Alessandro Astolfi · Lorenzo Marconi (Eds.)

Analysis and Design of Nonlinear Control Systems

Alessandro Astolfi · Lorenzo Marconi (Eds.)

Analysis and Design of Nonlinear Control Systems

In Honor of Alberto Isidori

With 68 Figures

Springer

Professor Alessandro Astolfi
Imperial College London
London SW7 2AZ, UK
and
Dipartimento di Informatica, Sistemi e Produzione
Università di Roma "Tor Vergata"
Via del Politecnico 1
00133 Roma, Italy
E-mail: a.astolfi@imperial.ac.uk

Professor Lorenzo Marconi
Dipartimento Elettronica Informatica e Sistemistica
C.A.SY./ D.E.I.S. – Università di Bologna
Viale Risorgimento 2
40136 Bologna, Italy
E-mail: lorenzo.marconi@unibo.it

Library of Congress Control Number: 2007934273

ISBN 978-3-540-74357-6 Springer Berlin Heidelberg New York

Springer is a part of Springer Science+Business Media

springer.com

Typesetting and production: LE-TEX Jelonek, Schmidt & Vöckler GbR, Leipzig
Cover design: eStudio Calamar S.L., F. Steinen-Broo, Girona, Spain

SPIN 11824350 89/3180/YL – 5 4 3 2 1 0 Printed on acid-free paper

to Alberto,
the Scientist, the Teacher, the Friend

Preface

This book is a tribute to

Prof. Alberto Isidori

on the occasion of his 65th birthday.

Prof. Isidori's prolific, pioneering and high-impact research activity has spanned over 35 years. Throughout his career, Prof. Isidori has developed ground-breaking results, has initiated research directions and has contributed towards the foundation of nonlinear control theory. In addition, his dedication to explain intricate issues and difficult concepts in a simple and rigorous way and to motivate young researchers has been instrumental to the intellectual growth of the nonlinear control community worldwide.

The volume collects 27 contributions written by a total of 52 researchers. The principal author of each contribution has been selected among the researchers who have worked with Prof. Isidori, have influenced his research activity, or have had the privilege and honour of being his PhD students. The contributions address a significant number of control topics, including theoretical issues, advanced applications, emerging control directions and tutorial works. The diversity of the areas covered, the number of contributors and their international standing provide evidence of the impact of Prof. Isidori in the control and systems theory communities.

The book has been divided into six parts: *System Analysis*, *Optimization Methods*, *Feedback Design*, *Regulation*, *Geometric Methods* and *Asymptotic Analysis*, reflecting important control areas which have been strongly influenced and, in some cases, pioneered by Prof. Isidori.

The first part "System Analysis" collects four contributions. In "Smooth Distributions Are Globally Finitely Spanned", H.J. Sussmann focuses on the foundations of differential geometry providing a characterization of smooth distributions. Then E.D. Sontag and Y. Wang, in "Uniformly Universal Inputs", prove the existence of universal inputs, uniformly for the observability of all analytic continuous-time systems. In "System Interconnection", J.C. Willems reviews and sheds new light on the classical notion of interconnection, with special emphasis on physically consistent formalizations. Finally, in "Reduced Order Modeling of Nonlinear Control Systems", A.J. Krener presents a method for nonlinear model reduction based on a normal form for the controllability and observability functions.

The second part "Optimization Methods" consists of four contributions. In "Nonholonomic Trajectory Optimization and the Existence of Twisted Matrix Logarithms", R.W. Brockett discusses an optimal control problem for bilinear systems as a representative example for a class of problems with a Lie group structure. Then A.B. Kurzhanski and P. Varaiya, in "The Hamilton-Jacobi Type Equations for Nonlinear Target Control and Their Approximation", present a comparison principle for first-order PDEs, of the Hamilton-Jacobi-Bellman type, arising in nonlinear target control synthesis and reachability analysis. In "Causal Coding of Markov Sources with Continuous Alphabets", S. Yüksel and T. Başar deal with the remote control problem for linear and nonlinear systems with quantization, by studying the structure of optimal causal encoders for kth-order Markov sources. Finally, in "Pseudospectral Optimal Control and Its Convergence Theorems", W. Kang, I.M. Ross and Q. Gong present convergence theorems for the pseudospectral methods of nonlinear optimal control with constraints.

The third part "Feedback Design" comprises six contributions. In "Event Based Control", K.J. Åström discusses the advantages of event-based control strategies over sampled-data theory in computer controlled systems. Then, A.P. Aguiar, J.P. Hespanha and P.V. Kokotović investigate, in "Zero Dynamics and Tracking Performance Limits in Nonlinear Feedback Systems", the tracking performance achievable for nonminimum-phase nonlinear systems by exploiting the concept of zero dynamics. In "A Nonlinear Model for Combustion Instability: Analysis and Quenching of the Oscillations", I.D. Landau, F. Bouziani and R.R. Bitmead study a model for combustion instability in gas-fueled turbo-machinery using the Krylov-Bogoliubov method. Then, M. Cao and A.S. Morse, in "Convexification of the Range-Only Station Keeping Problem", solve the three landmarks station keeping problem in the plane, by adopting concepts inherited from switched adaptive control. In "Control of Hydraulic Devices, an Internal Model Approach", K. Schlacher and K. Zehetleitner study the problem of controlling, by output feedback, nonlinear hydraulic devices to suppress periodic disturbances in steel rolling. Finally, in "Hybrid Zero Dynamics of Planar Bipedal Walking", J.W. Grizzle and E.R. Westervelt deal with the problem of designing stable periodic walking motions in bipedal robots by extending the concept of zero dynamics to hybrid systems.

Six contributions compose the fourth part "Regulation". In "Hybrid Systems: Limit Sets and Zero Dynamics with a View Toward Output Regulation", C. Cai, R. Goebel, R.G. Sanfelice and A.R. Teel investigate concepts such as limit sets and zero dynamics for hybrid systems and discuss their use in hybrid output regulation problems. Then, L. Marconi and L. Praly, in "Essential and Redundant Internal Models in Nonlinear Output Regulation", develop a few issues on the problem of semiglobal output regulation for nonlinear systems and, in particular, discuss the design of internal model-based regulators. In "Two Global Regulators for Systems with Measurable Nonlinearities and Unknown Sinusoidal Disturbances", R. Marino, G.L. Santosuosso and P. Tomei deal with the problem of global regulation for a class of possibly nonminimum-phase nonlinear systems in the presence of uncertainties on the system and the exosystem. Then A. Serrani, in "A Taxonomy for Time-Varying Immersions in Periodic Internal-Model Control", frames in the context of nonlinear output regulation the problem of classifying immersion mappings according to the observability properties of the steady-state generator system. In "Paving the Way Towards the Control of Wireless Telecommunication Networks", F. Delli Priscoli and A. Pietrabissa show how linear control methodologies can be used for the development of resource management procedures in communication networks. Finally, in "Nonlinear Synchronization of Coupled Oscillators: the Polynomial Case", J.-S. Kim and F. Allgöwer present a feedback method to achieve synchronization of coupled identical oscillators which are described by polynomial vector fields.

The fifth part "Geometric Methods" contains three chapters. In "Disturbance Decoupling for Open Quantum Systems: Quantum Internal Model Principle", N. Ganesan and T.J. Tarn explore the use of classical disturbance decoupling techniques to eliminate decoherence in quantum control systems. Then S. Monaco and D. Normand-Cyrot, in "Controller and Observer Normal Forms in Discrete-Time", study the problem of simplifying discrete-time nonlinear systems through feedback transformations and the use of output injection. Finally, in "A Geometric Approach to Dynamic Feedback Linearization", S. Battilotti and C. Califano characterize, from a geometric perspective, dynamically feedback linearizable systems and provide an algorithm for the computation of the linearizing dynamic controller.

The last part "Asymptotic Analysis" contains four contributions. In "The Steady-State Response of a Nonlinear Control System, Lyapunov Stable Attractors, and Forced Oscillations", C.I. Byrnes and D.S. Gilliam discuss the notion of steady-state response for nonlinear systems and its use in the study of forced oscillations. Then, in "Model Reduction by Moment Matching for Linear and Nonlinear Systems", A. Astolfi develops a theory of model reduction for nonlinear systems introducing a nonlinear enhancement of the notion of moment and exploiting the theory of the steady-state response of nonlinear systems. In "Adaptive Control of Nonlinear Systems with Unknown Parameters by Output Feedback: a Non-Identifier-Based Method", H. Lei and W. Lin solve the output feedback stabilization problem for a class of nonlinear systems with uncertain parameters. Finally, C. De Persis, in 'Hybrid Feedback Stabilization of Nonlinear Systems with Quantization Noise and Large Delays", illustrates the design of hybrid stabilizing controllers for nonlinear feedforward systems over finite-bandwidth networks with large delays.

The editors would like to thank all authors who have contributed to this exceptional book. We are also grateful to Michelle Hammond for her help in the preparation of the volume.

We complete the preface with a personal consideration. There are very few events in the life of a person that shape it in a unique way. For both of us, the encounter with Prof. Isidori has been one such event. It has determined the place we live and/or work, has directed our careers, inspired our work, set an example and influenced the way we work.

It is for us a great honour to celebrate Alberto's contributions to science and to our lives.

London, Rome, Bologna
June 2007

Alessandro Astolfi
Lorenzo Marconi

Contents

Part III Feedback Design

Part IV Regulation

List of Contributors

A. Pedro Aguiar
Instituto Superior Técnico
1049-001 Lisbon, Portugal.
pedro@isr.ist.utl.pt

Frank Allgöwer
University of Stuttgart
70550 Stuttgart, Germany
allgower@ist.uni-stuttgart.de

Alessandro Astolfi
Imperial College London
SW7 2AZ, London, UK
and
Università di Roma Tor Vergata
00133 Rome, Italy
a.astolfi@imperial.ac.uk

Karl J. Åström
Lund University
Lund, SE-221 00, Sweden
kja@control.lth.se

Tamer Başar
University of Illinois
Urbana, IL 61801 USA
tbasar@control.csl.uiuc.edu

Stefano Battilotti
Università di Roma "La Sapienza"
00184 Rome, Italy
battilotti@dis.uniroma1.it

Robert R. Bitmead
University of California, San Diego
La Jolla, CA 92093, USA
rbitmead@ucsd.edu

Fethi Bouziani
ENSIEG BP 46
38402 Saint-Martin d'Hères, France
fethi.bouziani@
lag.ensieg.inpg.fr

Roger W. Brockett
Harvard University
Cambridge, MA 02138, USA
brockett@deas.harvard.edu

Chris I. Byrnes
Washington University in St. Louis
St Louis, MO-63130, USA
chrisbyrnes@wustl.edu

Chaohong Cai
University of California
Santa Barbara, CA 93106, USA
cai@ece.ucsb.edu

Claudia Califano
Università di Roma "La Sapienza"
00184 Rome, Italy
claudia.califano@uniroma1.it

Ming Cao
Yale University
New Haven, CT 06520, USA
m.cao@yale.edu

Francesco Delli Priscoli
Università di Roma "La Sapienza"
00184 Rome, Italy
dellipriscoli@dis.uniroma1.it

Claudio De Persis
Università di Roma "La Sapienza"
00184 Rome, Italy
depersis@dis.uniroma1.it

Narayan Ganesan
Washington University in St. Louis
St Louis, MO-63130, USA
nganesan@wustl.edu

David S. Gilliam
Texas Tech University
Lubbock, TX 79409, USA
gilliam@math.ttu.edu

Rafal Goebel
3518 NE 42 St.,
Seattle, WA 98105, USA
rafal.k.goebel@gmail.com

Qi Gong
Univerisity of Texas at San Antonio
San Antonio, TX 78249, USA.
qi.gong@utsa.edu

Jessy W. Grizzle
University of Michigan
Ann Arbor, MI 48109, USA
grizzle@umich.edu

João P. Hespanha
University of California
Santa Barbara, CA 93106, USA
hespanha@ece.ucsb.edu

Wei Kang
Naval Postgraduate School
Monterey, CA 93943, USA
wkang@nps.edu

Jung-Su Kim
University of Stuttgart
70550 Stuttgart, Germany
kim@ist.uni-stuttgart.de

Petar V. Kokotović
University of California
Santa Barbara, CA 93106, USA
petar@ece.ucsb.edu

Arthur J. Krener
University of California
Davis, CA 95616, USA
and
Naval Postgraduate School
Monterey, CA 93943, USA
ajkrener@ucdavis.edu

Alexander B. Kurzhanski
Moscow State University
and
University of California at Berkeley
Berkeley, CA 94720, USA
kurzhans@eecs.berkeley.edu

Ioan D. Landau
ENSIEG BP 46
38402 Saint-Martin d'Hères, France
landau@lag.ensieg.inpg.fr

Hao Lei
Case Western Reserve University
Cleveland, OH 44106, USA
hao.lei@cace.edu

Wei Lin
Case Western Reserve University
Cleveland, OH 44106, USA
linwei@cwru.edu

Lorenzo Marconi
University of Bologna
40136 Bologna, Italy
lmarconi@deis.unibo.it

Riccardo Marino
Università di Roma Tor Vergata
00133 Rome, Italy
marino@ing.uniroma2.it

Salvatore Monaco
Università di Roma "La Sapienza"
00184 Rome, Italy
salvatore.monaco@uniroma1.it

A. Stephen Morse
Yale University,
New Haven, CT 06520, USA
as.morse@yale.edu

Dorothée Normand-Cyrot
CNRS, Supélec
91192 Gif-sur-Yvette, France
cyrot@lss.supelec.fr

Antonio Pietrabissa
Università di Roma "La Sapienza"
00184 Rome, Italy
pietrabissa@dis.uniroma1.it

Laurent Praly
École des Mines de Paris
77305 Fontainebleau, France.
Laurent.Praly@ensmp.fr

I. Michael Ross
Naval Postgraduate School
Monterey, CA 93943, USA
imross@nps.edu

Ricardo G. Sanfelice
University of California
Santa Barbara, CA 93106, USA
rsanfelice@ece.ucsb.edu

Giovanni L. Santosuosso
Università di Roma Tor Vergata
00133 Rome, Italy
santosuosso@ing.uniroma2.it

Kurt Schlacher
Johannes Kepler University
4040 Linz, Austria
kurt.schlacher@jku.at

Andrea Serrani
The Ohio State University
Columbus, OH 43206, USA
serrani@ece.osu.edu

Eduardo D. Sontag
Rutgers University
Piscataway, NJ 08854, USA
sontag@math.rutgers.edu

Hector J. Sussmann
Rutgers University
Piscataway, NJ 08854, USA
sussmann@math.rutgers.edu

Tzyh-Jong Tarn
Washington University in St. Louis
St Louis, MO-63130, USA
tarn@wuauto.wustl.edu

Andrew R. Teel
University of California
Santa Barbara, CA 93106, USA
teel@ece.ucsb.edu

Patrizio Tomei
Università di Roma Tor Vergata
00133 Rome, Italy
tomei@ing.uniroma2.it

Pravin Varaiya
University of California at Berkeley
Berkeley, CA 94720, USA
varaiya@eecs.berkeley.edu

Yuan Wang
Florida Atlantic University
Boca Raton, FL 33431, USA
ywang@math.fau.edu

Eric R. Westervelt
The Ohio State University
Columbus, OH 43210, USA
westervelt.4@osu.edu

Jan C. Willems
Katholieke Universiteit Leuven
B-3001 Leuven, Belgium
Jan.Willems@esat.kuleuven.be

Serdar Yüksel
Yale University
New Haven, CT 06511, USA
serdar.yuksel@yale.edu

Kurt Zehetleitner
Johannes Kepler University
4040 Linz, Austria
kurt.zehetleitner@jku.at

Short Curriculum Vitae of Alberto Isidori

Alberto Isidori, born in Rapallo (Italy) in 1942, obtained his *Laurea* degree in Electrical Engineering from the University of Rome in 1965 and the *Libera Docenza* in Automatic Control from the University of Rome in 1969. Since 1975, he has been Professor of Automatic Control at this University. He has held visiting positions in various leading Universities, which include the University of Illinois at Urbana-Champaign, the University of California at Berkeley and the ETH in Zurich. Since 1989 he has also been regularly collaborating with Washington University in St. Louis.

In 1996, at the opening of 13th IFAC World Congress in San Francisco, Dr. Isidori received the "Georgio Quazza Medal". This medal is the highest technical award given by the International Federation of Automatic Control, and is presented once every third year for lifetime contributions to automatic control science and engineering. The Georgio Quazza Medal was awarded to Dr. Isidori for "pioneering and fundamental contributions to the theory of nonlinear feedback control". He is also the recipient of the Ktesibios Award, from the Mediterranean Control Association (in 2000) and of the Bode Lecture Award, from the Control Systems society of IEEE (in 2001). In 1986 he was elected Fellow of IEEE and in 2005 he was elected Fellow of IFAC.

He was President of the European Union Control Association in the biennium 1995-1997, and he is currently President Elect of IFAC. He has been the organizer or co-organizer of several international conferences on the subject of feedback design for nonlinear systems. In particular, he was the initiator of a permanent series of IFAC Symposia on this topic.

Selected publications of Alberto Isidori

Books

[1] A. Isidori, *Sistemi di Controllo* (in Italian), Siderea, 1979.
[2] A. Ruberti and A. Isidori, *Teoria dei Sistemi* (in Italian), Boringhieri, 1979.
[3] A. Isidori, *Nonlinear Control Systems*, Springer Verlag, 1st ed. 1985, 2nd ed. 1989, 3rd. ed. 1995.
[4] A. Isidori, *Nonlinear Control Systems (volume II)*, Springer Verlag, 1999.
[5] H. Knobloch, A. Isidori, D. Flockerzi *Topics in Control Theory*, Birkhauser (Basel), 1993.
[6] C.I. Byrnes, F. Delli Priscoli, A. Isidori, *Output regulation of uncertain nonlinear systems*, Birkhauser (Boston), 1997.
[7] A. Isidori, L. Marconi, A. Serrani, *Robust Motion Control: an Internal-Model Approach*, Springer Verlag, 2003.

Selected Publications in Archival Journals

Automatica

[1] O.M. Grasselli, A. Isidori, F. Nicolò, Output regulation of a class of bilinear systems under constant disturbances, *Automatica*, 15, pp. 189–195 (1979).
[2] C.I. Byrnes, A. Isidori, On the attitude stabilization of rigid spacecraft, *Automatica*, 27, pp. 87–95 (1991). [*Recipient of the Best Paper Award*]
[3] C.I. Byrnes, F. Delli Priscoli, A. Isidori, Structurally stable output regulation of nonlinear systems, *Automatica*, 33, pp. 369–385 (1997).
[4] L. Marconi and A. Isidori, Mixed internal-model based and feed-forward control for robust tracking in nonlinear systems, *Automatica*, 36, pp. 993–1000, (2000).

[5] L. Marconi, A. Isidori, A. Serrani, Autonomous landing on a oscillating platform: an internal-model based approach, *Automatica*, 38, pp. 21–32, (2002).

[6] C. Bonivento, A. Isidori, L. Marconi, A. Paoli, Implicit fault tolerant control: application to induction motors, *Automatica*, 40, pp. 355–371, (2004). [*Recipient of the Best Paper Award*]

IEEE Transaction on Automatic Control

[1] C. Bruni, A. Isidori, A. Ruberti, A method of factorization of the impulse-response matrix, *IEEE Trans. Automatic Control*, 13, pp. 739–741 (1968).

[2] C. Bruni, A. Isidori, A. Ruberti, A method of realization based on the moments of the impulse-response matrix, *IEEE Trans. Automatic Control*, 14, pp. 203–204 (1969).

[3] A. Isidori, Direct construction of minimal bilinear realization's from nonlinear input-output maps, *IEEE Trans. Automatic Control*, 18, pp. 626–631 (1973).

[4] M.D. Di Benedetto, A.Isidori, Triangular canonical forms for bilinear systems, *IEEE Trans. Automatic Control*, 23, pp. 877–880 (1978).

[5] A.J. Krener, A. Isidori, C. Gori-Giorgi, S. Monaco, Nonlinear decoupling via feedback: a differential-geometric approach, *IEEE Trans. Automatic Control*, 26, pp. 331–345 (1981). [*Recipient of the Best Paper Award*]

[6] O.M. Grasselli, A. Isidori, An existence theorem for observers of bilinear systems, *IEEE Trans. Automatic Control*, 26, pp. 1299–1301 (1981).

[7] A. Isidori, The matching of a prescribed input-output behavior in a nonlinear system, *IEEE Trans. Automatic Control*, 30, pp. 258–265 (1985).

[8] D. Cheng, T.J. Tarn, A. Isidori, Global external linearization of nonlinear systems via feedback, *IEEE Trans. Automatic Control*, 30, pp. 808–811 (1985).

[9] A. Isidori, J.W. Grizzle, Fixed modes and nonlinear noninteracting control of nonlinear systems, *IEEE Trans. Automatic Control*, 33, pp. 907–914 (1988).

[10] S.S. Sastry, A. Isidori, Adaptive control of linearizable systems, *IEEE Trans. Automatic Control*, 34, pp. 1123–1131 (1989).

[11] A. Isidori, C.I. Byrnes, Output regulation of nonlinear systems, *IEEE Trans. Automatic Control*, 35, pp. 131–140 (1990). [*Recipient of the Best Paper Award*]

[12] C.I. Byrnes, A. Isidori, Asymptotic stabilization of minimum phase nonlinear systems, *IEEE Trans. Automatic Control*, 36, pp. 1122–1137 (1991).

[13] C.I. Byrnes, A. Isidori and J.C. Willems, Passivity, feedback equivalence, and the global stabilization of minimum phase nonlinear systems, *IEEE Trans. Automatic Control*, 36, pp. 1228–1240 (1991).

[14] A. Isidori, S.S. Sastry, P.V. Kokotovic and C.I. Byrnes, Singularly perturbed zero dynamics of nonlinear systems, *IEEE Trans. Automatic Control*, 37, pp. 1625–1631 (1992).

[15] A. Isidori, A. Astolfi, Disturbance attenuation and H_∞ control via measurement feedback in nonlinear systems, *IEEE Trans. Automatic Control*, 37, pp. 1283–1293 (1992).

[16] A. Isidori and W. Kang, H_∞ control via measurement feedback for general nonlinear systems, *IEEE Trans. on Automatic Control*, 40, pp. 466–472 (1995).

[17] A. Isidori and T.J. Tarn, Robust regulation for nonlinear systems with gain-bounded uncertainties, *IEEE Trans. on Automatic Control*, 40, pp. 1744–1754 (1995).

[18] A. Isidori, A remark on the problem of semiglobal nonlinear output regulation, *IEEE Trans. on Automatic Control*, 43, pp. 1734–1738 (1997).

[19] A. Isidori, B. Schwartz, and T.J. Tarn, Semi-global L_2 performance bounds for disturbance attenuation in nonlinear systems, *IEEE Trans. on Automatic Control*, 44, pp. 1535–1545 (1999).

[20] A. Isidori, A tool for semiglobal stabilization of uncertain non-minimum-phase nonlinear systems via output feedback, *IEEE Trans. on Automatic Control*, 45, pp. 1817–1827, (2000).

[21] R. De Santis and A. Isidori, On the output regulation for linear systems in the presence of input saturation, *IEEE Trans. on Automatic Control*, 46, pp. 156–160, (2001).

[22] C. De Persis and A. Isidori, A geometric approach to nonlinear fault detection and isolation, *IEEE Trans. on Automatic Control*, 46, pp. 853–865, (2001).

[23] A. Serrani, A. Isidori, L. Marconi, Semiglobal nonlinear output regulation with adaptive internal model, *IEEE Trans. on Automatic Control*, 46, pp. 1178–1194, (2001).

[24] A. Isidori, L. Marconi, A. Serrani, Robust nonlinear motion control of a helicopter, *IEEE Trans. on Automatic Control*, 48, pp. 413–426, (2003).

[25] C.I. Byrnes, A. Isidori, Limit sets, zero dynamics and internal models in the problem of nonlinear output regulation, *IEEE Trans. on Automatic Control*, 48, pp. 1712–1723, (2003).

[26] C.I. Byrnes, A. Isidori, Nonlinear internal models for output regulation, *IEEE Trans. on Automatic Control*, 49, pp. 2244–2247, (2004).

SIAM Journal on Control and Optimization

[1] P. d'Alessandro, A. Isidori, A. Ruberti, A new approach to the theory of canonical decomposition of linear dynamical systems, *SIAM J. Control*, 11, pp. 148–158 (1973).

[2] P. d'Alessandro, A. Isidori, A. Ruberti, Realization and structure theory of bilinear dynamical systems, *SIAM J. Control*, 12, pp. 517–535 (1974).

[3] M.D. Di Benedetto, A. Isidori, The matching of nonlinear models via dynamic state feedback, *SIAM J. Control and Optimization*, 24, pp. 1063–1075 (1986).

[4] F. Delli Priscoli, L. Marconi, A. Isidori, A new approach to adaptive nonlinear regulation, *SIAM J. Control and Optimization*, 45, pp. 829–855, (2006).
[5] L. Marconi, L. Praly, A. Isidori, Output stabilization via nonlinear Luenberger observers, *SIAM J. Control and Optimization*, 45, pp. 2277–2298, (2006).

Systems and Control Letters

[1] A. Isidori, A. Ruberti, A separation property of realizable Volterra kernels, *Systems and Control Lett.*, 1, pp. 309–311 (1981).
[2] A.J. Krener, A. Isidori, C. Gori-Giorgi, S. Monaco, Locally (f, g)-invariant distributions, *Systems and Control Lett.*, 1, pp. 12–15 (1981).
[3] A. Isidori, A.J. Krener, On feedback equivalence of nonlinear systems, *Systems and Control Lett.*, 2, pp. 118–121 (1982).
[4] A.J. Krener, A. Isidori, Linearization by output injection and nonlinear observers, *Systems and Control Lett.*, 3, pp. 47–52 (1983).
[5] A. Isidori, A. Ruberti, On the synthesis of linear input-output responses for nonlinear systems, *Systems and Control Lett.*, 4, pp.17–22 (1984).
[6] A. Isidori, S. Morse, State-feedback implementation of cascade compensators, *Systems and Control Lett.*, 8, pp. 63–68 (1986).
[7] C.I. Byrnes, A. Isidori, Local stabilization of minimum phase nonlinear systems, *Systems and Control Lett.*, 10, pp. 9–17 (1988).
[8] C.I. Byrnes, A. Isidori, New results and examples in nonlinear feedback stabilization, *Systems and Control Lett.*, 12, pp. 437–442 (1989).
[9] W. Zhan, T.J. Tarn and A. Isidori, A canonical dynamic extension for noninteraction with stability for affine nonlinear square systems, *Systems and Control Lett.*, 17, pp. 177–184 (1991).
[10] A. Isidori, A necessary condition for nonlinear H_∞ control via measurement feedback, *Systems and Control Lett.*, 23, pp. 169–177 (1994).
[11] A. Isidori, A note on almost disturbance decoupling for nonlinear minimum phase systems, *Systems and Control Lett.*, 27, pp. 191–194 (1996).
[12] A. Isidori, Global almost disturbance decoupling with stability for nonminimum phase single-input single-output nonlinear systems, *Systems and Control Lett.*, 28, pp. 115–122 (1996).
[13] A. Isidori, W. Lin, Global L_2-gain state feedback design for a class of nonlinear systems, *Systems and Control Lett.*, 34, pp. 395–302 (1998).
[14] A. Serrani, A. Isidori, Global robust output regulation for a class of nonlinear systems, *Systems and Control Lett.*, 39, pp. 133–139, (2000).
[15] A. Teel, L. Praly and A. Isidori, A note on the problem of semiglobal practical stabilization of uncertain nonlinear systems via dynamic output feedback, *Systems and Control Lett.*, 39, pp. 165–171, (2000).
[16] C. De Persis, A. Isidori, On the observability codistributions of a nonlinear system, with applications to state observation under unknown inputs, *Systems and Control Lett.*, 40, pp. 297–304, (2000).

[17] L. Marconi, A. Isidori, Robust global stabilization of a class of uncertain feedforward nonlinear systems, *Systems and Control Lett.*, 41, pp. 281–290, (2000).

[18] L. Marconi, A. Isidori, A. Serrani, Input disturbance suppression for a class of feedforward uncertain nonlinear systems, *Systems and Control Lett.*, 45, pp. 227–236, (2002)

[19] C. De Persis, A. Isidori, Global stabilizability by state feedback implies semiglobal stabilizability by encoded state feedback, *Systems and Control Lett.*, 53, pp 249–258, (2004).

[20] L. Marconi, A. Isidori, A. Serrani, Non-resonance conditions for uniform observability in the problem of nonlinear output regulation, *Systems and Control Lett.*, 53, pp 281–298, (2004).

[21] F. Delli Priscoli, L. Marconi, A. Isidori, Adaptive observers as nonlinear internal models, *Systems and Control Lett.*, 55, pp. 640–649, (2006).

[22] C. De Persis, A. Isidori, L. Marconi, Remote tracking via encoded information for nonlinear systems, *Systems and Control Lett.*, 55, pp. 809–818, (2006).

Part I

System Analysis

Smooth Distributions Are Globally Finitely Spanned

Hector J. Sussmann*

Rutgers University, Piscataway, NJ, USA

Dedicated to Alberto Isidori on his 65th birthday

Summary. A smooth distribution on a smooth manifold M is, by definition, a map that assigns to each point x of M a linear subspace $\Delta(x)$ of the tangent space T_xM, in such a way that, locally, there exist smooth sections $f_1, \ldots, f_d$ of Δ such that the linear span of $f_1(x), \ldots, f_d(x)$ is $\Delta(x)$ for all x. We prove that a much weaker definition of "smooth distribution," in which it is only required that for each $x \in M$ and each $v \in \Delta(x)$ there exist a smooth section f of Δ defined near x such that $f(x) = v$, suffices to imply that there exists a finite family $\{f_1, \ldots, f_d\}$ of smooth global sections of Δ such that $\Delta(x)$ is spanned, for every $x \in M$, by the values $f_1(x), \ldots, f_d(x)$. The result is actually proved for general singular subbundles E of an arbitrary smooth vector bundle V, and we give a bound on the number d of global spanning sections, by showing that one can always take $d = \operatorname{rank} E \cdot (1 + \dim M)$, where $\operatorname{rank} E$ is the maximum dimension of the fibers $E(x)$.

1 Introduction

"Smooth singular distributions" – or, simply, "smooth distributions" – on a smooth manifold M are often defined as maps $M \ni x \mapsto \Delta(x)$ such that

(i) *for each* $x \in M$, $\Delta(x)$ *is a linear subspace of* T_xM,
(ii) Δ *is locally finitely spanned.*

Condition (ii) can be assigned a precise meaning in several different ways. For example, we could take (ii) to mean

(LFS) *for every* $x_* \in M$ *there exist an open neighborhood* U *of* x_* *in* M *and a finite sequence* $(f_1, \ldots, f_d)$ *of smooth vector fields on* U *such that*

$$\Delta(x) = \operatorname{span}\{f_1(x), \ldots, f_d(x)\} \qquad \text{for all } x \in U\,.$$

* Research supported in part by NSF Grants DMS01-03901 and DMS-05-09930.

Naturally, it would also be possible to require a condition that in principle appears to be stronger, namely, that Δ is "globally finitely spanned," in the sense that the sequence $(f_1, \ldots, f_d)$ exist globally, that is

(GFS) *There exists a finite sequence* $(f_1, \ldots, f_d)$ *of smooth vector fields on* M *such that*

$$\Delta(x) = \mathrm{span}\{f_1(x), \ldots, f_d(x)\} \qquad \text{for all } x \in M\,.$$

(This is actually the requirement used by A. Isidori in [1], p. 14.)

Alternatively, one could require the seemingly much weaker condition that Δ is "determined by the values of its smooth sections," in the sense that

(DVSS) *For every* $x_* \in M$ *and every* $v \in \Delta(x_*)$ *there exists a smooth vector field* f *on some open neighborhood* U *of* x_* *such that* $f(x_*) = v$ *and* $f(x) \in \Delta(x)$ *for all* $x \in U$.

The purpose of this note is to prove that the three conditions are equivalent, and to give a bound for the number d of smooth global sections of Δ that are required to span $\Delta(x)$ for each x.

We will actually prove the result for a general smooth singular subbundle E of an arbitrary smooth vector bundle V over a smooth manifold M. The bound on the number d of global sections of E needed to span the space $E(x)$ at each point x of M will turn out to be $\rho(E) \cdot (1 + \dim M)$, where $\rho(E)$ is the rank of E. (In particular, a singular distribution Δ on a smooth manifold of dimension m is always spanned at every point by $m^2 + m$ smooth global sections.)

Acknowledgment. The author is grateful to Dmitry Roytenberg for bringing to his attention the question that is answered in this paper.

2 Statement and Proof of the Main Result

Throughout this note, the word "smooth" means "of class C^∞*", and "smooth manifold" means "smooth Hausdorff paracompact finite-dimensional manifold without boundary." If* M *is a smooth manifold, then* TM *is the tangent bundle of* M *and, for each* $x \in M$, T_xM *is the tangent space of* M *at* x. *If* S *is a subset of a real linear space* X, *then* $\mathrm{span}\, S$ *denotes the linear span of* S.

Our precise definitions are as follows.

Let V be a smooth vector bundle over a smooth manifold M. A ***singular subbundle*** of V on an open subset U of M is a map $M \ni x \mapsto E(x)$ such that $E(x)$ is a linear subspace of $V(x)$ for each $x \in M$. A ***smooth section*** of a singular subbundle E of V defined on U is a map $U \ni x \mapsto \sigma(x) \in E(x)$ which is smooth as a section of V.

Definition 1. *Let V be a smooth vector bundle over a smooth manifold M and let E be a singular subbundle of V over M. We say that E is* ***smooth*** *if for every $x \in M$ there exists a smooth section σ of E, defined on a neighborhood U of x, such that $v = \sigma(x)$.*

The ***rank*** *of a subbundlc E of V is thc maximum of thc dimcnsions of thc fibers of E.* □

We will use $\rho(E)$ to denote the rank of E.

In particular, a ***smooth distribution*** on M is a smooth singular subbundle of TM. (This means, that Δ is a map $M \ni x \mapsto \Delta(x)$ such that (i) and (ii) above hold, where (ii) is interpreted to mean "Condition DVSS holds.")

The following is then our main result.

Theorem 1. *Let V be a smooth vector bundle over a smooth manifold M of dimension m, and let E be a smooth singular subbundle of V in the sense of Definition 1. Let $d = \rho(E)(m+1)$. Then there exist d smooth global sections $\sigma_1, \ldots, \sigma_d$ of E such that the values $\sigma_1(x), \ldots, \sigma_d(x)$ linearly span $E(x)$ for every $x \in E$.*

Proof. Let $r = \rho(E)$. For each x, use $\delta(x)$ to denote the dimension of the space $E(x)$. Define $\Omega_k = \{x : \delta(x) \geq k\}$, for each $k \in \{0, 1, \ldots, r+1\}$. Then the Ω_k are open subsets of M. (Indeed, if $x \in \Omega_k$ then we may pick k linearly independent members $v_1, \ldots, v_k$ of $E(x)$, and smooth sections $\sigma_1, \ldots, \sigma_k$ of E, defined on a neighborhood U of x, such that $\sigma_j(x) = v_j$ for $j = 1, \ldots, k$. Since $\sigma_1(x), \ldots, \sigma_k(x)$ are linearly independent, it follows that $\sigma_1(x'), \ldots, \sigma_k(x')$ are linearly independent for all x' in some neighborhood U' of x. Then $\delta(x') \geq k$ for all $x' \in U'$, so $U' \subseteq \Omega_k$.)

Furthermore, it is clear that

$$M = \Omega_0 \supseteq \Omega_1 \supseteq \Omega_2 \supseteq \cdots \supseteq \Omega_{r-1} \supseteq \Omega_r \supseteq \Omega_{r+1} = \emptyset\,.$$

We will construct $m+1$-tuples $(\sigma_1^k, \ldots, \sigma_{m+1}^k)$ of smooth global sections of E, for each $k \in \{1, \ldots, r\}$, such that

$(*)$ *for every $k \in \{1, \ldots, r\}$ the following is true*

> $(\#_k)$ *for every $x \in \Omega_k$, the linear span of the set of $k(m+1)$ values $\sigma_i^j(x)$, $i \in \{1, \ldots, m+1\}$, $j \in \{1, \ldots, k\}$, is of dimension $\geq k$.*

It is then easy to see that, once this construction is carried out, the $r(m+1)$ sections σ_i^j, $i \in \{1, \ldots, m+1\}$, $j \in \{1, \ldots, r\}$, will have the desired property. (Indeed, suppose $x \in M$, and let $k = \delta(x)$. Then $x \in \Omega_k$, so the linear span of the $\sigma_i^j(x)$ for $i \in \{1, \ldots, m+1\}$ and $j \in \{1, \ldots, k\}$ is of dimension $\geq k$, and then *a fortiori* the linear span $L(x)$ of the $\sigma_i^j(x)$, for $i \in \{1, \ldots, m+1\}$ and $j \in \{1, \ldots, r\}$, is of dimension $\geq k$. But $E(x)$ is of dimension k, and $L(x) \subseteq E(x)$. So $L(x) = E(x)$.)

We construct $(\sigma_1^k, \ldots, \sigma_r^k)$ by induction on k, as follows. Suppose we have constructed the σ_i^j for $j < k$ and $i \in \{1, \ldots, m+1\}$, in such a way that

Conditions $(\#_j)$ are satisfied for all $j \in \{1, \ldots, k-1\}$. (In particular, this assumption is vacuously true if $k = 1$.)

For each $x \in \Omega_k$, let $A(x)$ be the linear span of the $\sigma_i^j(x)$ for $j < k$, $i \in \{1, \ldots, m+1\}$.

Now fix a point $x \in \Omega_k$. Then our inductive hypothesis implies that $\dim A(x) \geq k-1$. Since $\dim E(x) \geq k$ (because $x \in \Omega_k$), we may pick a vector $v \in E(x)$ such that the linear span of $A(x) \cup \{v\}$ has dimension $\geq k$, and then we may pick a smooth section ν^x of E, defined on some open neighborhood W_x of x, such that $\nu^x(x) = v$. After multiplying ν^x by a smooth function $\varphi : M \mapsto \mathbb{R}$ which is such that $\varphi(x) > 0$ and the support of φ is a compact subset of W_x, we may assume that ν^x is a global smooth section of E.

Since span $(A(x) \cup \{v\})$ has dimension $\geq k$, we may assume, after shrinking W_x, if necessary, that

(*) *the linear span of $A(x') \cup \{\nu^x(x')\}$ has dimension $\geq k$ for every $x' \in W_x$.*

This implies, in particular, that $W_x \subseteq \Omega_k$. Then $\mathcal{W} = \{W_x\}_{x \in \Omega_k}$ is an open covering of Ω_k.

We now come to the key step of our proof, namely, the construction of a refinement of $\mathcal{Z}$ of $\mathcal{W}$ with a very special property. (Recall that a ***refinement*** of an open covering $\mathcal{U} = \{U_\alpha\}_{\alpha \in A}$ of a topological space T is an open covering $\mathcal{V} = \{V_\beta\}_{\beta \in B}$ of T such that for every $\beta \in B$ there exists an $\alpha \in A$ such that $V_\beta \subseteq U_\alpha$.)

Lemma 1. *There exists $\mathcal{Z} = \{Z_\lambda\}_{\lambda \in L}$ such that*

- *$\mathcal{Z}$ is an open covering of Ω_k;*
- *$\mathcal{Z}$ is a refinement of $\mathcal{W}$;*
- *the index set L is a union*

$$L = L_1 \cup L_2 \cup \cdots \cup L_{m+1},$$

such that each L_i has the property that $Z_\lambda \cap Z_{\lambda'} = \emptyset$ whenever $\lambda \in L_i$, $\lambda' \in L_i$, and $\lambda \neq \lambda'$. (In other words, each $\mathcal{Z}_i = \{Z_\lambda\}_{\lambda \in L_i}$ is a family of pairwise disjoint open subsets of Ω_k.)

Proof. Let us prove the existence of a $\mathcal{Z}$ with the above properties. Equip M with a Riemannian metric. (This is possible because we are assuming that M is paracompact.) Then let $\mathcal{T}$ be a triangulation of Ω_k that refines $\mathcal{W}$, in the sense that every closed face F of $\mathcal{T}$ is a subset of W_x for some x. (The existence of a triangulation of Ω_k follows from the well known fact that every smooth manifold is triangulable, cf. [2]. Then a triangulation that refines the covering $\mathcal{W}$ can easily be constructed by successive barycentric subdivisions.)

For $j = 0, \ldots, m$, let T_j be the set of all j-dimensional open faces of $\mathcal{T}$, and let $|T_j|$ be the union of the members of T_j. Then each $F \in T_j$ is both relatively open and relatively closed in $|T_j|$. Given a face $F \in T_j$, let $C(F)$ be the union of the closures in Ω_k of all the faces $G \in T_j$ other than F. Then $C(F)$ is a relatively closed subset of Ω_k, because it is the union of a locally finite family of compact subsets of Ω_k. Furthermore, $F \cap C(F) = \emptyset$, because F is relatively open in T_j. If $x \in F$, then the Riemannian distance $\text{dist}(x, C(F))$ from x to $C(F)$ is strictly positive, because $x \notin C(F)$ and $C(F)$ is closed. Let B_x be the open ball in Ω_k with center x and radius $\frac{1}{2}\text{dist}(x, C(F))$. Let $B(F)$ be the union of the B_x for all $x \in F$.

We now prove that the $B(F)$, as F varies over T_j, are pairwise disjoint. To show this, let us assume that $F, F' \in T_j$ and $F \neq F'$, and let us show that $B(F) \cap B(F') = \emptyset$. Suppose $B(F) \cap B(F') \neq \emptyset$. Pick $y \in B(F) \cap B(F')$. Pick $x \in F$ such that $y \in B_x$ and $x' \in F'$ such that $y \in B_{x'}$. Let $\delta = \text{dist}(x, C(F))$, $\delta' = \text{dist}(x', C(F'))$. Assume, without loss of generality, that $\delta \geq \delta'$. Then

$$\text{dist}(x, x') \leq \text{dist}(x, y) + \text{dist}(y, x') < \frac{1}{2}\delta + \frac{1}{2}\delta' \leq \delta\,.$$

So $\text{dist}(x, x') < \delta$, which is a contradiction, since $\delta = d(x, C(F))$ and x' belongs to $C(F)$.

By construction, each $F \in T_j$ is a subset of W_x for some $x \in \Omega_k$. So we may pick, for each $F \in T_j$, a point $x_F \in \Omega_k$ such that $F \subseteq W_{x_F}$. We then define $\tilde{B}(F) = B(F) \cap W_{x_F}$. It is clear that

1. *for every $j \in \{0, \ldots, m\}$, and every $F \in T_j$, $\tilde{B}(F)$ is an open subset of Ω_k such that $F \subseteq \tilde{B}(F) \subseteq W_{x_F}$;*
2. *for every $j \in \{0, \ldots, m\}$, the sets $\tilde{B}(F)$, for $F \in T_j$, are pairwise disjoint;*
3. $\bigcup_{j=0}^{m} \bigcup_{F \in T_j} \tilde{B}(F) = \Omega_k$.

Now let $L_i = T_{i-1}$ for $i = 1, 2, \ldots, m+1$, and then let $L = L_1 \cup \cdots \cup L_{m+1}$. For $F \in L_i$, let $Z_F = \tilde{B}(F)$. Then $\mathcal{Z} = \{Z_F\}_{F \in L}$ is an open covering of Ω_k. Furthermore, $\mathcal{Z}$ is a refinement of $\mathcal{W}$, because if $F \in L$ then $Z_F \subseteq W_{x_F}$. Finally, each $\mathcal{Z}_i = \{Z_F\}_{F \in L_i}$ is clearly a family of pairwise disjoint sets. So $\mathcal{Z}$ has all the desired properties, and the proof of our lemma is complete. □

We now return to the proof of the main theorem. Let Z_i be the union of all the members of $\mathcal{Z}_i$. For each i and each member Z of $\mathcal{Z}_i$ pick a point $\bar{x}(i, Z) \in \Omega_k$ such that $Z \subseteq W_{\bar{x}(i,Z)}$. Then define a smooth section μ_i of E on Z_i by letting $\mu_i(x) = \nu^{\bar{x}(i,Z)}(x)$ if $x \in Z$, $Z \in \mathcal{Z}_i$. (This is possible because (a) if $x \in Z_i$ then there exists one and only one $Z \in \mathcal{Z}_i$ such that $x \in Z$, (b) $\nu^{\bar{x}(i,Z)}$ is a smooth section of E on $W_{\bar{x}(i,Z)}$, and (c) $Z \subseteq W_{\bar{x}(i,Z)}$.) It then follows from (*) that, whenever $i = 1, \ldots, m+1$, $Z \in \mathcal{Z}_i$, and $x \in Z$, then $\text{span}(A(x) \cup \{\mu_i(x)\})$ has dimension $\geq k$. In particular,

(**) *for every i, the linear span of $A(x) \cup \{\mu_i(x)\}$ has dimension $\geq k$ for every $x \in Z_i$.*

We now construct a smooth function $\varphi_i : M \mapsto \mathbb{R}$ such that

(1) $\varphi_i(x) > 0$ *for all* $x \in Z_i$,
(2) *if we let* $\theta_i(x) = \varphi_i(x)\mu_i(x)$ *for* $x \in Z_i$, *and* $\theta_i(x) = 0$ *for* $x \in M$, $x \notin Z_i$, *then* θ_i *is a smooth section of* V *on* M.

(The existence of φ_i is easy to prove. Indeed, it is clearly sufficient to construct φ_i on each connected component of M, so we may as well assume that M is connected. Since Z_i is open in M, we can express Z_i as a countable union $\bigcup_{\ell=1}^{\infty} K_\ell$ of compact sets such that $K_\ell \subseteq \operatorname{Int} K_{\ell+1}$ for each ℓ. For each ℓ, find a smooth nonnegative function $\psi_\ell : M \mapsto \mathbb{R}$ such that $\psi_\ell \equiv 1$ on K_ℓ and $\psi_\ell \equiv 0$ on $M \backslash K_{\ell+1}$. We construct φ_i by letting $\varphi_i = \sum_{\ell=1}^{\infty} \varepsilon_\ell \psi_\ell$, where the ε_ℓ are strictly positive numbers that converge to zero sufficiently fast. The maps $\theta_{i,\ell}$ given by $\theta_{i,\ell}(x) = \psi_\ell(x)\mu_i(x)$ are then smooth compactly supported global sections of V on M, and all we need is to find the ε_ℓ so that the series $\sum_{\ell=1}^{\infty} \varepsilon_\ell \psi_\ell$ and $\sum_{\ell=1}^{\infty} \varepsilon_\ell \theta_{i,\ell}$ converge to smooth limits φ_i, θ_i in the spaces $C^\infty(M, \mathbb{R})$, $C^\infty(M, V)$ – where $C^\infty(M, V)$ is the space of all smooth sections of V – endowed with the topology of uniform convergence on compact sets of all the derivatives of all orders. And it is a well known fact that such ε_ℓ always exist.)

We now let $\sigma_i^k = \theta_i$ for $i = 1, \ldots, m+1$. We have to verify that $(\#_k)$ holds. For this purpose, we pick $x \in \Omega_k$ and verify that the linear span $S(x)$ of the vectors σ_i^j, for $i \in \{1, \ldots, m+1\}$, $j \in \{1, \ldots, k\}$, has dimension $\geq k$. Since $\mathcal{Z}$ is a covering of Ω_k, we may pick an $F \in L$ such that $x \in Z_F$. Then F belongs to L_i for a unique i, and then $x \in Z_i$. It follows that $\operatorname{span}(A(x) \cup \{\mu_i(x)\})$ has dimension $\geq k$. Since $x \in Z_i$, we have $\varphi_i(x) > 0$, and then the fact that $\theta_i(x) = \varphi_i(x)\mu_i(x)$ implies that $\operatorname{span}(A(x) \cup \{\theta_i(x)\}) = \operatorname{span}(A(x) \cup \{\mu_i(x)\})$, so $\operatorname{span}(A(x) \cup \{\theta_i(x)\})$ also has dimension $\geq k$. In other words, we have shown that $\operatorname{span}(A(x) \cup \{\sigma_i^k(x)\})$ has dimension $\geq k$. Since $A(x)$ is the linear span of the vectors σ_i^j, for $i \in \{1, \ldots, m+1\}$, $j \in \{1, \ldots, k-1\}$ it is clear that $A(x) \cup \{\sigma_i^k(x)\} \subseteq S(x)$. Hence $\dim S(x) \geq k$ as desired, concluding our proof. □

References

1. A. Isidori. *Nonlinear Control Sytems.* Springer Verlag, New York, 2nd edition, 1989.
2. J. Munkres. *Elementary Differential Topology.* Princeton University Press, 1966.

Uniformly Universal Inputs

Eduardo D. Sontag[1*] and Yuan Wang[2**]

[1] Department of Mathematics, Rutgers University, Piscataway, NJ 08854, USA
[2] Department of Mathematical Sciences, Florida Atlantic University, Boca Raton, FL 33431, USA

Dedicated to Alberto Isidori on his 65th birthday

Summary. A result is presented showing the existence of inputs universal for observability, uniformly with respect to the class of all continuous-time analytic systems. This represents an ultimate generalization of a 1977 theorem, for bilinear systems, due to Alberto Isidori and Osvaldo Grasselli.

1 Introduction

One of the key concepts in control theory is that of a *universal input* for observability and parameter identification. Informally stated, an input u_0 is universal (for a given system) provided that the following property holds: if two internal states x_1 and x_2 are in principle distinguishable by any possible input/output experiment, then x_1 and x_2 can be distinguished by forcing the system with this particular input u_0 (and observing the corresponding output function). *Universal input theorem(s)* for distinguishability show that such inputs indeed do exist, and, furthermore, show that "generic" (in an appropriate technical sense) inputs have this property. Viewing unknown parameters as constant states, one may re-interpret the universal input property as one regarding parameter identifiability instead of observability.

In the seminal 1977 paper [5], Alberto Isidori (together with Osvaldo Grasselli) provided the first general result on existence of universal inputs for a wide class of nonlinear systems (bilinear systems). Motivated by this work [8] provided analogous results for discrete time systems as well as continuous-time analytic systems with compact state spaces, and this was extended to arbitrary continuous-time analytic systems in [14]. (See also the related work in [7]

* Supported in part by NSF Grant DMS-0504557
** Supported in part by NSF Grant DMS-0504296 and Chinese National Natural Science Foundation grant 60228003

for linear automata.) A different proof of the result in [14] was given in [21], where implications to the study of a nonlinear analog of "transfer functions" were discussed as well.

In the present paper, we provide an ultimate extension of the theorems for analytic continuous-time systems, showing that there are inputs that are universal with respect to *all* finite dimensional analytic systems, and, moreover, the set of such inputs is generic. A preliminary version of our result was presented at the 1994 IEEE Conference on Decision and Control [20] (see also [11]).

Besides their intrinsic theoretical appeal, universal input theorems help provide a rationale for systems identification when using information provided by "random" or unknown inputs. For example, in [16] universal inputs were used to justify the "dependent input" approach to the identification of molecular-biological systems, for which high complexity and the lack of sufficient quantitative measurements prevent the use of arbitrary test signals. The approach in [16], applied to measurements of nitrogen uptake fluxes in baker's yeast (*Saccharomyces cerevisiae*), was to view unmodeled dynamics (possibly due to mutations in the yeast strains being used) as generating fictitious "dependent inputs". In another direction, universal input theorems provide a basis for certain numerical methods for path planning in nonlinear systems, see for example [11, 10, 12].

2 Analytic Input/Output Operators

We first review some standard notions regarding analytic input/output operators. Let m be a fixed nonnegative integer. By an input we mean a Lebesgue measurable, essentially bounded function $u : [0, T] \to \mathbb{R}^m$ for some $T > 0$.

Consider a set $\Theta = \{X_0, X_1, \ldots, X_m\}$, whose elements will be thought as $m + 1$ non-commuting variables. We use Θ^* to denote the free monoid generated by Θ, where the neutral element of Θ^* is the empty word, and the product is concatenation. We define $\mathbb{R}[\Theta]$ to be the $\mathbb{R}$-algebra generated by Θ^*, that is, the set of all polynomials in the X_i's. By a *power series* in the variables $X_0, X_1, \ldots, X_m$ we mean a formal power series:

$$c = \sum_{w \in \Theta^*} \langle c,\ w \rangle w ,$$

where $\langle c,\ w \rangle \in \mathbb{R}$ for each $w \in \Theta^*$. We use $\mathbb{R}[[\Theta]]$ to denote the set of all power series in the X_i's. This is a vector space with "+" defined coefficient-wise. There is a linear duality between $\mathbb{R}[[\Theta]]$ and $\mathbb{R}[\Theta]$ provided by:

$$\langle c,\ d \rangle = \sum_{w \in \Theta^*} \langle c,\ w \rangle \langle d,\ w \rangle \tag{1}$$

for any $c \in \mathbb{R}[[\Theta]]$ and $d \in \mathbb{R}[\Theta]$.

A series $c \in \mathbb{R}[[\Theta]]$ is a *convergent* series if there is a positive (radius of convergence) ρ and a constant M such that

$$|\langle c,\, w\rangle| \le M\rho^l l!, \qquad \forall |w| = l\,, \tag{2}$$

where $|w|$ denotes the length of w, i.e., $|w| = l$ if $w = X_{i_1}X_{i_2}\cdots X_{i_l}$.

Let $L^m_{e,\infty}$ denote the set of measurable, locally essentially bounded functions $u : [0,\, \infty) \to \mathbb{R}^m$. For each $u \in L^m_{e,\infty}$ and $S_0 \in \mathbb{R}[[\Theta]]$, consider the initial value problem

$$\dot{S}(t) = \left(X_0 + \sum_{i=1}^{m} X_i u_i\right) S(t), \quad S(0) = S_0 \tag{3}$$

seen as a differential equation over $\mathbb{R}[[\Theta]]$. A solution is an absolutely continuous curve, where derivative is understood coefficient-wise. For any locally essentially bounded $u(\,\cdot\,)$, by the Peano-Baker formula, there is always a solution in $\mathbb{R}[[\Theta]]$ whose coefficients are iterated integrals of u. Furthermore, one can prove the uniqueness of the solutions successively by induction. In particular, the solution $C[u]$ with $C[u](0) = S_0 = 1$ defines the *generating* (or "Chen-Fliess") series of u (cf. [1, 2, 14]). Explicitly, For each u, the generating series $C[u]$ is given by

$$C[u](t) = \sum_w V_w[u](t)w\,,$$

where $V_w[u]$ is given recursively by $V_\phi[u](t) = 1$, and

$$V_{X_i w}[u](t) = \int_0^t u_i(s)V_w[u](s)\,ds\,, \qquad \forall w \in \Theta^*\,, \tag{4}$$

where $u_0 \equiv 1$. We say that a pair (T, r) of positive real numbers with $r \ge 1$ is *admissible* for a convergent series c if for some M and ρ as in (2) the following inequality holds:

$$Tr\rho(m+1) < 1.$$

For each pair (T, r) (where $r \ge 1$) that is admissible for a convergent series c, the series c defines an i/o operator $F_c^{T,r}$ on the set

$$\mathcal{V}_T(r) := \{u|\ u : [0,\, T] \to \mathbb{R}^m,\ \|u\|_\infty \le r\}$$

by means of the following formula:

$$F_c[u](t) = \langle c,\, C[u](t)\rangle = \sum_w \langle c,\, w\rangle V_w[u](t)\,. \tag{5}$$

It is known (c.f. [6]) that the series in (5) converges uniformly on $[0,\, T]$.

Note that, for every convergent series c, and for every two pairs (T_1, r_1) and (T_2, r_2) that are admissible for c, the functions $F_c^{T_1,r_1}$ and $F_c^{T_2,r_2}$ coincide on $V_r(T)$, where $T = \min\{T_1, T_2\}$ and $r = \min\{r_1, r_2\}$. Therefore, one may define a mapping F_c on the union of the sets $V_T(r)$ for all pairs (T, r) that are admissible for c, as an extension of the maps $F_c^{T,r}$. Such operators defined by convergent series have been extensively studied, c.f. [3, 6, 18, 19].

3 Uniformly Universal Inputs

In this section we study the distinguishability of operators by analytic input functions.

Let c and d be two convergent series. We say that c and d are *distinguishable by an input function* $u : [0, T_0] \to \mathbb{R}^m$, denoted by $c \not\sim_u d$, if for every $T \in (0, T_0]$ for which $(T, \max\{\|u\|_\infty, 1\})$ is admissible for both c and d, it holds that

$$F_c[u] \neq F_d[u]$$

as functions defined on $[0, T]$. Note here that "$c \not\sim_u d$" is stronger than merely requiring $F_c[u](t) \neq F_d[u](t)$ as functions over some interval. In our context, we require that $F_c[u] \neq F_d[u]$ as functions over *every* interval $[0, T]$ for which $(T, \max\{\|u\|_\infty, 1\})$ is admissible for both c and d.

An input u is called a *uniformly universal input* if $c \not\sim_u d$ for any convergent series c and d such that $c \neq d$. Note that an input u is a uniformly universal input if and only if u distinguishes c from 0 whenever $c \neq 0$.

For each $T > 0$, we consider $C^\infty[0, T]$, the set of all smooth functions from $[0, T]$ to $\mathbb{R}^m$, a topological space endowed with the Whitney topology. We will say that a subset S of a topological space is *generic* if S contains a countable intersection of open dense sets. Since $C^\infty[0, T]$ is a Baire space (cf.[4]), a generic subset of $C^\infty[0, T]$ is dense.

Let Ω^T denote the set of all uniformly universal inputs defined on $[0, T]$. The following is the main result.

Theorem 1. *For any fixed $T > 0$, the set Ω^T of uniformly universal inputs is a generic subset of $C^\infty[0, T]$.*

Theorem 1 asserts the existence of smooth uniformly universal inputs (and their genericity); however, there is no *analytic* uniformly universal input. To illustrate this fact, consider the following example.

Example 1. Take any fixed analytic function $\alpha : [0, \infty) \to \mathbb{R}$. For this function, consider the state space system:

$$\dot{x}_1 = 1\,, \quad \dot{x}_2 = 0\,, \quad \dot{x}_3 = (\alpha(x_1) - u)x_2\,, \quad y = x_3\,. \tag{6}$$

When writing the system as

$$\dot{x} = g_0(x) + g_1(x)u\,, \quad y = h(x)\,,$$

one has, in the standard coordinates of $\mathbb{R}^3$, $g_0(x) = (1,\, 0,\, \alpha(x_1)x_2)^\tau$, $g_1(x) = (0,\, 0,\, -x_2)^\tau$ and $h(x) = x_3$, where the superscript "τ" denotes the transpose.

For each $x \in \mathbb{R}^3$, let c_x be the generating series induced by the system with the initial state x, that is, c_x is given by

$$\langle c_x,\, X_{i_1} X_{i_2} \cdots X_{i_r} \rangle = L_{g_{i_r}} \cdots L_{g_{i_2}} L_{g_{i_1}} h(x)\,,$$

for all multi-indices $i_1 i_2 \ldots i_r$, and all $r \geq 0$. Then c_x is a convergent series, and for any initial state p, and each u, the corresponding output of (6) is given by the "Fliess fundamental formula" ([6]):

$$y(t) = F_{c_p}[u](t)\,.$$

Observe that for system (6), the two particular initial states $p = (0,0,0)$ and $q = (0,1,0)$ can always be distinguished by some input, i.e., $c_p \neq c_q$. (Indeed, whenever $p \neq q$ are two states such that $p_1 = q_1$, the input $u(t) = \alpha(p_1 + t) - 1$ distinguishes these initial states.) But the pair (p, q) cannot be distinguished by u, i.e., $F_{c_p}[u] = F_{c_q}[u]$, if $u(t) = \alpha(t)$. Hence, c_p and c_q cannot be distinguished by $\alpha(\,\cdot\,)$. This shows that for any analytic function $\alpha(\,\cdot\,)$, one can always find a pair (c_p, c_q) which α cannot distinguish but $c_p \neq c_q$. This shows that there is no uniformly universal input which is analytic. △

3.1 Universal Input Jets

For each $k \geq 1$, consider the polynomial $\mathfrak{d}_k(\mu)$ in $\mu = (\mu_0, \mu_1, \ldots, \mu_{k-1})$ given by

$$\mathfrak{d}_k(\mu) \;=\; \left.\frac{d^k}{dt^k}\right|_{t=0} C[u](t)\,, \tag{7}$$

where u is any input such that $u^{(i)}(0) = \mu_i$. Then one has the following formula for $k \geq 1$:

$$\frac{d^k}{dt^k} F_c[u](0) \;=\; \langle c,\; \mathfrak{d}_k(u(0), u'(0), \ldots, u^{(k-1)}(0))\rangle\,. \tag{8}$$

Let $\mathfrak{d}_0 = 1$. Then if $\mu = (\mu_0, \mu_1, \ldots)$ is such that $\langle c,\; \mathfrak{d}_k(\mu^k)\rangle \neq 0$ for some $k \geq 0$, then $c \not\sim_u 0$, for any $T > 0$ and any $u \in C^\infty[0, T]$ such that $u^{(i)}(0) = \mu_i$ for $0 \leq i \leq k-1$, where $\mu^k \in \mathbb{R}^{mk}$ is given by $\mu_i^k = \mu_i$ for $0 \leq i \leq k-1$.

Let $\mathbb{R}^{\mathrm{m},\infty} = \prod_{i=1}^\infty \mathbb{R}^m$ be endowed with the product topology whose basis of open sets consists of all sets of the form $\prod_{i=1}^\infty U_i$, where each U_i is an open subset of $\mathbb{R}^m$, and only finitely many of them are proper subsets of $\mathbb{R}^m$. Note that $\mathbb{R}^{\mathrm{m},\infty}$ is a Baire space, and hence, any generic subset of $\mathbb{R}^{\mathrm{m},\infty}$ is a dense set. For each $\mu \in \mathbb{R}^{\mathrm{m},\infty}$ and a series c, we let $\langle c,\; \mathfrak{d}(\mu)\rangle$ denote the sequence

$$\langle c,\; \mathfrak{d}_0\rangle, \langle c,\; \mathfrak{d}_1(\mu_0)\rangle,\; \langle c,\; \mathfrak{d}_2(\mu_0, \mu_1)\rangle,\; \langle c,\; \mathfrak{d}_3(\mu_0, \mu_1, \mu_2)\rangle,\; \ldots.$$

Let $\mathcal{J}$ be the subset of $\mathbb{R}^{\mathrm{m},\infty}$ defined by

$$\mathcal{J} = \{\mu :\; \langle d,\; \mathfrak{d}(\mu)\rangle \neq 0,\; \forall d \in \mathfrak{C},\; d \neq 0\}\,, \tag{9}$$

where $\mathfrak{C}$ stands for the set of all convergent series. Take $\mu \in \mathcal{J}$. It is easy to see from (8) that for any $u \in C^\infty$ with $u^{(i)}(0) = \mu_i$ for all i, u is a uniformly universal input. We call the elements in $\mathcal{J}$ universal input jets.

Theorem 2. *The set $\mathcal{J}$ of universal input jets is a generic subset of* $\mathbb{R}^{\mathrm{m},\infty}$.

4 Proofs of Theorems 1 and 2

To prove Theorems 1 and 2, we need to study some topological properties of the set $\mathfrak{C}$ of convergent series. This set can be identified with $\mathbb{R}^{\mathbb{N}}$, the set of all maps from $\mathbb{N}$ to $\mathbb{R}$, once the elements of Θ^* are linearly ordered; we again adopt the product topology on this set. With this topology, that a sequence $\{c_j\}$ converges to c means

$$\lim_{j\to\infty}\langle c_j,\ w\rangle = \langle c,\ w\rangle$$

for each $w \in \Theta^*$. Observe that a subset S of $\mathbb{R}[[\Theta]]$ is compact (in the product topology) if and only if S is closed, and for each w, there exists $M_w > 0$ such that for all $d \in S$,

$$|\langle d,\ w\rangle| \le M_w.$$

4.1 Equi-Convergent Families

A family S of convergent series is said to be *equi-convergent* if there exist $\rho,\ M > 0$ such that

$$|\langle d,\ w\rangle| \le M\rho^l l!\,, \qquad \forall |w| = l \tag{10}$$

holds for every $d \in S$. Clearly, every closed equi-convergent family is compact, and if S is equi-convergent, there exists some pair (T, r) that is admissible for every element of S. For such (T, r), we say that (T, r) is admissible for S.

For any convergent series c and $\mu \in \mathbb{R}^{mk}$, we let $\langle c,\ \mathfrak{d}_k(\mu)\rangle_k$ denote the k-vector

$$\Big(\langle c,\ \mathfrak{d}_0\rangle,\ \langle c,\ \mathfrak{d}_1(\mu_0)\rangle,\dots,\langle c,\ \mathfrak{d}_k(\mu_0,\dots,\mu_{k-1})\rangle\Big)\,.$$

For a set S of convergent series, we let

$$\mathcal{J}_S^k = \{\mu \in \mathbb{R}^{mk} :\ \langle d,\ \mathfrak{d}_k(\mu)\rangle_k \neq 0,\ \forall d \in S\}$$

(which maybe an empty set, e.g., in the case when S contains the zero series.) Let $\mu \in \mathbb{R}^{mk}$. We say that ν is a finite extension of μ if $\nu \in \mathbb{R}^{ml}$ for some $l \ge k$ such that $\nu_i = \mu_i$ for $0 \le i \le k-1$. For an equi-convergent family, we have the following conclusion.

Lemma 1. *Assume that S is compact and equi-convergent, and that $\mathcal{J}_S^l \neq \emptyset$ for some l. Then for any $k \ge 1$ and any $\mu \in \mathbb{R}^{mk}$, there exist K and a finite extension ν of μ such that $\nu \in \mathcal{J}_S^K$.*

To prove Lemma 1, we first discuss some continuity properties of the operators defined by the convergent series. Lemma 2.2 of [19] shows that if (T, r) is admissible for c, then the map $\mathcal{V}_T(r) \to C[0, T]$, $u \mapsto F_c[u]$ is continuous using the L_1 norm on $\mathcal{V}_T(r)$ in the special case when $r = 1$. The same proof can be used to prove the following result for equi-convergent families.

Lemma 2. *Assume that S is equi-convergent, and (T, r) is admissible for S. Then the map*

$$\mathcal{V}_T(r) \to C[0,T], \quad u \mapsto F_c[u]$$

is continuous with respect to the L_1 norm on $\mathcal{V}_T(r)$ and the C^0 norm on $C[0,T]$ uniformly for $c \in S$. □

This result can be strengthened further to the following, where the topology on $\mathcal{V}_T(r)$ is the L_1-topology, and the topology on $C[0,T]$ is the C^0 topology.

Lemma 3. *Let S be an equi-convergent family. Then, for any $r > 0$, there exists some $T_1 > 0$ such that for any $T < T_1$ the map*

$$\psi : S \times \mathcal{V}_T(r) \to C[0,T], \quad (c,u) \mapsto F_c[u]$$

is continuous.

Proof. Let S be compact and equi-convergent. Then there exists ρ such that

$$|\langle d,\ w\rangle| \le M\rho^k k! \qquad \forall |w| = k\,,\ \forall d \in S\,. \tag{11}$$

Let $T_1 = \frac{1}{r\rho(m+1)}$. Fix $T \in [0, T_1)$. Then F_d is defined on $\mathcal{V}_T(r)$ for each $d \in S$. For any $c, d \in S$, $u, v \in \mathcal{V}_T(r)$,

$$\|F_c[u] - F_d[v]\|_\infty \le \|F_c[u] - F_c[v]\|_\infty + \|F_c[v] - F_d[v]\|_\infty\,.$$

Hence, by Lemma 2, it is enough to show that the map

$$S \to C[0,T], \quad c \mapsto F_c[v] \tag{12}$$

is equi-continuous for $v \in \mathcal{V}_T(r)$, that is, for any $c \in S$, for any $\varepsilon > 0$, there exists a neighborhood $\mathcal{N}$ of c such that

$$\|F_c[v] - F_d[v]\|_\infty < \varepsilon$$

for all $d \in \mathcal{N}$ and all $v \in \mathcal{V}_T(r)$.

First note that for each $d \in S$ and $v \in \mathcal{V}_T(r)$, one has

$$|V_w[v](t)| \le \frac{r^k t^k}{k!} \qquad \forall\ |w| = k, \tag{13}$$

and therefore,

$$\left|\sum_{|w|\ge k} \langle d,\ w\rangle V_w[v](t)\right| \le \sum_{j=k}^{\infty} M\rho^j j!\,(m+1)^j \frac{r^j T^j}{j!} \le M \sum_{j\ge k} \frac{T^j}{T_1^j}$$

(where we have used the fact that there are at most $(m+1)^j$ elements in Θ^j). Since $0 < T < T_1$, it follows that for any $\varepsilon > 0$, there exists some $k > 0$ such that

$$|F_d[v](t) - F_{d_k}[v](t)| < \varepsilon \quad \forall t \in [0, T], \tag{14}$$

for all $v \in \mathcal{V}_T(r)$, all $d \in S$, where for each d, d_k is the polynomial given by

$$d_k = \sum_{|w| \le k} \langle c, w \rangle w \, .$$

Let $c \in S$ and $\varepsilon > 0$ be given. Choose k such that (14) holds for all $d \in S$ and $v \in \mathcal{V}_T(r)$ with ε replaced by $\varepsilon/4$. Then,

$$\begin{aligned} |F_c[v](t) - F_d[v](t)| &\le |F_{c_k}[v](t) - F_{d_k}[v](t)| + \varepsilon/2 \\ &= |F_{c_k - d_k}[v](t)| + \varepsilon/2 \le \sum_{|w| \le k} |\langle c - d, \, w \rangle V_w[v](t)| + \varepsilon/2. \end{aligned}$$

Let

$$R = \max_{0 \le j \le k} \left\{ \frac{r^j T^j}{j!} \right\}.$$

It follows from (13) that $\|V_w[v]\|_\infty \le R$ for all $v \in \mathcal{V}_T(r)$ and for all w with $|w| \le k$. Hence, there exists some $\delta > 0$ such that for any d satisfying $|\langle d, \, w \rangle - \langle c, \, w \rangle| < \delta$ for all $|w| \le k$,

$$\sum_{|w| \le k} |\langle c - d, \, w \rangle V_w[v](t)| < \varepsilon/2.$$

This means that there exists some neighborhood $\mathcal{N}$ of c such that for any $d \in \mathcal{N}$,

$$|F_c[v](t) - F_d[v](t)| < \varepsilon.$$

This shows that the map given in (12) is equi-continuous. □

Proof of Lemma 1. Let $\tilde{\mu} = (\tilde{\mu}_0, \ldots, \tilde{\mu}_{l-1}) \in \mathcal{J}_S^l$, and let $v \in C^\infty[0, 1]$ be given by

$$v(t) = \sum_{i=0}^{l-1} \tilde{\mu}_i \, \frac{t^i}{i!}, \qquad 0 \le t \le 1.$$

Let $r = 2\,\|v\|_\infty$. Without loss of generality, we assume that $r \ge 1$. Choose $0 < T < 1$ such that (T, r) is admissible for every $d \in S$.

By the assumption on $\tilde{\mu}$, it follows that $v \in \Omega_S^T$, where

$$\Omega_S^T := \{u \in C^\infty[0, T] : \; d \not\sim_u 0, \; \forall \, d \in S\}.$$

Hence, for any $c \in S$, there exists some $t_c \in [0, T]$ such that

$$|F_c[v](t_c)| = \tau_c > 0.$$

By the continuity property (c.f. Lemma 2), there exists a neighborhood $\mathcal{N}_c$ of c such that for any $d \in \mathcal{N}_c \cap S$,

$$|F_d[v](t_c)| \geq \tau_c/2\,.$$

Since S is compact, there exist $c_1, c_2, \ldots, c_n$ such that $S \subseteq \bigcup_{i=1}^{n} \mathcal{N}_{c_i}$. It then follows that for any $d \in S$, there exists some $1 \leq j \leq n$ such that

$$|F_d[v](t_j)| \geq \tau_{c_j}/2\,, \tag{15}$$

where $t_j = t_{c_j}$.

Let $\mu = (\mu_0, \mu_1, \ldots, \mu_{k-1}) \in \mathbb{R}^{mk}$ be given. Let $\{\omega_j\}$ be a sequence of analytic functions defined on $[0, T]$ such that

- $\omega_j^{(i)}(0) = \mu_i$ for $0 \leq i \leq k-1$, $j \geq 1$;
- $\omega_j \to v$ in the L_1 norm (as functions defined on $[0, T]$); and
- for some $M \geq 1$, $\|\omega_j\|_\infty \leq M$ for all $j \geq 1$.

(See Lemma A.3 in [21] for the existence of such sequences.) Reducing the value of T if necessary, one may assume that (T, M) is admissible for all $d \in S$.

Again, as it follows from the continuity property established in Lemma 2, one sees that for some n_0 large enough,

$$|F_d[\omega_{n_0}](t) - F_d[v](t)| \leq \tau/4 \quad \forall\, t \in [0,\, T],\ \forall\, d \in S, \tag{16}$$

where $\tau = \min\{\tau_{c_1}, \tau_{c_2},\, \ldots,\, \tau_{c_n}\}$. It follows from (15) and (16) that for each $d \in S$, there exists some $j > 0$ such that

$$|F_d[\omega_{n_0}](t_j)| \geq \tau/4 > 0\,,$$

from which it follows that $\omega_{n_0} \in \Omega_S^T$. As ω_{n_0} is analytic, it follows that $F_d[\omega_{n_0}]$ is also analytic (see Lemma 2.3 of [19]). This then implies that for any $d \in S$, there exists some $j_d \geq 1$ such that $y_d^{(j_d-1)}(0) \neq 0$, where $y_d(t) = F_d[\omega_{n_0}](t)$, and hence,

$$\langle d,\, \mathfrak{d}_{j_d}(\omega(0),\, \ldots,\, \omega^{(j_d-1)}(0))\rangle_{j_d} \neq 0\,,$$

where for simplicity, we have replaced ω_{n_0} by ω. Note then that this is equivalent to

$$\langle d_{j_d},\, \mathfrak{d}_{j_d}(\omega(0),\, \ldots,\, \omega^{(j_d-1)}(0))\rangle_{j_d} \neq 0\,.$$

Thus, for any $d \in S$, there exists a neighborhood $\mathcal{W}_d$ of d such that for any $\tilde{d} \in \mathcal{W}_d$,

$$\langle \tilde{d}_{j_d},\, \mathfrak{d}_{j_d}(\omega(0),\, \ldots,\, \omega^{(j_d-1)}(0))\rangle_{j_d} \neq 0\,,$$

and consequently,

$$\langle \tilde{d},\, \mathfrak{d}_{j_d}(\omega(0),\, \ldots,\, \omega^{(j_d-1)}(0))\rangle_{j_d} \neq 0\,.$$

Again, by compactness of S, there exists some $K \geq 1$ such that

$$\langle d,\, \mathfrak{d}_K(\omega(0),\, \ldots,\, \omega^{(K-1)}(0))\rangle_K \neq 0\,,$$

for any $d \in S$. Without loss of generality, one may assume that $K \geq k$. Let $\nu \in \mathbb{R}^{mK}$ be given by $\nu_i = \omega^{(i)}(0)$. Then $\nu \in \mathcal{J}_S^K$, and by the choice of $\{\omega_j\}$, ν is a finite extension of μ. □

4.2 Universal Jets for Equi-Convergent Families

For each element $w_0 \in \Theta^*$, and each integer $k > 0$, let $\mathfrak{C}_{w_0,k}$ be the set of all series satisfying:

$$|\langle c,\ w_0\rangle| \geq \frac{1}{k}\,, \tag{17}$$

and

$$|\langle c,\ w\rangle| \leq k^{n+1}n!\,, \qquad \forall |w| = n\,. \tag{18}$$

Clearly, each $\mathfrak{C}_{w,k}$ is compact, equi-convergent, and $d \neq 0$ for any $d \in \mathfrak{C}_{w,k}$. Moreover, it is easy to see that

$$\mathfrak{C} \setminus \{0\} = \bigcup_{w\in\Theta^*,k\geq 1} \mathfrak{C}_{w,k}\,. \tag{19}$$

We now let, for each w, k, and $T > 0$,

$$\Omega^T_{w,k} = \{u \in C^\infty[0, T] :\ c \not\sim_u 0,\ \forall c \in \mathfrak{C}_{w,k}\}\,.$$

Then it follows from (19) that

$$\Omega^T = \bigcap_{w,k} \Omega^T_{w,k}\,.$$

For a set S of convergent series, we define

$$\mathcal{J}_S = \{\mu \in \mathbb{R}^{\mathrm{m},\infty} :\ \langle d,\ \mathfrak{d}(\mu)\rangle \neq 0,\ \forall d \in S\}\,,$$

and we denote $\mathcal{J}_{\mathfrak{C}_{w,k}}$ by $\mathcal{J}_{w,k}$. Again, by (19), we have

$$\mathcal{J} = \bigcap_{w,k} \mathcal{J}_{w,k}\,.$$

Thus, to prove Theorem 2, it is enough to show that $\mathcal{J}_{w,k}$ is open dense in $\mathbb{R}^{\mathrm{m},\infty}$.

Lemma 4. *Let S be an equi-convergent and compact family so that $0 \notin S$. Then $\mathcal{J}_S$ is open and dense in $\mathbb{R}^{\mathrm{m},\infty}$.*

To prove Lemma 4, we first prove the following result which is stronger than Lemma 1 in that it is no longer a prior requirement that $\mathcal{J}^l_S \neq \emptyset$ for some l.

Lemma 5. *Let S be an equi-convergent and compact family so that $0 \notin S$. Then for any $j \geq 1$ and $\mu^j = (\mu_0, \ldots, \mu_{j-1}) \in \mathbb{R}^{mj}$, there exists a finite extension ν^k of μ^j such that $\nu^k \in \mathcal{J}^k_S$.*

Proof. Let $\mu_j \in \mathbb{R}^{mj}$ be given. Consider any fixed $c \in S$, $c \neq 0$. According to [17, Theorem 1] (see also Lemma A.4 in [21]), there are always some $l \geq j$ and finite extension $\nu_c \in \mathbb{R}^{ml}$ of μ^j such that

$$\langle c, \mathfrak{d}_l(\nu_c)\rangle_l \neq 0\,.$$

From here it follows that there exists some neighborhood $\mathcal{N}_c$ of c such that

$$\langle d, \mathfrak{d}_l(\nu_c)\rangle_l \neq 0\,,$$

for all $d \in \mathcal{N}_c \cap S$. Since S is Hausdorff and compact, one may assume that $\mathcal{N}_c$ is compact. Applying this argument for each c in S, and using compactness of S, one concludes that there are finitely many $c_1, c_2, \ldots c_n$ such that S is covered by $\cup_{i=1}^n \mathcal{N}_{c_i}$. Write $\mathcal{N}_{c_i} \cap S$ as $\mathcal{N}_i$. Then on each $\mathcal{N}_i$, there exists some finite extension $\nu_{c_i} \in \mathbb{R}^{ml_i}$ of μ^j such that

$$\langle d, \mathfrak{d}_{l_i}(\nu_{c_i})\rangle_{l_i} \neq 0\,,$$

for all $d \in \mathcal{N}_i$. In particular, note that, for each i, $\mathcal{N}_i$ is compact and $\mathcal{J}^{l_i}_{\mathcal{N}_i} \neq \emptyset$, so Lemma 1 can be applied to each such $\mathcal{N}_i$. We do this next, inductively.

Start by defining $s_1 = l_1$ and σ_1 as just ν_{c_1}. Then $\sigma_1 \in \mathbb{R}^{ms_1}$ is a finite extension of μ^j and $\sigma_1 \in \mathcal{J}^{s_1}_{\mathcal{N}_1}$. Consider $\mathcal{N}_2$. By Lemma 1, there exists some $s_2 \geq s_1$ and some finite extension σ_2 of σ_1 such that $\sigma_2 \in \mathcal{J}^{s_2}_{\mathcal{N}_2}$. Since σ_2 is an extension of σ_1, it follows that σ_2 is also in $\mathcal{J}^{s_2}_{\mathcal{N}_1}$, and it also a finite extension of μ^j. Repeating finitely many times, one concludes that there exists some finite extension $\sigma_n \in \mathbb{R}^{ms_n}$ of μ^j such that $\sigma_n \in \mathcal{J}^{s_n}_{\mathcal{N}_i}$ for all $1 \leq i \leq n$. Hence, $\sigma_n \in \mathcal{J}^{s_n}_S$. □

Proof of Lemma 4. Let S be an equi-convergent family so that $0 \notin S$. We first prove the density property of $\mathcal{J}_S$. Pick up any $\mu = (\mu_0, \mu_1, \ldots) \in \mathbb{R}^{\mathrm{m},\infty}$. Let W be a neighborhood of μ (in the product topology). Without loss of generality, one may assume that

$$W = W_0 \times W_1 \times \cdots \times W_{j-1} \times \mathbb{R}^m \times \mathbb{R}^m \times \cdots\,,$$

where W_i is an open subset of $\mathbb{R}^m$ for $0 \leq i \leq j-1$. By Lemma 5, there exists some finite extension ν^N of $\mu^j := (\mu_0, \ldots, \mu_{j-1})$ such that $\nu^N \in \mathcal{J}^N_S$. Note that every extension ν of ν^N is in $\mathcal{J}_S$ as well as in W since it is also an extension of μ^j. Hence, $W \bigcap \mathcal{J}_S \neq \emptyset$.

We now prove the openness property of $\mathcal{J}_S$. Pick $\mu = (\mu_0, \mu_1, \ldots) \in \mathcal{J}_S$. Then for each $c \in S$, there exists some $k \geq 0$ such that

$$\langle c, \mathfrak{d}_k(\mu)\rangle_k \neq 0\,. \tag{20}$$

By compactness of S, one can assume that k does not depend on c. Note that (20) involves only finitely many terms, so there are neighborhoods $\mathcal{N}_c$ of $c \in S$ and U_{c,μ^k} of μ^k in $\mathbb{R}^{mk}$ (where $\mu^k = (\mu_0, \ldots, \mu_{k-1})$) such that

$$\langle d, \mathfrak{d}_k(\nu)\rangle_k \neq 0$$

for all $d \in \mathcal{N}_c$ and all $\nu \in U_{c,\mu^k}$. Again, using compactness, one can show that there are finitely many $U_{c_1,\mu^k}, \ldots, U_{c_n,\mu^k}$, each of which is open, so that $S \subseteq \bigcup_{i=1}^n \mathcal{N}_{c_i}$, and $U_{c_i,\mu^k} \subseteq \mathcal{J}^k_{\mathcal{N}_{c_i}}$. Let

$$U_{\mu^k} = \bigcap_{i=1}^n U_{c_i,\mu^k}.$$

Then U_{μ^k} is a neighborhood of μ^k in $\mathbb{R}^{mk}$. Since $U_{\mu^k} \subseteq \mathcal{J}^k_{\mathcal{N}_{c_i}}$ for all $1 \le i \le n$, it follows that $U_{\mu^k} \subseteq \mathcal{J}^k_S$. Finally, let $U = U_{\mu^k} \times \mathbb{R}^{\mathrm{m},\infty}$. Then U is an open set containing μ. Furthermore, for any $\nu \in U$, the restriction ν^k of ν is in U_{μ^k}, and therefore, $\nu \in \mathcal{J}_S$. This shows that $U \subseteq \mathcal{J}_S$ and μ is an interior point of $\mathcal{J}_S$. □

4.3 Universal Inputs for Equi-Convergent Families

As discussed in Section 4.2, to prove Theorem 1, it is enough to show the following.

Lemma 6. *Let S be an equi-convergent and compact family so that $0 \notin S$. Then, for any $T > 0$, the set Ω_S^T is open and dense in $C^\infty[0,T]$.*

First of all, we make the following observation.

Remark 1. Suppose that $\Omega_S^{T_0}$ is open and dense in $C^\infty[0,T_0]$ for some T_0, then Ω_S^T is open and dense in $C^\infty[0,T]$ for every $T > T_0$. This can be shown in details as follows.

For each subset U of $C^\infty[0,T]$, let $U_{T_0} = \{v_{T_0} : v \in U\}$, where for $v \in C^\infty[0,T]$, v_{T_0} denotes the restriction of v to the interval $[0,T_0]$. Suppose U is open in $C^\infty[0,T]$, then U_{T_0} is open in $C^\infty[0,T_0]$, and every $u \in U_{T_0}$ can be smoothly extended to a function $\tilde{u} \in U$. Moreover, if $u \in \Omega_S^{T_0}$, then $\tilde{u} \in \Omega_S^T$. Hence, if $\Omega_S^{T_0} \bigcap U_{T_0} \neq \emptyset$, then $\Omega_S^T \bigcap U \neq \emptyset$. This shows the density property of Ω_S^T.

To show the openness property of Ω_S^T, let $u \in \Omega_S^T$. By definition, for any $c \in S$, there exists some $t_c \in [0,T_0]$ such that $F_c[u](t_c) \neq 0$, so $u_{T_0} \in \Omega_S^{T_0}$. By openness of $\Omega_S^{T_0}$, there is a neighborhood U of u_{T_0} in $C^\infty[0,T_0]$ such that $u_{T_0} \in U \subseteq \Omega_S^{T_0}$. Let

$$\tilde{U} = \{v \in C^\infty[0,T] : \; v_{T_0} \in U\}.$$

Then $\tilde{U}$ is a neighborhood of u in $C^\infty[0,T]$, and $\tilde{U} \subseteq \Omega_S^T$. This shows that every u in Ω_S^T is an interior element of Ω_S^T. ◁

Proof of Lemma 6. Assume that S is equi-convergent and compact. Let $T > 0$ be given. We first prove the density property of Ω_S^T. By Remark 1, one may assume that $T < 1/2$. Let $u \in C^\infty[0,T]$, and pick a neighborhood $\mathcal{W}$ of u. Again, without loss of generality, we may assume that

$$\mathcal{W} = \left\{ v \in C^\infty[0,T] : \left\| v^{(i)} - u^{(i)} \right\|_\infty < \delta,\ 0 \le i \le j-1 \right\}$$

for some $j \ge 1$ and some $\delta > 0$. Let $\mu = (\mu_0, \mu_1, \ldots)$ be given by $\mu_i = u^{(i)}(0)$. By Lemma 5, there exists some $K > j$ and a finite extension ν^K of μ^j such that $\nu^K \in \mathcal{J}_S^K$. By Lemma A.4 in [21], one sees that there exists some analytic function w_j such that $w_j^{(i)}(0) = \nu_{j+i} - \mu_{j+i}$ for $i = 0, \ldots, K-j-1$, and $\|w_j\|_{L_1} < \delta$. One then defines w_l inductively for $l = j-1, \ldots, 1, 0$ by

$$w_l(t) = \int_0^t w_{l+1}(s)\, ds.$$

It then can be seen that $w_{l+1}(t) = w_l'(t)$, $w_l(0) = 0$, and $\|w_l\|_\infty < \delta$ for $0 \le l \le j-1$. Consequently, $w_0 \in C^\infty[0,T]$ is a function such that $w_0^{(i)}(0) = 0$ for $0 \le i \le j-1$, and $w_0^{(i)}(0) = \nu_i - \mu_i$ for $j \le i \le K-1$, and $\left\| w_0^{(i)} \right\|_\infty < \delta$ for all $0 \le i \le j-1$.

Let $w(t) = u(t) + w_0(t)$. Then $w \in \mathcal{W}$. Also note that $w^{(i)}(0) = \nu_i$ for $0 \le i \le K-1$. Since $\nu^K \in \mathcal{J}_S^K$, it follows that $w \in \Omega_S^T$. This proves the density property of Ω_S^T.

Next we show the openness property of Ω_S^T. Let $u \in \Omega_S^T$. Again, by Remark 1, we may assume that (T, r) is admissible for every $c \in S$, where $r = \max\{\|u\|_\infty, 1\}$, and that $T < T_1$, where T_1 is defined as in Lemma 3. Since S is compact, there exists some $\delta > 0$ such that $\|F_c[u]\|_\infty \ge \delta$ for all $c \in S$. Observe that Lemma 3 still holds when $\mathcal{V}_T(r)$ is endowed with the Whitney topology. Hence, for each $c \in S$, there exist a neighborhood $\mathcal{N}_c$ of c and a neighborhood $U_c \subseteq \mathcal{V}_T(r)$ of u such that

$$\|F_c[v]\|_\infty > \delta/2$$

for all $c \in \mathcal{N}_c$, $v \in U_c$. By compactness of S, there are finitely many $c_1, c_2, \ldots, c_L$ such that $S \subseteq \bigcup_{i=1}^L \mathcal{N}_{c_i}$. Let $U = \bigcap_{i=1}^L U_{c_i}$. Then U is a neighborhood of u, and for each $v \in U$, $\|F_c[v]\|_\infty > \delta/2$ for all $c \in S$. It follows that $U \subseteq \Omega_S^T$. □

5 State Space Systems

Consider an *analytic system*

$$\Sigma : \begin{cases} x'(t) = g_0(x(t)) + \sum_{i=1}^m g_i(x(t)) u_i(t)\,, \\ y(t) = h(x(t))\,, \end{cases} \tag{21}$$

where for each t, $x(t) \in \mathcal{M}$, which is an analytic (second countable) manifold of dimension n, $h : \mathcal{M} \longrightarrow \mathbb{R}$ is an analytic function, and $g_0, g_1, \ldots, g_m$ are analytic vector fields defined on $\mathcal{M}$. Inputs are measurable essentially bounded maps $u : [0, T] \longrightarrow \mathbb{R}^m$ defined on $[0, T]$ for suitable choices of $T > 0$. In general, $\varphi(t, x, u)$ denotes the state trajectory of (21) corresponding to an input u and initial state x, defined at least for small t.

Fix any two states p, $q \in \mathcal{M}$ and take an input u. We say p and q are *distinguished by* u, denoted by $p \not\sim_u q$, if $h(\varphi(\cdot, p, u)) \neq h(\varphi(\cdot, q, u))$ (considered as functions defined on the common domain of $\varphi(\cdot, p, u)$ and $\varphi(\cdot, q, u)$); otherwise we say p and q cannot be distinguished by u, denoted by $p \sim_u q$. If p and q cannot be distinguished by *any* input u, then we say p and q are *indistinguishable*, denoted by $p \sim q$. If for any two states, $p \sim q$ implies $p = q$, then we say that system (21) is *observable*. (See [6] and [13].) See also [9] for other related notions as well as detailed concept of *generic local observability*.

For a given continuous time system Σ, let $\mathcal{F}$ be the subspace of functions $\mathcal{M} \longrightarrow \mathbb{R}$ spanned by the Lie derivatives of h in the directions of $g_0, g_1, \ldots, g_m$, i.e.,

$$\mathcal{F} := \operatorname{span}_{\mathbb{R}} \left\{ L_{g_{i_1}} L_{g_{i_2}} \cdots L_{g_{i_l}} h : \; l \geq 0, \; 0 \leq i_j \leq m \right\}. \tag{22}$$

This is the *observation space* associated to (21); see e.g. [13, Remark 5.4.2].

Now for any $\mu = (\mu_0, \mu_1, \ldots)$ in $\mathbb{R}^{m,\infty}$, we define

$$\psi_i(x, \mu) = \left.\frac{d^i}{dt^i}\right|_{t=0} h(\varphi(t, x, u)) \tag{23}$$

for $i \geq 0$, where u is any C^∞ input with initial values $u^{(j)}(0) = \mu_j$. The functions $\psi_i(x, \mu)$ can be expressed, – applying repeatedly the chain rule, – as polynomials in the $\mu_j = (\mu_{1j}, \ldots, \mu_{mj})$ whose coefficients are analytic functions.

For each fixed $\mu \in \mathbb{R}^{m,\infty}$, let $\mathcal{F}_\mu$ be the subspace of functions from $\mathcal{M}$ to $\mathbb{R}$ defined by

$$\mathcal{F}_\mu = \operatorname{span}_{\mathbb{R}} \{ \psi_0(\cdot, \mu), \psi_1(\cdot, \mu), \psi_2(\cdot, \mu), \ldots \}, \tag{24}$$

and let $\mathcal{F}_\mu(x)$ be the space obtained by evaluating the elements of $\mathcal{F}_\mu$ at x for each $x \in \mathcal{M}$.

For system (21), we consider the series c_p, for each $p \in \mathcal{M}$, defined by

$$\langle c_p, X_{i_1} X_{i_2} \cdots X_{i_l} \rangle = L_{g_{i_l}} \cdots L_{g_{i_2}} L_{g_{i_1}} h(p). \tag{25}$$

According to [15, Lemma 4.2], this is always a convergent series. Note then that $p \not\sim q$ if and only if $c_p \neq c_q$ (see [6, 17]). Also, for each $i \geq 0$, it holds that

$$\psi_i(p, \mu) = \langle c_p, \mathfrak{d}_i(\mu_0, \ldots, \mu_{i-1}) \rangle,$$

where $\mathfrak{d}_i$ is still the same as defined in (7). For each $\mu \in \mathbb{R}^{\mathrm{m},\infty}$, we denote

$$\Psi_\mu(p) = (\psi_0(p,\mu), \psi_1(p,\mu), \psi_2(p,\mu), \ldots), \quad p \in \mathcal{M}.$$

Consider the set

$$\mathcal{J}_\Sigma := \{\mu \in \mathbb{R}^{\mathrm{m},\infty} : \quad \Psi_\mu(p) \neq \Psi_\mu(q), \ \forall p \not\sim q\},$$

and the set

$$\mathfrak{J} := \bigcap_\Sigma \mathcal{J}_\Sigma,$$

where the intersection is taken over the collection of all analytic systems with m inputs as in (21). Clearly, $\mathfrak{J} \supseteq \mathcal{J}$, and hence, the following is an immediate consequence of Theorem 2.

Corollary 1. *The set $\mathfrak{J}$ is a generic subset of $\mathbb{R}^{\mathrm{m},\infty}$.*

Using Corollary 1, one recovers the existence of universal inputs for analytic systems previously established in [14], but in a stronger form, uniformly on all state space systems of all dimensions with input functions taking values in $\mathbb{R}^m$.

References

1. K.T. Chen. Iterated path integrals. *Bulletin AMS*, 83:831–879, 1987.
2. M. Fliess. Réalisation locale des systèmes non linéaires, algèbres de Lie filtrées transitives et séries génératrices non commutatives. *Invent. Math.*, 71:521–537, 1983.
3. M. Fliess and C. Reutenauer. Une application de l'algèbre differentielle aux systèmes réguliers (ou bilinéaires). In A. Bensoussan and J. L. Lions, editors, *Analysis and Optimization of Systems*, Lecture Notes in Control and Information Science, pages 99–107. Springer-Verlag, Berlin, 1982.
4. M. Golubitsky and V. Guillemin. *Stable Mapping and Their Singularities.* Springer-Verlag, New York, 1973.
5. O.M. Grasselli and A. Isidori. Deterministic state reconstruction and reachability of bilinear control processes. *Proc. Joint Autom. Control Conf.*, 71:1423–1427, 1977.
6. A. Isidori. *Nonlinear Control Sytems.* Springer Verlag, New York, 3rd edition, 1995.
7. A.A. Muchnik. General linear automata. In I. Cox and G. Wilfong, editors, *Systems Theory Research, A.A. Lyapunov*, volume 23, pages 179–218. Consultants Bureau, 1973.
8. E.D. Sontag. On the observability of polynomial systems. I. Finite-time problems. *SIAM J. Contr. Optimization*, 17(1):139–151, 1979.
9. E.D. Sontag. A concept of local observability. *Systems & Control Letters*, 5(1):41–47, 1984.

10. E.D. Sontag. Universal nonsingular controls. *Systems & Control Letters*, 19(3):221–224, 1992.
11. E.D. Sontag. Spaces of observables in nonlinear control. *Proceedings of the International Congress of Mathematicians*, 1:1532–1545, 1994.
12. E.D. Sontag. Control of systems without drift via generic loops. *IEEE Trans. on Automat. Contr.*, 40(7):1210–1219, 1995.
13. E.D. Sontag. *Mathematical Control Theory: Deterministic Finite Dimensional Systems.* Springer-Verlag, New York, 2nd edition, 1998.
14. H.J. Sussmann. Single-input observability of continuous-time systems. *Mathematical Systems Theory*, 12:371–393, 1979.
15. H.J. Sussmann. Lie brackets and local controllability: A sufficient condition for scalar-input systems. *SIAM J. Contr. Optimization*, 21:686–713, 1983.
16. N.A.W. van Riel and E.D. Sontag. Parameter estimation in models combining signal transduction and metabolic pathways: The dependent input approach. *IEE Proc. Systems Biology*, 153:263–274, 2006.
17. Y. Wang and E.D. Sontag. On two definitions of observation spaces. *Systems & Control Letters*, 13:279–289, 1989.
18. Y. Wang and E.D. Sontag. Algebraic differential equations and rational control systems. *SIAM J. Contr. Optimization*, 30(6):1126–1149, 1992.
19. Y. Wang and E.D. Sontag. Generating series and nonlinear systems: Analytic aspects, local realizability and I/O representations. *Forum Mathematicum*, 4:299–322, 1992.
20. Y. Wang and E.D. Sontag. Orders of I/O equations and uniformly universal inputs. *CDC94*, 13:1270–1275, 1994.
21. Y. Wang and E.D. Sontag. Orders of input/output differential equations and state space dimensions. *SIAM J. Contr. Optimization*, 33(4):1102–1126, 1995.

System Interconnection

Jan C. Willems

Katholieke Universiteit Leuven, B-3001 Leuven, Belgium

Summary. Viewing a system as an architecture of subsystems is of central importance, both in modeling and in design. The aim of this article is to put forward a language for discussing the interconnection of dynamical systems. Under the influence of feedback control and signal processing, it has become customary to regard interconnections as output-to-input assignment. It is argued that this picture is not appropriate for physical systems, where it more logical to view interconnection as variable sharing. The modeling philosophy that emerges from this vantage point is tearing, zooming, and linking. This is formalized in the context of the notions from the behavioral approach, and illustrated by means of a number of examples.

1 Introduction

It is a pleasure to contribute an essay to this volume dedicated to Alberto Isidori on the occasion of his 65-th birthday. As the topic of my article, I chose an issue which is at the core of systems thinking, namely the formalization and the mathematization of system interconnection. This pertains to linear and nonlinear systems alike. In view of Alberto's early interest in foundational aspects of system theory, especially in the context of linear systems, and his later concentration on control problems for nonlinear systems, it is my intention to make this article a fitting tribute to his wide ranging scientific interests and to the influence that his work had in the field of systems and control theory.

Systems, physical and man-made alike, usually consist of interconnections of interacting subsystems. This feature is crucial in modeling, analysis, and synthesis. The notion of a dynamical system that took shape in the field of systems and control throughout the 20-th century is input/output based. This statement ignores the notion of a dynamical system as a 'flow', as used in mathematics, since we consider this setting totally inadequate as a general vantage point for modeling. The statement also ignore developments in computer science, were very subtle types of interactions have been put forward.

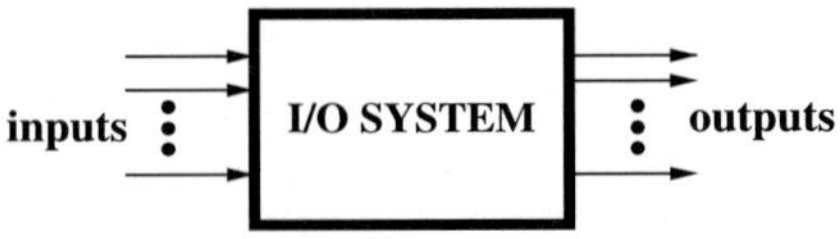

Fig. 1. Input/output system

The central idea in input/output thinking is that the environment acts on a system by imposing inputs, and that the system reacts by imposing outputs on its environment (see Figure 1). This mental image also suggests the functioning of interconnections, by assigning outputs of one system to function as inputs to another system (see Figure 2). These are very appealing ideas indeed, but the question should be examined if this is the mechanism by which the environment interacts with a system and if this is the way interconnections function in reality, physical and otherwise.

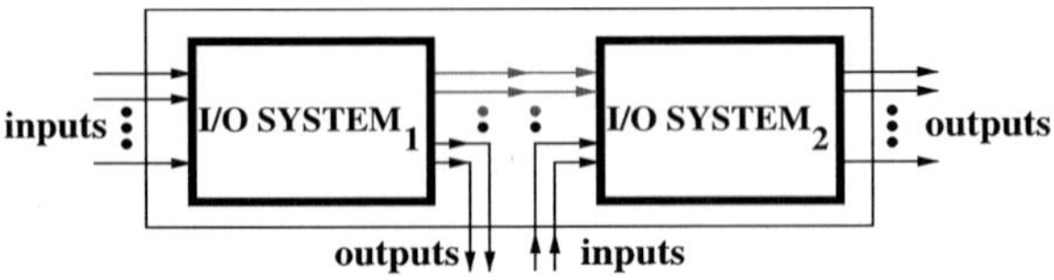

Fig. 2. Input/output interconnection

Early on, a system was regarded as an input/output *map*. This view is especially prevalent in signal processing, in classical control theory, and in Wiener's work. However, in all but the simplest examples, a dynamical input/output system is simply not a *map*. One can go a certain distance with this 'map' idea in the context of linear systems, say by assuming both the input and the output to be zero in the far past, or by restricting to square integrable inputs and outputs. But already this is very awkward, for example in connection with feedback or with unstable systems. However, the 'map' aspect is a totally inappropriate, indeed basically an impossible, starting point for nonlinear or discrete event systems.

Later on, state space systems came in vogue. By taking into consideration initial conditions, the state space point of view gives a much better vantage point to discuss dynamical models. Thus we arrived at

$$\dot{\mathrm{x}} = f(\mathrm{x}, \mathrm{u}, \mathrm{t}), \;\; \mathrm{y} = h(\mathrm{x}, \mathrm{u}, \mathrm{t})$$

as the mathematical structure on which system and control theory is based since the introduction of state models around 1960.

Both the classical input/output maps, as well as the modern input/state/output version, consider a system as an cause/effect relation. The input/output point of view led to signal flow graphs, and to system interconnection as *output-to-input assignment* (see Figure 2). In control and in signal processing (and, but to a lesser extent, in circuit theory), signal flow diagrams combining series, parallel, and feedback connections are viewed as the standard way to deal with interconnections. Since an adder can be viewed as a input/output system, with two inputs and their sum as output, we end up with output-to-input assignment as the basic operation by which systems are interconnected. Since this also fits the classical picture of control as *feedback* so very well, it is this view that came to dominate the field of systems and control.

2 Tearing, Zooming, and Linking

From an applications point of view, the input/output framework is much more restrictive than one is often led to believe. The architecture formalized by the signal flow arrows is often viewed as essential for describing the interaction of a system with its environment. But, the opposite is actually the case, especially for the description of physical systems and for describing their interconnections. In many situations, signal flow graphs are unphysical, a figment of the imagination, cumbersome, and unnecessary. Sharing common variables is a much more key idea for system interconnection than output-to-input connection.

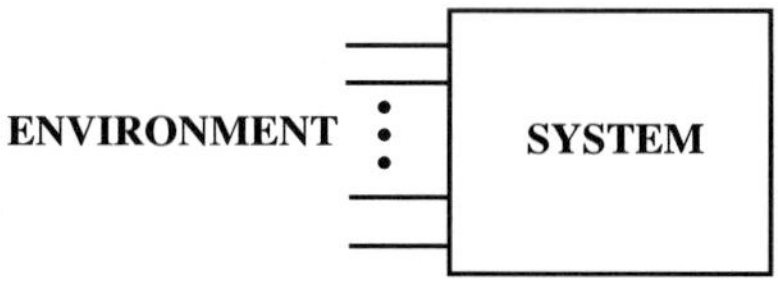

Fig. 3. Blackbox

A typical modeling task can be viewed as follows. Our aim is to model the dynamic behavior of a number of related variables. This is visualized by means of a blackbox (see Figure 3) with a number of terminals. One should think of these terminals as 'places' where the variables which we set out to model 'live'. Sometimes one should take these terminals literally, sometimes not. In first instance, this only means that the modeler has declared what the variables of interest are: the terminals are merely a visualization. Often, thou, the terminals are real, and the aim is to model the variables associated with physical terminals through which a system interacts with its environment. When dealing with interconnections, it is natural to assume
(i) that these terminals and their variables are real physical entities, and
(ii) that there are usually *many* physical variables collectively and indivisibly associated with *one and the same* terminal.

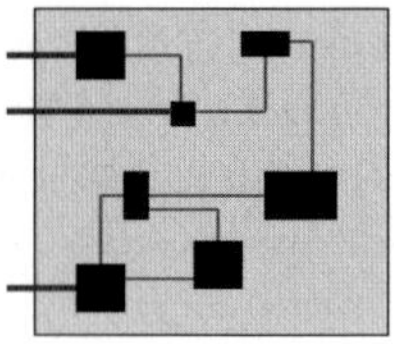

Fig. 4. Greybox

Most systems consist of interacting components. In order to discover these interactions, we look inside the blackbox of Figure 3, where we find an interconnection architecture of 'smaller' blackboxes that interact through terminals of their own (see Figure 4). Modeling then proceeds by examining the smaller blackboxes and their interactions.

This modeling process is called *tearing, zooming, and linking* .

1) *Tearing* refers to viewing a system as an interconnection of smaller subsystems.
2) *Zooming* refers to modeling the subsystems.
3) *Linking* refers to modeling the interconnections.

There is an obvious hierarchical structure in this modeling process. Indeed, zooming involves modeling the (dynamic) laws that govern the variables on the terminals of a subsystem. This subsystem may in turn consist of interacting sub-sub-systems. Modeling the subsystem then again involves tearing, zooming, and linking. This goes on until we meet components whose model specification follows from first principles, or a subsystem whose model has been stored in a database, or where system identification is the modeling procedure that is called for.

The question which we examine in this paper is what actually happens when subsystems are interconnected. Our theme is that this does not (usually) imply *input-to-output assignment*, but rather *variable sharing.* However, in order to put these ideas in their proper setting, we briefly backtrack to the mathematical description of dynamical systems outside the input/output setting.

3 Behavioral Systems

Over the last two decades, a framework for the study of systems has been developed that does not take the input/output structure as its starting point. The 'behavioral approach', as this has been called, simply identifies the dynamics of a system with a family of trajectories, called the *behavior*, and develops systems theory (including control) from there.

The behavioral framework views modeling as follows. Assume that we have a phenomenon that we wish to describe mathematically. Nature (that is, the reality that governs this phenomenon) can produce certain events (also called outcomes). The totality of possible events (*before* we have modelled the phenomenon) forms a set $\mathbb{U}$, called the *universum*. A *mathematical model* of the phenomenon restricts the outcomes that are declared possible to a subset $\mathcal{B}$ of $\mathbb{U}$; $\mathcal{B}$ is called the *behavior* of the model. We refer to $(\mathbb{U}, \mathcal{B})$ (or to $\mathcal{B}$ by itself, since $\mathbb{U}$ usually follows from the context) as a mathematical model.

As an example, consider the *ideal gas law*, which poses $PV = kNT$ as the relation between the pressure P, the volume V, the number N of moles, and the temperature T of an ideal gas, with k a universal physical constant. The universum $\mathbb{U}$ is $(\mathbb{R}_+)^4$, and the behavior $\mathcal{B} = \{(P, V, N, T) \in (\mathbb{R}_+)^4 \mid PV = kNT\}$.

In the study of dynamical systems we are more specifically interested in situations where the events are signals, trajectories, i.e. maps from a set of *independent variables* (time, in the present paper) to a set of *dependent variables* (the values taken on by the signals). In this case the universum is the collection of all maps from the set of independent variables to the set of dependent variables. It is convenient to distinguish these sets explicitly in the notation: $\mathbb{T}$ for the set of independent variables, and $\mathbb{W}$ for the set of dependent variables. $\mathbb{T}$ suggests 'time', the case of interest in the present article. Whence a (dynamical) *system* is defined as a triple

$$\Sigma = (\mathbb{T}, \mathbb{W}, \mathcal{B})$$

with $\mathcal{B}$, *the behavior*, a subset of $\mathbb{W}^{\mathbb{T}}$ ($\mathbb{W}^{\mathbb{T}}$ is standard mathematical notation for the set of all maps from $\mathbb{T}$ to $\mathbb{W}$). The behavior is the central object in this definition. It formalizes which signals $w : \mathbb{T} \to \mathbb{W}$ are possible, according to the model: those in $\mathcal{B}$, and which are not: those not in $\mathcal{B}$. The behavioral framework treats a model for what it is: an exclusion law. Of course, in applications, the behavior $\mathcal{B}$ must be specified somehow, and it is here that differential equations (and difference equations for discrete-time systems) enter the scene.

In the equations describing a behavior, very often other variables appear in addition to those the model aims at. The origin of these auxiliary variables varies from case to case. They may be state variables (as in flows, automata, and input/state/output systems); they may be potentials (as in the well-known expressions for the solutions of Maxwell's equations); most frequently and most germane for the purposes of the present article, they are interconnection variables. It is important to incorporate these auxiliary variables in our modeling language *ab initio*, and to distinguish clearly between the variables whose behavior the model aims at, and the auxiliary variables introduced in the modeling process. The former are called *manifest* variables and the latter *latent* variables.

A *mathematical model with latent variables* is defined as a triple $(\mathbb{U}, \mathbb{L}, \mathcal{B}_{\text{full}})$, with $\mathbb{U}$ the universum of manifest variables, $\mathbb{L}$ the universum of latent variables, and $\mathcal{B}_{\text{full}} \subseteq \mathbb{U} \times \mathbb{L}$ the *full behavior*. It induces (or *represents*) the *manifest model* $(\mathbb{U}, \mathcal{B})$, with $\mathcal{B} = \{w \in \mathbb{U} \mid \text{there exists } \ell \in \mathbb{L} \text{ such that}$

$(w, \ell) \in \mathcal{B}_{\text{full}}\}$. A (dynamical) *system with latent variables* is defined completely analogously as

$$\Sigma_{\text{full}} = (\mathbb{T}, \mathbb{W}, \mathbb{L}, \mathcal{B}_{\text{full}})$$

with $\mathcal{B}_{\text{full}} \subseteq (\mathbb{W} \times \mathbb{L})^{\mathbb{T}}$. The notion of a system with latent variables is the natural end-point of a modeling process and hence a very natural starting point for the analysis and synthesis of systems. More details and examples of behavioral systems may be found in [3, 4].

The procedure of modeling by *tearing, zooming, and linking* is an excellent illustration of the appropriateness of the behavioral approach. We assume throughout finiteness, i.e., a finite number of subsystems are interconnected, each with a finite number of terminals. Our view of interaction through terminals is certainly not the end point of the development of formalizing the interaction of systems. There are many interactions between subsystems that do not fit this 'terminal' paradigm: actions at a distance (as gravity), rolling and sliding, mixing, components that are interconnected through distributed surfaces, etc. Interconnecting systems through terminals fits very well lumped electrical and mechanical systems, many hydraulic systems, some thermal systems, etc. Interconnection via terminals also serves as a useful paradigm for more complex situations.

4 Formalization

In this section an outline is given of a formal procedure for obtaining a model by viewing a system (a blackbox) as an interconnection of subsystems (smaller blackboxes). The idea is to formalize the picture shown in Figure 2: a finite number of systems are interconnected through terminals by to other subsystems. This suggests a graph with the subsystems in the nodes, and the interconnections in the edges. As we shall see, this formalism uses the notions of a behavior and of latent variables in an effective way.

The ingredients are:

1) terminals,
2) (parameterized) modules,
3) the interconnection architecture,
4) the module embedding, and
5) the manifest variable assignment.

4.1 Terminals

A *terminal* is specified by its *type*. Giving the type of a terminal identifies the kind of a physical terminal that we are dealing with. The type of terminal implies a universum of *terminal variables*. These variables are physical quantities that characterize the possible 'signal states' on the terminal, it specifies

Table 1. Examples of terminals

Type of terminal	Variables	Universum
electrical	(voltage, current)	$\mathbb{R} \times \mathbb{R}$
1-D mechanical	(force, position)	$\mathbb{R} \times \mathbb{R}$
2-D mechanical	(position, attitude, force, torque)	$\mathbb{R}^2 \times [0, 2\pi) \times \mathbb{R}^2 \times \mathbb{R}$
thermal	(temperature, heat-flow)	$\mathbb{R}_+ \times \mathbb{R}$
fluidic	(pressure, mass-flow)	$\mathbb{R} \times \mathbb{R}$
input	u	$\mathbb{R}$
output	(y)	$\mathbb{R}$
etc.	etc.	etc.

how the module interacts with the environment through this terminal. Some examples of terminals are given below.

4.2 Modules

A *module* is specified by its *type*, and its *behavior*. Giving the type of a module identifies the kind of physical system that we are dealing with. Giving a *behavior specification* of a module implies giving a *representation* and the values of the *parameters* associated with a representation. Combined, these specify the behavior of the variables on the terminals of the module. The type of a module implies an ordered set of terminals. Since each of the terminals comes equipped with a universum of terminal variables, we thus obtain an ordered set of variables associated with that module. The module behavior then specifies what time trajectories are possible for these variables. Thus a module defines a dynamical system $(\mathbb{R}, \mathbb{W}, \mathcal{B})$ with $\mathbb{W}$ the Cartesian product over the terminals of the universa of the terminal variables.

However, there are very many ways to specify a behavior (for example, as the solution set of a differential equation, as the image of a differential operator, through a latent variable model, through a transfer function, and many other ways). The behavioral representation picks out one of these. These representations will in first instance contain unspecified parameters (for example, the coefficients of the differential equation, or the rational functions in a transfer function). Giving the parameter values specifies their numerical values, and completes the specification of the behavior of the signals that are possible on the terminals of a module.

We give two examples. The first is a simple 3 Ohm resistor (see Figure 5). The module type is *ohmic resistor*. This means that it has two terminals, both of 'electrical' type, and that it is parameterized by a non-negative real number (the value of the resistor in Ohms). Since the terminals are electrical, there are two variables, a voltage and a current (counted positive when it runs into the device), on each terminal. This yields in total four real variables associated with a resistor: (V_1, I_1) and (V_2, I_2). From the fact that we

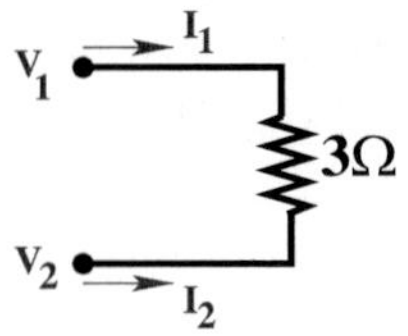

Fig. 5. A resistor

have an ohmic resistor, we know that the relation between these variables is

$$V_1 - V_2 = RI_1, \ I_1 + I_2 = 0.$$

Giving in addition the value of the parameter $R = 3$ of the Ohmic resistor leads to the behavioral equations

$$V_1 - V_2 = 3I_1, \ I_1 + I_2 = 0.$$

These equations completely specify the behavior of the terminal variables of a 3 ohm resistor.
Our second example of a module is a transfer function. The module type is *multivariable proper transfer function.* Its parameters are $(\mathtt{m}, \mathtt{p}, G)$, with $\mathtt{m}, \mathtt{p} \in \mathcal{N}$ and G a $\mathtt{p} \times \mathtt{m}$ matrix of proper real rational functions. This means that we have a system with $\mathtt{m} + \mathtt{p}$ terminals, the first $\mathtt{m}$ of 'input' type, the last $\mathtt{p}$ of 'output' type, and behavior described by the controllable input/state/output system

$$\tfrac{d}{dt}x = Ax + Bu, \ y = Cx + Du,$$

with A, B, C, D such that $G(\xi) = D + C(I\xi - A)^{-1}B$. The behavior of this system consists of all $(u, y) : \mathbb{R} \to \mathbb{R}^{\mathrm{m}} \times \mathbb{R}^{\mathrm{p}}$ for which there exists $x : \mathbb{R} \to \mathbb{R}^{\mathrm{n}}$ such that these equations are satisfied: the state serves as a latent variable. Of course, to be precise, we would have to add some smoothness, but we will slide over these technical points, since they are not germane to the purposes of this article.

This representation of the module behavior requires specification of the numerical value of the state space system parameter matrices A, B, C, D. We have identified 'transfer function' with controllable linear time-invariant differential system. In this case there are very many other ways of translating this specification into dynamic equations. For example, by using left or right polynomial co-prime factorizations of the transfer function, we obtain differential equation representations in kernel or image form. By using factorizations with rational functions, we can obtain proper stable rational functions as parametrization. This class of systems, linear time-invariant differential systems, have been dealt with extensively in the literature.

Some general examples of modules types with their terminals and of behavioral specifications are given in the tables above.

Table 2. Examples of modules

Type of module	Terminals	Type of terminals
resistor	(terminal1, terminal2)	(electrical, electrical)
transistor	(collector, emitter, base)	(electrical, idem,idem)
mass, 2 applicators	(appl1, appl2)	(3-D mechanical, idem)
2-inlet vessel	(inlet1, inlet2)	(fluidic, fluidic)
heat exchanger	(inlet, outlet)	(fluidic-thermal, idem)
signal processor	(in, out)	(input, output)
etc.	etc.	etc.

Table 3. Examples of module specifications

Type of module	Specification	Parameter
resistor	default	R in Ohms
n-terminal circuit	transfer impedance	$G \in \mathbb{R}^{\mathtt{n}\times\mathtt{n}}(\xi)$
n-port circuit	i/s/o admittance	(A, B, C, D)
bar, 2 applicators	Lagrangian equations	mass and length
2-inlet vessel	default	geometry
signal processor	kernel representation	$R \in \mathbb{R}[\xi]^{\bullet\times\bullet}$
signal processor	latent variable	(R, M)
etc.	etc.	etc.

A module Σ of a given type with T terminals yields the signal space $\mathbb{W} = \mathbb{W}_1 \times \mathbb{W}_2 \times \cdots \times \mathbb{W}_T$, with $\mathbb{W}_k$ the universum associated with the k-th terminal. The behavioral specification yields the behavior $\mathcal{B} \subseteq \mathbb{W}^{\mathbb{R}}$. If $(w_1, w_2, \ldots, w_T) \in \mathcal{B}$, then we think of the w_k's as signals $w_k : \mathbb{R} \to \mathbb{W}_k$ that can occur on the k-th terminal.

4.3 The Interconnection Architecture

The next element in the specification of a model is the *interconnection architecture* (or *interconnection graph*). This is defined as a graph with leaves. Recall that a *graph* is defined as $\mathcal{G} = (\mathbb{V}, \mathbb{E}, \mathcal{A})$, with $\mathbb{V}$ the set of *vertices*, $\mathbb{E}$ the set of *edges*, and $\mathcal{A}$ the *adjacency map*. $\mathcal{A}$ associates with each edge $e \in \mathbb{E}$ an unordered pair $\mathcal{A}(e) = [v_1, v_2]$ with $v_1, v_2 \in \mathbb{V}$, in which case e is said to be *adjacent* to v_1 and v_2. A graph with leaves is a graph in which some of the 'edges' are adjacent to only one vertex. These special 'edges' are called 'leaves'. Formally, a *graph with leaves* is defined as $\mathcal{G} = (\mathbb{V}, \mathbb{E}, \mathbb{L}, \mathcal{A})$, with $\mathbb{V}$ the set of *vertices*, $\mathbb{E}$ the set of *edges*, $\mathbb{L}$ the set of *leaves*, and $\mathcal{A}$ the *adjacency map*. $\mathcal{A}$ associates with each edge $e \in \mathbb{E}$ an unordered pair $\mathcal{A}(e) = [v_1, v_2]$ with $v_1, v_2 \in \mathbb{V}$, and with each leaf $\ell \in \mathbb{L}$ an element $\mathcal{A}(\ell) = v \in \mathbb{V}$, in which case e is said to be *adjacent* to v_1 and v_2, and ℓ to v.

4.4 The Module Embedding

The *module embedding*
(i) associates with each vertex of the interconnection architecture a module, and
(ii) specifies for every vertex a $1 \leftrightarrow 1$ assignment between the edges and leaves adjacent to the vertex and the terminals of the module that has been associated with this vertex.
This is illustrated in the figure below.

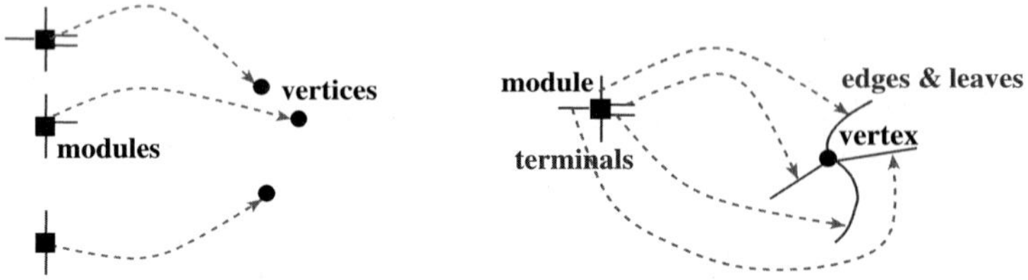

Fig. 6. Terminal assignment

Since each edge is adjacent to two vertices, each edge is associated by the module embedding with 2 terminals. It is assumed that this assignment results in terminals that are of the same type if the type is physical (both electrical, or mechanical, or hydraulic, or thermal, etc.), or of opposite type (one input, one output) if the terminals are of logical type. In other words, if the edge e is adjacent to vertices v_1 and v_2, then the module embedding makes v_1 and v_2 either of the same physical type, or of opposite logical type. In this way, each edge and leaf is labelled by a terminal type, and each vertex is labelled as a module.

Consider again a few examples. The first is the electrical circuit shown in Figure 7. The goal is to model the behavior of the voltage and current in the external port.

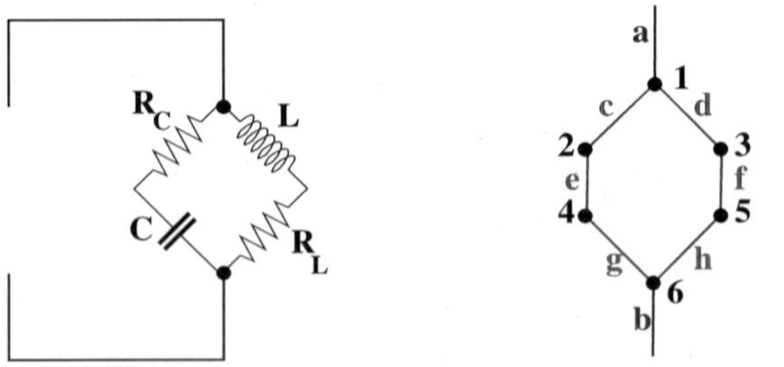

Fig. 7. RLC circuit

This circuit has 6 modules. Two resistors with parameter values R_C and R_L respectively, one capacitor with parameter value C, one inductor with parameter value L, and two connectors with parameter value 3 (meaning that it connects 3 terminals). All terminals are of electrical type, the resistors, capacitor, and inductor each have 2 terminals, and the connectors both have 3. The interconnection architecture is shown on the right side of Figure 7. There are 6 vertices, labelled 1 to 6, 6 edges, labelled c, d, e, f, g, h, and 2 leaves labelled a, b. The module embedding consists of

$$R_C \mapsto 2, R_L \mapsto 5, C \mapsto 4, L \mapsto 3, \text{connector}_1 \mapsto 1, \text{connector}_2 \mapsto 6.$$

Because of the special symmetries that are valid for the electrical elements used, we need not specify how the terminals of the modules are exactly associated with the edges. If, for example, there would have been a diode in edge 2, we would have had to specify if its current blocking direction is associated with edge c or with edge e.
The second example is the classical feedback system shown in Figure 8. The interconnection architecture is the graph with vertices A_1, A_2, G_1, and G_2, edges $3, 4, 5, 6$, and leaves $1, 2$, shown on the right side of the figure. The mod-

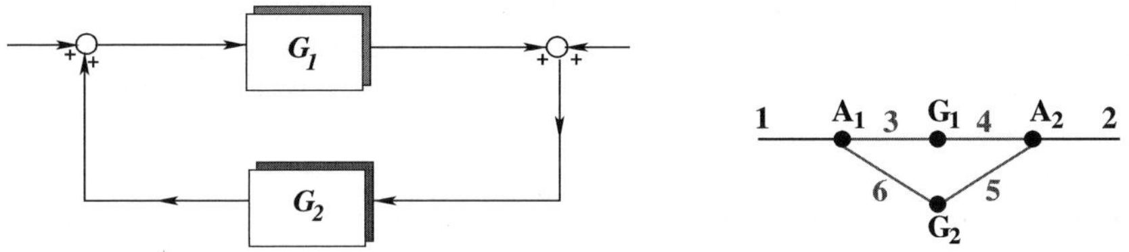

Fig. 8. Feedback system

ules are two adders, associated with vertices A_1 and A_2, each with 2 inputs and 1 output, and two input/output systems, associated with vertices G_1 and G_2. The module embedding requires that the appropriate input-to-output directions are respected.

4.5 Interconnection Equations

The edges of the interconnection architecture specify how terminals of modules are linked. *Assume that there are universal rules that specify relations among the variables on the terminals that are linked.* Pairing of terminals by the edges of the interconnection architecture implies an *interconnection law.* Some examples of interconnection laws are shown in the table below.

Proceeding this way leads to a complete set of behavioral equations:
(i) For each vertex of the interconnection architecture, we obtain a behavior relating the variables that 'live' on the terminals of the module that is associated with this vertex. These behavioral equations are called the *module equations.*

Table 4. Examples of interconnection laws

Pair of terminals	Variables terminal 1	Variables terminal 2	Interconnection constraints
electrical	(V_1, I_1)	(V_2, I_2)	$V_1 = V_2, I_1 + I_2 = 0$
1-D mechanical	(F_1, q_1)	(F_2, q_2)	$F_1 + F_2 = 0, q_1 = q_2$
thermal	(Q_1, T_1)	(Q_2, T_2)	$Q_1 + Q_2 = 0, T_1 = T_2$
fluidic	$(p_1, f_1$	(p_2, f_2)	$p_1 = p_2, f_1 + f_2 = 0$
information processing	input u	output y	$u = y$
etc.	etc.	etc.	etc.

(ii) For each edge of the interconnection architecture, we obtain behavioral equations relating the variables that 'live' on the terminals and that are linked by this edge. These behavioral equations are called the *interconnection equations.* Note that no interconnection equations result from the leaves, but the associated terminal variables do enter in the module equations.
These equations together specify the behavior of all the variables on all the terminals involved. Each vertex of the interconnection graph is in the end labelled as a module, and each edge as a terminal type: we have systems in the vertices, and interconnections in the edges. This stands in contrast to conventional electrical circuit theory, which has the elements (i.e. modules) in the edges, and the interconnections in the vertices. The interconnection equations are usually very simple (see the table above). Typically they equate certain variables and put the sum of other variables to zero. We therefore think of interconnection as *variable sharing.*

For the examples discussed earlier, and with the obvious notation, we obtain the following specification of the behavior.
1. For the circuit, we obtain the module equations

$$\begin{aligned} V_{c''} - V_{e'} &= R_C I_{c''},\ I_{c''} = I_{e'}; \\ V_{f''} - V_{h'} &= R_L I_{f''},\ I_{f''} = I_{h'}; \\ C\tfrac{d}{dt}\left(V_{c''} - V_{e'}\right) &= I_{c''},\ I_{c''} = I_{e'}; \\ L\tfrac{d}{dt} I_{d''} &= \left(V_{d''} - V_{f'}\right),\ I_{d''} = I_{f'}; \\ V_a = V_{c'} &= V_{d'},\ I_a + I_{c'} + I_{d'} = 0; \\ V_b = V_{g''} &= V_{h''},\ I_b + I_{g''} + I_{h''} = 0, \end{aligned}$$

and the interconnection equations:

$$\begin{aligned} V_{c'} &= V_{c''},\ I_{c'} + I_{c''} = 0; \\ V_{d'} &= V_{d''},\ I_{d'} + I_{d''} = 0; \\ V_{e'} &= V_{e''},\ I_{e'} + I_{e''} = 0; \\ V_{f'} &= V_{f''},\ I_{f'} + I_{f''} = 0; \\ V_{g'} &= V_{g''},\ I_{g'} + I_{g''} = 0; \\ V_{h'} &= V_{h''},\ I_{h'} + I_{h''} = 0. \end{aligned}$$

2. For the feedback system, we obtain the module equations

$$(u_3, y_4) \in \mathcal{B}_{G_1}; \ (u_5, y_6) \in \mathcal{B}_{G_2};$$
$$y_2 = u_1 + u_6; \ y_4 = u_2 + u_4,$$

and the interconnection equations

$$u_3 = y_3; \ u_4 = y_4; \ u_5 = y_6; \ u_6 = y_6.$$

Here $\mathcal{B}_{G_1}$ and $\mathcal{B}_{G_2}$ denote the behavior of the input/output systems in respectively the forward loop and the feedback loop of the feedback system.

4.6 The Manifest Variable Assignment

The final step consists of the *manifest variable assignment.* This is a mapping that assigns the manifest variables as a function of the terminal variables. *The terminal variables are henceforth considered as latent variables.*

For the circuit example the manifest variable assignment consists of the maps that defines the port voltage and port current as a function of the terminal voltages and currents:

$$V_{\text{external}} = V_a - V_b, \ I_{\text{external}} = I_a.$$

For the feedback system, the manifest variable assignment consists of

$$u_{\text{external}} = (u_1, u_2), \ y_{\text{external}} = (y_6, y_5).$$

The behavioral equations (the combination of the module equations and the interconnection equations), combined with the manifest variable assignment, define the full behavior of the system that is being modelled. It is the end result of the modeling process based on tearing ($\cong$ the interconnection architecture), zooming ($\cong$ obtaining the module equations and the manifest variable assignment), and linking ($\cong$ setting up the interconnection equations).

This tearing-zooming-linking modeling methodology has many virtues: it is *systematic* and *modular*, it is *adapted to computer assisted modeling* (with module equations in parametric form stored in a database, and with the interconnection equations also stored in a database), it is *hierarchical* (once a model of a system has been obtained, it can be used as a subsystem-module on a higher level). A good model library will have items that are *re-useable, extendable, modifiable, flexible,* etc.).

Disadvantages are that it immediately involves many variables. This can be alleviated by the (partial) elimination of latent variables when possible. For example, the interconnection equations often immediately lead to elimination of half of the interconnection variables. There are situations where the special structure of the modules and the interconnections allow a more direct elimination of some of the variables. For example, modeling of mechanical systems using Lagrangians, modeling of electrical circuits using ports instead of terminals, etc.

Our philosophy is to keep the interconnections highly standardized and simple, and to deal with complex models in the modules. For example, in the electrical circuit, a multi-terminal connector was viewed as a module, not as a connection. Also, in mechanical systems, joints, hooks and hinges should be viewed as modules, not as connections.

As a caveat, we should emphasize that not all interconnections or interactions of subsystems fit the framework described above. In particular, distributed interconnections were not considered, for example mechanical systems that are interconnected by sharing a surface, or heat conduction along a surface. Also, terminals do not capture interconnections along virtual terminals, as action at a distance. Finally, interactions as rolling, sliding, bouncing, mixing, etc., also require a different setting.

The resulting graph structure of an interconnected system has the modules in the nodes and the interconnections as the branches. This is faithful to the physics, and should be contrasted with the graph structure pursued in electrical circuit theory, which has the modules in the branches and the connectors in the nodes. This works fine with 2-terminal elements, but is awkward otherwise, and is difficult to generalize to other, non-electrical, domains.

5 Interconnected Behavior

We now formalize the interconnected system. The most effective way to proceed is to specify it as a latent variable system, with as manifest variables the variables specified in the manifest variable assignment, and as latent variables all the terminal variables associated with all the modules. Its full behavior behavior then consists of the behavior as specified by each of the modules, combined by the interconnection laws obtained by the interconnection architecture. A first principles model of an interconnected system obtained this way always contains many latent variables. This is one of the main motivations to introduce latent variables in our modeling language *ab initio*. It also underscores the importance of the *elimination theorem* [1, 2, 4].

The modeling procedure described above has some similarity with modeling using *bondgraphs*. However, there are important differences, especially that in our setting it is not required that the interconnection variables have any relation to the energy that is transmitted along the interconnection terminals. The comparison with bondgraphs is taken up in detail in [5].

6 Conclusions

Modeling interconnected via the above method of *tearing, zooming, and linking* provides the prime example of the usefulness of behaviors and the inadequacy of input/output thinking. Even if our system, after interconnection, allows for a natural input/output representation, it is unlikely that this will be the case of the subsystem and of the interconnection architecture. It is only when considering the more detailed signal flow graph structure of a system that input/output thinking becomes useful. Signal flow graphs are useful building blocks for interpreting information processing systems, but physical systems need a more flexible framework.

Acknowledgments

This research is supported by the Research Council KUL project CoE EF/05/006 (OPTEC), Optimization in Engineering, and by the Belgian Federal Science Policy Office: IUAP P6/04 (Dynamical systems, Control and Optimization, 2007–2011).

References

1. T. Cotroneo. *Algorithms in Behavioral Systems Theory.* Doctoral dissertation, Faculty of Mathematics and Physical Sciences, University of Groningen, 2001.
2. T. Cotroneo and J.C. Willems. The simulation problem for high order differential equations. *Applied Mathematics and Computation*, 145:821–851, 2003.
3. J.W. Polderman and J.C. Willems. *Introduction to Mathematical Systems Theory: A Behavioral Approach.* Springer-Verlag, 1998.
4. J.C. Willems. Paradigms and puzzles in the theory of dynamical systems. *IEEE Trans. on Automat. Contr.*, 36:259–294, 1991.
5. J.C. Willems. Modeling open and interconnected systems. *IEEECSM*, 2007. To appear.

Reduced Order Modeling of Nonlinear Control Systems

Arthur J. Krener[1,2]

[1] Department of Mathematics, University of California, Davis, CA 95616-8633, USA
[2] Department of Applied Mathematics, Naval Postgraduate School, Monterey, CA 93943-5216, USA

This paper is dedicated to my good friend and esteemed colleague Alberto Isidori on the occasion of his 65th birthday.

Summary. We consider the problem of model reduction for nonlinear control systems. We introduce the concept of input normal form of degree d and show that a sufficiently smooth nonlinear control system can always be brought to this form by local changes of state coordinates. The changes of coordinates are not uniquely defined but the resulting normal form of the controllability and observability functions are if $d \leq 6$. The parameters in this normal form indicates the relative importance of the state coordinates to the input output map of the control system. Then we offer a new interpretation of linear balanced truncation and show how it can be extended to nonlinear system. Finally we offer new estimates of error between the full and reduced Hankel maps.

1 Introduction

The theory of model reduction for linear control systems was initiated by B. C. Moore [9]. His method, called balanced truncation, is applicable to controllable, observable and exponentially stable linear systems in state space form. The reduction is accomplished by making a linear change of state coordinates to simultaneously diagonalize the controllability and observability gramians and make them equal. Such a state space realization is said to be balanced. The diagonal entries of the gramians are the singular values of the Hankel map from past inputs to future outputs. The balanced reduction is accomplished by Galerkin projection onto the states associated to the largest singular values. The method is intrinsic in that the reduced order model depends only on the dimension of the reduced state space.

Scherpen [12] extended Moore's method to locally asymptotically stable nonlinear systems. She defined the controllability and observability functions which are the nonlinear analogs of the controllability and observability gramians. Scherpen made a change of state coordinates that took the system into input normal form where the controllability function is one half of the sum of squares of the state coordinates. She then made additional changes of state coordinates that preserved the input normal form while diagonalizing the observability function where the diagonal entries, which she called the singular value functions, are state dependent. She reduced the system by Galerkin projection onto coordinates with the largest singular value functions.

The reduction technique of Scherpen is not intrinsic. The singular value functions themselves are not unique [6]. Moreover the resulting reduced order system depends on the changes of coordinates that are used to achieve it and these are not unique.

The goal of this paper is to present a more intrinsic method of nonlinear model reduction. Our approach differs from Scherpen in that we analyze the changes of coordinates degree by degree and give a normal form of the controllability and observability functions for each degree. Generically this normal form of the controllability and observability functions is unique up through terms of degree 7 and it is diagonalized in some sense. There are many changes of coordinates that achieve the normal form and this choice can affect the lower order model.

2 Input Normal Form of Degree d

Suppose we have an n dimensional system of the form

$$\begin{aligned} \dot{x} = f(x,u) &= Fx + Gu + f^{[2]}(x,u) + \ldots + f^{[d]}(x,u) + O(x,u)^{d+1} \\ y = h(x) &= Hx + h^{[2]}(x) + \ldots + h^{[d]}(x) + O(x)^{d+1} \end{aligned} \tag{1}$$

where $f(x,u),\ h(x)$ are $C^{d+1}, d \geq 1$ in some neighborhood of the equilibrium $x = 0, u = 0$. The superscript $^{[j]}$ denotes a function that is homogeneous and polynomial of degree j in its arguments so the right sides of the above are the Taylor series expansions of $f,\ h$ around $x = 0,\ u = 0$ with remainders of degree $d + 1$.

Following Moore [9] we assume that F is Hurwitz, i.e., all its eigenvalues lie in the open left half plane, F, G is a controllable pair and and H, F is an observable pair. Scherpen [12] defined the controllability and observability functions of the system. The controllability function $\pi_c(x)$ is the solution of the optimal control problem

$$\pi_c(x^0) = \inf_{u(-\infty:0)} \frac{1}{2} \int_{-\infty}^{0} |u|^2 dt \tag{2}$$

subject to the system (1) and the boundary conditions

$$\begin{aligned} x(-\infty) &= 0 \\ x(0) &= x^0. \end{aligned}$$

The notation $u(-\infty : 0)$ denotes a function in $L^2((-\infty, 0), I\!R^m)$. Loosely speaking $\pi_c(x)$ is the minimal "input energy" needed to excite the system from the zero state to x over the time interval $(-\infty, 0]$.

If $\pi_c(x)$ exists and is smooth then it and the optimal control $u = \kappa(x)$ satisfy the Hamilton-Jacobi-Bellman equations

$$0 = \frac{\partial \pi_c}{\partial x}(x) f(x, \kappa(x)) - \frac{1}{2}|\kappa(x)|^2, \qquad 0 = \frac{\partial \pi_c}{\partial x}(x) \frac{\partial f}{\partial u}(x, \kappa(x)) - \kappa'(x) \tag{3}$$

locally around $x = 0$ where $'$ denotes transpose. The negative signs in front of second terms in the above equations occur because we are considering an optimal control problem on $(-\infty, 0]$ rather than the more usual $[0, \infty)$.

Because F is Hurwitz and F, G is a controllable pair then from [1], [8] we know there there exists a unique local solution of these equations around $x = 0$ where $\pi_c(x)$ is C^{d+2} and $\kappa(x)$ is C^{d+1}. Moreover the Taylor series of this solution can be computed term by term,

$$\begin{aligned} \pi_c(x) &= \tfrac{1}{2} x' P_c^{-1} x + \pi_c^{[3]}(x) + \ldots + \pi_c^{[d+1]}(x) + O(x)^{d+2} \\ \kappa(x) &= Kx + \kappa^{[2]}(x) + \ldots + \kappa^{[d]}(x) + O(x)^{d+1} \end{aligned} \tag{4}$$

where $P_c > 0$ is controllability gramian, i.e., the unique solution to linear Lyapunov equation

$$0 = P_c F + F' P_c + GG' \tag{5}$$

and the linear part of the feedback is

$$K = G' P_c^{-1} . \tag{6}$$

The controllability gramian is finite because F, G is a controllable pair and positive definite because F is Hurwitz. The higher degree terms of $\pi(x)$, $\kappa(x)$ can be computed degree by degree following the method of Al'brecht [1].

The observability function $\pi_o(x)$ is defined by

$$\pi_o(x^0) = \frac{1}{2} \int_0^\infty |y(t)|^2 dt$$

subject to the system (1) and the initial condition

$$x(0) = x^0 .$$

Since F is Hurwitz we are assured that if x^0 is small enough then $x(t) \to 0$ exponentially fast as $t \to \infty$ so $y(0 : \infty) \in L^2((0, \infty), I\!R^p)$. Again speaking loosely $\pi_c(x)$ is the "output energy" that is released by the system over the time interval $[0, \infty)$ when it is started at $x(0) = x$ and the input is zero.

The observability function satisfies the nonlinear Lyapunov equation

$$0 = \frac{\partial \pi_o}{\partial x}(x) f(x, 0) + \frac{1}{2}|h(x)|^2. \tag{7}$$

Because F is Hurwitz and H, F is an observable pair then there exists a unique C^{d+2} solution of this equation defined locally around $x = 0$. The Taylor series of this solution can also be computed term by term,

$$\pi_o(x) = \frac{1}{2}x' P_o x + \pi_o^{[3]}(x) + \ldots + \pi_o^{[d+1]}(x) + O(x)^{d+2}$$

where $P_o > 0$ is the observability gramian, i.e., the unique solution to the linear Lyapunov equation

$$P_o F' + F P_o + H' H = 0.$$

The observability gramian is finite because F is Hurwitz and positive definite because $H,\ F$ is an observable pair.

From [9], [12] we know that we can choose a linear change of coordinates so that in the new coordinates also denoted by x

$$\pi_c(x) = \frac{1}{2}|x|^2 + \pi_c^{[3]}(x) + O(x)^4 \tag{8}$$

$$\pi_o(x) = \frac{1}{2}\sum \tau_i x_i^2 + \pi_o^{[3]}(x) + O(x)^4 \tag{9}$$

where the squared singular values $\tau_1 \geq \tau_2 \geq \ldots \geq \tau_n > 0$ are the ordered eigenvalues of $P_o P_c$. When (8) holds then we say that the system is in *input normal form of degree one.*

Instead we could have made a linear change of coordinates to make $P_o = I$ and P_c a diagonal matrix. Then we say that the system is in *output normal form of degree one.* Throughout this paper we shall concentrate on systems that are in input normal form but there are analogous results for systems that are in output normal form.

The linear part of the system is said to be balanced [9] if the state coordinates have been chosen so that P_c and P_o are diagonal and equal. The diagonal entries $\sigma_1 \geq \ldots \geq \sigma_n > 0$ are called the Hankel singular values of the linear part of the system and they are related to the squared singular values $\tau_i = \sigma_i^2$.

Definition 1. *A system with distinct squared singular values $\tau_1 > \tau_2 > \ldots > \tau_n$ is in input normal form of degree d if*

$$\pi_c(x) = \frac{1}{2}\sum_{i=1}^{n} x_i^2 + O(x)^{d+2}, \qquad \pi_o(x) = \frac{1}{2}\sum_{i=1}^{n} \tau_i^{[0:d-1]}(x_i) x_i^2 + O(x)^{d+2} \tag{10}$$

where $\tau_i^{[0:d-1]}(x_i) = \tau_i + \ldots$ is a polynomial in x_i with terms of degrees 0 through $d-1$. They are called the squared singular value polynomials of degree $d-1$.

Input normal form of degree d is similar to a normal form of Scherpen [12]. She showed that, for nonlinear systems with controllable, observable and exponentially stable linear part, state coordinates x can be found such that

$$\pi_c(x) + \frac{1}{2}\sum_{i=1}^{n} x_i^2, \qquad \pi_o(x) = \frac{1}{2}\sum_{i=1}^{n} \tau_i(x)x_i^2 \tag{11}$$

where Scherpen called $\tau_i(x)$ the singular value functions.

Grey and Scherpen [6] have shown that the singular value functions $\tau_i(x)$ are not unique except at $x = 0$ where they equal the squared singular values of the linear part of the system $\tau_i(0) = \tau_i = \sigma_i^2$. For example, suppose $1 \leq i < j \leq n$ and we define for any $c \in \mathbb{R}$

$$\begin{aligned}
\bar{\tau}_i(x) &= \tau_i(x) + cx_j^2 \\
\bar{\tau}_j(x) &= \tau_j(x) - cx_i^2 \\
\bar{\tau}_l(x) &= \tau_l(x) \qquad\qquad \text{otherwise}
\end{aligned}$$

then

$$\pi_o(x) = \frac{1}{2}\sum_{l=1}^{n} \tau_l(x)x_l^2 = \frac{1}{2}\sum_{l=1}^{n} \bar{\tau}_l(x)x_l^2 \,.$$

Moreover there are many local coordinate systems around zero in which the controllability and observability functions of the system are in the normal form of Scherpen (11), see [6].

The differences between Scherpen's normal form and input normal form of degree d are threefold. First the former is exact while the latter is only approximate through terms of degree $d+1$. The second difference is that, in the former, the parameters $\tau_i(x)$ can depend on all the components of x, while, in the latter, when the Hankel singular values are distinct, the ith parameter $\tau_i^{[0:d-1]}(x_i)$ only depends on x_i. Finally and most importantly, the singular value functions $\tau_i(x)$ of the former are not unique except at $x = 0$ while the squared singular value polynomials $\tau_i^{[0:d-1]}(x_i)$ of the latter are unique if $d \leq 6$ and the Hankel singular values are distinct. If the system is odd, i.e., $f(-x,-u) = -f(x,u)$, $h(-x) = -h(x)$ then the squared singular value polynomials $\tau_i^{[0:d-1]}(x_i)$ are unique if $d \leq 12$.

Recently Fujimoto and Scherpen [3] have shown the existence of a normal form where π_c is one half the sum of squares of the state coordinates and

$$\frac{\partial \pi_o}{\partial x_i}(x) = 0 \ \text{ iff } \ x_i = 0. \tag{12}$$

But the normal form of Fujimoto and Scherpen [3] is not unique while the input normal form of degree $d \leq 6$ is unique.

While writing this paper we became aware of a earlier paper of Fujimoto and Scherpen [2] that claims the following. Suppose the linear part of the system is controllable, observable and Hurwitz and the Hankel singular values are distinct. Then there exists a local change of coordinates such that the controllability and observability functions are of the form

$$\pi_c(x) = \frac{1}{2}\sum_{i=1}^{n} x_i^2 \qquad \pi_o(x) = \frac{1}{2}\sum_{i=1}^{n} \left(\rho_i(x_i)x_i\right)^2 . \tag{13}$$

Unfortunately there appears to be a gap in their proof. Such a result would be an extremely useful generalization of Morse's Lemma.

Notice that if a system with distinct squared singular values $\tau_i = \tau_i(0)$ is in input normal form of degree d then its controllability and observability functions are "diagonalized" through terms of degree $d+1$. They contain no cross terms of degree less than or equal to $d+1$ where one coordinate multiplies a different coordinate. This is reminiscent of the balancing of linear systems by B. C. Moore [9].

For linear systems the squared singular value τ_i is a measure of the importance of the coordinate x_i. The "input energy" in the state x is $\pi_c(x)$ and the "output energy" is $\pi_o(x)$. The states that are most important are those with the most "output energy" for fixed "input energy". Therefore in constructing the reduced order model, Moore kept the subspace of states with largest τ_i for they have the most "output energy" per unit "input energy".

In Scherpen's generalization [12] of Moore, the singular value functions $\tau_i(x)$ measure the importance of the state x_i. To obtain a reduced order model, she assumed $\tau_i(x) > \tau_j(x)$ whenever $1 \le i \le k < j \le n$ and x is in a neighborhood of the origin. Then she kept the states $x_1, \ldots, x_k$ in the reduced order model. But the $\tau_i(x)$ are not unique so this approach is not uniquely defined.

For nonlinear systems in input normal form of degree d, the polynomial $\tau_i^{[0:d-1]}(x_i)$ is a measure of the importance of the coordinate x_i for moderate sized x. We shall show that if the τ_i are distinct and $d \le 6$ then $\tau_i^{[0:d-1]}(x_i)$ is unique. The leading coefficient of this polynomial is the squared singular value τ_i so in constructing a reduced order model we will want to keep the states with the largest τ_i. But τ_i can be small yet $\tau_i^{[0:d-1]}(x_i)$ can be large for moderate sized x_i. If we are interested in capturing the behavior of the system for moderate sized inputs, we may also want to keep such states in the reduced order model. We shall return to this point when we discuss reduced order models in a Section 4.

Theorem 1. *Suppose the system (1) is C^r, $r \ge 2$ with controllable, observable and exponentially stable linear part . If the squared singular values $\tau_1, \ldots, \tau_n$ are distinct and if $2 \le d < r-1$ then there is at least one change of state coordinates that takes the system into input normal form of degree d (10). The controllability and observability functions of a system in input normal form of degree $d \le 6$ are unique. But the system and a change of coordinates that achieves input normal form are not necessarily unique even to degree d. If the system is odd then the controllability and observability functions of a system* *in input normal form of degree $d \leq 12$ are unique but again the system and a change of coordinates that achieves it are not necessarily unique.*

Proof. We shall prove the first part by induction. Moore has shown the existence of input normal form of degree $d = 1$ so assume that we have shown the existence of input normal form of degree $d-1$. Then there are state coordinates x and polynomials $\tau_i^{[0:d-2]}(x_i)$ of degree 0 through $d-2$ such that

$$\pi_c(x) = \frac{1}{2}\sum_{i=1}^{n} x_i^2 + \pi_c^{[d+1]}(x) + O(x)^{[d+2]}$$

$$\pi_o(x) = \frac{1}{2}\sum_{i=1}^{n} \tau_i^{[0:d-2]}(x_i)x_i^2 + \pi_o^{[d+1]}(x) + O(x)^{[d+2]}.$$

A *near identity change of coordinates of degree* $d > 1$ is one of the form $x = z + \phi^{[d]}(z)$.. For brevity we refer to it as a change of coordinates of degree d. Notice that a change of coordinates of degree d does not change the expansions of π_c and π_o through terms of degree d but it can change terms of degrees greater than d. We shall show that there is a degree d change of coordinates that will bring a system from input normal form of degree $d-1$ to input normal form of degree d. In fact there may be several such degree d changes of coordinates.

Suppose we have a degree $d+1$ monomial

$$x_i x_j x_{k_1} \cdots x_{k_{d-1}} \tag{14}$$

with at least two distinct indices, say $i \neq j$. Let γ_c and γ_o be the coefficients of this monomial in $\pi_c^{[d+1]}(x)$ and $\pi_o^{[d+1]}(x)$

After the degree d change of coordinates

$$\begin{aligned} \phi_i^{[d]}(z) &= a_i z_j z_{k_1} \cdots z_{k_{d-1}} \\ \phi_j^{[d]}(z) &= a_j z_i z_{k_1} \cdots z_{k_{d-1}} \\ \phi_l^{[d]}(z) &= 0 \qquad\qquad \text{otherwise} \end{aligned} \tag{15}$$

we have

$$\pi_c(z) = \frac{1}{2}\sum_{i=1}^{n} z_i^2 + \pi_c^{[d+1]}(z) + (a_i + a_j) z_i z_j z_{k_1} \cdots z_{k_{d-1}}$$
$$+O(z)^{[d+2]}$$

$$\pi_o(z) = \frac{1}{2}\sum_{i=1}^{n} \tau_i^{[0:d-2]}(z_i) z_i^2$$
$$+\pi_o^{[d+1]}(z) + (\tau_i a_i + \tau_j a_j) z_i z_j z_{k_1} \cdots z_{k_{d-1}}$$
$$+O(z)^{[d+2]}.$$

We would like to choose a_i and a_j so as to cancel the monomial $z_i z_j z_{k_1} \cdots z_{k_{d-1}}$ from both $\pi_c^{[d+1]}(z)$ and $\pi_o^{[d+1]}(z)$ so they must satisfy

$$\begin{bmatrix} 1 & 1 \\ \tau_i & \tau_j \end{bmatrix} \begin{bmatrix} a_i \\ a_j \end{bmatrix} = - \begin{bmatrix} \gamma_c \\ \gamma_o \end{bmatrix} \tag{16}$$

Since $i \neq j$ then $\tau_i \neq \tau_j$ and this is always possible .

We proceed in this way to cancel all monomials in $\pi_c^{[d+1]}(z)$ and $\pi_o^{[d+1]}(z)$ with at least two distinct indices and so all that are left are monomials with all indices the same $i = j = k_1 = \cdots = k_{d-1}$. For such a monomial the degree d change of coordinates

$$\begin{aligned} \phi_i^{[d]}(z) &= -\gamma_c z_i^{d+1} \\ \phi_l^{[d]}(z) &= 0 \qquad\qquad \text{otherwise} \end{aligned} \tag{17}$$

can be used to cancel the monomial z_i^{d+1} from $\pi_c^{[d+1]}(z)$ but nothing can be done about the same monomial in $\pi_o^{[d+1]}(z)$. Hence it is added to $\tau_i^{[0:d-2]}(z_i)$ to form $\tau_i^{[0:d-1]}(z_i)$.

Next we show that if $d \leq 6$ the normal form is unique. Let γ_c and γ_o be the coefficients the monomial $x_i x_j x_k$ in $\pi_c^{[3]}$ and $\pi_o^{[3]}$. If $i = j = k$ then there is only one change of coordinates (17) that cancels the monomial from $\pi_c^{[3]}$. If there are only two distinct indices among $i,\ j,\ k$ then there is only one change of coordinates (15, 16) that cancels the monomial from $\pi_c^{[3]}$ and $\pi_o^{[3]}$.

Assume that the indices are distinct, $i < j < k$. Then there is a one parameter family of degree two transformations that cancel this monomial from $\pi_c^{[3]},\ \pi_o^{[3]}$,

$$\begin{aligned} x_i &= z_i + a_i z_j z_k \\ x_j &= z_j + a_j z_i z_k \\ x_k &= z_k + a_k z_i z_j \\ x_l &= z_l, \qquad \text{otherwise.} \end{aligned} \tag{18}$$

The coefficients $a_i,\ a_j,\ a_k$ must satisfy

$$\begin{bmatrix} 1 & 1 & 1 \\ \tau_i & \tau_j & \tau_k \end{bmatrix} \begin{bmatrix} a_i \\ a_j \\ a_k \end{bmatrix} = - \begin{bmatrix} \gamma_c \\ \gamma_o \end{bmatrix}$$

Since $\tau_i > \tau_j > \tau_k$ we can choose any a_i and adjust $a_j,\ a_k$ to satisfy this constraint. This freedom propagates to the higher order remainders of π_c and π_c in three ways.

The first way is that it introduces terms like $z_j^2 z_k^2,\ z_i^2 z_k^2,\ z_i^2 z_j^2$ with coefficients that are not unique because they depend on a_i. But all these contain two distinct indices and so all can be cancelled. For example we would cancel the $z_j^2 z_k^2$ terms with a degree three change of coordinates of the form

$$\begin{aligned} z_j &= \xi_j + b_j \xi_j \xi_k^2 \\ z_k &= \xi_k + b_k \xi_j^2 \xi_k. \end{aligned}$$

This introduces nonunique terms like $\xi_j^2\xi_k^4$ and $\xi_j^4\xi_k^2$ but these are easily cancelled because they contain two distinct indices. The coordinate transformations that cancel them introduce nonunique terms of degree 12 that we don't care about.

Here is another way that (18) can nonuniquely change the higher remainders. Suppose the monomial x_i^3 appears in π_o in the input normal form of degree $d > 2$. Then after (18) it is replaced by

$$z_i^3 + 3a_i z_i^2 z_j z_k + 3a_i^2 z_i z_j^2 z_k^2 + a_i^3 z_j^3 z_k^3.$$

The first nonunique term $3a_i z_i^2 z_j z_k$ contains three distinct indices so it is easily cancelled by a change of coordinates of degree 3 that introduce nonunique terms of degree 6 with at least two distinct indices which in turn are easily cancelled by a changes of coordinates of degree 5 which introduce nonunique terms of degree 10 that we don't care about. The second nonunique term $3a_i^2 z_i z_j^2 z_k^2$ also contains three distinct indices so it is easily cancelled by a change of coordinates of degree 4 that introduce extra terms of degree 8 that we don't care about. The last nonunique term $a_i^3 z_j^3 z_k^3$ has two distinct indices so it can be cancelled by a change of coordinates that introduce nonunique terms of degree 12 that we don't care about.

The last way that (18) can nonuniquely change the higher remainders is as follows. Suppose the monomial $x_i x_{l_1} x_{l_2} x_{l_3}$ appears in the quartic remainders of π_c, π_o Then after (18) it is replaced by

$$z_i z_{l_1} z_{l_2} z_{l_3} + a_i z_j z_k z_{l_1} z_{l_2} z_{l_3} + \dots$$

The nonunique term $a_i z_j z_k z_{l_1} z_{l_2} z_{l_3}$ contains at least two distinct indices $j \neq k$ so it can be cancelled by a change of coordinates of degree 4 which introduces nonunique terms of degree 8 that we don't care about.

But if if $l_1 = l_2 = l_3 = k$ then the change of coordinates that cancels the nonunique term $a_i z_j z_k^4$ is of the form

$$\begin{aligned} z_j &= \xi_j + b_j \xi_k^4 \\ z_k &= \xi_k + b_k \xi_j \xi_k^3 \end{aligned}$$

and the first of these introduces a nonunique term $b_j^2 \xi_k^8$ that contains only one distinct index and so it cannot be cancelled from $\pi_o^{[8]}$. This is why input normal form is not unique for $d \geq 7$.

If the system is odd then it is easy to see that π_c, π_o are even functions

$$\pi_c(x) = \pi_c(-x), \qquad \pi_o(x) = \pi_o(-x)$$

so there Taylor expansions contain only even terms. A slight modification of the above argument shows that input normal of degree $d \leq 12$ is unique. □

The *normal change of coordinates of degree* d that achieves input normal form of degree d is constructed as follows. For each monomial (14) let $i,\ j$ be the pair of distinct indices that are furthest apart. Then we choose $a_i,\ a_j$ in the change of coordinates (15) to cancel this monomial in $\pi_c^{[d+1]}(z)$ and $\pi_o^{[d+1]}(z)$. If there are not two distinct indices then we choose the change of coordinates (17) to cancel the monomial in $\pi_c^{[d+1]}(z)$. Then form the composition of all such changes of coordinates as one ranges over all monomials of degree $d+1$ and throw away the part of composition of degree greater than d. The result does not depend on the order of the composition and it is *the unique normal change of coordinates of degree* d. The rational behind using the normal change of coordinates of degree d is that if $i,\ j$ are as far apart as possible then so are $\tau_i,\ \tau_j$. The coefficients $a_i,\ a_j$ that are used to cancel the monomial (14) in both $\pi_c^{[d+1]}$ and $\pi_o^{[d+1]}$ satisfy the pair of linear equations (16). The determinant of the matrix on the left is $\tau_j - \tau_i$ and we would like to make its magnitude as large as possible to minimize the effect of numerical errors in solving these linear equations. Hence we choose $i,\ j$ as far apart as possible.

While writing this paper we became aware of a paper of Fujimoto and Tsubakino [4] that discusses the term by term computation of a change of coordinates that takes a system into input normal form of degree d. They show that at each degree the coefficients of the change of coordinates must satisfy a set of linear equations that is underdetermined, there are more coordinates than there are constraints in the normal form. But they don't show that the set of linear equations is always solvable as we have above.

Next we drop the assumption that the squared singular values are distinct.

Definition 2. *The system is in input normal form of degree d if*

$$\begin{aligned}\pi_c(x) &= \tfrac{1}{2}\sum_{i=1}^n x_i^2 + O(x)^{d+2}\\ \pi_o(x) &= \tfrac{1}{2}\sum_{i=1}^n \tau_i^{[0:d-1]}(x)x_i^2 + O(x)^{d+2}\end{aligned} \tag{19}$$

where $\tau_i^{[0:d-1]}(x)$ *is a polynomial of degrees* 0 *through* $d-1$ *in the variables* $\{x_j : \tau_j = \tau_i\}$ *and with constant term*

$$\tau_i^{[0:d-1]}(0) = \tau_i.$$

Theorem 2. *Suppose the system (1) is* C^r $r \geq 2$ *with controllable, observable and exponentially stable linear part . If* $2 \leq d < r-1$ *then there is a change of state coordinates that takes the system into input normal form of degree d (19).*

Proof. A slight extension of the proof of the previous theorem yields the existence of the input normal form of degree d. One uses changes of coordinates of the form (15) where $\tau_i \neq \tau_j$ to cancel the monomial $z_i z_j z_{k_1}\cdots z_{k_{d-1}}$ from $\pi_c^{[d+1]}(z)$ and $\pi_c^{[d+1]}(z)$.

If $\tau_i = \tau_j = \tau_{k_1} = \ldots = \tau_{k_{d-1}}$ then the monomial $z_i z_j z_{k_1} \cdots z_{k_{d-1}}$ can be canceled from $\pi_c^{[d+1]}(z)$ using the change of coordinates

$$\begin{aligned} \phi_i^{[d]}(z) &= c_i z_i z_j z_{k_1} \cdots z_{k_{d-1}} \\ \phi_l^{[d]}(z) &= 0 \qquad \text{otherwise.} \end{aligned} \tag{20}$$

But this change of coordinates is not uniquely determined unless $i = j = k_1 = \ldots, k_{d-1}$. For example if $i \neq j$ we could as well use the change of coordinates

$$\begin{aligned} \bar{\phi}_j^{[d]}(z) &= c_j z_i z_j z_{k_1} \cdots z_{k_{d-1}} \\ \bar{\phi}_l^{[d]}(z) &= 0 \qquad \text{otherwise} \end{aligned} \tag{21}$$

to cancel the monomial $z_i z_j z_{k_1} \cdots z_{k_{d-1}}$ from $\pi_c^{[d+1]}(z)$. Or we could use a combination of them both

$$\begin{aligned} \tilde{\phi}_i^{[d]}(z) &= c_i z_i z_j z_{k_1} \cdots z_{k_{d-1}} \\ \tilde{\phi}_j^{[d]}(z) &= c_j z_i z_j z_{k_1} \cdots z_{k_{d-1}} \\ \tilde{\phi}_l^{[d]}(z) &= 0 \qquad \text{otherwise.} \end{aligned} \tag{22}$$

□

If the squared singular values $\tau_1, \ldots, \tau_n$ are not distinct then input normal form of any degree $d > 1$ is not unique. For example we could make a block diagonal change of coordinates

$$x_i = z_i + \phi_i(z)$$

where $\phi_i(z)$ only depends on those z_j such that $\tau_j = \tau_i$. By the an argument similar to the above we see that input normal form of degree $d \leq 6$ is unique up to such block diagonal changes of coordinates.

3 Linear Model Reduction

Moore's method [9] of obtaining a reduced order model of a linear system is called balanced truncation. One chooses so-called balanced linear state coordinates z where the controllability and observability gramians are diagonal and equal,

$$P_c = P_o = \text{diagonal}(\sigma_1, \ldots, \sigma_n)\,.$$

If $\sigma_1 \geq \sigma_2 \geq \ldots \geq \sigma_k >> \sigma_{k+1} \geq \cdots \geq \sigma_n > 0$ then a k dimensional reduced order model is obtained by Galerkin projection onto the subspace of the first k balanced coordinates.

An equivalent method is to chose linear coordinates x so that the system is in input normal form of degree 1,

$$P_c = I \qquad P_o = \text{diagonal}(\tau_1, \ldots, \tau_n)$$

where $\tau_i = \sigma_i^2$. A k dimensional reduced order model is obtained by Galerkin projection onto the subspace of the first k input normal coordinates. The two sets of coordinates and the reduced order models are related by $z_i = \pm\sigma_i^{\frac{1}{2}} x_i$.

It is convenient to let $\mathbf{x}_1 = (x_1, x_2, \dots, x_k)$ and $\mathbf{x}_2 = (x_{k+1}, \dots, x_n)$ then in these coordinates the full order linear system is

$$\begin{aligned} \dot{\mathbf{x}}_1 &= F_{11}\mathbf{x}_1 + F_{12}\mathbf{x}_2 + G_1 u \\ \dot{\mathbf{x}}_2 &= F_{21}\mathbf{x}_1 + F_{22}\mathbf{x}_2 + G_2 u \\ y &= H_1\mathbf{x}_1 + H_2\mathbf{x}_2 \end{aligned} \tag{23}$$

and the reduced order model is

$$\begin{aligned} \dot{\mathbf{x}}_1 &= F_{11}\mathbf{x}_1 + G_1 u \\ y &= H_1\mathbf{x}_1 \,. \end{aligned} \tag{24}$$

How does one justify balanced truncation? Consider a linear system

$$\begin{aligned} \dot{x} &= Fx + Gu \\ y &= Hx \,. \end{aligned} \tag{25}$$

If F is Hurwitz then it defines an input-output map

$$\begin{aligned} \mathcal{IO} &: L^2((-\infty,\infty), I\!R^m) \to L^2(-\infty,\infty, I\!R^p) \\ \mathcal{IO} &: u(-\infty:\infty) \mapsto y(-\infty:\infty) \end{aligned}$$

given by

$$y(t) = \int_{-\infty}^{t} He^{F(t-s)}Gu(s)\, ds \,.$$

Ideally want would like to choose the reduced order model to minimize over all models of state dimension k the norm of the difference between the input-output maps of the full and reduced models. Balanced truncation does not achieve this goal.

The input-output map of a linear system is not a compact operator which causes mathematical difficulties. There is a closely related map which is of finite rank hence compact. It is the Hankel map from past inputs to future outputs which factors through the current state

$$\begin{aligned} \mathcal{H} &: L^2(-\infty,0], I\!R^m) \to L^2([0,\infty), I\!R^p) \\ \mathcal{H} &: u(-\infty:0) \mapsto y(0:\infty) \end{aligned}$$

given by

$$\begin{aligned} x(0) &= \int_{-\infty}^{0} e^{-Fs}Gu(s)\, ds \\ y(t) &= He^{Ft}x(0) \,. \end{aligned}$$

Unfortunately balanced truncation does not minimize the difference of norm between the Hankel maps of the full and reduced models over all reduced models of state dimension k.

So how does one justify balanced truncation and how can it be generalized to nonlinear systems? Newman and Krishnaprasad [11] have given a stochastic way. Here is another way. We start by restricting our attention to reduced order models that can be obtained by Petrov Galerkin projection of (25). A Petrov Galerkin requires two linear maps

$$\begin{aligned} \Psi : I\!R^k \to I\!R^n, &\qquad \Psi : z \mapsto x = \Psi z \\ \Phi : I\!R^n \to I\!R^k, &\qquad \Phi : x \mapsto z = \Phi x \end{aligned}$$

such that $\Phi\Psi z = z$ and$(\Psi\Phi)^2 = \Psi\Phi$. The reduced order model of (25) is then

$$\begin{aligned} \dot{z} &= \Phi\left(F\Psi z + Gu\right) \\ y &= H\Psi z \,. \end{aligned}$$

Balanced truncation is a Galerkin projection in balanced coordinates where

$$\Phi = \Psi' = \begin{bmatrix} I \; 0 \end{bmatrix} . \tag{26}$$

But in the original coordinates it is a Petrov Galerkin projection. How were Ψ and Φ chosen?

Intuitively to obtain a reduced order model of dimension k of the linear system (25) we should Ψ so that the states in its range have the largest output energy $\pi_o(x)$ for given input energy $\pi_c(x)$. More precisely, the range of Ψ should be the k dimensional subspace through the origin where $\pi_o(x)$ is maximized for given $\pi_c(x)$. If the linear system is in input normal coordinates the clearly this subspace is given by $x_{k+1} = \cdots = x_n = 0$ and a convenient choice of Ψ is (26).

We choose Φx so that the norm of the difference in the outputs starting from x and $\Psi\Phi x$ is as small as possible. To do this we define the co-observability function

$$\pi_{oo}(x, \bar{x}) = \frac{1}{2}\int_0^\infty |y(t) - \bar{y}(t)|^2 \, dt \tag{27}$$

where $y(0:\infty)$, $\bar{y}(0:\infty)$ are the outputs of the linear system (25) starting from x, $\bar{x}$ at $t = 0$ with $u(0:\infty) = 0$. Then we choose Φx to minimize

$$\pi_{oo}(x, \Psi\Phi x) \,.$$

Because of the system is linear, $\pi_{oo}(x, \bar{x})$ is a quadratic form in $(x, \bar{x})$ and

$$\pi_{oo}(x, \bar{x}) - \pi_o(x - \bar{x}) = \frac{1}{2}\sum \tau_i(x_i \quad \bar{x}_i)^2 \,.$$

If the system is in input normal coordinates then the minimizing Φ is given by (26). This explains choices of Ψ, Φ that are made in balanced truncation.

4 Nonlinear Model Reduction

We would like to generalize linear balanced truncation to nonlinear systems of the form

$$\begin{aligned} \dot{x} &= f(x,u) \\ y &= h(x)\,. \end{aligned} \tag{28}$$

To do so we restrict our attention to reduced order models that are constructed by a nonlinear Galerkin projection. A nonlinear Galerkin projection of (28) is defined by two maps, an embedding ψ from the lower dimensional state space into the higher one and a submersion ϕ of the higher dimensional state space onto the lower dimensional one. These state spaces could be manifolds but, since we will focus on local methods, we shall assume that they are neighborhoods of the origin in $I\!R^k$, $I\!R^n$. To simplify notation we just let $I\!R^k$, $I\!R^n$ stand for these neighborhoods.

So we seek

$$\begin{aligned} \psi : I\!R^k &\rightarrow I\!R^n, \qquad & \psi : z \mapsto x \\ \phi : I\!R^n &\rightarrow I\!R^k, \qquad & \phi : x \mapsto z \end{aligned}$$

such that $\phi(\psi(z)) = z$ and $(\psi \circ \phi)^2(x) = \psi \circ \phi(x)$

Motivated by the interpretation of linear balanced truncation given above we would like to choose ψ so that the submanifold that is its range maximizes output energy $\pi_o(x)$ for fixed input energy $\pi_c(x)$. But if $k > 1$ and the system is nonlinear then this submanifold is not well-defined.

Suppose $k = 1$ then for each small positive constant c we can maximize $\pi_o(x)$ subject to $\pi_c(x) = c$. Since the quadratic parts of these functions are positive definite, for small $c > 0$, $\pi_o(x)$ will have two local maxima on each level set $\pi_c(x) = c$. The locus of these local maxima form a one dimensional submanifold through the origin which we can take as the state space of our one dimensional reduced order model.

But if $k > 1$ then the k dimensional submanifold that maximizes $\pi_o(x)$ for $\pi_c(x) = c$ is not well-defined. When the system is linear and $\pi_o(x)$, $\pi_c(x)$ are quadratic forms then this submanifold is assumed to be a subspace and hence is well-defined. It is the subspace spanned by the k leading eigenvectors of P_o when the system is in input normal form $P_c = I$.

Suppose that the squared singular values are distinct and that the nonlinear system has been brought to input normal form (19) of degree d by changes of coordinates up to degree d. The minimum input energy necessary to excite the system to state x is

$$\pi_c(x) = \frac{1}{2}|x|^2 + O(x)^{d+2}.$$

The output energy generated by the system relaxing from the state x is

$$\pi_o(x) = \frac{1}{2}\sum_{j=1}^{n} \tau_j^{[0:d-1]}(x_j)x_j^2 + O(x)^{d+2}.$$

Suppose further that there is a gap in the squared singular value polynomials over the range of states of interest $|x| \leq c$,

$$\tau_i^{[0:d-1]}(x_i) >> \tau_j^{[0:d-1]}(x_j) \tag{29}$$

for $1 \leq i \leq k < j \leq n$ and $|x_i| \leq c,\ |x_j| \leq c$.

Then a k dimensional submanifold that "approximately maximizes" $\pi_o(x)$ for given $\pi_c(x)$ is given by $x_{k+1} = \cdots = x_n = 0$ and we define

$$\psi(z_1, \ldots, z_k) = x = (z_1, \ldots, z_k, 0, \ldots, 0)\,. \tag{30}$$

We find the submersion ϕ as before. Define the co-observability function $\pi_{oo}(x, \bar{x})$ as before (27) except that $y(0 : \infty)$, $\bar{y}(0 : \infty)$ are the outputs of the nonlinear system (28) starting from x, $\bar{x}$ with $u(0 : \infty) = 0$. It is not hard to see that $\pi_{oo}(x, \bar{x})$ satisfies the Lyapunov PDE

$$0 = \left[\frac{\partial \pi_{oo}}{\partial x}(x, \bar{x}) \ \frac{\partial \pi_{oo}}{\partial \bar{x}}(x, \bar{x}) \right] \begin{bmatrix} f(x, 0) \\ f(\bar{x}, 0) \end{bmatrix} + \frac{1}{2}|h(x) - h(\bar{x})|^2$$

and this can be easily solved term by term,

$$\pi_{oo}(x, \bar{x}) = \frac{1}{2} \sum \tau_i (x_i - \bar{x}_i)^2 + \pi_{oo}^{[3]}(x, \bar{x}) + \pi_{oo}^{[4]}(x, \bar{x}) + \ldots$$

We choose $\phi(x)$ to minimize $\pi_{oo}(x, \psi(\phi(x)))$. Assume that the system is in input normal form of degree d and ψ has been chosen as above . Then a straightforward calculation leads to

$$\begin{aligned} \phi_i(x) &= x_i \\ &+ \tfrac{1}{\tau_i} \left(\tfrac{\partial \pi_{oo}^{[3]}}{\partial \bar{x}_i}(x, (\phi(x), 0)) + \tfrac{\partial \pi_{oo}^{[4]}}{\partial \bar{x}_i}(x, (\phi(x), 0)) + \ldots \right) \end{aligned} \tag{31}$$

for $i = 1, \ldots, k$ which can be solved by repeated substitution.

The reduced order nonlinear model is

$$\begin{aligned} \dot{z} = a(z, u) &= \frac{\partial \phi}{\partial x}(\psi(z)) f(\psi(z), u) \\ y = c(z) &= h(\psi(z))\,. \end{aligned} \tag{32}$$

Here is our algorithm for nonlinear model reduction to degree d.

1) Compute the controllability and observability functions $\pi_c(x)$, $\pi_o(x)$ to degree $d + 1$ by solving the HJB and Lyapuov equations (3, 7) term by term.
2) Make normal changes of coordinates of degrees 1 through d to bring the system into input normal form of degree d, (10)
3) Examine the squared singular value polynomial $\tau_i^{[0:d-1]}(x_i)$ to see if there is a gap (29) for some k over the range of states of interest.
4) Define the embedding ψ by (30).
5) Find the submersion ϕ by solving (31) to degree d.
6) The degree d reduced order model is given by the truncation of (32) to terms of degree less than or equal to d.

5 Linear Error Estimates

K. Glover [5] has given an important error bound for the norm of the difference between the input-output map of the full linear system $\mathcal{IO}_n$ and the input-output map of its balanced truncation $\mathcal{IO}_k$,

$$\|\mathcal{IO}_n - \mathcal{IO}_k\| \le 2 \sum_{j=k+1}^{n} \sigma_j$$

where the norm is the induced L^2 norm.

We know that the corresponding Hankel maps $\mathcal{H}_n,\ \mathcal{H}_k$ satisfy

$$\sigma_{k+1} \le \|\mathcal{H}_n - \mathcal{H}_k\|$$

so we have for linear balanced truncation

$$\sigma_{k+1} \le \|\mathcal{H}_n - \mathcal{H}_k\| \le \|\mathcal{IO}_n - \mathcal{IO}_k\| \le 2 \sum_{j=k+1}^{n} \sigma_j \,.$$

Unfortunately we do not have similar estimates for nonlinear model reduction. But there is a new error estimate for linear systems that can be extended to nonlinear systems. To each state x of the full order model there is an optimal open loop control $u_x(-\infty : 0)$ that excites the system from state 0 at $t = -\infty$ to state x at $t = 0$. It is the solution of the optimal control problem (2). These optimal controls form a n dimensional subspace of the space of inputs $L^2([-\infty : 0], I\!R^m)$ to the Hankel map. We expect that these optimal controls are typical of those that are used in the full order model and hence we are interested in the size of errors when they are used in the reduced order model. The error of the Hankel maps can be readily bounded as follows.

The optimal open loop controls are generated by the optimal feedback (6). If the linear system is in input normal coordinates then the optimal gain is $K = G'$. We drive both the full model (23) and the reduced model (24) obtained by balanced truncation by this feedback to get the combined system

$$\begin{bmatrix} \dot{x} \\ \dot{z} \end{bmatrix} = \begin{bmatrix} F + GK & 0 \\ G_1 K & F_{11} \end{bmatrix} .$$

For balanced truncation $-(F + GK)$ and F_{11} are Hurwitz so there is an unstable subspace $z = Tx$ where T satisfies the Sylvester equation

$$T(F + GK) - F_{11}T = G_1 K \,. \tag{33}$$

The meaning of T is that if we excite the reduced order system with the optimal control $u_x(-\infty : 0)$ that excites the full order system to x then $z(0) = Tx$.

Next we define the cross-observability function

$$\rho(x,z) = \frac{1}{2}\int_0^\infty |y_f(t) - y_r(t)|^2\,dt \tag{34}$$

where $y_f(0:\infty)$ is the output of the full order model starting at $x(0) = x$ and $y_r(0:\infty)$ is the output of the reduced order model starting at $z(0) = z$ with $u(0:\infty) = 0$. Because the systems are linear, ρ is a quadratic form

$$\rho(x,z) = \frac{1}{2}\begin{bmatrix} x \\ z \end{bmatrix}' \begin{bmatrix} Q_{11} & Q_{12} \\ Q_{21} & Q_{22} \end{bmatrix} \begin{bmatrix} x \\ z \end{bmatrix}$$

where Q satisfies the Sylvester equation

$$0 = \begin{bmatrix} F & 0 \\ 0 & F_{11} \end{bmatrix}' \begin{bmatrix} Q_{11} & Q_{12} \\ Q_{21} & Q_{22} \end{bmatrix} + \begin{bmatrix} Q_{11} & Q_{12} \\ Q_{21} & Q_{22} \end{bmatrix} \begin{bmatrix} F & 0 \\ 0 & F_{11} \end{bmatrix} + \begin{bmatrix} H'H & H'H_1 \\ H_1'H & H_1'H_1 \end{bmatrix}. \tag{35}$$

Clearly Q_{11} is the observability gramian. If the system is in balanced or input normal coordinates then Q_{22} is the upper left $k \times k$ block of Q_{11}.

If we use the optimal control $u_x(-\infty : 0)$ then the norm of error between the full and reduced Hankel maps is

$$2\rho(x,Tx) = x' \begin{bmatrix} I \\ T \end{bmatrix}' \begin{bmatrix} Q_{11} & Q_{12} \\ Q_{21} & Q_{22} \end{bmatrix} \begin{bmatrix} I \\ T \end{bmatrix} x\,.$$

If the system is in input normal form then the maximum squared norm of error between the Hankel maps restricted to optimal inputs of the full system is the largest eigenvalue of

$$\begin{bmatrix} I \\ T \end{bmatrix}' \begin{bmatrix} Q_{11} & Q_{12} \\ Q_{21} & Q_{22} \end{bmatrix} \begin{bmatrix} I \\ T \end{bmatrix}. \tag{36}$$

This error estimate is always greater than or equal to the largest neglected squared singular value τ_{k+1} because the right singular vectors of the Hankel map of the full system are a basis for the space of optimal controls. In the few examples that we have computed we found that this error estimate is much closer to τ_{k+1} than to the square of Glover's bound.

One can compute the maximum norm of error between the Hankel maps restricted to optimal inputs of the reduced system in a similar fashion.

6 Nonlinear Error Estimates

Unfortunately for nonlinear model reduction we don't have an error bound on the input-output maps similar to Glover. Furthermore we don't have a lower bound on the norm of the Hankel error like the first neglected singular value. But we can generalize the error bounds of the Hankel maps restricted to optimal inputs of the full or reduced system. We present the bound for the optimal inputs of the full system. The other is very similar.

Suppose we have a full order system (1) which is in input normal form of degree d and its reduced order model

$$\begin{aligned} \dot{z} = a(z,u) &= F_{11}z + G_1 u + a^{[2]}(z,u) + \ldots + a^{[d]}(z,u) \\ y = c(z) &= H_1 z \qquad\qquad + c^{[2]}(z) \quad + \ldots + c^{[d]}(x) \end{aligned} \tag{37}$$

found by the method above or a similar method.

For each $x \in I\!R^n$ there is an optimal open loop control $u_x(-\infty : 0)$ that excites the system from state 0 at $t = -\infty$ to state x at $t = 0$. It is the solution of the optimal control problem (2). These optimal controls form a n dimensional submanifold of the space of inputs $L^2([-\infty : 0], I\!R^m)$ of the Hankel map. Again we expect that these optimal controls are typical of those that are used in the full order model and hence we are interested in the errors when they are used in the reduced order model.

The optimal controls are generated by the feedback (4) which can be computed term by term. We plug this feedback into the combined system

$$\begin{aligned} \dot{x} &= f(x,\kappa(x)) = (F + GK)x + \ldots \\ \dot{z} &= a(z,\kappa(x)) = F_{11}z + G_1 Kx + \ldots \end{aligned}$$

Again $-(F + GK)$ and F_{11} are Hurwitz so there exists an unstable manifold $z = \theta(x)$ which satisfies the PDE

$$a(\theta(x),\kappa(x)) = \frac{\partial \theta}{\partial x}(x) f(x,\kappa(x)) \,.$$

This PDE can be solved term by term and the linear coefficient is the T satisfying (33).

The cross-obervability function $\rho(x,z)$ is defined as before (34) except now the full (1) and reduced (37) systems are nonlinear and the input is zero. The cross-obervability function satisfies the PDE

$$0 = \begin{bmatrix} \frac{\partial \rho}{\partial x}(x,z) & \frac{\partial \rho}{\partial z}(x,z) \end{bmatrix} \begin{bmatrix} f(x,0) \\ a(z,0) \end{bmatrix} + \frac{1}{2}|h(x) - c(z)|^2 .$$

This also has a series solution

$$\rho(x,z) = \frac{1}{2} \begin{bmatrix} x \\ z \end{bmatrix}' \begin{bmatrix} Q_{11} & Q_{12} \\ Q_{21} & Q_{22} \end{bmatrix} \begin{bmatrix} x \\ z \end{bmatrix} + \rho^{[3]}(x,z) + \ldots$$

where Q satisfies the Sylvester equation (35).

Then the squared norm of the error between the full and reduced Hankel maps using the optimal control $u_x(-\infty : 0)$ is

$$2\rho(x,\theta(x))$$

and good estimate of the maximum relative squared norm of the error is

$$\sup \frac{\rho(x, \theta(x))}{\pi_c(x)}.$$

Suppose the system is in input normal form of degree d, so that

$$\pi_c(x) = \frac{1}{2}|x|^2 + O(x)^{d+2}. \tag{38}$$

Then we can make a linear orthogonal change of coordinates to diagonalize the quadratic part (36) of $\rho(x, \theta(x))$. If the diagonal entries are distinct, then as with the transformation to input normal form of degree d we can make further changes of state coordinates of degrees 2 through d that leave $\pi_c(x)$ as above (38) and bring $\rho(x, \theta(x))$ into the normal form

$$\rho(x, \theta(x)) = \frac{1}{2} \sum_i \epsilon_i^{[0:d-1]}(x_i) x_i^2.$$

The $\epsilon_i^{[0:d-1]}(x_i)$ are polynomials of degrees 0 through $d-1$ and are called the squared error polynomials. They are unique if $d \leq 6$ ($d \leq 12$ for odd systems). They measure how fast the squared error between the Hankel maps grows restricted to optimal controls $u_x(-\infty : 0)$ as x grows.

7 Example

We consider three linked rods connected by planar rotary joints with springs and dampening hanging from the ceiling. The input is a torque applied to the top joint and the output is the horizontal displacement of the bottom. Each rod is uniform of length 2, mass 1, with spring constant 3, dampening constant 0.5 and gravity constant 0.5.

We approximated the nonlinear system by its Taylor series through terms of degree 5. The Taylor series of controllability and observability functions $\pi_c(x)$, $\pi_o(x)$ were computed through terms of degree 6. The system was brought into input normal form of degree 5 by a changes of state coordinates of degrees 1 through 5. The Hankel singular values of the linear part of the system are 15.3437, 14.9678, 0.3102, 0.2470, 0.0156, 0.0091. Apparently only two dimensions are linearly significant.

Figure 1 is a semilog plot of the squared singular value polynomials $\tau_i^{[0:4]}$ Notice the difference in scale and how flat they are. Apparently only two dimensions are nonlinearly significant.

Let $u_x(-\infty : 0)$ be the optimal input that excites the full system to x. Let $\mathcal{H}_n, \mathcal{H}_k$ be the Hankel of the full order model ($n = 6$) and the reduced order model ($k = 2$). The error between them satisfies

$$|\mathcal{H}_n(u_x(-\infty : 0)) - \mathcal{H}_k(u_x(-\infty : 0))|^2 \leq 0.0965|x|^2 - 0.0009|x|^4 + \ldots$$

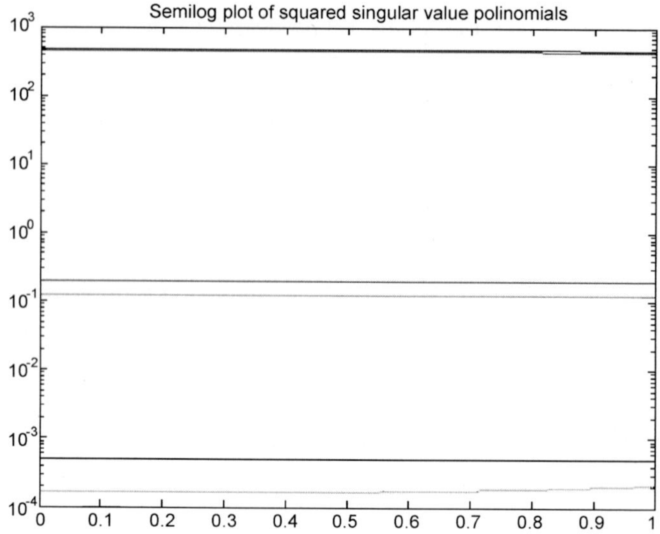

Fig. 1. Squared singular value polynomials

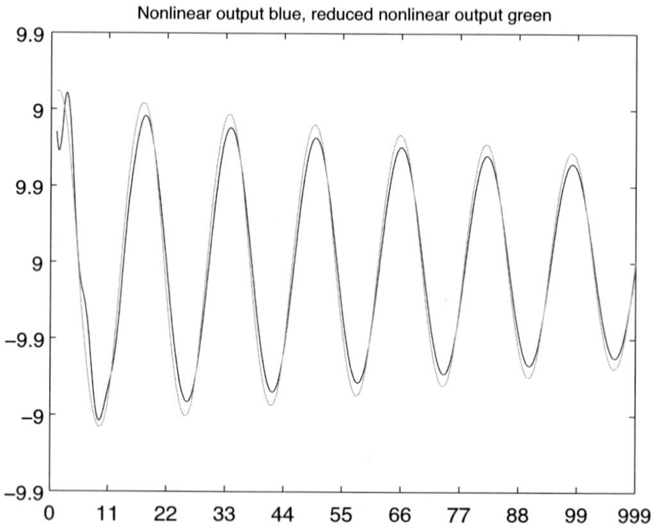

Fig. 2. Optimal response

By way of comparison, the square of the third Hankel singular value is 0.0962 so this estimate is tight. Figure 2 shows the outputs of the Hankel maps of the full and reduced systems excited by an optimal control $u_x(-\infty : 0)$ for random x.

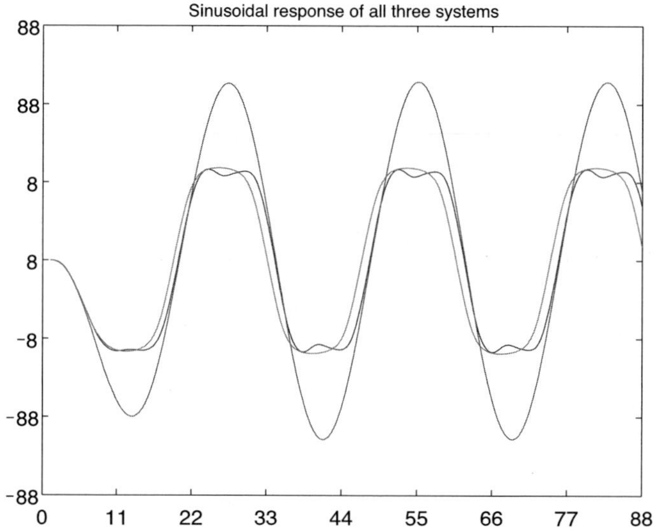

Fig. 3. Sinusoidal response

Figure 3 shows the responses of the full nonlinear model, the reduced nonlinear model and the linear part of the full model to a sinusoidal input. The linear response has the largest amplitude and exceeds the total length of the three rods. The full nonlinear response has a small secondary oscillation that does not appear in the reduced nonlinear response.

8 Conclusion

We have presented a new normal form for the controllability and observability functions of a nonlinear control system. This normal form is valid through terms of degree $d + 1$ where d is an integer chosen by the user and less than the degree of smoothness of the system. There are several advantages to this normal form. It can be computed term by term from the Taylor expansion of the nonlinear system. It is essentially unique for $d \leq 6$ and therefore gives an unambiguous measure of the relative importance of the different components of the state. A reduced order model can be constructed by projection onto the most important coordinates. One nice property of the reduced order model is that its controllability and observability functions are almost the restrictions of the controllability and observability functions of the full order model. Also the state space of the reduced order model almost achieves the intuitive goal of maximizing the observability function while holding the controllability function constant. Our methods readily extend to other forms of nonlinear model reduction such as LQG [7], [14] and H_∞ [10], [13].

References

1. E.G. Al'brecht. On the optimal stabilization of nonlinear systems. *PMM-J. Appl. Math. Mech.*, 25:1254–1266, 1961.
2. K. Fujimoto and J.M.A. Scherpen. Nonlinear balanced realizations based on singular value analysis of Hankel operators. *Proc. of the 42nd IEEE Conf. on Decision and Contr.*, 2003.
3. K. Fujimoto and J.M.A. Scherpen. Nonlinear input-normal realizations based on the differential eigenstructure of Hankel operators. *IEEE Trans. on Automat. Contr.*, 50:2–18, 2005.
4. K. Fujimoto and D. Tsubakino. On computation of nonlinear balanced realization and model reduction. *Proc. of the 2006 Amer. Contr. Conf.*, 2006.
5. K. Glover. All optimal Hankel-norm approximations of linear multivariable systems and their l^∞ error bounds. *Int. J. of Control*, 39:1150–1193, 1984.
6. W.S. Grey and J.M.A. Scherpen. On the nonuniqueness of singular value functions and balanced nonlinear realizations. *Systems & Control Letters*, 44:219–232, 2001.
7. E.A. Jonckheere and L.M. Silverman. A new set of invariants for linear systems-application to reduced order compensator design. *IEEE Trans. on Automat. Contr.*, 28:953–964, 1983.
8. D.L. Lukes. Optimal regulation of nonlinear dynamical systems. *SIAM J. Contr. Optimization*, 7:75–100, 1969.
9. B.C. Moore. Principle component analysis in linear systems: controllability, observability and model reduction. *IEEE Trans. on Automat. Contr.*, 26:17–32, 1981.
10. D. Mustafa and K. Glover. Controller reduction by h_∞ balanced truncation. *IEEE Trans. on Automat. Contr.*, 36:668–682, 1991.
11. A.J. Newman and P.S. Krishnaprasad. Computation for nonlinear balancing. *Proc. of the 37th IEEE Conf. on Decision and Contr.*, pages 4103–4104, 1998.
12. J.M.A. Scherpen. Balancing for nonlinear systems. *Systems & Control Letters*, 21:143–153, 1993.
13. J.M.A. Scherpen. h_∞ balancing for nonlinear systems. *Int. J. of Robust and Nonlinear Control*, 6:645–668, 1996.
14. J.M.A. Scherpen and A.J. van der Schaft. Normalized coprime factorizations and balancing for unstable nonlinear systems. *Int. J. of Control*, 60:1193–1222, 1994.

Part II

Optimization Methods

Nonholonomic Trajectory Optimization and the Existence of Twisted Matrix Logarithms

Roger W. Brockett

Division of Engineering and Applied Sciences, Harvard University, Cambridge, MA, USA

Dedicated to Alberto Isidori, a leader in the field of systems engineering.

Summary. The problem studied here is a shortest path problem of the type encountered in sub-Riemannian geometry. It is distinguished by special structures related to its Lie group setting and the $\mathbb{Z}_2$ graded structure on the relevant Lie algebra. In spite of the fact that the first order necessary conditions lead to differential equations that are integrable in terms of elementary functions, in this case there remain questions related to the existence of appropriate values for the parameters which appear. In this paper we treat the problem in some generality but establish the existence of suitable parameter values only in the case of the general linear group of dimension two.

1 Introduction

In the early 1970's Alberto Isidori worked with Antonio Ruberti and others in the control group in Rome to help define the subject of bilinear systems. They showed that various structural results on linear time-invariant systems generalized nicely to the bilinear case (see, e.g., their paper in [5]) and work in this area provided a pattern for subsequent results on more general classes of nonlinear systems, some of which figure prominently in his widely used book [7].

In this paper we return to the bilinear model, focusing on an optimal control problem which illustrates remarkable complexity. It is but one example of a class of problems having a Lie group theoretic flavor and which find application in quantum control [8], nonholonomic mechanics [4], etc.

We are concerned here with a problem which has a deceptively simple description. We want to transfer a nonsingular matrix $X(0)$ to a second nonsingular matrix $X(1)$ under the assumption that the matrix evolves according to

$$\dot{X} = UX$$

with $U(t) = U^T(t)$. This system is controllable on the space of nonsingular matrices with positive determinant and in our earlier work [4], we observed that the first order necessary conditions associated with minimizing

$$\eta = \int_0^T ||U(t)||\, dt$$

subject to the condition that U should steer the system from $X(0) = X_0$ to $X(1) = X_1$ imply that U should take the form

$$U(t) = e^{\Omega t} H e^{-\Omega t}$$

with $\Omega = -\Omega^T$ and $H = H^T$, both constant. A standard argument involving the rescaling of time shows that η does not depend on the available time, T, as long as $T > 0$ and that the optimal control remains the same if $||U(t)||$ in the integrand defining η is replaced with $||U(t)||^2$. The trajectories satisfying the first order necessary conditions take the explicit form

$$X(t) = e^{\Omega t} e^{(H-\Omega)t} .$$

Writing this in the alternative way

$$X = e^{\frac{1}{2}(A^T - A)} e^A$$

suggests that, on one hand, finding A is something like finding a logarithm of X, but on the other, something like finding a polar representation of X.

This optimization problem makes contact with several rather distinct lines of work of which we mention three.

Expressing matrices as products of positive definite matrices: Notice that because the exponential of a real symmetric matrix is necessarily positive definite, when $X(0) = I$ and the differential equation is approximated by a discrete time system, we need to express $X(T)$ as the product of symmetric positive definite matrices This is a problem with some history. A focus of interest has been on the problem of determining the fewest number of positive definite factors required to represent an arbitrary matrix, see Ballantine [2]. Our problem can be thought of as a variant in which it is not the number of factors but the sum of the norms of the logarithms of the factors that is being minimized.

Sub-Riemannian Distance Problems: Baillieul [1], Gaveau [6] and the author [4, 3] have studied model problems in sub-Riemannian geometry of various types. One such involves the minimization of

$$\eta = \int_0^T (u^2 + v^2)^{1/2}\, dt$$

subject to the condition that u, v should drive the system

$$\dot{x} = u \;\; ; \;\; \dot{y} = v \;\; ; \;\; \dot{z} = xv - yu$$

from a given value of (x, y, z) at $t = 0$ to a second value at $t = T$. If we write the matrix system as

$$\dot{X} = U + U(X - I)$$

we see that in a neighborhood of the identity, the two dimensional problem takes the form

$$\frac{d}{dt}\begin{bmatrix} x_{11} & x_{12} \\ x_{21} & x_{22} \end{bmatrix} = \begin{bmatrix} u & v \\ v & -u \end{bmatrix} + \begin{bmatrix} u & v \\ v & -u \end{bmatrix}\begin{bmatrix} x_{11} - 1 & x_{12} \\ x_{21} & x_{22} - 1 \end{bmatrix}$$

so that in terms of $a = x_{11} - x_{22}$, $b = x_{12} + x_{21}$, $c = x_{12} - x_{21}$ we have the approximation

$$\dot{a} = u \ ; \ \dot{b} = v \ ; \ \dot{c} = av \, .$$

Symmetric Spaces: The geometry of certain classes of Riemannian manifolds can be studied in unusual detail by virtue of their representation as a quotients of Lie groups. The explicit solvability of our optimal control equations in terms of matrix exponentials is a consequence of $\mathbb{Z}_2$ graded structure on the Lie algebra [4], which is the algebraic counterpart of the idea of a symmetric space. An important tool in this theory is the so-called Iwasawa decomposition of the relevant group–i.e., the representation of the elements of the group in the form $g = k\,a\,n$, where k belongs to a compact subgroup, a is from an Abelian subgroup and n is from a nilpotent subgroup. This decomposition is often more useful than the familiar polar form. For example, in numerical linear algebra where factorizations involving upper (or lower) triangular matrices play a role, one can see the effectiveness of this point of view.

Here we bring out some of the relationships with sub-Riemannian geometry and the appearance of cut points and conjugate points in that subject.

2 Some Preliminaries

Although every nonsingular complex matrix has a complex logarithm and every real orthogonal matrix and every real symmetric positive definite matrix has a real logarithm, not every real nonsingular matrix has a real logarithm. If M is to be expressed as e^A then the eigenvalues of M must be the exponentials of the eigenvalues of A. But if the eigenvalues of M are $\lambda_1, \lambda_2, ..., \lambda_n$ with each being real and negative then the eigenvalues of A must be

$$\ln|\lambda_1|e^{\pi i + n_1 2\pi i}, \ln|\lambda_2|e^{\pi i + n_2 2\pi i}, ..., \ln|\lambda_n|e^{\pi i + n_n 2\pi i}$$

for some choice of integers $n_1, n_2, ..., n_n$. On the other hand, if A is to be real then these must occur in complex conjugate pairs but this is only possible if the values of the real and negative eigenvalues are repeated with even multiplicity. Thus, for example, because the negative definite matrix

$$X_1 = \begin{bmatrix} -(2\sqrt{2}+3) & 0 \\ 0 & 2\sqrt{2}-3 \end{bmatrix}$$

has two distinct negative eigenvalues it does not have a real logarithm. However, and the significance of this has already been hinted at, X_1 can be expressed as $e^{\Omega}e^{H-\Omega}$. One possibility being

$$X_1 = exp\begin{bmatrix} 0 & \frac{3}{2}\pi \\ -\frac{3}{2}\pi & 0 \end{bmatrix} exp\begin{bmatrix} 0 & \sqrt{2}\pi - \frac{3}{2}\pi \\ \sqrt{2}\pi + \frac{3}{2}\pi & 0 \end{bmatrix}.$$

As will be explained, this representation is critical in showing, for example, that the minimum value of

$$\eta = \int_0^1 ||U||dt$$

required to steer the system $\dot{X} = UX$ from I to X_1 with $U(t)$ symmetric is no more than 2π. Likewise, the representation

$$-I = exp\begin{bmatrix} 0 & 2\pi \\ -2\pi & 0 \end{bmatrix} exp\begin{bmatrix} 0 & \sqrt{3}\pi - 2\pi \\ \sqrt{3}\pi + 2\pi & 0 \end{bmatrix}$$

is critical in showing that the minimum value of η required to steer the same system from I to $-I$ is $\sqrt{6}\pi$. (Notice that the cost would have been just $\sqrt{2}\pi$ if the admissible controls had included the skew-symmetric matrices.)

Definition 1. *If X_1 and X_2 are nonsingular matrices with positive determinants let $d(X_1, X_2)$ be the minimizing value of η relative to all measurable choices of U that steer the system from X_1 at $t = 0$ to X_2 and $t = 1$.*

The following facts are more or less immediate. If I is the identity matrix then

$$\text{i)}\ \mathrm{d(X_1, X_2) = d(I, X_2X_1^{-1})}$$

$$\text{ii)}\ \mathrm{d(I, X) = d(I, X^T) = d(I, X^{-1})}$$

and if Ψ is orthogonal then

$$iii)\ d(I, X) = d(I, \Psi^T X \Psi^T).$$

The first two statements are immediate. Observe that if Ψ is an orthogonal matrix and if we define $\Omega^* = \Psi^T \Omega \Psi$ and $H^* = \Psi^T H \Psi$ then

$$\Psi^T e^{\Omega} e^{H-\Omega} \Psi = e^{\Omega^*} e^{H^* - \Omega^*}.$$

Of course the Frobenius norm of H and that of H^* are the same.

This kind of minimum distance problem plays a central role in our paper [4] and it was shown there that in addition to the obvious fact that every symmetric positive definite matrix can be expressed as $e^{\Omega}e^{H-\Omega}$ simply by letting $\Omega = 0$, every orthogonal matrix can be expressed as $e^{\Omega}e^{H-\Omega}$ as follows from the explicit solution for two-by-two blocks

$$\begin{bmatrix} \cos\psi & \sin\psi \\ -\sin\psi & \cos\psi \end{bmatrix} = \begin{bmatrix} \cos(\psi+\pi) & \sin(\psi+\pi) \\ -\sin(\psi+\pi) & \cos(\psi+\pi) \end{bmatrix} e^{\begin{bmatrix} 0 & b-\psi-\pi \\ b+\pi+\psi & 0 \end{bmatrix}}$$

with $b = \sqrt{2\pi\psi + \psi^2}$. Here one can limit ψ to be between π and $-\pi$.

An important question left open in [4] was that of determining if every real square matrix with positive determinant can be expressed as $e^{\Omega}e^{H-\Omega}$ with $\Omega = -\Omega^T$ and $H = H^T$ real. This is significant because unless this is the case the above variational problem will necessarily require broken extremals.

Definition 2. *If M can be expressed as $M = e^{\frac{1}{2}(A^T-A)}e^A$ as above we will say that A is a* twisted logarithm *of M.*

The following remarks maybe helpful in gaining an initial understanding of the problem of finding twisted logarithms.

Remark 1. Suppose that s is a real positive constant. If A is a twisted logarithm of M then $A + (\log a)I$ is a twisted logarithm of aM. ◁

Remark 2. If A is a twisted logarithm of M and if Θ is orthogonal, then $\Theta^T A\Theta$ is a twisted logarithm of $\Theta^T M\Theta$.

Clearly if $M = e^{\frac{1}{2}(A^T-A)}e^A$ then $\Theta^T M\Theta = \Theta^T e^{\frac{1}{2}(A^T-A)}e^A\Theta$. Using the fact that $\Theta e^A\Theta^{-1} = e^{\Theta A\Theta^{-1}}$ leads to the remark. ◁

3 Some Properties of the Linearization

In this section we investigate some properties of the map defining the twisted logarithm and their implications for the optimal control problem. Recall the well-known integral representation of the linear term in the Taylor series expansion of $e^{A+\epsilon B}$

$$e^{A+\epsilon B} = e^{At}\left(I + \epsilon\int_0^1 e^{-At}Be^{At}dt\right) + hot\,.$$

This may be proven by starting with the differential equation $\dot{x} = (A+\epsilon B)x$, changing variables to $z(t) = e^{-At}x(t)$ to get

$$\dot{z} = \epsilon e^{-At}Be^{At}z$$

then observing that from the Peano-Baker series

$$z(1) \approx \left(I + \epsilon\int_0^1 e^{-At}Be^{At}dt\right) z(0)$$

and finally $x(1) = e^A z$.

Notation. We denote by ψ the map that takes a square matrix A and maps it to the twisted exponential,

$$\psi : A \mapsto e^{\frac{1}{2}(A^T - A)} e^A \,.$$

Let $gl(n)$ denote the set of all n-by-n matrices and $\mathcal{G}l(n)$ the set of all nonsingular n-by-n matrices.

Theorem 1. *Consider the map $\psi : gl(n) \to \mathcal{G}l(n)$ defined above. The linearization of ψ at any point A with $A = A^T$ and A having distinct eigenvalues is of full rank. More specifically, in a coordinate system in which $A = D$, with D diagonal, the derivative with respect to ϵ of $e^{\epsilon \Delta} e^{D + \epsilon(E - \Delta)}$ with $\Delta = -\Delta^T$ and $E = E^T$ is*

$$\frac{d}{d\epsilon} e^{\epsilon \Delta} e^{D + \epsilon(E - \Delta)} \Big|_0 = \Delta e^D + e^D \int_0^1 e^{-Dt}(E - \Delta) e^{Dt} dt = F e^D$$

with the ij^{th} entry of F being given by

$$f_{ij} = e_{ij} \left(\frac{e^{(d_i - d_j)} - 1}{d_i - d_j} \right) e^{d_j} + \delta_{ij} \left(1 - \frac{e^{(d_i - d_j)} - 1}{d_i - d_j} \right) e^{d_j}$$

with the understanding that when $i = j$ the coefficient of e_{ij} is interpreted as being one.

Proof. First we verify the statements about the derivative assuming that A is diagonal. Because we have assumed that the d_i are distinct, the quantity $(e^{d_i - d_j} - 1)/(d_i - d_j)$ is never zero. This means that the coefficients of the e_{ij} are never zero. As for the coefficients of δ_{ij}, we need only consider the case $i \neq j$ because Δ is skew-symmetric. For $i \neq j$ the coefficients of the skew-symmetric ω_{ij} are never zero. Thus we see that when A is diagonal with unrepeated eigenvalues the Jacobian of the map ψ is diagonal with nonzero eigenvalues. In the general case we can find Θ such that $A = \Theta^T D \Theta$ and Θ simply acts on E and Ω in an invertible way. □

When used with the implicit function theorem this result shows that there exist open sets of real nonsingular matrices that have real twisted logarithms. However, it is also true that at many values of Ω, H the linearization of the map ψ is singular.

Lemma 1. *If $e^{\frac{1}{2}(A^T - A)} e^A = M$ and if there exists a skew-symmetric matrix S such that $[S, M] = 0$ but $[S, A] \neq 0$ then the Jacobian of ψ is singular at A.*

Proof. It is easily verified that, under the hypothesis, the matrix $[S, A]$ is mapped to 0 by the Jacobian. □

4 Main Theorem

We now turn to the proof of the fact that every real two-by-two matrix with positive determinant has a real twisted logarithm. It seems very likely that the same result holds in all dimensions but the proof given here does not seem to be extendable. In general solutions are not unique and our proof involves a nonconstructive homotopy argument. We will give some further information on location of solutions in a later section.

Theorem 2. *Every real two-by-two matrix with positive determinant can be expressed as*

$$M = e^{\Omega} e^{H-\Omega}$$

with $H = H^T$, $\Omega = -\Omega^T$ and both real.

Proof. We begin by establishing some properties of a family of t-parametrized curves associated with a special case of $e^{\Omega t} e^{(H-\Omega)t}$, for which H is zero on the diagonal. Consider

$$X_{b,c}(t) = exp \begin{bmatrix} 0 & ct \\ -ct & 0 \end{bmatrix} exp \begin{bmatrix} 0 & (b-c)t \\ (b+c)t & 0 \end{bmatrix}$$

under the assumption that $c > b > 0$. This matrix product evaluates to

$$X_{b,c}(t) = \begin{bmatrix} \cos ct \cos \omega t + \frac{c+b}{\omega} \sin ct \sin \omega t & \frac{b-c}{\omega} \cos ct \sin \omega t + \sin ct \cos \omega t \\ -\frac{b+c}{\omega} \sin ct \cos \omega t + \cos ct \sin \omega t & \cos ct \cos \omega t - \frac{c+b}{\omega} \sin ct \sin \omega t \end{bmatrix}$$

where $\omega = \sqrt{c^2 - b^2}$.

Introduce the notation

$$\Omega_0 = \begin{bmatrix} 0 & -1 \\ 1 & 0 \end{bmatrix}$$

and make the definitions

$$\alpha(t) = \frac{1}{2} \mathrm{tr} \left(\mathrm{e}^{\Omega} \mathrm{e}^{\mathrm{H}-\Omega} \right)$$

and

$$\beta(t) = \frac{1}{2} \mathrm{tr} \left(\Omega_0 \left(\mathrm{e}^{\Omega} \mathrm{e}^{\mathrm{H}-\Omega} \right) \right) .$$

For the given $X_{c,b}$ it is not difficult to verify that

$$\begin{bmatrix} \alpha(t) \\ \beta(t) \end{bmatrix} = \begin{bmatrix} \cos ct & \sin ct \\ -\sin ct & \cos ct \end{bmatrix} \begin{bmatrix} 1 & 0 \\ 0 & \frac{c}{\omega} \end{bmatrix} \begin{bmatrix} \cos \omega t & -\sin \omega t \\ \sin \omega t & \cos \omega t \end{bmatrix} \begin{bmatrix} 1 \\ 0 \end{bmatrix} = e^{-c\Omega_0 t} D e^{\omega \Omega_0 t} e_1$$

where D is the diagonal matrix with entries $1, c/\omega$ and e_1 is the first standard basis vector in $\mathbb{R}^2$. More explicitly,

$$\alpha(t) = \cos ct \cos \omega t + \frac{c}{\omega} \sin ct \sin \omega t$$

$$\beta(t) = -\sin ct \cos \omega t + \frac{c}{\omega} \cos ct \sin \omega t \,.$$

As we have seen, we can scale the determinant of $e^{\Omega} e^{H-\Omega}$ to one by adding a multiple of the identity to H. This insures that the trace of H is then zero. In this case we can we can apply an orthogonal transformation to H, Ω and X making the on-diagonal elements of H both zero. Thus, establishing the theorem is equivalent to showing that we can solve the $\alpha - \beta$ equations. However,

$$\alpha^2(t) + \beta^2(t) = \cos^2 \omega t + \frac{c^2}{\omega^2} \sin^2 \omega t = 1 + \frac{b^2}{\omega^2} \sin^2 \omega t$$

so that we actually only need to solve them for any (α^*, β^*) lying outside the unit disk. Moreover,

$$\frac{d}{dt} \tan^{-1} \frac{\beta(t)}{\alpha(t)} = \frac{\dot\beta\alpha - \dot\alpha\beta}{\alpha^2 + \beta^2} = -\langle \begin{bmatrix} \alpha(t) \\ \beta(t) \end{bmatrix}, \Omega_0 \begin{bmatrix} \dot\alpha(t) \\ \dot\beta(t) \end{bmatrix} \rangle \frac{1}{\alpha^2(t) + \beta^2(t)} \,.$$

This expression evaluates to

$$\frac{d}{dt} \tan^{-1} \frac{\beta(t)}{\alpha(t)} = \frac{1}{\alpha^2(t) + \beta^2(t)} e_1^T e^{-\Omega_0 \omega t} D e^{\Omega_0 ct} \Omega_0 e^{-\Omega_0 ct} \left(-c\Omega_0 D + \omega D \Omega_0\right) e^{\Omega_0 \omega t} e_1 \,.$$

After some some simplification, the numerator can be seen to be a quadratic form in $e^{\Omega_0 \omega t} e_1$ defined by the symmetric matrix

$$D\Omega_0 \left(c\Omega_0 D + \omega D \Omega_0\right) = -cD^2 + \omega (D\Omega_0)^2 = (c - \omega) \begin{bmatrix} -1 & 0 \\ 0 & -\frac{c^2}{\omega^2} \end{bmatrix}$$

Because this is negative definite, we see that $\tan^{-1}(\beta/\alpha)$ is monotone decreasing with t.

Thus, for c and b in the ranges indicated, and for $0 \le t \le \pi/\omega$, the locus of points in $\mathbb{R}^2$ with coordinates $(\alpha(t), \beta(t))$ swept out as t increases from zero, starts at $\alpha(0), \beta(0)) = (1, 0)$ and rotates clockwise with an increasing radius which reaches a local maximum of b^2/ω^2 when $t = \pi/2\omega$ and then decreases monotonically until $t = \pi/\omega$ when it meets the unit circle. Moreover, the radius exceeds $b^2/2\omega^2$ for $t \in [\pi/2\omega - \pi/4\omega, \pi/2\omega + \pi/4\omega]$

Now suppose that we are given a particular value of (α, β), say (α^*, β^*), and that we want to show that we can find a value of $(ct, \omega t)$ satisfying the above equations. Because $\alpha^2(t) + \beta^2(t) > c^2/2\omega^2$ for a range of ct that exceeds 2π we see that the curve encircles a disk of radius $c^2/2\omega^2$. If we chose c/ω so that $c/\omega > \sqrt{2(\alpha^*)^2 + 2(\beta^*)^2}$ then the curve will encircle the given point. With each pair (c, ω) we associate a closed curve beginning and ending at $(\alpha, \beta) = (1, 0)$. The curve consists of two parts, the first is the curve described above which begins at $(1, 0)$ and ends on the unit circle. The second part begins

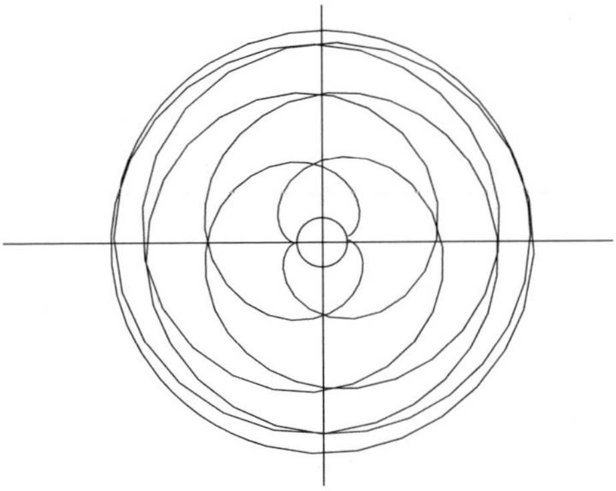

Fig. 1. An example of a plot of the $\alpha-\beta$ curve discussed in the proof. The *horizontal axis* is α, the *vertical* β and the *curve* winds clockwise, starting from $(1,0)$. The values of c and ω are 5 and 4.96, respectively. The *circle* in the center is the locus $\alpha^2+\beta^2=1$.

at the end point of the first and continues around the circle in a clockwise sense until it reaches $(1,0)$. This curve encircles the point (α^*,β^*) one or more times. Now consider the continuous deformation of this curve defined by reducing $(c^2-\omega^2)/\omega^2$ to one while keeping $c-\omega$ positive. This shrinks the maximal radius of the closed curve from its original value to one. Thus, for some value of c the curve must pass through (α^*,β^*) and the theorem is proven. □

This analysis suggests that a convenient locally convergent numerical algorithm for finding a solution can be obtained by adjusting (c,ω). We rewrite the equations for (α,β) as

$$\begin{bmatrix}0\\0\end{bmatrix}=\begin{bmatrix}1&0\\0&\omega\end{bmatrix}\begin{bmatrix}\cos ct&-\sin ct\\\sin ct&\cos ct\end{bmatrix}\begin{bmatrix}\alpha\\\beta\end{bmatrix}-\begin{bmatrix}1&0\\0&c\end{bmatrix}\begin{bmatrix}\cos\omega t&\sin\omega t\\-\sin\omega t&\cos\omega t\end{bmatrix}\begin{bmatrix}1\\0\end{bmatrix}.$$

Let $\phi(c,\omega)$ denote the norm squared of the right-hand side of these equations and consider

$$\begin{bmatrix}\dot c\\\dot\omega\end{bmatrix}=-\begin{bmatrix}\frac{\partial}{\partial c}\\\frac{\partial}{\partial\omega}\end{bmatrix}\phi(c,\omega).$$

Corollary 1. *Real matrices with positive determinants that have real twisted logarithms include*

1) all nonsingular symmetric matrices (positive definiteness not required);
2) all orthogonal matrices;
3) all nonsingular skew-symmetric matrices.

Proof. Given any nonsingular symmetric matrix there exists an orthogonal matrix Θ such that $\Theta^T M \Theta$ is diagonal with the the diagonals ordered such that the 11 and 22 elements have the same sign, the 33 and 44 elements have the same sign, etc. In this case theorem two can be applied one block at a time to give the result. A similar reduction to the two -by-two case is possible for the other cases as well. □

5 The Nilpotent Approximation

Known results on the nilpotent problem

$$\dot{x} = u \quad ; \quad \dot{Z} = xu^T - ux^T$$

show that the entire set of points of the form $(x, Z) = (0, Z_1)$ are conjugate to the point $(x, Z) = (0, 0)$. The optimal controls that generate any such transfer in one unit of time are all periodic of period 1 but there are controls of smaller period that satisfy the first order necessary conditions. In the case where x is two-dimensional these can be organized into families according to the least period of u, which takes the form $1/n$ with n an integer. Even in higher dimensions there are nicely characterized submanifolds of points conjugate to $(0, 0)$. Different points on the submanifolds correspond to different amplitudes of the control.

The problem under consideration here has a nilpotent approximation of the following form. Introduce the symmetric matrix Q and the skew-symmetric matrix Ω. Let U be symmetric and consider

$$\dot{Q} = U \quad ; \quad \dot{\Omega} = [U, Q] \, .$$

If the matrices in question are of size n-by-n then this can be thought of being the restriction of a system of the above form with x being of dimension $n(n+1)/2$ and Z being a skew-symmetric matrix with the corresponding number of rows and columns.

This is a nilpotent approximation to $\dot{X} = UX$ but there are very strong differences between the system and its nilpotent approximation. For example, as we have seen, conjugate to $X = I$ is any X_1 such that for some skew-symmetric matrix S we have $[S, X_1] = 0$ but X_1 is expressible as $e^{\Omega} e^{H - \Omega}$ with $[S, \Omega]$ or $[S, H]$ nonzero. In this case the controls that steer X to a conjugate point need not be periodic. As can be seen from the explicit formulas given above, every orthogonal matrix is conjugate to I but in the two by two case only those of the form $\pm I$ are generated by periodic controls. Moreover, the optimizing control to reach $-I$ is periodic with least period $1/2$, not 1. Thus we see that the deviation between the system $\dot{X} = UX$ and its nilpotent approximation is quite important.

6 Shortest Paths

We consider in this final section a direct demonstration of the optimality of certain paths which satisfy the first order necessary conditions. Such arguments are of interest because, as we have seen, in some cases it is not yet clear that there is any solution to the first order necessary conditions and, in cases where we know there is one, there may be many. Moreover, because the underlying space is not simply connected there will be cut points, beyond which a minimizing trajectory will cease to be minimizing. Here is a problem which can be treated in a elementary way and which provides a direct confirmation of optimality in certain cases. Observe that for $\dot{X} = UX$, any particular column of X satisfies a vector equation. For example, in the two by two case we have

$$\frac{d}{dt}\begin{bmatrix} x_{11} \\ x_{21} \end{bmatrix} = \begin{bmatrix} u & v \\ v & -u \end{bmatrix}\begin{bmatrix} x_{11} \\ x_{21} \end{bmatrix} = \begin{bmatrix} x_{11} & x_{21} \\ -x_{21} & x_{11} \end{bmatrix}\begin{bmatrix} u \\ v \end{bmatrix}.$$

It follows that

$$\frac{d}{dt}\tan^{-1}\left(\frac{x_{21}}{x_{11}}\right) = \frac{\dot{x}_{21}x_{11} - \dot{x}_{11}x_{21}}{x_{11}^2 + x_{21}^2} = \frac{v(x_{11}^2 - x_{21}^2) - 2ux_{11}x_{21}}{x_{11}^2 + x_{21}^2}$$

$$\frac{d}{dt}x_{11}^2 + x_{21}^2 = 2u(x_{11}^2 - x_{21}^2) + 2vx_{11}x_{21}\,.$$

These equations have an important property best revealed by writing them in terms of $\theta = \tan^{-1}(x_{21}/x_{11})$ and $\rho = \sqrt{x_{11}^2 + x_{21}^2}$. These equations take the form

$$\dot{\theta} = \langle f_1, \begin{bmatrix} u \\ v \end{bmatrix}\rangle \ ; \ \dot{\rho} = \langle f_2, \begin{bmatrix} u \\ v \end{bmatrix}\rangle$$

where f_1 and f_2 can be evaluated from the given equations. What is important is that f_1 and f_2/ρ define an orthonormal basis for $\mathbb{R}^2$. This makes it clear that the optimal control for advancing θ must align (u, v) with f_1. Thus we see, for example, that to transfer from the identity to any matrix of the form

$$M = \begin{bmatrix} \alpha & \beta \\ \gamma & \frac{1+\beta\gamma}{\alpha} \end{bmatrix}$$

requires at least

$$\eta* = \tan^{-1}\frac{\gamma}{\alpha} + \frac{1}{2}|\ln(\alpha^2 + \gamma^2)|\,.$$

The resulting trajectory in θ, ρ coordinates is

$$\theta(t) = \theta(0) + (\theta(1) - \theta(0))\,t \ ; \ \rho(t) = e^{\ln(\alpha^2+\gamma^2)t}\,.$$

Remark 3. To reach any matrix of the form

$$M = \begin{bmatrix} -e & s \\ 0 & -e^{-1} \end{bmatrix}$$

requires at least $\eta = \pi + 1$. ◁

References

1. J. Baillieul. *Some Optimization Problems in Geometric Control Theory.* PhD thesis, Harvard University, Cambridge MA, 1965.
2. C.S. Ballantine. Products of positive definite matrices. *Linear Algebra and Appl.*, 3:79–114, 1970.
3. R.W. Brockett. Control theory and singular Riemannian geometry. In P. Hilton and G. Young, editors, *New Directions in Applied Mathematics*, pages 11–27. Springer-Verlag, New York, 1981.
4. R.W. Brockett. Explicitly solvable control problems with nonholonomic constraints. *Proc. of the 38th IEEE Conf. on Decision and Contr.*, pages 13–16, 1999.
5. P. d'Alessandro, A. Isidori, and A. Ruberti. Structure analysis of linear and bilinear systems. In R. R. Mohler and A. Ruberti, editors, *Theory and Analysis of Variable Structure Systems*, pages 25–35. Academic Press, New York, 1972.
6. B. Gaveau. Principe de moindre action propagation de la chaleur et estimees sons elliptiques sur certains groupes nilpotents. *Acta Mathematica*, 139:95–153, 1977.
7. A. Isidori. *Nonlinear Control Systems.* Springer Verlag, London, UK, 3rd edition, 1995.
8. N. Khaneja, B. Luy, and S.J. Glaser. Boundary of quantum evolution under decoherence. *Proceedings of National Academy of Sciences*, 100(23):13162–13166, 2003.

The Hamilton-Jacobi Type Equations for Nonlinear Target Control and Their Approximation

Alexander B. Kurzhanski[1] and Pravin Varaiya[2]

[1] Moscow State (Lomonosov) University and Electrical Engineering and Computer Science University of California at Berkeley
[2] Electrical Engineering and Computer Science University of California at Berkeley

We dedicate this paper to Professor Alberto ISIDORI on the occasion of his birthday, in recognition of his outstanding contributions to nonlinear control theory.

Summary. This paper gives a comparison principle for first-order PDEs of the Hamilton-Jacobi-Bellman type that arise in problems of nonlinear target control synthesis and reachability analysis under hard bounds on the controls. The emphasis is is on treating backward reachability sets for a system with moving target sets which may also turn to be forward reachability tubes of another system. The target sets are to be reached within preassigned intervals. The exact solutions to this problem, given in set-theoretic terms, are expressed as level sets to the solutions of some specific types of the HJB equation. But these solutions require fairly complicated calculation. The present paper presents an alternative approach that avoids exact solutions in favor of their upper and lower bounds, which in many cases may suffice for solving the required problems. For systems with linear structure ellipsoidal estimates are given, which ensure accurate approximations of nonconvex reachability sets through unions of ellipsoids.

1 Introduction

A central issue in control theory is to investigate nonlinear control systems, [11]. Among these are problems of reachability and control synthesis. A related topic is how to calculate "weakly invariant" sets of points, namely, those from which it is possible to reach a given terminal set under feedback control, [13], [16]. The knowledge of these, also known as backward reachability sets, is crucial for designing feedback control strategies, constructing safety zones in motion planning and other related problems.

The present paper deals with the indicated problems under additional hard bounds on the controls, under complete noise-free measurements of the state space variable and under the assumption that the target set is moving and that the set may be reached within a given time interval rather than at given time. The target is therefore a set, possibly nonconvex, taken in the product space "time × state". This yields backward reach sets that are nonconvex even for linear systems.

It is well known that such problems for systems described by ODEs may be in principle reduced to the investigation of first order PDEs of the Hamilton-Jacobi-Bellman (HJB) type and their modifications, [13, 16, 3, 5, 19, 4]. However, solutions to equations of the HJB type are rather difficult to calculate, and the design of related computational algorithms is still under development [22, 24]. Nevertheless, for many applied problems one may often be satisfied with approximate solutions that impose a smaller computational burden and may be achieved by substituting the original HJB equations by variational inequalities due to certain *comparison principles* ([16, 6, 9]). This paper provides comparison theorems for the HJB equations of this paper, introducing upper and lower estimates for their solutions, which may used for checking sufficient conditions for verification problems of reachability analysis. In the case of linear systems this approach may lead to effective algorithms based on ellipsoidal techniques which also allow to accurately approximate nonconvex reach sets through unions of ellipsoids.

2 Backward Reachability from Moving Target Set

We begin with ordinary systems without disturbances. Consider equation

$$\dot{x} = f(t, x, u),\ t \in [t_0, t_1] = T(t_0), \tag{1}$$

where $x \in \mathbb{R}^n$ is the state space variable, $u \in \mathcal{P}(t) \subset \mathbb{R}^m$ is the control. Function $f(t, x, u)$ is assumed continuous in all variables and satisfying standard conditions of existence, uniqueness and prolongability of solutions within the intervals under consideration; $\mathcal{P}(t)$ – is a set-valued function with compact values, Hausdorff-continuous ([14], [1]).

Introduce the notation $X[t] = X(t, t_0, X^0)$ for the *forward reach set, at time* t, from the set-valued position $\{t_0, X^0\}$, where $X[t_0] = X^0$ is the starting set, and

$$\mathcal{H}(t, x, p) = \min\{(p, f(t, x, u)) | u \in \mathcal{P}(t)\}$$

for the *Hamiltonian* of system (1). Given *moving target set* $\mathcal{M}(s),\ s \in [t, \tau]$, - a compact set-valued function, we have to find $W_m[t] = W(t, \mathcal{M}_\alpha^\beta(\cdot))$ – the set of points $x = x(t)$, *from which* it is possible to reach set $\mathcal{M}(\tau)$ *for some* $\tau \in [\alpha, \beta] \subseteq T(t),\ t \leq \alpha,\ t_1 > \beta$. This is the *weakly invariant* or *backward reachability set* relative to moving target set $\mathcal{M}_\alpha^\beta(\cdot) = \mathcal{M}(\tau),\ \tau \in [\alpha, \beta]$, [12], [16].

The value function which grasps the desired property is

$$V_m(t,x) = \min_u\{\min_\tau\{d^2(x(\tau),\mathcal{M}(\tau)) \mid \tau \in [\alpha,\beta], u \in \mathcal{P}(\cdot)\},\ x(t)=x\}\,. \quad (2)$$

Here $d(x,\mathcal{X}) = \inf\{(x-z,x-z)^{1/2} \mid z \in \mathcal{X}\}$ is the Euclid distance from point x to set $\mathcal{M}$. Then (see [16])

$$W_m[t] = \{x : V_m(t,x) \le 0\}, \quad (3)$$

where

$$V_m(t,x) = \min_\tau\{V(t,\tau,x) \mid \tau \in [\alpha,\beta]\},$$

and

$$V(t,\tau,x) = \min_u\{d^2(x(\tau),\mathcal{M}(\tau)) \mid u \in \mathcal{P}(\cdot),\ x(t)=x\}\,. \quad (4)$$

In its turn, $V(t,\tau,x)$ satisfies the "backward" HJB equation

$$V_t + \mathcal{H}(t,x,V_x) = 0,\ V(\tau,\tau,x) = d^2(x(\tau),\mathcal{M}(\tau))\,. \quad (5)$$

Introducing the level sets $W(t,\tau) = \{x : V(t,\tau,x) \le 0\}$, we come to the following property.

Lemma 1. *The relation*

$$W_m[t] = \cup\{W(t,\tau) \mid \tau \in [\alpha,\beta]\},$$

where $W(t,\tau)$ is the zero-level set of the function $V(t,\tau,x)$ – the solution to the HJB equation (5), is true.

The last property is independent of whether V was obtained through classical or generalized viscosity solutions of equation (5). In the last case (5) is a formal notation where the partial derivatives have to be substituted by generalized subdifferentials or Dini derivatives, [8], [7], [2], [25]. The next item to discuss is the *strongly invariant set* relative to $\mathcal{M}_\alpha^\beta(\cdot)$. This is the set of points $W_s[t] = W(t,\mathcal{M}_\alpha^\beta(\cdot))$, $x = x(t)$ from each of which *the whole forward reach set $X(\tau,t,x) \subseteq \mathcal{M}(\tau)$ for some* time $\tau \in [\alpha,\beta]$, $t \le \alpha$. The value function which grasps the last property is

$$V_s(t,x) = \max_u \min_\tau\{d^2(x(\tau),\mathcal{M}(\tau)) \mid \tau \in [\alpha,\beta], u \in \mathcal{P}(\cdot),\ x(t)=x\}, \quad (6)$$

so that $W_s[t] = \{x : V_s(t,x) \le 0\}$. Then the formulae for calculating this set are similar to (2) – (5), but with $\min_u$ substituted for $\max_u$. Thus the exact description of the set $W[t]$ requires to solve the first-order PDE (5) followed by a minimization problem in τ. Such a problem is in general difficult to solve as the reachability sets for nonlinear systems may turn out to have a very peculiar form (see [21], [10]).

We shall therefore seek for the upper and lower estimates of functions $V_m(t,x)$, $V_s(t,x)$, and as a consequence, also the external and internal estimates of sets $W[t], W_s[t]$. Following are two theorems on internal external approximations (note the difference from [16] and similar results).

Consider the solution $V(t, \tau, x)$ of equation (5). The next theorem gives a lower approximation of this function.

Assumption 1. *The functions $H(t, x, p)$ and $w^+(t, \tau, x) \in C_1, \mu(t) \in L_1$, satisfy the inequalities*
(i) $\mathcal{H}(t, x, p) \geq H(t, x, p), \ \ \forall \{t, x, p\}$,
(ii) $w_t^+ + H(t, x, w_x^+) \geq \mu(t)$,
(iii) $V(\tau, \tau, x) \geq w^+(\tau, \tau, x)$.

Theorem 1. *Let Assumption 1 be true. Then, for all x*

$$w^+(t, \tau, x) + \int_t^\tau \mu(s)ds \leq V(t, \tau, x). \tag{7}$$

Proof. Let τ be fixed and the pair $\{\tau, x^*(\tau) = x^* \in \mathcal{M}\}$ generate the trajectory $x^*(s), \ s \in [t, \tau]$, which solves problem (4), with given $x^*(t) = x$. Then

$$dw^+(t, \tau, x)/dt|_{x=x^*(t)} =$$

$$= w_t^+(t, \tau, x^*) + \mathcal{H}(t, x^*, w_x^+) \geq w_t^+(t, \tau, x^*) + H(t, x^*, w_x^+) \geq \mu(t).$$

The last relations imply

$$dw^+(t, \tau, x)/dt|_{x=x^*(t)} \geq \mu(t).$$

Integrating this inequality from t to τ, we have

$$w^+(t, \tau, x^*(t)) + \int_t^\tau \mu(s)ds \leq w^+(\tau, \tau, x^*(\tau)) \leq V(\tau, \tau, x^*(\tau)) = V(t, \tau, x^*(t)).$$

This proves the theorem, due to Assumption 1 and the fact that $V(t, \tau, x) = V(t, \tau, x^*(t)) = V(\tau, \tau, x^*(\tau))$, due to equation (5). □

We look for upper approximations of $V(t, \tau, x)$.

Assumption 2. *The functions $H_-(t, x, p)$ and $w^-(t, \tau, x) \in C_1, \ \ \nu(t) \in L_1$ satisfy the inequalities*
(i) $\mathcal{H}(t, x, p) \leq H_-(t, x, p), \ \ \forall \{t, x, p\}$,
(ii) $w_t^- + H_-(t, x, w_x^-) \leq \nu(t)$,
(iii) $V(\tau, \tau, x) \leq w^-(\tau, \tau, x)$.

Theorem 2. *Let Assumption 2 be true. Then*

$$w^-(t, \tau, x) + \int_t^\tau \nu(s)ds \geq V(t, \tau, x). \tag{8}$$

The proof is similar to the above.

We may now introduce the lower and upper approximations of the function $V_m(t,x)$. Since (7) is true for all $t \le \tau$ and all $\tau \in [\alpha, \beta]$, we have

$$w_m^+(t,x) = \min_\tau\{w^+(t,\tau,x) + \int_t^\tau \mu(s)ds \mid \tau \in [\alpha,\beta]\} \le V_m(t,x). \quad (9)$$

Since (8) is true for all $t \le \tau$ and all $\tau \in [\alpha, \beta]$, we also have

$$w_m^-(t,x) = \min_\tau\{w^-(t,\tau,x) + \int_t^\tau \nu(s)ds \mid \tau \in [\alpha,\beta]\} \ge V_m(t,x)\,. \quad (10)$$

The last two relations lead to the next *comparison theorem.*

Theorem 3. *The function $V_m(t,x)$ possesses lower and upper bounds*

$$w_m^+(t,x) \le V_m(t,x) \le w_m^-(t,x).$$

Passing to level sets

$$W_m^+(t,\tau) = \{x : w_m^+(t,\tau,x) \le 0\},$$

$$W_m^-(t,\tau) = \{x : w_m^-(t,\tau,x) \le 0\},$$

we come to the next conclusion.

Theorem 4. *The backward reach set $W_m[t]$ for a moving target $\mathcal{M}_\alpha^\beta(\,\cdot\,)$ satisfies the relations*

$$\cup\{W_m^-(t,\tau) \mid \tau \in [\alpha,\beta]\} \subseteq W_m[t] \subseteq \cup\{W_m^+(t,\tau) \mid \tau \in [\alpha,\beta]\}.$$

A similar theorem may be proved for the set $W_s[t]$.

3 The Verification Problem for Moving Target Sets

We now introduce some rules to check whether a given point does belong to the sets from which, *within a given time interval* $[\alpha,\beta]$, one may ensure:

(a) the intersection $X(\tau,t,x) \cap \mathcal{M}(\tau) \ne \emptyset$ *for some* $t \in [\alpha,\beta]$ – a collision with target set,
(b) the condition $X(\tau,t,x) \cap \mathcal{M}_\varepsilon(\tau) = \emptyset$ *for all* $\tau \in [\alpha,\beta]$ and some $\varepsilon > 0$, – an evasion from $\mathcal{M}(\tau)$,
(c) the inclusion $X(\tau,t,x) \subseteq \mathcal{M}(\tau)$ *for all* $\tau \in [\alpha,\beta]$, - an immersion in $\mathcal{M}$ *for some* $\tau \in [\alpha,\beta]$ and *for all* $\tau \in [\alpha,\beta]$.

Here $\mathcal{M}_\varepsilon$ stands for an ε - neighborhood of $\mathcal{M}$ taken in some metric. The schemes for verifying the desired properties will usually consist in checking some inequalities or solving some optimization problems over respective value functions or their approximations.

Lemma 2. *Consider set $W_m[t]$ of the previous Section. In order that $x^* \in W_m[t]$ it is*

- *necessary and sufficient that $V_m(t,x^*) \leq 0$,*
- *sufficient that $w_m^-(t,x^*) \leq 0$.*

Lemma 3. *In order that $X(\tau,t,x^*) \cap \mathcal{M}_\varepsilon(\tau) = \emptyset$ for some $\varepsilon > 0$ it is*

- *necessary and sufficient that $V_m(t,x^*) > 0$,*
- *sufficient that $w_m^+(t,x^*) > 0$.*

Denote

$$V_s(t,\tau,x) = \max_u\{d^2(x(\tau),\mathcal{M}(\tau)) \mid u \in \mathcal{P}(\cdot), x(t) = x\},$$

and $w_s^-(t,\tau,x)$ to be the upper estimate of $V_s(t,\tau,x)$, similar to $w_m^-(t,\tau,x)$ – the upper estimate of $V_m(t,\tau,x)$.

Lemma 4.

(i) In order that $X(\tau,t,x^*) \subseteq \mathcal{M}(\tau)$ for some $\tau \in [\alpha,\beta]$, *it is*
  - *necessary and sufficient that* $\min_\tau V_s(t,\tau,x^*) \leq 0$,
  - *sufficient that* $\min_\tau w_s^-(t,\tau,x^*) \leq 0$.

(ii) In order that $X(\tau,t,x^*) \subseteq \mathcal{M}(\tau)$ for all $\tau \in [\alpha,\beta]$, *it is*
  - *necessary and sufficient that* $\max_\tau V_s(t,\tau,x^*) \leq 0$,
  - *sufficient that* $\max_\tau w_s^-(t,\tau,x^*) \leq 0$.

Consider now the particular situation when $\mathcal{M}(\tau) = Y[\tau]$ and $Y[\tau] = Y(\tau,t_0,y^0)$ is the reach set for another system of type (1), namely

$$\dot{y} = g(t,y,v),\ t \in [t_0,t_1] = T(t_0), \tag{11}$$

with $y \in \mathbb{R}^n$, $v(t) \in \mathcal{Q}(t)$ and the properties of $g(t,y,v)$, $\mathcal{Q}(\cdot)$ are similar to $f(t,x,u)$, $\mathcal{P}(\cdot)$. Then (see [16]), taking

$$V^T(\tau,y) = \min_v\{d^2(y(t_0),y^0) \mid v(\cdot) \in \mathcal{Q}(\cdot),\ y(\tau) = y\},$$

one gets $Y[\tau] = \{y : V^T(\tau,y) \leq 0\}$. Redefine the functions $V(\tau,x,t), V_s(\tau,x,t)$ of (4), (6) as

$$V(t,\tau,x) = \min_u\{V^T(\tau,x(\tau)) \mid x(t) = x\} \tag{12}$$

and

$$V_s(t,\tau,x) = \max_u\{V^T(\tau,x(\tau)) \mid x(t) = x\}, \tag{13}$$

taking $V^T(\tau,x)$ instead of $d^2(x(\tau),Y[\tau])$. Here $x(\tau)$ is the state generated by system (1) from $x(t) = x$, with some control $u(\cdot)$ and $y(\tau)$ is the state generated by system (11) from $y(t_0) = y$, with some control $v(\cdot)$. Relations (12), (13) may serve to define sets $W_m[t], W_s[t]$ as well as the former (4), (6).

Thus we have

$$V(t,\tau,x) = \min_u\{\min_v\{d^2(y(t_0),y^0) \mid v(\cdot) \in \mathcal{Q}(\cdot),\ y(\tau) = x(\tau)\} \mid x(t) = x\},$$

and

$$V_s(t,\tau,x) = \max_u\{\min_v\{d^2(y(t_0),y^0) \mid v(\cdot) \in \mathcal{Q}(\cdot),\ y(\tau) = x(\tau)\} \mid x(t) = x\},$$

which brings us, in the second case, to *a new type* of minmax problem. We have thus solved the functions for verifying conditions when the intersection $\mathcal{Z}(\tau) = X(\tau,t,x) \cap Y[\tau] \neq \emptyset$ or $\mathcal{Z}(\tau) = \emptyset$, or $\mathcal{Z}(\tau) = X(\tau,t,x) \subseteq Y[\tau]$ *at given time* τ Denoting $V(t,\tau,x) = V(t,\tau,x \mid t_0,y^0)$, $V_s(t,\tau,x) = V_s(t,\tau,x) \mid t_0,y^0)$, we now have to solve the same problem *within the interval* $[\alpha,\beta]$. Defining $V_m(t,x), V_s(t,x)$ and their level sets $W_m[t], W_s[t]$ as in the above, we come to verification rules similar to Lemmas 2.1-2.3. Problems of such type are relevant for decision support in organizing safe air and water traffic and related problems. Note that as a rule, the sets $W_m[t], W_s[t]$ are nonconvex. For systems with original linear dynamics constructive solutions are nevertheless available.

4 Linear Systems – Approximating Nonconvex Reachability Sets

We shall now illustrate how to use the results of the above for linear systems of type

$$\dot{x} = A(t)x + B(t)u, \tag{14}$$

with

$$u(t) \in \mathcal{E}(q(t),Q(t)), \quad \mathcal{M}(t) = \mathcal{E}(m(t),M(t)) \tag{15}$$

where

$$\mathcal{E}(q(t),Q(t)) = \{u : (u - q(t), Q^{-1}(t)(u - q(t))) \leq 1\}$$

stands for an ellipsoid with support function

$$\max\{(l,x) | x \in \mathcal{E}(q,Q)\} = (l,q) + (l,Ql)^{1/2}.$$

Then, applying Theorem 2.1, omitting the plus in $w^+(t,\tau,x)$, we have

$$\begin{aligned} dw/dt &= w_t + \mathcal{H}(t,x,w_x) \\ &= (w_x, A(t)x) - (w_x, B(t)Q(t)B'(t)w_x)^{1/2} + (w_x, B(t)q(t)) \\ &\geq w_t + (w_x, A(t)x) - \mu(t) - \\ &\quad (4\mu(t))^{-1}(w_x, B(t)Q(t)B'(t)w_x) + (w_x, B(t)q(t)) \\ &= w_t + H(t,w,w_x), \end{aligned} \tag{16}$$

for any $\mu(t) > 0$, with equality attained if

$$\mu(t) = \mu_e(t,w_x) = -\frac{1}{2}(w_x, B(t)Q(t)B'(t)w_x)^{1/2}.$$

We look for $w(t, \tau, x)$ as a quadratic form

$$w(t, \tau, x) = (x - x^*(t), P(t)(x - x^*(t))) - 1$$

and require that $w + 1$ satisfies the PDE

$$w_t + (w_x, A(t)x) - (4\mu(t))^{-1}(w_x, B(t)Q(t)B'(t)w_x) + (w_x, B(t)q(t)) = 0 \quad (17)$$

with boundary condition

$$w(\tau, \tau, x) + 1 \equiv (x - m, M(x - m)). \quad (18)$$

Then, after integrating the inequality $dw/dt \geq -\mu(t)$, from $\{t, x\}$ to $\{\tau, x(\tau)\}$, in view of (18), we come to

$$(x - m, M(x - m)) - 1 + \int_t^\tau \mu(s)ds = w(\tau, x(\tau)) + \int_t^\tau \mu(s)ds \geq w(t, x). \quad (19)$$

Here $P(t)$ may be obtained through a standard procedure of solving equation (17) by introducing a resulting Riccati equation

$$-\dot{P} = A(t)P + PA'(t) - \frac{1}{\mu(t)}(P, B(t)Q(t)P'(t)), \quad (20)$$

where $P(\tau) = M$ and equation

$$\dot{x}^* = A(t)x^* + B(t)q(t), \; x^*(\tau) = m\,. \quad (21)$$

The transformations

$$\dot{P}^{-1} = -P^{-1}\dot{P}P^{-1}, \quad \mathcal{P}_+(t) = (1 + \int_t^\tau \mu(s)ds)P^{-1}(t),$$

convert into

$$\dot{\mathcal{P}}_+ = \mathcal{P}_+A(t) + A'(t)\mathcal{P}_+ - \frac{1}{\pi(t)}B(t)Q(t)B'(t) - \pi(t)\mathcal{P}_+, \;\; \mathcal{P}_+(\tau) = M^{-1}\,. \quad (22)$$

Here $\pi(t) = \mu(t)/(1 + \int_t^\tau \mu(s)ds)$. These equations now coincide with those of Refs. [15, 17] derived through *inductive* procedures in contrast with present formulas derived *deductively*. Here

$$w(t, \tau, x) \leq V(t, \tau, x), \; \forall x \in \mathbb{R}^n, \tau \geq t,$$

so that the zero-level sets for function $w(t, \tau, x) = w(t, \tau, x \mid \pi(t)))$ are ellipsoids of type $\mathcal{E}(x^*(t), \mathcal{P}(t))$. This is all the more true for ellipsoids of type $\mathcal{E}(x^*(t), \mathcal{P}_+(t), \; \mathcal{P}_+(t)) = \mathcal{P}_+^\pi(t))$ that depend on parametrizing functions π. We thus arrive at the proposition below.

Theorem 5. *The backward reachability set* $W[t,\tau]$ *and the value function* $V(t,\tau,x)$ *possess the following bounds*

$$\mathcal{E}(x^*,\mathcal{P}_+^\pi(t)) = \{x : w(t,\tau,x \mid \pi(\cdot)) \le 1\} = \mathcal{E}_\pi^+[t] \supseteq W(t,\tau), \qquad (23)$$

where $w(t,\tau,x \mid \pi(\cdot)) = (x - x^*(t), (\mathcal{P}_+^\pi)^{-1}(t)(x - x^*(t))) - 1 \le V(t,\tau,x)$.

Note that in [15], [17] it was proved that

$$W[t,\tau] = \cap\{\mathcal{E}_\pi^+[t] \mid \pi(\cdot) \ge 0\} \qquad (24)$$

which gives *exact* representation of the backward reachability set. Selecting k functions $\pi_i(\cdot)$, we come to the estimate

$$W[t,\tau] \in \cap\{\mathcal{E}_{\pi_i}^+[t] \mid 1 \le i \le k\}. \qquad (25)$$

An appropriate selection of $\pi_i(\cdot)$ ensures that ellipsoids $\mathcal{E}_{\pi_i}^+[t]$ are tight in the sense that there is no other external ellipsoid of type (25) that could be squeezed in between $\mathcal{E}_{\pi_i}^+[t]$ and $W[t,\tau]$, ([17]). Denote

$$\min_\tau\{w(t,\tau,x \mid \pi(\cdot)) \mid \tau \in [\alpha,\beta]\} = w_\pi^+(t,x).$$

Then we come to the conclusion.

Lemma 5.

(i) The inequality $w_\pi^+(t,x) \le W(t,x)$ *is true for all* $\pi(s) > 0, s \in [t,\beta]$.
(ii) The equality $w(t,x) = \max\{w_\pi^+(t,x) \mid \pi(\cdot)\} = V(t,x)$ *is true.*

Here the first assertion follows from Theorem 5 while the second from (25). Recall that functions $w_\pi(t,\tau,x \mid \pi(\cdot))$ are parametrized quadratic forms. Their minimization over τ yields a lower estimate $w_\pi(t,x)$ for function $V(t,x)$ whose zero-level set $W[t]$ is nonconvex in general. A further maximization over parameter $\pi(\cdot)$ leads to the equality $w(t,x) = V(t,x)$. We now pass to upper estimates for $V(t,\tau,x)$ and internal estimates for $W(t,\tau)$. Applying Theorem 2.2, we have for $w^-(t,\tau,x)$ the equation

$$\begin{aligned}\frac{dw^-}{dt} &= w_t^- + \mathcal{H}(t,x,w_x^-)\\ &= w_t^- + (w_x^-, A(t)x) - (w_x^-, B(t)Q(t)B'(t)w_x^-)^{1/2} + (w_x^-, B(t)q(t))\\ &\le w_t^- + (w_x^-, A(t)x + B(t)q(t)) -\\ &\quad -(K^{-1}(t)w_x^-, K^{-1}(t)w_x^-)^{-1/2}(K^{-1}(t)w_x^-, S(t)(B(t)Q(t)B'(t))^{1/2}w_x^-)\\ &\le 0,\end{aligned} \qquad (26)$$

where $S(t)$ is any continuous matrix, whose values $S(t)$ are orthogonal matrices with $S(t)S'(t) = I$, and $K(t) = K'(t) > 0$. Equality is reached here under collinearity of vectors $S(t)(B(t)Q(t)B'(t))^{1/2}p$ and $K^{-1}(t)p,\ w_x = p$. We shall look for $w^-(t, \tau, x)$ to be quadratic, with $w^-(t, \tau, x) = (x - x^*, K(t)(x - x^*))$ -1, where $K(t) = K'(t) > 0$ is differentiable.

Note that in the domain

$$D(r) = \{\{t, x\} : (x - x^*(t), K(t)(x - x^*(t)) < r^2,\ t \in [t_0, t_1],\ t_0 > \alpha, t_1 < \tau\}\,,$$

with r sufficiently large, we have

$$\begin{aligned}
&(w_t^- + \mathcal{H}(t, x, w_x) \leq (x - x^*(t), \dot{K}(t)(x - x^*(t)))- \\
&-2(\dot{x}^*(t), K(t)(x - x^*(t)) + 2(K(x - x^*(t)), A(t)x + B(t)q(t))- \\
&-2r^{-1}((x - x^*(t)), S(t)(B(t)Q(t)B'(t))^{1/2}K(x - x^*(t))) \leq 0.
\end{aligned} \tag{27}$$

Therefore, in this domain we demand that the next equality be true

$$\begin{aligned}
&(x - x^*(t), \dot{K}(t)(x - x^*(t))) - 2(\dot{x}^*(t), K(t)(x - x^*(t))+ \\
&2(K(t)(x - x^*(t), A(t)(x - x^*(t)) + A(t)x^*(t) + B(t)q(t))- \\
&-2r^{-1}((x - x^*(t), S(t)(B(t)Q(t)B'(t))^{1/2}K(t)(x - x^*(t)) = 0.
\end{aligned}$$

Equalizing to zero the terms with multipliers of second order in $x - x^*$, then those of first order in the same variable, we observe that the last equality will be fulfilled if and only if, in the domain $D(r)$, the following equations are true $((B(t)Q(t)B'(t))^{1/2} = \mathbf{B(t)})$,

$$\dot{K} = -K'A(t) - A'(t)K - r^{-1}(KS(t)\mathbf{B}(t) + \mathbf{B}(t)S'(t)K) = 0, \tag{28}$$

$$\dot{x}^* = A(t)x^*(t) + B(t)q(t), \tag{29}$$

with boundary condition

$$K(\tau) = M^{-1}, x^*(\tau) = m.$$

Under the last equations (28), (29), the relation (27) yields the inequality

$$dw^-/dt \leq 0,$$

hence, integrating it from t to τ, we obtain the condition

$$w^-(t, \tau, x) \geq w^-(\tau, \tau, x) = V(\tau, \tau, x) = (x - m, M^{-1}(x - m)) - 1, \tag{30}$$

where $w^-(t, x) = (x - x^*(t), K(t)(x - x^*(t))$, and $K(t)$ and $x^*(t)$ are defined through equations (28) and (29), with

$$w(\tau, \tau, x) + 1 = (x - x^*(t), M^{-1}(x - x^*(t)))\,.$$

Note that for the inverse $K^{-1}(t) = \mathcal{K}(t)$ we have

$$\begin{aligned}
\dot{\mathcal{K}} &= A(t)\mathcal{K} + \mathcal{K}A'(t) + r^{-1}(S(t)\mathbf{B}(t)\mathcal{K} + \mathcal{K}\mathbf{B}(t)S'(t)) = 0, \\
\mathcal{K}(\tau) &= K^{-1}(\tau) = M(\tau)\,.
\end{aligned} \tag{31}$$

Thus $\{x : w(t, \tau, x) \leq 0\} = \mathcal{E}(x^*, \mathcal{K}(t)) = \mathcal{E}_S^-[t] \subseteq W(t, \tau)$. Summarizing the above, we have the following statement.

Theorem 6.

(i) The function

$$w^-(t,\tau,x|S(\cdot)) \geq V(t,\tau,x),\ \forall S(\cdot).$$

(ii) The inclusion

$$\mathcal{W}(t,\tau) \subseteq \mathcal{E}_S^-[t],$$

is true, where

$$\mathcal{E}_S^-[t] = \{x : (x - x^*(t), K(t)(x - x^*(t))) \leq 1\}$$

and $\mathcal{K}(t) = K^{-1}(t)$ *and* $x^*(t)$ *are defined by equations (29), (31).*

Finally, let

$$\min_{\tau}\{w - (t,\tau,x|S(\cdot)) \mid \tau \in [\alpha,\beta]\} = w^-(t,x \mid S(\cdot))\}\,.$$

Then we come to the conclusion.

Lemma 6.

(i) The inequality $w^-(t,x \mid S(\cdot)) \geq W(t,x)$ *is true for all* $S(s) > 0, s \in [t,\beta]$.
(ii) The equality $w(t,x) = \min\{w^-(t,x \mid S(\cdot)) \mid S(\cdot)\} = V(t,x)$ *is true.*

It now remains to pass to the approximation of the functions

$$V_s(t,x) = \min_{\tau} V_s(t,\tau,x), \quad V_{ss}(t,x) = \max_{\tau} V_s(t,\tau,x),$$

where

$$V_s(t,\tau,x) = \max_{u}\{d^2(x(\tau),\mathcal{M}(\tau)) \mid u(\cdot) \in Q(\cdot)\}.$$

The difference of the last function $V_s(t,\tau,x)$ from $V(t,\tau,x)$ is that in the definition of the former we have a "max" rather than a "min". This difference is reflected in the schemes for quadratic upper and lower approximations of $V_s(t,\tau,x)$ and for external and internal ellipsoidal approximations of sets $W_s(t,\tau)$, which may be repeated along the same lines as before with obvious changes. As a result we have the next propositions.

Theorem 7. *The following inclusions are true*

$$W_s(t,\tau) \supseteq \mathcal{E}(x^*(t),\mathcal{P}_s(t)\ |\gamma(\cdot)),$$

where $x^*(t)$ *is defined by (21) and* $\mathcal{P}_s$ *is defined as*

$$\dot{\mathcal{P}}_s = \mathcal{P}_s A(t) + A'(t)\mathcal{P}_s + \frac{1}{\gamma(t)}B(t)Q(t)B'(t) + \gamma(t)\mathcal{P}_s,\ \ \mathcal{P}_s(\tau) = M^{-1}(\tau), \quad (32)$$

where

$$\gamma(t) = \mu(t)/(1 - \int_t^\tau \mu(s)ds) > 0.$$

Here $\mathcal{E}(x^*(t), \mathcal{P}_s(t) \,|\gamma(\cdot))$ depends upon a positive parameter $\gamma(\cdot)$. The set

$$\mathcal{E}(x^*(t), \mathcal{P}_s(t) \mid \gamma(\cdot))$$

becomes empty when $\mathcal{P}_s(t)$ departs from being positive definite.

Theorem 8. *The following inclusions are true*

$$W_s(t,\tau) \subseteq \mathcal{E}(x^*(t), \mathcal{K}_s(t) \,|S(\cdot)),\ \forall S(\cdot),$$

where $x^*(t)$ is defined by (21) and $\mathcal{K}_s$ is defined by

$$\begin{aligned} \dot{\mathcal{K}} &= A(t)\mathcal{K} + \mathcal{K}A'(t) - r^{-1}(S(t)\mathbf{B}(t)\mathcal{K} + \mathcal{K}\mathbf{B}(t)S'(t)) = 0, \\ \mathcal{K}(\tau) &= K^{-1}(\tau) = M(\tau), \end{aligned}$$

where $S(t)$ is any orthogonal matrix-valued function, $S'(t)S(t) \equiv I$.

Lemmas 5, 6 allow us to verify the sufficient conditions of Lemmas 2.1-2.3 by using approximations of value functions through parametrized families of quadratic forms and approximations of nonconvex sets through unions of parametrized ellipsoids. For example, suppose for the linear system of this Section that

$$w_m^-(t,x) = \min_\tau\{w^-(t,\tau,x \mid S(\cdot)\}) \mid \tau \in [\alpha,\beta]\} = w^-(t,x \mid S(\cdot)),$$

where $w(t,\tau,x \mid S(\cdot))$ is a parametrized quadratic form received due to Theorem 6. Also suppose that for x given, there exists *at least one* parametrizing function $S(\cdot)$ that yields $w^-(t,\tau,x \mid S(\cdot)) < 0$ for *at least one* τ. Then this $x \in W_m[t]$! For another example, suppose for the linear system of this Section that

$$w_m^+(t,x) = \min_\tau\{w^+(t,\tau,x \mid \pi(\cdot)) \mid \tau \in [\alpha,\beta]\} = w_\pi^+(t,x),$$

where $w^+(t,\tau,x \mid \pi(\cdot))$ is a parametrized quadratic form received due to Theorem 5. Also suppose that for x given, there exists *at least one* parametrizing function $\pi(\cdot)$ that yields $w^+(t,\tau,x \mid \pi(\cdot)) > 0$ *for all* τ. Then there exists an ε such that $X(t,\tau,x) \cap \mathcal{M}_\varepsilon = \emptyset$. In a similar way, to check the sufficient condition of Lemma 4, one has to use quadratic approximations of $V_s(t,\tau,x)$ described in Theorems 3.3, 3.4. Approximation by parametrized quadratic functions allows to effectively approximate verification problems for moving targets that are reach sets $Y[t]$ of another system, as indicated at the end of the previous Section. A detailed description of such algorithms lies beyond the scope of this paper.

5 Conclusion

This paper indicates Hamilton-Jacobi-Bellman equations for problems of backward reachability for nonlinear systems under moving target sets. Some comparison principles for approximating the solutions to these equations from above and below are further indicated. For systems with original linear dynamics the types of backward reach sets discussed here turn out to be nonconvex, despite the original linearity. Nevertheless the application of parametrized quadratic forms and related families of ellipsoids in principle allow in nondegenerate cases to approximate these nonconvex sets by unions of ellipsoids. Among other approaches to the problem under discussion are those in publications [26], [18], [23], [20].

References

1. J.P. Aubin and H. Frankowska. *Set Valued Analysis.* Birkhäuser, Boston, 1990.
2. M. Bardi and I. Capuzzo Dolcetta. *Optimal Control and Viscosity Solutions of Hamilton-Jacobi-Bellman Equations.* Birkhäuser, Boston, 1997. SCFA.
3. R. Bellman and R. Kalaba. *Dynamic Programming and Modern Control Theory.* London math. society monographs, London, 1965.
4. D.P. Bertsekas. *Dynamic Programming and Optimal Control.* Dynamic Systems and Mycrophysics. Athena Scientific, v.I,II,, Belmont, Mass., 1995.
5. A. Blaquière and G. Leitmann, editors. *Optimality and Reachability via Feedback Controls.* Math. Theory of Dyn. Systems and Microphysics. Academic Press, 1982.
6. F.H. Clarke, Y.S. Ledyaev, R.J. Stern, and P.R. Wolenski. *Nonsmooth Analysis and Control Theory,* volume 178 of *GTM.* Springer Verlag, 1998.
7. M.G. Crandall, L.C. Evans, and P.L. Lions. Some properties of viscosity solutions of Hamilton-Jacobi equations. *Transaction American Mathematical Society,* 282(2):487–502, 1984.
8. W.H. Fleming and H.M. Soner. *Controlled Markov Processes and Viscosity Solutions.* Springer-Verlag, New York, 1993.
9. V.I. Gurman. *The Extension Principle in Problems of Control.* Fizmatlit, Moscow, 1997.
10. H.I. Guseinov, A.N. Moiseev, and V.N. Ushakov. On the approximation of reachability sets for control systems. *PMM – Applied Mathematics and Mechanics,* pages 179–186, 1998.
11. A. Isidori. *Nonlinear Control Sytems.* Springer Verlag, 3rd edition, 1995.
12. N.N. Krasovski. *Rendezvous Game Problems.* Nat. Tech. Inf. Service. Springfield, Virginia, 1971.
13. N.N. Krasovski and A.N. Subbotin. *Game-theoretical Control Problems.* Springer-Verlag, 1988.
14. A.B. Kurzhanski. *Control and Observation Under Uncertainty.* Nauka, Moscow, 1977.
15. A.B. Kurzhanski and I. Vályi. *Ellipsoidal Calculus for Estimation and Control.* Birkhäuser, Boston, 1997. SCFA.
16. A.B. Kurzhanski and P. Varaiya. Dynamic optimization for reachability problems. *J. Opt. Theory and Applications,* 108(2):227–251, 2001.
17. A.B. Kurzhanski and P. Varaiya. *Ellipsoidal Techniques for Reachability Analysis: Part I: External Approximations. Part II: Internal Approximation. Box-valued Constraints,* volume 17 of *Optimization Methods and Software.* Taylor and Francis, 2002.

18. J. Lygeros. *Lecture Notes on Hybrid Systems.* ENSIETA, University of Patras, Greece, 2004.
19. J. Lygeros, C.J. Tomlin, and S.S. Sastry. Controllers for reachability specifications for hybrid systems. *Automatica*, pages 349–370, 1999.
20. I. Mitchell, A.M. Bayen, and C.J. Tomlin. Validating a hamilton-jacobi approximation to hybrid system reachable sets. In M. DiBenedetto and A. San-Giovanni-Vincentelli, editors, *Hybrid Systems: Computation and Control*, LNCS N. 2034, pages 418–432. Springer-Verlag, 2001.
21. V.S. Patsko, S.G. Pyatko, and A.A. Fedotov. Three-dimensional reachability set for a nonlinear control system. *Journal of Computer and Syst. Sc. Intl*, 42(3):320–328, 2003.
22. Osher. S. and R. Fedkiw. *Level Set Methods and Dynamic Implicit Surfaces*, volume 153 of *AMS*. Springer Verlag, 2003.
23. P. Saint-Pierre. Approximation of viability kernels and capture basins for hybrid systems. *Proc. of the 2001 European Control Conference*, pages 2776–2783, 2001.
24. J.A. Sethian. *Level Set Methods and Fast Marching Methods.* Cambridge Univ. Press, 2nd edition, 1999.
25. A.I. Subbotin. *Generalized Solutions of First-order PDE's. The Dynamical Optimization Perspective.* Birkhäuser, Boston, 1995. SCFA.
26. C.J. Tomlin, G.J. Pappas, and S.S. Sastry. Conflict resolution for air traffic management: a case study in multi-agent hybrid systems. *IEEE Trans. on Automat. Contr.*, 43(4):509–521, 1998.

Causal Coding of Markov Sources with Continuous Alphabets

Serdar Yüksel[1] and Tamer Başar[2]

[1] Yale University. Dept. of Electrical Engineering, 10 Hillhouse Avenue, New Haven, CT 06511, USA.
[2] Coordinated Science Laboratory, University of Illinois, 1308 West Main Street, Urbana, IL 61801-2307 USA.

This chapter is dedicated to Professor Alberto Isidori on the occasion of his 65th birthyear.

Summary. We study, for kth-order Markov sources, the structure of optimal causal encoders minimizing total rate subject to a mean-square distortion constraint over a finite horizon. The class of sources considered is general alphabet sources, and the encoder is allowed to be variable-rate. This leads to an optimization problem in an infinite-dimensional space, for which a solution exists. We show that the optimal causal encoder for a kth-order Markov source uses only the most recent k source symbols and the information available at the receiver. We also consider the infinite-horizon version, but for $k = 1$, provide an existence result for an optimal stationary solution, discuss the effect of randomization on performance, and show that a recurrence-based time sharing of at most two deterministic quantizers is optimal. We further argue that for real-valued processes innovation encoding is not in general optimal, but for coding of a stable linear or non-linear source, the quantization of the innovation is an almost optimal scheme in the limit of low-distortion. This class of problems has natural applications in remote control of linear and non-linear systems with quantization.

1 Introduction

In real-time applications such as remote control of time-sensitive processes [6], live streaming and voice over Internet, causality in encoding and decoding is a natural limitation. For instance, there is a natural causal ordering of events in a controlled process, consisting of observation, estimation, and actuation [23]. All these events need to take place in real time and with no arbitrary delay. Such a requirement rules out the use in control systems of many of the traditional communication and encoder schemes which use long block codes, since their designs build on the asymptotic analysis required by information theory and rate-distortion theory.

Practical systems require fast and easily implementable schemes, which may not perform satisfactorily. Scalar quantizers, and in particular uniform scalar quantizers, are the most widely used encoding schemes in many applications, and in particular in control systems. Uniform scalar quantizers, in addition to their simplicity, are also approximately optimal in the limit of high-rate distortion [8] with respect to the mean-square distortion measure. Furthermore, they are the optimal quantizers for the asymptotic distortion minimization for noiseless linear time-invariant scalar systems [25]. However, when one deviates from the limit of low-rate distortion or asymptotic analysis, in general, the structures of the quantizers get more complicated, and little is known about the optimal encoders.

Causal coding has been studied in the literature in different contexts and with different assumptions on the classes of sources and encoder types. Some of the major results on the topic are the following: When the elements of a sequence $\{x_i\}$, taking values on a discrete alphabet, are independent and identically distributed, and if the decoder is allowed to perform with delay, then the optimal memoryless coder is the optimal causal encoder minimizing the data rate subject to a distortion constraint [18]. The delayless encoding scheme was studied in [7], which demonstrated the optimality of memoryless fixed-rate encoders when the sources belong to a discrete and finite set, and the encoding is fixed-rate. We note that delayless encoding (zero-delay encoding) is stronger than that of causality in that both encoding and decoding are instantaneous. If the source is kth-order Markov, then the optimal causal fixed-rate coder minimizing any measurable distortion uses only the last k source symbols and the current state at the receiver's memory [24]. The results of [24] were extended in [21] to systems without feedback, under the assumption of a fixed decoder structure. The problem of optimal transmission over noisy channels with perfect causal feedback was considered in [22] for the case when the source belongs to a finite and discrete set. In the limit of low distortion (high rate), [14] studied stationary encoding of a stable stationary process and showed that memoryless quantizer followed by a conditional entropy coder is at most 0.25 bits worse than any noncausal encoder. A relevant problem in causal rate-distortion theory was studied in [20], where under the criterion of *directed mutual information* minimization subject to a distortion constraint, and with availability of feedback, optimal causal conditional coding laws are obtained.

If one allows variable rate coding, the optimization problem for the evaluation of the optimal quantizer becomes somewhat complicated since it involves possibly infinite-dimensional spaces. Entropy, as a lower bound on error-free transmission for a discrete source, can be attained fairly closely by means of entropy coding techniques or via block coding. We note that in variable rate coding, the delays in the transmission of higher length symbols might affect the overall delay. Nonetheless, in variable-rate coding, the system is still causal in that no future data is allowed to be used in encoding and decoding. Furthermore, using tools from variable rate encoding, a bound on the maximum length code can be enforced and entropy coding schemes can still be applied under such a restriction (note that such a restriction necessarily leads to finite number of codewords). Various studies [10], [11] have looked at the problem of *entropy-constrained* quantization for the design of variable rate quantizers.

Stochastic control can provide a useful tool in the characterization of optimal coding schemes, as has been shown in [24]. An important work in this direction is [4], which studied infinite horizon quantization of a partially observed source generated by a stationary stable Markov process by also allowing variable rate coding, but under the restriction of finitely many codewords. The existence of a solution in a general, infinite-horizon average performance criterion Markov decision process (MDP) problem was studied in several publications, such as [16], [12].

1.1 Main Results

We provide in this chapter extensions of the available results on causal coding/decoding in finite horizon to Markov sources of arbitrary degree (which include i.i.d. sources) and in general state spaces, and to the case where the quantization is variable rate. We also provide extensions to the infinite-horizon case with a first-order Markov source. Specifically we give the following results.

1) We obtain existence results for optimal causal deterministic and randomized quantizers under general conditions.
2) We provide the structure of the optimal causal encoders, and show that the optimal causal coder minimizing the total data rate subject to a distortion constraint over a finite horizon for a kth-order Markov source uses the last k source outputs and the information available at the receiver.
3) For the infinite-horizon problem, and for a first-order Markov source, we obtain existence results, and characterize the optimal solution.
4) We show that performance can be improved through randomization, where both the transmitter and the receiver have access to the realized outcome of the randomized choice. Randomization can be restricted to two-point distributions, without any loss of generality.
5) For the infinite horizon case, even without sharing randomization information, an optimal time-sharing policy based on recurrence properties of the Markov dynamics attains the optimal performance.
6) For various source dynamics, including non-linear systems, we provide bounds on the performance of innovation coding. We show that innovation coding is in general suboptimal, but for coding of stable linear as well as non-linear sources, innovation coding is almost optimal in the limit of low distortion.

2 Preliminaries

We first introduce the notion of a quantizer.

Definition 1. *A quantizer, Q, for a scalar continuous variable is a Borel-measurable mapping from the real line to a finite or countable set, characterized by corresponding bins $\{\mathcal{B}_i\}$ and their reconstruction levels $\{q^i\}$, such that $\forall i$, $Q(x) = q^i$ if and only if $x \in \mathcal{B}_i$.*

We assume that the quantization bins are regular [10]. Thus, (for scalar quantization) $\mathcal{B}_i$ can be taken to be nonoverlapping semiopen intervals, $\mathcal{B}_i = (\delta_i, \delta_{i+1}]$, with $\delta_i < \delta_{i+1}, i = 0, \pm 1, \pm 2, \ldots$, such that δ_0 is the one closest to the origin, where $\{\delta_i\}$ are termed "bin edges". Here, we also have $q^i \in \mathcal{B}_i$.

In a dynamic, discrete-time setting, the construction of a quantizer at any time t could depend on the past quantizer values. To make this precise, let $\mathcal{X}$ be the input space, $\hat{\mathcal{X}}$ be the output space, and $\mathcal{Q}_t$ be the set of quantizer reconstruction values, q_t, at time t, $t = 1, 2, \ldots$. Then, the quantizer at time t, to be denoted by f_t, is a mapping from $\mathcal{X}^t$ to $\mathcal{Q}_t$, where $\mathcal{X}^t$ is the t-product of $\mathcal{X}$. Such a quantizer is said to be *causal* (that is, it depends only on the past and present values of the input process), in addition to being *dynamic*. A quantizer is assumed to be followed by an encoder, which provides the binary representation of the quantization outputs. However, hereafter, we will use the term encoder to refer to the ensemble of the quantizer and the encoder. Finally, we also introduce $g_t : \mathcal{Q}_1 \times \mathcal{Q}_2 \times \ldots \mathcal{Q}_t \to \hat{\mathcal{X}}$ as the *decoder function*, which again has causal access to the past received values.

The class of quantizers we have introduced above are so-called deterministic quantizers, in the sense that for each fixed t, and given f_t and history $x_1^t := \{x_s,\ s = 1, \ldots, t\}$, the quantizer value induced, $q_t = f_t(x_1^t)$, is a uniquely defined element of $\mathcal{Q}_t$. Let $\mathcal{T}$ be the space of such deterministic quantizers, and let $\sigma(\mathcal{T})$ be its σ-algebra. A more general class quantizers is the *randomized* ones, which assign a probability measure to selection of bins for each fixed $x_1^\infty \in \mathcal{X}^\infty$. More precisely, and by a slight abuse of notation, quantizer policy $q(.|x)$ is randomized if, for each $x \in \mathcal{X}^\infty$, $q(.|x)$ is a probability measure on $\sigma(\mathcal{T})$, and if, for every fixed $D \in \sigma(\mathcal{T})$, $q(D|.)$ is a well-defined function on $\mathcal{X}^\infty$, whose restriction to the interval $[1, t]$ agrees with a deterministic quantizer f_t.

In the development of this chapter, we will first work with deterministic quantizers, and subsequently extend the analysis to randomized policies, utilizing in both cases entropy analysis and dynamic programming. We will also be using Markov Decision Process (MDP) tools; an overview as relevant to the development here is provided in the Appendix.

3 Problem Formulation

Let v_i^m, $i \leq m$, denote the ith through mth components, $\{v_i, \ldots, v_m\}$, of a vector v. Consider a sequence: $x_t \in \mathcal{R}, t = 1, 2, \ldots, N$. The encoder causally encodes this sequence, $f_t(x_1^t)$, $1 \le i \le N$, and a delayless causal decoder generates the estimates $\hat{x}_t = g_t(f_1(x_1), \dots, f_t(x_1^t))$, for $1 \le t \le N$. Let x_1^N be generated by an i.i.d. or a kth order Markov process, $k \ge 1$, where $x_t, 1 \le t \le N$, takes values in $\mathcal{R}$, and $E[x_t^2] < \infty$ with the marginal density functions $p(x_t)$ continuous. Let $H(f_1^N)$ denote, with a slight abuse of notation, the entropy of the quantizer outputs $q_1^N := \{q_1, \dots, q_N\}$ under the quantization policies $f_i(\cdot)$, $i = 1, \dots, N$, that is, with $p_i(q_i)$ denoting the probability of q_i,

$$H(f_1^N) = -\sum_{i=1}^{N} E[p_i(q_i) \log_2 p_i(q_i)] \,, \quad q_i = f(x_1^i) \,, \; i = 1, \dots, N \,.$$

We study the following constrained minimization problem: For a given positive integer N,

$$\inf_{f_1^N} \frac{1}{N} H(f_1^N) \tag{1}$$

subject to

$$\frac{1}{N} \sum_{t=1}^{N} E[(x_t - \hat{x}_t(f_1^t))^2] \le D, \tag{2}$$

for some finite $D > 0$, where $\hat{x}_t(f_1^t)$ is (as the output, $g_t(\cdot)$, of the decoder) the conditional mean of x_t given the quantizer policy f_s, $s \le t$, and the output of the quantizer, q_s, $s \le t$. We will consider also the infinite-horizon case, when $N \to \infty$. Further, we will study the improvement in the value of (1) when randomization is allowed on quantizer policies.

In the causal coding literature, the underlying optimization problem has generally been restricted to finite dimensional spaces. The analysis then builds on the fact that a continuous function over a compact set attains a minimum. However, when there is an entropy constraint, as opposed to a fixed length rate constraint, the optimization problem is one of infinite dimension and the optimal quantizer could then have infinitely many quantization levels [9]. Then, the appropriate framework in this case is one of infinite-dimensional optimization using weak topology. Some background on this is provided in the Appendix.

4 The Structure of Optimal Deterministic Quantizers

We use the notation $|.|$ to denote the Euclidean norm (absolute value) of a scalar quantity and $||.||_2$ to denote the Euclidean norm of a vector.

Since it is quite inconvenient to work with the set of quantization functions, we will work with the set of quantization output distributions on integer indexed bins. Thus, one could carry out the analysis via well-studied topological results on the space of probability distribution functions. We assume the input probability distributions to be absolutely continuous with respect to the Lebesgue measure, and define the (Radon-Nikodym) derivative of the distributions with respect to the Lebesgue measure as the probability density function, μ. We also let, by slight abuse of notation, the conditional density (or conditional probability mass function, as appropriate) of the random variable x_t given $f_s(x_1^s) = q_s$, $1 \le s \le t$, for a fixed quantizer policy f_1^t by $P(x_t|f_1^t)$; note that this quantity depends not only on the observed quantizer outputs q_1^t (suppressed in this description) but also on the functional form of the quantizer policy, f_1^t. A similar interpretation also applies to $P(f_i|f_1^{i-1})$ (probability mass function of $q_i = f_i(x_1^i)$ given $f_s(x_1^s) = q_s$, $1 \le s \le i-1$), as well as the conditional entropy, $H(f_i|f_1^{i-1})$, which is that of the quantizer output at time $t = i$ given all previous quantizer outputs and all quantizer functions (past and present).

Our main results for the finite-horizon case are now given below, first for memoryless sources, and then for Markov sources with memory. In both cases, a solution to (1)–(2) exists because it is minimization of a weak* continuous function, $H(f_1^N)$, over a weak* compact set, as determined by (2)[3].

Theorem 1. *Suppose $\{x_t\}$ are i.i.d. random variables which can be discrete or continuous valued. Then, the optimal deterministic encoder uses only the current symbol.*

Proof. We use dynamic programming. The underlying constrained optimization problem has an associated Lagrange multiplier, $\lambda > 0$ ([25], Proof of Lemma 3.3). Let us introduce the Lagrangian

$$J_\lambda = \sum_{i=1}^{N} H(f_i|f_1^{i-1}) + \lambda \sum_{i=1}^{N} E_{x_i}[(x_i - g(f_i, f_1^{i-1}))^2 | x_1^i, f_1^{i-1}].$$

The cost at time N can be written as

$$J_\lambda^N = \lambda E_{x_N}[(x_N - g(f_N, f_1^{N-1}))^2 | x_1^N, f_1^{N-1}] + H(f_N|f_1^{N-1}),$$

which is identical to

$$J_\lambda^N = \lambda E_{x_N}[(x_N - g(f_N, f_1^{N-1}))^2 | x_N, f_1^{N-1}] + H(f_N|f_1^{N-1}),$$

since $x_N - g(f_N, f_1^{N-1})^2$ does not depend on x_1^{N-1}, which follows from the observation that the encoder has perfect access to f_1^{N-1} and x_N.

Define the scalar quantizer and the decoder for the last stage as

$$f_{f_1^{N-1}}(x_N) := f_N(x_N, f_1^{N-1}),$$
$$g_{u^N}(f_{f_1^{N-1}}(x_N)) := g_N(f_1^{N-1}, f_{f_1^{N-1}}(x_N)).$$

[3] Weak* compactness of the constraint set follows from an argument used in [25] for a similar (but not identical) set.

Thus,

$$J_\lambda^N = \lambda E_{x_N}[[(x_N - g_{u^N}(f_{f_1^{N-1}}(x_N)))^2 | x_N, f_1^{N-1}] + H(f_{f_1^{N-1}}(x_N)|f_1^{N-1}) .$$

We have, since the source is memoryless and the quantizer is deterministic,

$$H(f_{f_1^{N-1}}(x_N))|f_1^{N-1}) = H(f_{f_1^{N-1}}(x_N)).$$

The optimal decoder policy does not affect the entropy component of the cost, and therefore it generates the mean-square minimizing estimate:

$$\hat{x}_N = g_{u^N}(f_{f_1^{n-1}}(x_N)) = E[x_N|f_1^{N-1}, f_N] = E[x_N|f_N] ,$$

since

$$\begin{aligned} &p(x_N|f_1^{N-1}, f_N) \\ &= \frac{p(f_{f_1^{N-1}}(x_N)|x_N, f_1^{N-1})p(x_N|f_1^{N-1})p(f_1^{N-1})}{p(f_1^{N-1}, f_{f_1^{N-1}}(x_N))} \\ &= \frac{p(f_{f_1^{N-1}}(x_N)|x_N, f_1^{N-1})p(x_N)p(f_1^{N-1})}{p(f_{f_1^{N-1}}(x_N)|f_1^{N-1})p(f_1^{N-1})} \\ &= p(x_N|f_{f_1^{N-1}}(x_N)). \end{aligned} \tag{3}$$

Thus,

$$E_{x_N}[g_{u^N}(f_{f_1^{N-1}}(x_N))|f_1^{N-1}] = E_{x_N}[g_{u^N}(f_{f_1^{N-1}}(x_N))] ,$$

and we have

$$J_\lambda^N = \lambda E_{x_N}[(x_N - E_{x_N}[g_{u^N}(f_{f_1^{N-1}}(x_N))])^2) + H(f_{f_1^{N-1}}(x_N))] .$$

Now, regarding $f_{f_1^{N-1}}(x_N)$ as a function to be optimized over, it is evident that the optimal $f_{f_1^{N-1}}$ is only a function of the last state symbol x_N. This characterizes the structure for the last stage.

At time $N-1$, we have the problem of minimization of:

$$\begin{aligned} J_\lambda^{N-1} = \lambda E_{x_{N-1}}[(x_{N-1} - g_{N-1}(f_{N-1}, f_1^{N-2}))^2 \\ + H(f_{N-1}|f_1^{N-2}) + E[J^N|x_1^N, f_1^{N-1}]|x_1^{N-1}, f_1^{N-2}]. \end{aligned}$$

We thus know that J^N is independent of both x_1^N and the encoder outputs, and hence the last term can be taken out of the expectation. The structure of the cost, J_λ^{N-1}, then becomes identical to that in the problem for the Nth stage, and the quantizer in the $(N-1)$th stage is also memoryless. The recursions for the other stages follow similar reasoning. Thus, memoryless scalar quantization is optimal within the class of deterministic quantizer policies. □

Theorem 2. *For a kth-order Markov source, the finite-horizon optimal causal deterministic encoder at stage t, $0 \leq t \leq N-1$ use the most recent (available)* $\min(t, k)$ *symbols and the information available at the receiver. The optimal deterministic encoder for the last stage, N, uses only the last symbol and the information available at the receiver.*

Proof. Let us introduce the Lagrangian

$$J_\lambda = \sum_{i=1}^{N} H(f_i|f_1^{i-1}) + \lambda \sum_{i=1}^{N} E_{x_i}[(x_i - g(f_i, f_1^{i-1}))^2 | x_1^i, f_1^{i-1}].$$

The cost at time N can be written as

$$J_\lambda^N = \lambda E_{x_N}[(x_N - g(f_N, f_1^{N-1}))^2 | x_1^N, f_1^{N-1}] + H(f_N|f_1^{N-1}),$$

which is identical to

$$J_\lambda^N = \lambda E_{x_N}[(x_N - g(f_N, f_1^{N-1}))^2 | x_N, f_1^{N-1}] + H(f_N|f_1^{N-1}),$$

since $x_N - g(f_N, f_1^{N-1})$ does not depend on x_1^{N-1}, which follows from the observation that the encoder has perfect access to f_1^{N-1} and x_N.

Define the scalar quantizer and the decoder for the last stage as

$$f_{f_1^{N-1}}(x_N) := f_N(x_N, f_1^{N-1}),$$
$$g_{u^N}(f_{f_1^{N-1}}(x_N)) := g_N(f_1^{N-1}, f_{f_1^{N-1}}(x_N)).$$

Thus,

$$J_\lambda^N = \lambda E_{x_N}[[(x_N - g_{u^N}(f_{f_1^{N-1}}(x_N)))^2 | x_N, f_1^{N-1}] + H(f_{f_1^{N-1}}(x_N) | f_1^{N-1}).$$

Unlike the memoryless case, we cannot take the conditioning out in the expression for $H(f_{f_1^{N-1}}(x_N)) | f_1^{N-1})$.

The optimal decoder policy does not affect the entropy component of the cost, and therefore it generates the mean-square minimizing estimate:

$$\hat{x}_N = g_{u^N}(f_{f_1^{n-1}}(x_N)) = E[x_N | f_1^N, f_N].$$

Thus, we have

$$J_\lambda^N = \lambda E_{x_N}[(x_N - E_{x_N}[g_{u^N}(f_{f_1^{N-1}}(x_N))])^2)$$
$$+ H(f_{f_1^{N-1}}(x_N)) | f_1^{N-1})],$$

from which it is evident that the optimal $f_{f_1^{N-1}}$ is only a function of the last state symbol x_N and $f_{1^{N-1}}$. This characterizes the structure for the last stage.

Upon observing the functional dependence of $J^N(\lambda)$, we write the functional to be minimized at stage $N-1$ as,

$$J_\lambda^{N-1} = \lambda E_{x_{N-1}}[(x_{N-1} - g_{N-1}(f_{N-1}, f_1^{N-2}))^2 | x_1^{N-1}, f_1^{N-2}]$$
$$+ H(f_{N-1}|f_1^{N-2}) + E_{x_N}[J^N | x_{(N-1)-k+1}^N, f_1^{N-1}].$$

It should be observed that, J_λ^{N-1} is a function of $(x_{(N-1)-k+1}^{N-1}, f_1^{N-1})$, due to the fact that the last stage cost is a function of $x_{(N-1)-k+1}^{N-1}$. This follows from the observation that the statistics of x_N is completely characterized by $x_{(N-1)-k+1}^{N-1}$. Thus, we write $J_\lambda^{N-1}(x_{N-1-k+1}^{N-1}, f_1^{N-1})$ to show the explicit dependence on its arguments.

For the time stage $N-2$, we have

$$J_\lambda^{N-2} = \lambda E_{x_{N-2}}[(x_{N-2} - g_{N-2}(f_{N-2}, f_1^{N-3}))^2$$
$$+H(f_{N-2}|f_1^{N-3}) + \{E_{x_{N-1}}[J_\lambda^{N-1}(x_{N-1-k+1}, f_1^{N-2})$$
$$+E_{x_N}[J^N(x_{(N-1)-k+1}^N, f_1^{N-1})]]|x_1^{N-2}, f_1^{N-2}]\}.$$

Here, we have an expectation over x_{N-1} and x_{N-2}. Since x_{N-1} depends on x_{N-1-k}^{N-2}, and these terms are measurable, they affect the expectation for the cost-to-go, i.e.,

$$E[J^N | x_{(N-1)-k+1}^N, f_1^{N-1}]$$

and are to be used in the optimization. An inductive argument proves the result for t, such that $k \le t \le N-3$.

For $t \le k$, one uses only the available source values, x_1^t and the information at the receiver. □

5 The Infinite-Horizon Solution

Here we restrict the source to be first-order Markov, to simplify the analysis and the convergence result below. For such a source, the problem in the infinite-horizon case is the minimization of the quantity

$$\limsup_{N\to\infty} \frac{1}{N} \sum_{t=1}^{N} H(f_t | f_1^{t-1}) \tag{4}$$

subject to the average distortion constraint

$$\limsup_{N\to\infty} \frac{1}{N} \sum_{t=1}^{N} E[(x_t - \hat{x}_t(f_1^t))^2] \le D. \tag{5}$$

A version of the infinite horizon problem has been studied in [4]. Our setup here is different, however, in that we do not assume that the number of bins is finite.

We first need a non-linear filter for the evolution of the conditional density, which was developed in [4]. First, it follows from the property of total probability that

$$\int_{x_{n-1}} P(x_n, x_{n-1}, f_n, f_1^{n-1})\, dx_{n-1} = P(x_n|f_1^n)P(f_1^n)\,.$$

Using properties of conditional probability,

$$P(x_n, x_{n-1}, f_n|f_1^{n-1}) = P(f_n|x_n)P(x_n|x_{n-1})P(x_{n-1}|f_1^{n-1})\,,$$

which leads to the following expression for $P(x_n|f_1^n)$:

$$\frac{\int P(x_{n-1}|f_1^{n-1})P(f_n|x_n)P(x_n|x_{n-1})dx_{n-1}}{\int\int P(x_{n-1}|f_1^{n-1})P(f_n|x_n)P(x_n|x_{n-1})dx_n dx_{n-1}}. \tag{6}$$

The entropy and the distortion constraint can be written as a function of this conditional density for all time stages. Following the terminology of [4], we define

$$\pi_n(x) := P(x_n = x|f_1^n).$$

Let $\mathcal{P}$ be the set of probability distributions for $\pi_n(x), n \geq 1$. Then the conditional density and the quantization output process , $(\pi_n(x), f_{n+1})$, form a joint Markov process in $\mathcal{P} \times \mathcal{Q}$. We refer the reader to [5] and [13] for an analysis of the conditions for the ergodicity of such filtering processes which involve deterministic observation dynamics.

Here, we have the minimization of

$$\limsup_{N\to\infty} \frac{1}{N} \sum_{t=1}^{N} \int H(f_t|\pi_{t-1})\pi_{t-1}(x_{t-1})dx_{t-1} =: r(\pi, f) \tag{7}$$

subject to the constraint

$$\begin{aligned} s(\pi, f) := \limsup_{N\to\infty} \Bigg[\frac{1}{N}\sum_{t=1}^{N} \int \Big(\int (x_t - \hat{x}_t(f_t, \pi_{t-1}))^2 \\ \cdot P(x_t|x_{t-1})dx_t\Big)\pi_{t-1}(x_{t-1})dx_{t-1}\Bigg] \leq D\,, \end{aligned} \tag{8}$$

where $\hat{x}_t(f_t, \pi_{t-1})$ denotes the estimation at the receiver at time t, using the conditional density and the received codeword.

Theorem 3. *For an irreducible Markov sequence $\{x_t\}$, suppose that $(x_t)^2$ and $p_t \log_2(p_t)$ are uniformly integrable over the set of distributions. If x_t converges almost surely to a random variable x with $E[x^2] < \infty$, then there exists an optimal stationary randomized quantizer solving (7)–(8).*

Proof. Following [12], Theorem 3.2, we have the following.

1) *The set* of feasible quantizers that lead to satisfaction of the distortion constraint *is nonempty.* This follows from the observation that a sub-optimal strategy would be to encode the source without using the past information. There exists such a scheme that yields an arbitrarily small distortion subject to a finite entropy rate (such as memoryless high-rate uniform quantization of the source). Hence there exists at least one solution.
2) $r(\pi, f)$ *is nonnegative and weak* continuous.* This follows from the fact that entropy of a discrete variable is nonnegative and that entropy is a weak* continuous function of both the input distribution and the quantizer. Since entropy is differentiable in the quantizers (see [25]), continuity follows directly. Following a similar argument, it can be shown that entropy is also weak* continuous in the conditional densities.
3) $s(\pi, f)$ *is nonnegative and weak* continuous.* Expected value of the Euclidean distance norm is clearly nonnegative. We need to show that it is weak* continuous. This follows from the fact that an arbitrarily small change in the distance can only be caused by an arbitrarily small change in the quantizer, since this distance metric is differentiable [25] . The same argument applies to the case of a fixed quantizer when the input densities are arbitrarily perturbed.
4) *The transition function for* f_n *is weak* continuous.* For this, we refer to Lemma 3.1 of [4]. It follows from the properties of the nonlinear filter that

$$\pi_n(x_n) = \frac{\int \pi_{n-1}(x_{n-1})P(q_n|x_n)P(x_n|x_{n-1})dx_{n-1}}{\int\int \pi_{n-1}(x_{n-1})P(q_n|x_n)P(x_n|x_{n-1})dx_n dx_{n-1}},$$

and the fact that the conditional densities $\pi_n(x_n)$ belong to a tight set for all time stages.

In view of 1-4 above, there exists an optimal solution. □

6 The Structure of Optimal Randomized Quantizers

Randomization might improve performance in a constrained optimization problem. Toward this end we have the following result.

Theorem 4. *For both the finite-horizon and the infinite-horizon problems, performance can be improved using a randomized causal quantizer. The optimal randomized stationary policies are convex combinations of at most two stationary deterministic policies, with the randomization information at the encoder made available to the decoder.*

Proof. The fact that only two-point randomization is needed follows from the fact that there is only one constraint in the optimization problem. This follows arguments in convex optimization to meet Kuhn-Tucker conditions. Suppose there was an n-point randomization, where randomization information is available at the decoder. Suppose we use quantizer f^i with probability p^i, with entropies $H(f^i)$ and distortions $D(f^i)$, $1 \leq i \leq n$. In this case, we minimize, with respect to $\{p_i\}$, the Lagrangian

$$\sum_i p^i H(f^i) + \lambda p^i D(f^i) + \lambda_2 \sum_i p^i,$$

where λ_2 is needed to enforce that p^i's fare probabilities, and λ is the Lagrange multiplier that was introduced earlier.

Taking partial derivatives with respect to all the elements, we obtain

$$H(f^i) + \lambda D(f^i) + \lambda_2 = 0\,.$$

This suggests that the candidate $H(f^i), D(f^i)$ pairs must all be on a line. However, if there are more than two such candidate elements, then one could, without any loss of generality, take one of them out and place its mass on one of the remaining quantizers, since the constraint $\lambda p^i D(f^i) = D$ can be met with only two such quantizers. □

The reason why the randomization information has to be available at both the decoder and the encoder is that otherwise randomization adds additional entropy to the process, which in turn hurts the performance.

It could be difficult to implement randomized quantizers since these require common randomness between the encoder and the decoder. However, for the infinite-horizon problem, one might achieve the optimal performance without the assumption of such common randomness. This can be achieved via time-sharing, which has to exploit the recurrence properties of Markov chains.

The set of probability density functions, unlike discrete sets, requires a more involved analysis for recurrence properties. Normally, one would want to have a state in the set of probability densities, which would be visited with probability 1, hence an atom. In this case, this atom would be the state where after a sufficient number of visits, the policy change would be implemented. However, for general state spaces, such an atom might not be possible [17]. If the transition of the Markov evolution of $P(x_t|f_1^{t-1})$ were to form an irreducible Markov chain, one could still be able to use this set as the policy-sharing instant; however, such a notion of irreducibility is not directly applicable to the space of probability distributions since such distributions are non-atomic. Toward this end, we use the notion of a small set [17]. Let $\mathcal{P}$ be the set of probability distributions. A set, C in $\mathcal{P}$, is small if there exists a natural number z and a probability distribution ν, such that, for all $P(x_t|q_1^{t-1}) \in C$,

$$P(P(x_{t+z}|q_1^{t+z-1}) \in A | P(x_t|q_1^{t-1})) > \nu(A),$$

for all $A \subset \mathcal{P}$. If this holds, then one can obtain a recurrence set by enlarging the probability space.

Proposition 1. *Suppose that there exists a set C which is recurrent under each of the deterministic policies used in the randomized stationary quantizer. Then, there exists an ϵ-optimal time-sharing scheme of two deterministic policies corresponding to the optimal randomized stationary policy.*

Proof. Suppose the randomization rate η is rational, such that there exists m and n so that $\eta = m/n$, and that quantization policy f' is used with probability η, and another quantization policy f'' is used with probability $1-\eta$.

There exists an invariant distribution such that the conditional mass distribution has a limiting distribution under both policies. Let $\pi_{\infty,f'}(x)$ be the invariant conditional density under policy f' and let $\pi_{\infty,f''}(x)$ be the invariant density under f''. By assumption under both policies there exists a small set C.

Now, apply the policy f' in the first m cycles between the visits to the recurrent state C, and apply f'' in the remaining $n-m$ successive visits to the same state. Such a time-sharing will achieve the optimal performance [19].

If η is not rational, then for any $\delta > 0$ there exist large enough m', n' such that $|\eta - (m'/n')| < \delta$. Applying the above scheme for m' and $n' - m'$ cycles will arbitrarily approximate the optimal randomization policy. □

Remark 1. For the memoryless discrete source case, it was shown in [18] that the optimal memoryless encoder time-shares between two scalar quantizers. This result, as observed above, is applicable to dynamic encoding as well, with the difference here being in the additional analysis required in the recurrence properties of the chain. In such a case, the policy is not stationary. ◁

7 On Innovation Encoding

Linear differential coding, or innovation encoding, is a common coding scheme used for various applications. However, except for Gaussian sources [2], with transmission over Gaussian channels, it is not apparent whether innovation encoding is the best scheme.

In fact, following the analysis on the non-linear filter derived by [4], one immediately sees that, innovation encoding is in general not optimal for a continuous source, since the conditional density information cannot be sufficiently represented by the mean value. An exception of interest is the use of uniform quantization for a linear noise-free source which has a uniform initial distribution [25]. Linear innovation coding for a Gaussian source is another rare case where innovation coding is optimal.

We consider here a linear source with the dynamics

$$x_{t+1} = ax_t + w_t \tag{9}$$

where $|a| < 1$ (that is the process is stable), and w_t is an i.i.d. Gaussian driving process.

For the coding problem, there does exist a solution by our earlier argument. We now consider a scheme where the innovation process, $e_t =$

$x_t - aE[x_{t-1}|q_1^{n-1}]$, is quantized. The process $\{e_t\}$ is Markov if the quantizer is time invariant. We note that uniformly quantizing the innovation at a high rate is fairly efficient, and this property in fact holds for any value of the parameter a (that is, for the unstable case as well). The proof of the result below in part builds on the findings in [14].

Theorem 5. *For causal coding of the source given in (9), suppose innovations are uniformly quantized. In the limit of low distortion, such an encoder is at most 0.254 bits worse than any (possibly noncausal) encoder.*

Proof. We provide and compare the performances based on some lower and upper bounds. The lower bound is obtained via the Shannon lower bound. We have

$$\begin{aligned}
\lim_{n\to\infty}(1/n)H(f_1^n) &\geq \lim_{n\to\infty}(1/n)\sum_{t=1}^{n} I(x_t; f_t|f_1^{t-1}) \\
&= \lim_{n\to\infty}(1/n)\sum_{t=1}^{n} H(x_t|f_1^{t-1}) - H(x_t|f_1^t) \\
&= \lim_{n\to\infty}(1/n)\sum_{t=1}^{n} H(a\, x_{t-1} + w_{t-1}|f_1^{t-1}) - H(x_t|f_1^t) \\
&\geq \lim_{n\to\infty}(1/n)\sum_{t=1}^{n} H(a\, x_{t-1} + w_{t-1}|x_1^{t-1}) - H(x_t|f_1^t) \\
&= \lim_{n\to\infty}(1/n)\sum_{t=1}^{n} H(w_{t-1}) - H(x_t - \hat{x}_t|f_1^t) \\
&\geq \lim_{n\to\infty}(1/n)\sum_{t=1}^{n} H(w_{t-1}) - H(x_t - \hat{x}_t) \\
&= (1/2)\log_2(\sigma^2/D)\,.
\end{aligned}$$

For the upper bound, we quantize the innovation as described above. We first argue that there exists an invariant density for the error process. This follows from the observation that, there exists an $\epsilon > 0$ and a bounded set $C = \{x : |x|^2 \leq E[w_0^2/(1-a^2)]\}$ such that (see [17])

$$E[e_{t+1}^2|e_t] \leq (1-\epsilon)e_t^2 + E[w_0^2]1_{e_t \in C}\,,$$

where $1_{(.)}$ is the indicator function. Accordingly, we uniformly quantize e_t. The rate as a function of the distortion is given by the Gish-Pierce formula [8]:

$$\begin{aligned}
R &= \frac{1}{2}\log_2(2H) - \frac{1}{2}\log(12D) \\
&\leq \frac{1}{2}\log_2((2\pi e(a^2 D + \sigma^2)/12D))\,.
\end{aligned}$$

Hence, as $D \to 0$, the difference is $(1/2)\log_2(2\pi e/12) = 0.254$ bits. □

When the source is unstable, that is $|a| \geq 1$, (10) seems to still hold and one might argue that the same rate loss as above applies. However, for the unstable case, if one uses fixed-length encoding, there does not exist an invariant density for the state. One then needs to use non-trivial feedback or memory policies.

A similar approach as in Theorem 5 applies to the case when the source evolves as a non-linear process. Coding for control of non-linear systems has been studied by Isidori and De Persis [6] among others. We have the following, the proof of which follows that of Theorem 5.

Theorem 6. *Consider optimal causal coding of a stable source given by*

$$x_{t+1} = f(x_t) + w_t \tag{10}$$

where $f(x_t)$ is a stable Lipschitz function (for example, a contraction), and w_t is an i.i.d. Gaussian driving process. Suppose that innovations are uniformly quantized. Then, in the limit of low distortion, such an encoder is at most 0.254 bits worse than any (possibly noncausal) encoder.

8 Concluding Remarks

In this chapter, we studied optimal causal coding for a rate minimization problem subject to a distortion constraint. We provided an existence result for an optimal solution. We then derived for the finite-horizon case the structures of the optimal encoders for Markov sources of arbitrary finite order, including memoryless sources. We provided an existence result for the infinite-horizon problem. We showed that optimal causal quantizers can be randomized, with randomization between at most two deterministic policies. For the infinite horizon problem, this randomization performance can be arbitrarily closely attained via time-sharing and recurrence policies. We also investigated the performance of innovation coding of linear and non-linear sources.

9 Acknowledgments

Research reported here has been supported in part by the NSF grant CCR 00-85917 ITR. We are thankful to Sean Meyn for various insightful discussions on the topic of and results presented in this chapter.

A Weak Topology

In finite dimensional spaces, compactness of a set is equivalent to closedness and boundedness. However, this ceases to be true in infinite dimensional spaces. Due to the strict definition of strong convergence, in optimization problems, it usually suffices to work with a weaker notion, that of weak* convergence: A sequence of functions, f_n, converges in weak* topology to f, if and only if

$$\int g(x)f_n(x)dx \to \int g(x)f(x)dx$$

for all $g \in C_b(X)$, where X is the space under consideration, which is Polish, that is, a topological space whose topology is metrizable by some metric such that X is complete and separable (contains a countably dense set) and $C_b(X)$ denotes the set of continuous and bounded functions on X. With this definition of convergence, one can introduce the notion of weak* compactness: a set of functions is weak* compact if every sequence in the set has a subsequence which converges in the weak* topology.

When applying these notions to probability distributions, there exists a simpler way of checking compactness, which is given by the Prohorov theorem, which states that tightness and relative weak* compactness are identical when X is Polish (such as $\mathcal{R}^n$ for finite n). A set of probability distribution functions $\mathcal{P}$ is tight if for every arbitrarily small $\epsilon > 0$, there exists a compact set C, such that

$$p(x \in C) \geq 1 - \epsilon$$

for all $p \in \mathcal{P}$.

A relevant result is the Arzela-Ascoli theorem, which states that, a set of continuous functions mapping a compact space to another one is compact if and only if each of the functions is uniformly equicontinuous [15]. Some further details on weak topology and relevant results can be found in [3].

B Markov Decision Processes

Given a metric space S, the $\sigma-$algebra generated by $\mathcal{S}$ is called the *Borel σ-algebra* of set $\mathcal{S}$, and is denoted by $\mathcal{B}(\mathcal{S})$. A Borel set is any measurable subset of $\mathcal{S}$. Thus, a Borel set is an element of $\mathcal{B}(\mathcal{S})$.

Given two Borel spaces $\mathcal{S}$ and $\mathcal{T}$, a *stochastic kernel* on $\mathcal{S}$ given $\mathcal{T}$ is a function $f : \mathcal{T} \times \mathcal{B}(\mathcal{S}) \to \mathcal{R}$ such that $f(., B)$ is a measurable function and $f(x, .)$ is a probability measure, for each $(x, B) \in \mathcal{T} \times \mathcal{B}(\mathcal{S})$.

A constrained Markov decision model is a sequence $(\mathcal{X}, \mathcal{A}, A, \mathcal{Q}, c, d)$, where

1) $\mathcal{X}$ is the state space, which is a Borel space.
2) $\mathcal{A}$ is the action space, which is Borel.
3) $A : \mathcal{X} \to \mathcal{B}(\mathcal{A})$ is the action function .
4) Q is a stochastic kernel on $\mathcal{X}$ given $\mathcal{X} \times \mathcal{A}$. This kernel is the dynamics of the system.
5) c is a cost function.
6) d is a constraint function.

A policy, $\{\pi_n\}$ is a sequence of stochastic kernels in $\mathcal{A}$ given the history, such that

$$\pi_n(A(x_n)|h_n) = 1,$$

where the history process is

$$h_n = (x_1, a_1, x_2, a_2, \ldots, x_{n-1}, a_{n-1}, x_n),$$

for integer valued $n \geq 1$.

A policy is *randomized stationary* if there exists a stochastic kernel ϕ such that $\pi_n(.|h_n) = \phi(x_n)$ for each history h_n. A policy is called *deterministic stationary* is there exists a function h such that $\pi_n(.|h_n)$ is the Dirac measure at $h(x_n)$ for all h_n. For details, see for instance [12] and [1].

References

1. E. Altman. Constrained Markov decision processes. *Chapman & Hall/CRC*, 1999.
2. R. Bansal and T. Başar. Simultaneous design of measurement and control strategies for stochastic systems with feedback. *Automatica*, 25:679–694, 1989.
3. P. Billingsley. *Convergence of Probability Measures.* John Wiley & Sons, New York, 1968.
4. V.S. Borkar, S.K. Mitter, and S. Tatikonda. Optimal sequential vector quantization of Markov sources. *SIAM J. Contr. Optimization*, 40:135–148, 2001.
5. P. Chigansky and R. Liptser. Stability of nonlinear filters in non-mixing case. *Annals App. Prob.*, 14:2038–2056, 2004.
6. C. De Persis and A. Isidori. Stabilizability by state feedback implies stabilizability by encoded state feedback. *Systems & Control Letters*, 53:249–258, 2004.
7. N. Gaarder and D. Slepian. On optimal finite-state digital transmission systems. *IEEE Transaction Information Theory*, 28:167–186, 1982.
8. H. Gish and J.N. Pierce. Asymptotically efficient quantization. *IEEE Transaction Information Theory*, 14:676–683, 1968.
9. A. György and T. Linder. Optimal entropy-constrained scalar quantization of a uniform source. *IEEE Transaction Information Theory*, 46:2704–2711, 2000.
10. A. György and T. Linder. On the structure of optimal entropy constrained scalar quantizers. *IEEE Transaction Information Theory*, 48:416–427, 2002.
11. A. György, T. Linder, P.A. Chou, and B.J. Betts. Do optimal entropy constrained quantizers have a finite or infinite number of codewords? *IEEE Transaction Information Theory*, 49:3031–3037, 2003.
12. O. Hernandez-Lerma, J. Gonzales-Hernandez, and R.R. Lopez-Martinez. Constrained average cost Markov control processes in Borel spaces. *SIAM J. Contr. Optimization*, 42:442–468, 2003.
13. T. Kaijser. A limit theorem for partially observed Markov chains. *Annals of Probability*, 4:677–696, 1975.
14. T. Linder and R. Zamir. Causal coding of stationary sources and individual sequences with high resolution. *IEEE Transaction Information Theory*, 52(2):662–680, 2006.

15. D.G. Luenberger. *Optimization by Vector Space Methods.* John Wiley & Sons, New York, 1969.
16. S.P. Meyn. The policy iteration algorithm for average reward Markov decision processes with general state space. *IEEE Trans. on Automat. Contr.*, 42:1663–1680, 1997.
17. S.P. Meyn and R. Tweedie. *Markov Chains and Stochastic Stability.* Springer Verlag, London, 1993.
18. D.L. Neuhoff and R.K. Gilbert. Causal source codes. *IEEE Transaction Information Theory*, 28:701–713, 1982.
19. K.W. Ross. Randomized and past-dependent policies for Markov decision processes with multiple constraints. *Operations Research*, 37:474–477, 1989.
20. S. Tatikonda. *Control Under Communication Constraints.* PhD thesis, MIT, Cambridge, 2000.
21. D. Teneketzis. Real-time encoding-decoding of Markov sources in noisy environments. *Proc. of Mathematical Theory of Networks and Systems*, 2004.
22. J.C. Walrand and P. Varaiya. Optimal causal coding-decoding problems. *IEEE Transaction Information Theory*, 19:814–820, 1983.
23. H.S. Witsenhausen. Separation of estimation and control for discrete time systems. *Proceedings of the IEEE*, 59:1557–1566, 1971.
24. H.S. Witsenhausen. On the structure of real-time source coders. *Bell Syst. Tech. J.*, 58:1437–1451, 1979.
25. S. Yüksel and T. Başar. Minimum rate coding for LTI systems over noiseless channels. *IEEE Trans. on Automat. Contr.*, 51(12):1878–1887, 2006.

Pseudospectral Optimal Control and Its Convergence Theorems*

Wei Kang[1], I. Michael Ross[1], and Qi Gong[2]

[1] Naval Postgraduate School, Monterey, CA 93943, USA.
[2] Univ. of Texas at San Antonio, San Antonio, TX 78249, USA.

Dedicated to Professor Alberto Isidori on the occasion of his 65th birthday.

Summary. During the last decade, pseudospectral (PS) optimal control methods have emerged as demonstrable efficient methods for computational nonlinear optimal control. Some fundamental problems on the feasibility and convergence of the Legendre PS method are addressed. In the first part of this paper, we summarize the main results published separately in a series of papers on these topics. Then, a new result on the feasibility and convergence is proved. Different from existing results in the literature, in this new theorem neither the invertibility of necessary conditions nor the existence of limit points is assumed.

1 Introduction

Optimal feedback control is a fundamental problem in control theory and control system engineering. For a very limited set of problems, optimal feedback solutions can be obtained explicitly either through the Hamilton-Jacobi-Bellman equation or the Minimum Principle. For general problems with state- and control constraints, and nonlinear dynamics, achieving explicit solutions is quite impossible. An alternative approach is to develop efficient numerical methods and generate feedback by way of real-time computation, and idea that goes back to Pontryagin et al. [18]. For many years, the computational approach had been widely considered as being too slow for real-time applications of highly nonlinear problems. In recent years, a new class of methods known as pseudospectral (PS) methods have emerged as demonstrable candidates for real-time computation [1, 11, 22, 24, 25, 26, 29]. As a matter fact, feedback control via real-time computation has been demonstrated not merely in simulation but also in

* The research was supported in part by the Air Force Office of Scientific Research under AFOSR Grant F1ATA0-60-6-2G002.

practice such as the ground test of the PS attitude control system of NPSAT1 [26, 29], an experimental spacecraft scheduled for launch in 2007. The advent of practical real-time computation requires a new theoretical framework for optimal control theory. In this paper, we focus on convergence theorems related to discrete approximations arising from an application of PS methods.

Currently, there are two approaches to analyze the convergence of discrete approximations. One is based on the theory of consistent approximations [17] and the other based on Robinson's implicit function theorem [20, 21]. In the theory of consistent approximations, sufficient conditions are constructed to prove that the limit point of a sequence of discrete optimal solutions must be the optimal solution of the original optimal control problem, provided that a limit point exists. Such an approach has been used since the 1960s and modern results for Runge-Kutta approximations are described in [28]. The other approach is based on the invertibility of the discrete necessary conditions (i.e. the Karush-Kuhn-Tucker conditions) [13]. Following this line, one can prove both convergence and convergence rate at a price of stronger assumptions. For state-constrained problems, it is necessary to impose significant conditions for a proof of convergence even when the discrete approximations are based on Eulerian approximations [5]. In this paper, we develop convergence theorems for PS approximations for a special family of control systems, namely feedback linearizable systems. In this work, we rely on exploring the approximation theory from spectral analysis in conjunction with the structure of feedback linearizable systems. This allows us to state stronger results in which certain fundamental consistency type of assumptions previously required on this topic are removed. The results in this paper represent a first success beyond some of the consistency cornerstones on the convergence theory of PS methods. In particular, the proof presented in this paper is independent of the discrete-time or continuous-time necessary conditions, which is a fundamental difference from most existing proofs on the convergence of discrete approximations.

Pseudospectral methods were largely developed in the 1970s for solving partial differential equations arising in fluid dynamics and meteorology [3]. During the 1990s, PS methods were introduced for solving optimal control problems [7, 6, 9, 8]; and since then, have gained considerable attention [10, 14, 16, 19, 30, 31]. One of the main reasons for the popularity of PS methods is that they demonstrably offer an exponential convergence rate for the approximation of analytic functions while providing Eulerian-like simplicity. Although PS methods are easy to apply, proofs of existence and convergence of approximations is a difficult problem and currently an active area of research for general nonlinear systems. Significant progress has been made during the last few years for the family of feedback linearizable systems with either continuous or discontinuous optimal control. In the next few sections, we will first summarize the main results published separately in a series of papers on the existence and convergence of discrete approximations using PS optimal control methods. These results are formulated in a unified framework so that one can easily analyze the differences and similarities. Then, a new result is proved in which neither invertibility of necessary conditions nor the existence of limit points is assumed.

2 Problem Definition

In this paper, we address the following Bolza problem of control systems in the feedback linearizable normal form.

Problem B: Determine the state-control function pair $(x(t), u(t))$, $x \in \mathbb{R}^r$ and $u \in \mathbb{R}$, that minimizes the cost function

$$J(x(\cdot), u(\cdot)) = \int_{-1}^{1} F(x(t), u(t))\, dt + E(x(-1), x(1)) \tag{1}$$

subject to

$$\dot{x}_1 = x_2, \ \cdots, \ \dot{x}_{r-1} = x_r, \ \dot{x}_r = f(x) + g(x)u \ \text{(state equations)} \tag{2}$$

$$e(x(-1), x(1)) = 0 \ \text{(endpoint conditions)} \tag{3}$$

$$h(x(t), u(t)) \leq 0 \ \text{(state-control constraints)} \tag{4}$$

where $x \in \mathbb{R}^r$, $u \in \mathbb{R}$, and $F : \mathbb{R}^r \times \mathbb{R} \to \mathbb{R}$, $E : \mathbb{R}^r \times \mathbb{R}^r \to \mathbb{R}$, $f : \mathbb{R}^r \to \mathbb{R}$, $g : \mathbb{R}^r \to \mathbb{R}$ $e : \mathbb{R}^r \times \mathbb{R}^r \to \mathbb{R}^{N_e}$ and $h : \mathbb{R}^r \times \mathbb{R}^r \to \mathbb{R}^s$ are all Lipschitz continuous functions with respect to their arguments. In addition, we assume $g(x) \neq 0$ for all x. Throughout the paper we make extensive use of Sobolev spaces, $W^{m,p}$, that consists of functions, $\xi : [-1, 1] \to \mathbb{R}$ whose j-th weak derivative, $\xi^{(j)}$, lies in L^p for all $0 \leq j \leq m$ with the norm,

$$\| \xi \|_{W^{m,p}} = \sum_{j=0}^{m} \| \xi^{(j)} \|_{L^p} .$$

We limit our discussions to those optimal control problems that have at least one optimal solution $(x^*(t), u^*(t))$. The problem of convergence is addressed in three different situations. In the first case, we assume $x_r^*(t)$ has bounded m-th order weak derivative with $m \geq 2$, i.e. $x_r^*(t)$ is in $W^{2,\infty}$. This condition implies that the optimal control $u^*(t)$ is continuous. In this case, uniform convergence is guaranteed under consistent type of assumptions. In the second case, the optimal control is allowed to be discontinuous; however, due to the fundamental limitation in global polynomial approximations, convergence is proved in a week sense rather than uniform convergence. We hasten to note that PS methods are not limited to global polynomial approximations, and that numerical experiments with PS knotting methods [23] based on non-global polynomials suggest that uniform convergence is possible; however, a theoretical framework for a proof of this result is an open area of research. In the third case, we assume $x_r^*(t)$ is in $W^{m,\infty}$ with $m \geq 3$. Under this assumption of additional regularity of the optimal trajectory, we are able to remove the strong consistent type of assumptions made in the other two cases.

In this paper, we focus on the Legendre PS method for optimal control. The ideas are applicable to other PS methods as well but we limit our discussions to the Legendre method for the purpose of clarity in presentation. In Legendre PS optimal control method, the state, $x(t)$, is approximated by N-th order Lagrange polynomials based on the interpolation at the Legendre-Gauss-Lobatto (LGL) quadrature nodes. The LGL nodes, $t_0 = -1 < t_1 < \cdots < t_N = 1$, are defined by

$$\begin{array}{l} t_0 = -1, \;\; t_N = 1, \text{ and} \\ \text{for } k = 1, 2, \ldots, N-1, t_k \text{ are the roots of } \dot{L}_N(t) \end{array}$$

where $\dot{L}_N(t)$ is the derivative of the N-th order Legendre polynomial $L_N(t)$. It has been proven in computational mathematics that the interpolation at the LGL nodes is an extremely efficient method in the approximation of smooth functions. In the discretization, $x(t_k)$ is approximated by the vector $\bar{x}^{Nk} \in \mathbb{R}^r$,

$$\bar{x}^{Nk} = \left[\bar{x}_1^{Nk} \; \bar{x}_2^{Nk} \; \cdots \; \bar{x}_r^{Nk} \right]^T .$$

Similarly, $\bar{u}^{Nk}$ is the approximation of $u(t_k)$. Thus, a discrete approximation of the function $x_i(t)$ is the row vector

$$\bar{x}_i^N = \left[\bar{x}_i^{N1} \; \bar{x}_i^{N2} \; \cdots \; \bar{x}_i^{NN} \right] .$$

A continuous-time approximation of the state $x_i(t)$ is defined by its polynomial approximation denoted as $x_i^N(t)$, i.e.,

$$x_i(t) \approx x_i^N(t) = \sum_{k=0}^{N} \bar{x}_i^{Nk} \phi_k(t), \tag{5}$$

where $\phi_k(t)$ is the Lagrange interpolating polynomial. Instead of polynomial interpolation, the control input is approximated by the following non-polynomial interpolation

$$u^N(t) = \frac{\dot{x}_r^N(t) - f(x^N(t))}{g(x^N(t))} . \tag{6}$$

The notation used in this paper is summarized as follows. The discrete-time variables are denoted by letters with an upper bar, such as $\bar{x}_i^{Nk}$ and $\bar{u}^{Nk}$, where N represents the number of LGL nodes and k represents the kth node. If k in the superscript and/or i in the subscript are missing, it represents the corresponding vector or matrix inwhich the indices run from their minimum to the maximum. The index k generates row vectors, and i generates column vectors. For example,

$$\bar{x}_i^N = \left[\bar{x}_i^{N0}\ \bar{x}_i^{N1} \cdots \bar{x}_i^{NN}\right], \quad \bar{x}^{Nk} = \left[\bar{x}_1^{Nk}\ \bar{x}_2^{Nk} \cdots \bar{x}_r^{Nk}\right]^T$$

$$\bar{x}^N = \begin{bmatrix} \bar{x}_1^{N0} & \bar{x}_1^{N1} & \cdots & \bar{x}_1^{NN} \\ \vdots & \vdots & \vdots & \vdots \\ \bar{x}_r^{N0} & \bar{x}_r^{N1} & \cdots & \bar{x}_r^{NN} \end{bmatrix}.$$

Given a discrete approximation of a continuous function, its interpolation polynomial is denoted by the same notation without the upper bar. For example, $x_i^N(t)$ in (5), $u^N(t)$ in (6). The derivative of $x_i^N(t)$ at the LGL node t_k can be easily computed by a matrix multiplication

$$\left[\dot{x}_i^N(t_0)\ \dot{x}_i^N(t_1) \cdots \dot{x}_i^N(t_N)\right]^T = D(\bar{x}_i^N)^T \tag{7}$$

where D is the $(N+1)\times(N+1)$ differentiation matrix. An explicit formula for its computation is given in [3]. The cost functional $J[x(\cdot), u(\cdot)]$ is approximated by the Gauss-Lobatto integration rule (8), in which w_k are the LGL weights [3]. Now, we can define Problem $\mathrm{B^N}$, a PS discretization of Problem B as follows.

Problem $\mathbf{B^N}$: Find $\bar{x}^{Nk} \in \mathbb{R}^r$ and $\bar{u}^{Nk} \in \mathbb{R}$, $k = 0, 1, \ldots, N$, that minimize

$$\bar{J}^N(\bar{x}^N, \bar{u}^N) = \sum_{k=0}^{N} F(\bar{x}^{Nk}, \bar{u}^{Nk})w_k + E(\bar{x}^{N0}, \bar{x}^{NN}) \tag{8}$$

subject to

$$D(\bar{x}_1^N)^T = (\bar{x}_2^N)^T,\ D(\bar{x}_2^N)^T = (\bar{x}_3^N)^T,\ \cdots,\ D(\bar{x}_{r-1}^N)^T = (\bar{x}_r^N)^T$$

$$D(\bar{x}_r^N)^T = \begin{bmatrix} f(\bar{x}^{N0}) + g(\bar{x}^{N0})\bar{u}^{N0} \\ \vdots \\ f(\bar{x}^{NN}) + g(\bar{x}^{NN})\bar{u}^{NN} \end{bmatrix} \tag{9}$$

$$\|e(\bar{x}^{N0}, \bar{x}^{NN})\|_\infty \leq (N - r - 1)^{-m+\delta} \tag{10}$$

$$h(\bar{x}^{Nk}, \bar{u}^{Nk}) \leq (N - r)^{-m+\delta} \cdot \mathbf{1}, \qquad \text{for all } 0 \leq k \leq N \tag{11}$$

$$\underline{\boldsymbol{b}} \leq \begin{bmatrix} \bar{x}^{Nk} \\ \bar{u}^{Nk} \end{bmatrix} \leq \bar{\boldsymbol{b}}, \qquad \text{for all } 0 \leq k \leq N \tag{12}$$

where δ is a positive number to be determined later.

Comparing Problem B^N to Problem B, constraints (10) and (11) are relaxed by a small margin that approaches zero as N is increased. This is critical because, without the relaxation, a counter example in [12] shows that Problem $\mathrm{B^N}$ may have no feasible trajectory. From a practical view point, the relaxation makes sense because of the finite precision in computer hardware and tolerances in optimization solvers. Problem B^N has an additional constraint (12) which is not in Problem B. We assume $\underline{b}$ and $\bar{b}$ are large enough so that the optimal solution of Problem B is contained in this region. Without this additional constraint, it is possible for Problem B^N to have feasible trajectories but no optimal trajectories. It will be proved later that this additional constraint do not change the final optimal solution because it becomes inactive for large N. Furthermore, (12) reduces the search region for the optimal solution and helps speeding up computation. In the PS optimal control, we use the optimal solution $(\bar{x}^{*N}, \bar{u}^{*N})$ of Problem B^N and its interpolation to approximate the optimal solution of Problem B. However, the seemingly straightforward approach belies the danger of infeasible discrete constraints and divergence of the optimal solutions. Examples have been found [12] in which a PS discretization does not have any feasible trajectory even though the original optimal control problem has infinitely many feasible continuous-time trajectories. In another example, the discretization of Problem B has feasible trajectories but no optimal trajectories. Such complications on the existence and convergence of approximations are not limited to PS methods, rather they are intrinsic to optimal control [13, 4]. Nonetheless, as noted earlier, theories developed for the convergence of discrete approximations for one method are not quite portable to an analysis of convergence of approximations of another method. Given that PS methods are quite different in the constructions of discrete approximations, it is evident that we need a first-principles approach to analyze the convergence of its approximations. In this spirit, we focus on the following fundamental problems of PS optimal control methods.

Question 1. Does there exist a feasible trajectory $(\bar{x}^N, \bar{u}^N)$ to Problem B^N?
Question 2. Under what condition does a sequence of optimal solutions of Problem B^N converge to an optimal solution of Problem B as N increases?

3 Problems with Continuous Optimal Control

In [12], Questions 1 and 2 were answered for the case of continuous optimal control. In this case, Problem B^N is defined for $\delta = \frac{3}{2}$. Convergence is proved on the basis of the following assumption.

Assumption 1. *There is a subsequence* $\{N_j\}_{j=1}^{\infty}$ *of the sequence* $\{N\}_{N=1}^{\infty}$ *such that* $\left\{\bar{x}^{N_j 0}\right\}_{j=1}^{\infty}$ *converges as* $N_j \to \infty$. *Additionally, there exists a continuous function* $q(t)$ *such that* $\dot{x}_r^{N_j}(t)$ *converges to* $q(t)$ *uniformly in* $[-1, 1]$.

The following theorem summarizes the key results in [12] on the existence and convergence of the PS optimal control.

Theorem 1. *Suppose Problem B has a feasible trajectory* $(x(t), u(t))$ *satisfying* $x_r(t) \in W^{m,\infty}$, $m \geq 2$.

(i) *There exists a positive integer* N_1 *such that, for any* $N > N_1$, *Problem* B^{N} *has a feasible trajectory,* $(\bar{x}^N, \bar{u}^N)$. *Furthermore, it satisfies*

$$\begin{aligned} ||x(t_k) - \bar{x}^{Nk}||_\infty &\le L(N-r)^{1-m} \\ |u(t_k) - \bar{u}^{Nk}| &\le L(N-r)^{1-m} \end{aligned}$$

for all $k = 0, \ldots, N$, *where* L *is a positive constant independent of* N.

(ii) *Suppose* $\{(\bar{x}^N, \bar{u}^N)\}_{N=N_1}^\infty$ *be a sequence of feasible trajectories of Problem* B^{N} *and suppose the sequence satisfies Assumption 1. Then, there exists* $(x^\infty(t), u^\infty(t))$ *satisfying (2)–(4) such that the following limit converges uniformly on* $[-1, 1]$.

$$\lim_{N_j \to \infty} (x^{N_j}(t) - x^\infty(t)) = 0 \tag{13}$$

$$\lim_{N_j \to \infty} (u^{N_j}(t) - u^\infty(t)) = 0 \tag{14}$$

$$\lim_{N_j \to \infty} \bar{J}^{N_j}(\bar{x}^{N_j}, \bar{u}^{N_j}) = J(x(\cdot), u(\cdot)) \tag{15}$$

$$\lim_{N_j \to \infty} J(x^{N_j}(\cdot), u^{N_j}(\cdot)) = J(x(\cdot), u(\cdot)) \tag{16}$$

(iii) *If* $\{(\bar{x}^N, \bar{u}^N)\}_{N=N_1}^\infty$ *in (ii) is a sequence of optimal solutions to Problem* B^{N}, *then* $(x^\infty(t), u^\infty(t))$ *in (ii) must be an optimal solution to Problem B.*

This theorem answers Questions 1 and 2 raised in Section 2. It is a slightly generalized version of the key results in [12]. The proof is omitted but interested readers are referred to [12].

4 Problems with Discontinuous Optimal Control

A critical assumption in Theorem 1 is $x_r(t) \in W^{m,\infty}$ for some $m \ge 2$. This implies that $\dot{x}_r(t)$ and $u(t)$ are continuous. However, in many optimal control problems, $u(t)$ is discontinuous as in the case of a bang-bang controller. In this section, we extend the results of Theorem 1 to problems with piecewise C^1 optimal control. Therefore, in Problem B^{N} we assume $m = 1$ and $\delta = \frac{3}{4}$.

Definition 1. *A function* $\psi(t) : [-1, 1] \to \mathbb{R}^k$ *is called piecewise* C^1 *if there exist finitely many points* $\tau_0 = -1 < \tau_1 < \cdots < \tau_{s+1} = 1$ *such that, on every subinterval* (τ_i, τ_{i+1}), $i = 0, \ldots, s$, $\psi(t)$ *is continuously differentiable and both* $\psi(t)$ *and its derivative,* $\dot{\psi}(t)$, *are bounded.*

In the following, we need an assumption that is similar to Assumption 1.

Assumption 2. *Given a sequence of discrete feasible trajectories, namely* $\{\bar{x}^N, \bar{u}^N\}_{N=N_1}^\infty$, *there exists a subsequence* $\{N_j\}_{j=1}^\infty$ *of* $\{N\}_{N=1}^\infty$ *such that (a) for all* $1 \le i \le r$, $\{\bar{x}_i^{N_j 0}\}_{N_j=N_1}^\infty$ *converges as* $N_j \to \infty$; *(b)* $\dot{x}_r^{N_j}(t)$ *is uniformly* *bounded for $N_j \geq N_1$ and $t \in [-1,1]$; and, (c) there exists a piecewise C^1 function $q(t)$ such that, for any fixed $\epsilon > 0$, $\dot{x}_r^{N_j}(t)$ converges to $q(t)$ uniformly on the interval I_ϵ, where*

$$I_\epsilon = [-1,1] \setminus \bigcup_{j=1}^{s} (\tau_j - \epsilon, \tau_j + \epsilon) \tag{17}$$

and $-1 < \tau_1 < \cdots < \tau_s < 1$ represent the discontinuous points of $q(t)$.

Theorem 2. *Assume that the optimal state $x_r^*(t)$ is continuous and piecewise C^1; the optimal control $u^*(t)$ is piecewise C^1; and the set $\{(x,u) | h(x,u) \leq 0\}$ is convex. Let $(x(t), u(t))$ be any feasible trajectory of Problem B. Then we have the following properties.*

(i) There exists a positive integer N_1 such that, for any $N > N_1$, Problem B^{N} has a feasible trajectory, $(\bar{x}^N, \bar{u}^N)$. Furthermore, the feasible trajectory satisfies

$$\|x(t_k) - \bar{x}^{Nk}\|_\infty \leq (N-r)^{-\frac{1}{4}}, \quad 0 \leq k \leq N \tag{18}$$

$$|u(t_k) - \bar{u}_k^N| \leq (N-r)^{-\frac{1}{4}}, \quad \forall\, t_k \in I_\rho \tag{19}$$

where I_ρ is defined by (17) with $\rho = (N-r)^{-\frac{1}{2}}$.

(ii) Suppose $\{(\bar{x}^N, \bar{u}^N)\}_{N=N_1}^{\infty}$ be a sequence of feasible trajectories of Problem B^{N} and suppose the sequence satisfies Assumption 2. Then, there exists $(x^\infty(t), u^\infty(t))$ satisfying (2)–(4) such that

$$\lim_{N_j \to \infty} (x^{N_j}(t) - x^\infty(t)) = 0 \quad \textit{uniformly on } [-1,1]$$

$$\lim_{N_j \to \infty} (u^{N_j}(t) - u^\infty(t)) = 0 \quad \textit{uniformly on any closed set } I_\epsilon, \epsilon > 0$$

$$\lim_{N_j \to \infty} \bar{J}^{N_j}(\bar{x}^{N_j}, \bar{u}^{N_j}) = J(x^\infty(\cdot), u^\infty(\cdot))$$

$$\lim_{N_j \to \infty} J(x^{N_j}(\cdot), u^{N_j}(\cdot)) = J(x(\cdot), u(\cdot))$$

(iii) If $\{(\bar{x}^N, \bar{u}^N)\}_{N=N_1}^{\infty}$ in (ii) is a sequence of optimal solutions of Problem B^{N}, then $(x^\infty(t), u^\infty(t))$ in (ii) is an optimal solution of Problem B.

Due to the discontinuity in the optimal control, the proof of this theorem calls for highly involved algebraic manipulations and inequality estimations. The reader is referred to [15] for its proof. The importance of Theorem 2 is self-evident. The theorem guarantees that Problem B^{N} is well-posed with a nonempty set of feasible discrete-time trajectories around any trajectory of Problem B, even if the input is discontinuous. Furthermore, (18) and (19) imply that the feasible discrete-time trajectories can be arbitrarily close to the continuous-time trajectories.

Compared to Theorem 1, Theorem 2 is significantly different in two key aspects that are beyond a generalization of allowing for a discontinuous control input for Problem B. First, the concept of convergence in this section is different from that of the last section. In fact, due to the fundamental limitation of global polynomial approximations to discontinuous functions, it is impossible to prove the uniform convergence of the discrete solutions like in Theorem 1. If the optimal control is discontinuous, the convergence is proved for an interval I_ρ in which an open neighborhood around the discontinuities must be removed. The second major difference lies in the assumption that the state-control constraint must be convex, which is not required in Theorem 1.

5 A Convergence Theorem without Assumptions 1 and 2

The goal of this section is to prove a similar convergence result without Assumptions 1 and 2. To remove these assumptions, we found it necessary to add more constraints to Problem $\mathrm{B^N}$; however, we note that these additional constraints are not necessarily required in practical problem solving. As a matter of fact, in all of our numerical experimentations, these additional constraints were not required. With this perspective in mind, we let D be the differentiation matrix at the LGL nodes [3]. For any integer $m_1 > 0$, let $\{a_0^{m_1N}, a_1^{m_1N}, \cdots, a_{N-r-m_1+1}^{m_1N}\}$ denotes the sequence of spectral coefficients for the interpolation polynomial of the vector $\bar{x}_r^N(D^T)^{m_1}$. There are only $N-r-m_1+2$ coefficients because the order of $x_r^N(t)$ is at most $N-r+1$ (see Lemma 1). These coefficients depend linearly on $\bar{x}_r^N$ [2],

$$\begin{bmatrix} a_0^{m_1N} \\ \vdots \\ a_{N-r-m_1+1}^{m_1N} \end{bmatrix} = \begin{bmatrix} \frac{1}{2} & & \\ & \ddots & \\ & & N-r-m_1+1+\frac{1}{2} \end{bmatrix} \times$$
$$\begin{bmatrix} L_0(t_0) & \cdots & L_0(t_N) \\ & \vdots & \\ L_{N-r-m_1+1}(t_0) & \cdots & L_{N-r-m_1+1}(t_N) \end{bmatrix} \begin{bmatrix} w_0 & & \\ & \ddots & \\ & & w_N \end{bmatrix} D^{m_1} \begin{bmatrix} \bar{x}_r^{N0} \\ \vdots \\ \bar{x}_r^{NN} \end{bmatrix}. \tag{20}$$

We now modify Problem $\mathrm{B^N}$ by adding a set of linear inequality constraints as follows.

Problem $\mathbf{B^{N+}}$: Problem $\mathrm{B^{N+}}$ is defined to be Problem $\mathrm{B^N}$ plus the following additional constraints

$$\underline{\boldsymbol{b}}_j \leq \begin{bmatrix} 1 & 0 & \cdots & 0 \end{bmatrix} D^j (\bar{x}_r^N)^T \leq \bar{\boldsymbol{b}}_j, \quad 1 \leq j \leq m_1 - 1 \tag{21}$$

$$\sum_{n=0}^{N-r-m_1+1} |a_n^{m_1N}| \leq d \tag{22}$$

where m_1 is a positive integer, $2 \leq m_1 \leq m-1$, and d is a number sufficiently large.

For Problem $\mathbf{B^{N+}}$, the selection of m_1 does not change the convergence result proved in this section; however, it is proved in a separate paper that the selection of m_1 determines the rate of convergence. The discrete optimal solution converges faster if m_1 is selected around $\frac{m-1}{2}$. In Problem $\mathrm{B^N}$, we assume $m \geq 3$ and $\frac{3}{2} < \delta < m$. Let $\underline{\boldsymbol{b}}_j$ and $\bar{\boldsymbol{b}}_j$ be the lower and upper bounds so that the jth order derivative of $x_r^*(t)$ of the optimal trajectory is contained in the interior of the region; and, d is an upper bound larger than

$$\frac{6}{\sqrt{\pi}}(U(x_r^{*\,(m_1+1)}) + V(x_r^{*\,(m_1+1)}))\zeta(3/2) \tag{23}$$

where $\zeta(s)$ is the ζ function, $x_r^*(t)$ is the rth component of the optimal trajectory, $U(x_r^{*\,(3)})$ is the upper bound of the third order derivative of $x_r^*(t)$ and $V(x_r^{*\,(3)})$ is the total variation of $x_r^{*\,(3)}(t)$. In practice, the quantities, $\underline{\boldsymbol{b}}$, $\bar{\boldsymbol{b}}$, $\underline{\boldsymbol{b}}_j$, $\bar{\boldsymbol{b}}_j$, and d are unknown and must be estimated based upon experience or other information about the system; however, as noted earlier, numerical experiments reveal that the sequence of optimal solutions converge in most tested examples without implementing the constraints (21) and (22). Consequently, they must be viewed as practical safeguards against pathological cases rather than as a burden on problem solving.

Theorem 3. Suppose Problem B has an optimal solution $(x^*(t), u^*(t))$ in which $(x_r^*(t))^{(m)}$ has a bounded variation for some $m \geq 3$. In Problem $\mathrm{B^N}$ we assume $\delta = m_1 + \delta_1$, where $2 \leq m_1 \leq m-1$ and $\frac{1}{2} < \delta_1 < m - m_1$. Then the following hold.

(i) There exists an $N_1 > 0$ such that Problem $\mathrm{B^{N+}}$ has feasible trajectories for all $N \geq N_1$, i.e. there always exist at least one pair $(\bar{x}^N, \bar{u}^N)$ for each $N \geq N_1$ that satisfies all the constraints in Problem $\mathrm{B^{N+}}$.

(ii) Let $\{(\bar{x}^{*N}, \bar{u}^{*N})\}_{N=N_1}^{\infty}$ be a sequence of optimal solutions to Problem $\mathrm{B^{N+}}$. Then, there exists a subsequence, $\{(\bar{x}^{*N_j}, \bar{u}^{*N_j})\}_{j\geq 1}^{\infty}$ and an optimal solution, $(x^*(t), u^*(t))$, of Problem B so that the following limits converge uniformly:

$$\begin{aligned} \lim_{N_j\to\infty}(x^{*N_j}(t) - x^*(t)) &= 0 \\ \lim_{N_j\to\infty}(u^{*N_j}(t) - u^*(t)) &= 0 \\ \lim_{N_j\to\infty} \bar{J}^{N_j}(\bar{x}^{*N_j}, \bar{u}^{*N_j}) &= J(x^*(\cdot), u^*(\cdot)) \\ \lim_{N_j\to\infty} J(x^{*N_j}(\cdot), u^{*N_j}(\cdot)) &= J(x^*(\cdot), u^*(\cdot)) \end{aligned} \tag{24}$$

where $(x^{*N_j}(t), u^{*N_j}(t))$ is an interpolant of $(\bar{x}^{*N}, \bar{u}^{*N})$.

To prove this theorem we need several lemmas. The first lemma is on a one-to-one mapping between the trajectory sets of (9) and (2).

Lemma 1. *(i) For any trajectory, $(\bar{x}^N, \bar{u}^N)$, of the dynamics (9), the pair $(x^N(t), u^N(t))$ defined by (5) and (6) satisfies the differential equations defined in (2). Furthermore,*

$$\bar{x}^{Nk} = x^N(t_k),\ \bar{u}^{Nk} = u^N(t_k),\ \text{for } k = 0, 1, \cdots, N. \tag{25}$$

(ii) For any pair $(x^N(t), u^N(t))$ in which $x^N(t)$ consists of polynomials of degree less than or equal to N and $u^N(t)$ is a function. If $(x^N(t), u^N(t))$ satisfies the differential equations in (2), then $(\bar{x}^N, \bar{u}^N)$ defined by (25) satisfies the discrete equations in (9).

(iii) If $(\bar{x}^N, \bar{u}^N)$ satisfies (9), then the degree of $x_i^N(t)$ is less than or equal to $N - i + 1$.

The proof of this lemma follows (6), (7), (9), and some basic ideas from spectral analysis. For the purposes of brevity, the details are omitted.

Lemma 2. *Suppose $\{(\bar{x}^N, \bar{u}^N)\}_{N=N_1}^{\infty}$ is a sequence satisfying (9), (21) and (22). Then,*

$$\left\{ ||(x^N(t))^{(l)}||_\infty \middle| N \geq N_1,\ l = 1, \cdots, m_1 \right\} \tag{26}$$

is bounded. If $f(x)$ and $g(x)$ are C^{m_1-1}, then

$$\left\{ ||(u^N(t))^{(l)}||_\infty \middle| N \geq N_1,\ l = 1, \cdots, m_1 - 1 \right\} \tag{27}$$

is bounded.

Proof. Consider $(x_r^N(t))^{(m_1)}$. From Lemma 1, it is a polynomial of degree less than or equal to $N - r - m_1 + 1$. Therefore,

$$(x_r^N(t))^{(m_1)} = \sum_{n=0}^{N-r-m_1+1} a_n^{m_1 N} L_n(t)$$

where $L_n(t)$ is the Legendre polynomial of degree n. It is known that $|L_n(t)| \leq 1$. Therefore, (22) implies that $||(x_r^N(t))^{(m_1)}||_\infty$ is bounded by d for all $N \geq N_1$. Then, the integrations of $(x_r^N(t))^{(m_1)}$ over $[-1, 1]$ are bounded, which implies the boundedness of (26). Then, using (6), we can prove the boundedness of (27). □

Lemma 3. *Let $\{(\bar{x}^N, \bar{u}^N)\}_{N=N_1}^{\infty}$ be a sequence satisfying (9)–(12). Assume the set*

$$\{ ||\ddot{x}_r^N(t)||_\infty | N \geq N_1 \} \tag{28}$$

is bounded. Then, there exists $(x^\infty(t), u^\infty(t))$ satisfying (2)–(4) and a subsequence $\{(\bar{x}^{N_j}, \bar{u}^{N_j})\}_{N_j \geq N_1}^{\infty}$ such that (13), (14), and (15) hold. Furthermore, if $\{(\bar{x}^N, \bar{u}^N)\}_{N=N_1}^{\infty}$ is a sequence of optimal solutions to Problem B^N, then $(x^\infty(t), u^\infty(t))$ must be an optimal solution to Problem B.

Proof. Let $x_r^N(t)$ be the interpolating polynomial of $\bar{x}_r^N$. Because (28) is a bounded set, we know that the sequence of functions $\{\dot{x}_r^N(t)|N \geq N_1\}$ is uniformly equicontinuous. Then, Lemma 3 is a corollary of the Arzelà-Ascoli Theorem and Theorem 1. □

Given any function $h(t)$ defined on $[-1, 1]$. In the following, $U(h)$ represents an upper bound of $h(t)$ and $V(h)$ represents the total variation. In the following, $K(N) = N - r - m_1 + 1$.

Lemma 4. *Let $(x(t), u(t))$ be a solution of the differential equation (2). Suppose $x_r^{(m)}(t)$ has bounded variation for some $m \geq 3$. Let m_1 be an integer satisfying $2 \leq m_1 \leq m - 1$. Then, there exist constants $M > 0$ and $N_1 > 0$ so that for each integer $N \geq N_1$ the differential equation (2) has a solution $(x^N(t), u^N(t))$ in which $x^N(t)$ consists of polynomials of degree less than or equal to N. Furthermore, the pair $(x^N(t), u^N(t))$ satisfies*

$$||x_i^N(t) - x_i(t)||_\infty \leq \frac{MV(x_r^{(m)}(t))}{K(N)^{(m-m_1)-1/2}}, \quad i = 1, 2, \cdots, r \tag{29}$$

$$||(x_r^N(t))^{(l)} - (x_r(t))^{(l)}||_\infty \leq \frac{MV(x_r^{(m)}(t))}{K(N)^{(m-m_1)-1/2}}, \quad l = 1, 2, \cdots, m_1 \,. \tag{30}$$

Furthermore, the spectral coefficients of $(x_r^N)^{(m_1)}(t)$, denoted by $a_n^{m_1 N}$, satisfy

$$|a_n^{m_1 N}| \leq \frac{6(U(x_r^{(m_1+1)}) + V(x_r^{(m_1+1)}))}{\sqrt{\pi} n^{3/2}}, \quad n = 1, 2, \cdots, N - r - 1 \,. \tag{31}$$

If $f(x)$ and $g(x)$ have Lipschitz continuous Lth order partial derivatives for some $L \leq m_1 - 1$, then

$$||(u^N(t))^{(l)} - (u(t))^{(l)}||_\infty \leq \frac{MV(x_r^{(m)}(t))}{K(N)^{(m-m_1)-1/2}}, \quad l = 0, 1, \cdots, L \,. \tag{32}$$

Proof. Consider the Legendre series

$$(x_r)^{(m_1)}(t) \sim \sum_{n=0}^{N-r-m_1+1} a_n^{m_1 N} L_n(t)$$

A sequence of polynomials $x_1^N(t), \cdots, x_{r+m_1}^N(t)$ is defined as follows,

$$x_{r+m_1}^N(t) = \sum_{n=0}^{K(N)} a_n^{m_1 N} L_n(t), \; x_{r+m_1-1}^N(t) = (x_r)^{(m_1-1)}(-1) + \int_{-1}^{t} x_{r+m_1}^N(s) ds$$

$$\cdots, \; x_{r+1}^N(t) = \dot{x}_r(-1) + \int_{-1}^{t} x_{r+2}^N(s) ds, x_r^N(t) = x_r(-1) + \int_{-1}^{t} x_{r+1}^N(s) ds, \cdots$$

Define $x^N(t) = \begin{bmatrix} x_1^N(t) & \cdots & x_r^N(t) \end{bmatrix}^T$ and define $u^N(t)$ by (6). It is obvious that $x_i^N(t)$ is a polynomial of degree less than or equal to N. And $(x^N(t), u^N(t))$

satisfies the differential equation (2). Because we assume $V(x_r^{(m)}) < \infty$, it is known [3] that

$$||x_r^{(m_1)}(t) - x_{r+m_1}^N(t)||_\infty = ||x_r^{(m_1)}(t) - \sum_{n=0}^{N-r-m_1+1} a_n^{m_1 N} L_n(t)||_\infty$$
$$\leq C_1 V(x_r^{(m)})(N-r-m_1+1)^{-(m-m_1)+1/2}$$

for some constant $C_1 > 0$. Therefore,

$$|x_{r+m_1-1}^N(t) - (x_r)^{(m_1-1)}(t)| \leq \int_{-1}^{t} |x_{r+m_1}^N(s) - (x_r)^{(m_1)}(s)| ds$$
$$\leq 2C_1 V(x_r^{(m)})(N-r-m_1+1)^{-(m-m_1)+1/2}$$

Similarly, we can prove (29) and (30).

To prove (31), note that the spectral coefficient $a_n^{m_1 N}$ of $(x_r^N)^{(m_1)}(t)$ is the same as the spectral coefficients of $(x_r)^{(m_1)}(t)$. From Jackson's Theorem [27],

$$|a_n^{m_1 N}| < \frac{6}{\sqrt{\pi}}(U(x_r^{(m_1+1)}) + V(x_r^{(m_1+1)}))\frac{1}{n^{3/2}}.$$

Because f and g are Lipschitz continuous. In a bounded set around $x(t)$, $g(x) > \alpha > 0$ for some $\alpha > 0$. Therefore, the function $\frac{s-f(x)}{g(x)}$ is Lipschitz in a neighborhood of (x, s), i.e.

$$|u^N(t) - u(t)| = \left| \frac{x_{r+1}^N(t) - f(x^N(t))}{g(x^N(t))} - \frac{\dot{x}_r(t) - f(x(t))}{g(x(t))} \right|$$
$$\leq C_2(|x_{r+1}^N(t) - \dot{x}_r(t)| + |x_1^N(t) - x_1(t)| + \cdots + |x_r^N(t) - x_r(t)|) \quad (33)$$

for some C_2 independent of N. Hence, (32) follows (29), (30) and (33) when $l = 0$. Similarly, we can prove (32) for $l \leq L$. □

Proof of Theorem 3. (i) Because we assume Problem B has an optimal solution, there must exist a feasible solution $(x(t), u(t))$ satisfying (2)–(4). Let $(x^N(t), u^N(t))$ be the pair in Lemma 4 that satisfies (2). Define

$$\bar{x}^{Nk} = x^N(t_k), \; \bar{u}^{Nk} = u^N(t_k) \quad (34)$$

for $0 \leq k \leq N$. From Lemma 1, we know that $\{(\bar{x}^N, \bar{u}^N)\}$ satisfies (9).

Next we prove that the mixed state-control constraint (11) is satisfied. Because h is Lipschitz continuous and because of (29) and (32), there exists a constant C independent of N so that

$$||h(x(t), u(t)) - h(x^N(t), u^N(t))|| \leq C(\sum_{j=1}^{r} |x_j(t) - x_j^N(t)| + |u(t) - u^N(t)|)$$
$$\leq CMV(x_r^{(m)}(t))(r+1)(N-r-m_1+1)^{-(m-m_1)+1/2}.$$

Hence

$$h(x^N(t), u^N(t)) \leq CMV(x_r^{(m)}(t))(r+1)(N-r-m_1+1)^{-(m-m_1)+1/2}.$$

Since $m - m_1 \geq 1$ and $\delta > \frac{1}{2}$, there exists a positive integer N_1 such that,

$$CMV(x_r^{(m)}(t))(r+1)(N-r-m_1+1)^{-(m-m_1)+1/2} \leq (N-r-1)^{-(m-m_1)+\delta}$$

for all $N > N_1$. Therefore $x_1^N(t_k), \ldots, x_r^N(t_k), u^N(t_k), k = 0, 1, \ldots, N$, satisfy the mixed state and control constraint (11) for all $N > N_1$.

By a similar procedure, we can prove that the endpoint condition (10) is satisfied. Because $x_r^N(t)$ is a polynomial of degree less than or equal to N, and because of (7) and (34), we know $(x_r^N(t))^{(j)}$ equals the interpolation polynomial of $\bar{x}_r^N(D^T)^j$. So, $\begin{bmatrix} 1 & 0 & \cdots & 0 \end{bmatrix} D^j(\bar{x}_r^N)^T$ equals $(x_r^N(t))^{(j)}|_{t=-1}$. Therefore, (30) implies (21) if the interval between $\underline{\boldsymbol{b}}_j$ and $\boldsymbol{b}_j$ is large enough. In addition, the spectral coefficients of $\bar{x}_r^N(D^T)^{m_1}$ is the same as the spectral coefficients of $(x_r^N(t))^{(m_1)}$. From (31), we have

$$\sum_{n=0}^{N-r-m_1+1} |a_n^{m_1 N}| \leq d$$

if d is large enough. So, $\{(\bar{x}^N, \bar{u}^N)\}$ satisfies (22). Because we select $\underline{\boldsymbol{b}}$ and $\bar{\boldsymbol{b}}$ large enough so that the optimal trajectory of the original continuous-time problem is contained in the interior of the region, we can assume that $(x(t), u(t))$ is also bounded by $\underline{\boldsymbol{b}}$ and $\bar{\boldsymbol{b}}$. Then, (29) and (32) imply (12) for N large enough. Thus, $(\bar{x}_k^N, \bar{u}_k^N)$ is a discrete feasible solution satisfying all the constraints in Problem $\mathrm{B}^{\mathrm{N}+}$.

(ii) To prove the second part of the theorem, consider $\{(\bar{x}^{*N}, \bar{u}^{*N})\}_{N=N_1}^{\infty}$, a sequence of optimal solutions of Problem $\mathrm{B}^{\mathrm{N}+}$. From Lemma 2, the set (28) is bounded. By applying Lemma 3, it follows that there exists a subsequence of $\{(x^{*N}(t), u^{*N}(t))\}_{N=N_1}^{\infty}$ and an optimal solution of Problem B so that the limits in (24) converge uniformly. □

6 Conclusion

By focusing on optimal control problems subject to feedback linearizable systems, and an appropriate differentiability assumption, it is proved that the PS discretization, Problem $\mathrm{B}^{\mathrm{N}+}$, is always feasible and that its solutions converge to the optimal solution of Problem B as $N \to \infty$. For Problem B with a discontinuous optimal control, convergence is proved under Assumption 2.

References

1. K. Bollino, I. M. Ross, and D. Doman. Optimal nonlinear feedback guidance for reentry vehicles. *Proc. of AIAA Guidance, Navigation and Control Conference*, 2006.

2. J.P. Boyd. *Chebyshev and Fourier Spectral Methods.* Dover, 2nd edition, 2001.
3. C. Canuto, M.Y. Hussaini, A. Quarteroni, and T.A. Zang. *Spectral Method in Fluid Dynamics.* Springer-Verlag, New York, 1998.
4. J. Cullum. Finite-dimensional approximations of state constrainted continuous optimal problems. *SIAM J. Contr. Optimization*, 10:649–670, 1972.
5. A.L. Dontchev and W.W. Hager. The Euler approximation in state constrained optimal control. *Mathematics of Computation*, 70:173–203, 2000.
6. G. Elnagar and M.A. Kazemi. Pseudospectral Chebyshev optimal control of constrained nonlinear dynamical systems. *Computational Optimization and Applications*, 11:195–217, 1998.
7. G. Elnagar, M.A. Kazemi, and M. Razzaghi. The pseudospectral Legendre method for discretizing optimal control problems. *IEEE Trans. on Automat. Contr.*, 40:1793–1796, 1995.
8. F. Fahroo and I.M. Ross. Computational optimal control by spectral collocation with differential inclusion. *Proc. of the 1999 Goddard Flight Mechanics Symposium*, pages 185–200, 1999. NASA/CP-1999-209235.
9. F. Fahroo and I.M. Ross. Costate estimation by a Legendre pseudospectral method. *J. of Guidance, Control, and Dynamics*, 24(2):270–277, 2001.
10. F. Fahroo and I.M. Ross. Direct trajectory optimization by a Chebyshev pseudospectral method. *J. of Guidance, Control, and Dynamics*, 25(1):160–166, 2002.
11. F. Fahroo and I.M. Ross. Pseudospectral methods for infinite-horizon nonlinear optimal control problems. *Proc. of AIAA Guidance, Navigation and Control Conference*, 2005.
12. Q. Gong, W. Kang, and I.M. Ross. A pseudospectral method for the optimal control of constrained feedback linearizable systems. *IEEE Trans. on Automat. Contr.*, 51(7):1115–1129, 2006.
13. W.W. Hager. Runge-Kutta methods in optimal control and the transformed adjoint system. *Numerische Mathematik*, 87:247–282, 2000.
14. S.I. Infeld and W. Murray. Optimization of stationkeeping for a libration point mission. *AAS Spaceflight Mechanics Meeting*, 2004. AAS 04-150.
15. W. Kang, Q. Gong, and I.M. Ross. Convergence of pseudospectral methods for a class of discontinuous optimal control. *Proc. of the 44rd IEEE Conf. on Decision and Contr.*, 2005.
16. P. Lu, H. Sun, and B. Tsai. Closed-loop endoatmospheric ascent guidance. *J. of Guidance, Control, and Dynamics*, 26(2):283–294, 2003.
17. E. Polak. *Optimization: Algorithms and Consistent Approximations.* Springer-Verlag, Heidelberg, 1997.
18. L.S. Pontryagin, V.G. Boltyanskii, R.V. Gamkrelidze, and E.F. Mischenko. *The Mathematical Theory of Optimal Processes.* Wiley-Interscience, New York, 1962.
19. J. Rea. Launch vehicle trajectory optimization using a Legendre pseudospectral method. *Proc. of AIAA Guidance, Navigation and Control Conference*, 2003. Paper No. AIAA 2003-5640.
20. S.M. Robinson. Strongly regular generalized equations. *Mathematics of Operations Research*, 5:43–62, 1980.
21. S.M. Robinson. An implicit function theorem for a class of nonsmooth functions. *Mathematics of Operations Research*, 16:292–309, 1991.
22. I.M. Ross and F. Fahroo. A unified framework for real-time optimal control. *Proc. of the 42nd IEEE Conf. on Decision and Contr.*, 2003.

23. I.M. Ross and F. Fahroo. Pseudospectral knotting methods for solving optimal control problems. *J. of Guidance, Control, and Dynamics*, 27(3):397–405, 2004.
24. I.M. Ross and F. Fahroo. Issues in the real-time computation of optimal control. *Mathematical and Computer Modelling, An International Journal*, 43(9–10):1172–1188, 2006.
25. I.M. Ross, Q. Gong, F. Fahroo, and W. Kang. Practical stabilization through real-time optimal control. *Proc. of the 2006 Amer. Contr. Conf.*, 2006.
26. I.M. Ross, P. Sekhavat, A. Fleming, and Q. Gong. Pseudospectral feedback control: foundations, examples and experimental results. *Proc. of AIAA Guidance, Navigation and Control Conference*, 2006. AIAA-2006-6354.
27. G. Sansone, A.H. Diamond, and E. Hille. *Orthogonal Functions.* Robert E. Krieger Publishing Co., Huntington, New York, 1977.
28. A. Schwartz and E. Polak. Consistent approximations for optimal control problems based on Runge-Kutta integration. *SIAM J. Contr. Optimization*, 34:1235–1269, 1996.
29. P. Sekhavat, A. Fleming, and I.M. Ross. Time-optimal nonlinear feedback control for the NPSAT1 spacecraft. *Proc. of the 2005 IEEE/ASME International Conference on Advanced Intelligent Mechatronics*, 2005.
30. S. Stanton, R. Proulx, and C. D'Souza. Optimal orbit transfer using a Legendre pseudospectral method. *Proc. of AAS/AIAA Astrodynamics Specialist Conference*, 2003. AAS-03-574.
31. P. Williams, C. Blanksby, and P. Trivailo. Receding horizon control of tether system using quasi-linearization and Chebyshev pseudospectral approximations. *Proc. of AAS/AIAA Astrodynamics Specialist Conference*, 2003. AAS-03-534.

Part III

Feedback Design

Event Based Control

Karl J. Åström

Department of Automatic Control LTH, Lund University, Box 118, SE-221 00 Lund, Sweden

Summary. In spite of the success of traditional sampled-data theory in computer control it has some disadvantages particularly for distributed, asynchronous, and multi-rate system. Event based sampling is an alternative which is explored in this paper. A simple example illustrates the differences between periodic and event based sampling. The architecture of a general structure for event based control is presented. The key elements are an event detector, an observer, and a control signal generator, which can be regarded as a generalized data-hold. Relations to nonlinear systems are discussed. Design of an event based controller is illustrated for a simple model of a micro-mechanical accelerometer.

1 Introduction

Computer controlled systems are traditionally based on periodic sampling of the sensors and a zero order hold of the actuators, see [31], [48], and [52]. When controlling linear time invariant systems the approach leads to closed loop systems that are linear but periodic. The periodic nature can be avoided by considering only the behavior at times that are synchronized with the sampling resulting in the *stroboscopic model.* The system can then be described by difference equations with constant coefficients. Traditional sampled data theory based on the stroboscopic model is simple, and has been used extensive in practical applications of computer controlled systems. Periodic sampling also matches the time-triggered model of real time software which is simple and easy to implement safely, see [10], [37].

Since standard sampled-data theory only considers the behavior of the system at the sampling instants, it is necessary to ensure that the inter-sample behavior is satisfactory. A simple way to do this is to also sample not only the system but also the continuous time loss function, see [1], [2] and [6]. This approach, which is called *loss function sampling*, is equivalent to minimising the continuous time behavior subject to the constraint that the control signal is piece-wise constant, see [1]. Analysis and design reduces to calculations for a time-invariant discrete system. Another approach, called *lifting*, is to describe the behavior of the the state over a whole sampling interval, see [12], [56]. Lifting also gives a description of the system which is time-invariant but there are technical difficulties because the state space is infinite dimensional.

Even if traditional sampled data control is the standard tool for implementing computer control, there are severe difficulties when dealing with systems having multiple sampling rates or systems with distributed computing. With multi-rate sampling the complexity of the system depends critically on the ratio of the sampling rates. For distributed systems the theory requires that the clocks are synchronized. In networked distributed systems it has recently been a significant interest in the effects of sampling jitter, lost samples and delays on computer controlled systems, see [11].

Event based sampling is an alternative to periodic sampling. Signals are then sampled only when significant events occurs, for example, when a measured signal exceeds a limit or when an encoder signal changes. Event based control has many conceptual advantages. Control is not executed unless it is required, control by exception, see [36]. For a camera based sensor it could be natural to read off the signal when sufficient exposure is obtained. Event based control is also useful in situations when control actions are expensive, for example when controlling manufacturing complexes, and when when it is expensive to acquire information like in computer networks. Event based control is also the standard form of control in biological systems, see [55]. A severe drawback with event based systems is that very little theory is available.

All sampled systems, periodic as well as event based share a common property that the feedback is intermittent and that control is open loop between the samples. After an event the control signal is generated in an open loop manner and applied to the process. In traditional sampled-data theory the control signal is simply kept constant between the sampling instants, a scheme that is commonly called a zero order hold (ZOH). In event based systems the generation of the open loop signal is an important issue and the properties of the closed loop system depends critically on how the signal is generated.

There has not been much development of theory for systems with event based control. There are early results on discontinuous systems, [20], [53], [54], and impulse control, see [8] and [7]. Event based systems can also be regarded as special cases of hybrid control, where the system runs open loop between the regions, [15].

This paper gives an overview of systems with event based control. Section 2 presents a number of examples of were event based control is beneficial. Section 3 which is based on [4] and [5] gives a detailed discussion of a simple example that illustrates several issues about event based control. The example shows the benefits of event based control compared with conventional periodic sampling with impulse holds and zero order holds. Section 4 presents new ideas on analysis of design of event based control. A general system structure is given, and the different subsystems are discussed. Particular emphasis is given to the design of the control signal generator which can be viewed as a generalized hold circuit. It is shown that control signal generator can be chosen so that the event based system is equivalent to a nonlinear control system. This implies that techniques for nonlinear control can be applied, [28]. The design procedure is illustrated by a simple version of a MEMS accelerometer in Section 5.

2 Examples

Event based control occurs naturally in many situations from simple servo system to large factory complexes and computer networks. It is also the dominating control principle in biological systems. Encoders are primarily event based sensors. Relay systems with on-off control and satellite thrusters are event based [20], [16]. Systems with pulse-width of pulse-frequency modulation are other examples [50], [43], [22], [21], [49]. In this case the control signal is restricted to be a positive or negative pulse of given size. The controller decides when the pulses should be applied, what sign they should have, and the pulse lengths.

Event based control is easy to implement and was therefore used in many early feedback systems. Accelerometer and gyros with pulse feedback were used in systems for inertial navigation, see [41], [17]. A typical accelerometer consists of a pendulum which is displaced by acceleration. The displacement is compensated by force sensors and a feedback which keeps the pendulum in its neutral position. The restoring force was generated by pulses that moved the pendulum towards the center position as soon as a deviation was detected. Since all correcting impulses have the same form, the velocity can be obtained simply by adding pulses. Much work on systems of this type was done in the period 1960-1990.

Systems with relay feedback [20], [53] are other examples of event based control. Here feedback occurs when certain signals exceed given limits. The sigma-delta modulator or the one-bit AD converter, [42] [9], which is commonly used in audio and mobile telephone system, is another example. It is interesting to note that in spite of their wide spread use there does not exist a comprehensive theory for design of sigma-delta modulators. Design is currently done based on extensive simulations and experiments.

When controlling automotive engines it is natural to base sampling on crank angle position rather than on time, see [24], [23] and [13]. In ship control it is natural to base control on distance travelled rather than time. It is also much more natural for the ships captain to deal with quantities like turning radius instead of turning rate. Similarly in control of rolling mills it is natural to sample based on length rather than time. Other examples of event based process control are given in [38].

A typical plant in the process industry has many production units separated by buffer tanks for smoothing production variations. It is highly desirable not too change production rates too frequently because each change will cause upsets. There are always disturbances in the production. It may also be necessary to change production when the levels in the storage tanks are close to upper and lower limits. Controlling a large production facility can be approached via event based control. Nothing is done unless there are severe upsets or when storage tanks are approaching the limits. An early attempt with event based control of a paper mill is given in [46], a more recent project is described in [47].

Control of networks are other examples of event based control. In the Internet there are a large number of routers that just forward messages. The end-to-end transmission is controlled by the transmission control protocol (TCP) that ensures that the full message is received. To achieve this some information must be added to the message as well as mechanism for resending the message and for controlling the transmission rate. The TCP protocol is thus an example of event based control, see [29] and [34].

In biological systems the neurons interact by sending pulses. Electrical stimuli changes ion concentrations in the neuron and a pulse is emitted when the potential reaches a certain level, see [33], [27]. Several efforts have been made to mimic these systems. Implementations of silicon neurons are found in [40] and the paper [14] shows how they can be used to implement simple control systems. An interesting feature is the ease of interfacing and the possibility of constructing very reliable systems by duplicating the neurons and simply adding pulses.

3 Comparison of Periodic and Event Based Control

To provide insight into the differences between periodic and event based control we will first consider a simple regulation problem where all calculations can be performed analytically. The results are based on [4] and [5], where many additional details are given. Consider a system to be controlled described by the equation

$$dx = udt + dv, \tag{1}$$

where the disturbance $v(t)$ is a Wiener process with unit incremental variance and u the control signal. It is assumed that the state x is measured without error. The object is to control the system so that the state is as close to the origin as possible. To be specific we will minimize the mean square variations

$$V = \frac{1}{T} E \int_0^T x^2(t)dt. \tag{2}$$

Conventional periodic sampling with different holds will be compared with event based control where control actions are taken only when the output is outside the interval $-a < x < a$. The effects of different data holds will also be explored. We will compare the distribution of $x(t)$ and the variances of the outputs for both control schemes.

Periodic Sampling with Impulse Hold

We will first consider the case of impulse hold. The control signal is then an impulse applied when an event occurs. Since the process dynamics is an integrator it is possible to reset the state to zero instantaneously at each event. Let t_k be the time when the event occurs, the control law then becomes

$$u(t) = -x(t_k)\delta(t - t_k), \tag{3}$$

where δ is the delta function. The control law (3) implies that an impulse which resets the control to zero is applied at each sampling interval. After the impulse the closed loop system is governed by $dx = dv$ and the variance then grows linearly until the next sampling interval occurs. Since the incremental variance is 1 the average variance over a sampling interval is $h/2$ and the minimal loss function (2) for periodic sampling and impulse hold becomes

$$V_{PIH} = \frac{1}{2}h. \tag{4}$$

Periodic Sampling with Zero Order Hold

To illustrate that the data hold influences the results we will consider the standard situation with periodic sampling and a first order hold. Let h the sampling period, the sampled system is then

$$x(t + h) = x(t) + hu(t) + e(t), \tag{5}$$

which a special case of a standard discrete system with $\Phi = 1$ and $\Gamma = h$ [6]. The mean square variance over one sampling period is

$$\begin{aligned} V &= \frac{1}{h}\int_0^h Ex^2(t)\,dt = \frac{1}{h}J_e(h) \\ &\quad + \frac{1}{h}\left(Ex^TQ_1(h)x + 2x^TQ_{12}(h)u + u^TQ_2(h)u\right) \\ &= \frac{1}{h}(R_1(h)S(h) + J_e(h)), \end{aligned} \tag{6}$$

where $Q_1(h) = h,\ Q_{12}(h) = h^2/2,\ Q_2(h) = h^3/3,\ R_1(h) = h$ and

$$J_e(h) = \int_0^h Q_{1c}\int_0^t R_{1c}\,d\tau\,dt = h^2/2 \tag{7}$$

see [2]. The function $S(h)$ is the positive solution of the Riccati equation

$$S = \Phi^TS\Phi + Q_1 - L^TRL, \quad L = R^{-1}(\Gamma^TS\Phi + Q_{12}^T), \quad R = Q_2 + \Gamma^TS\Gamma,$$

where the argument h has been dropped to get cleaner equations. The Riccati equation has the solution $S(h) = h\sqrt{3}/6$, and the controller which minimizes the loss function (2) is

$$u = -Lx = \frac{1}{h}\frac{3+\sqrt{3}}{2+\sqrt{3}}x.$$

The minimum variance with periodic sampling and a zero order hold is thus

$$V_{PZOH} = \frac{3+\sqrt{3}}{6}h. \tag{8}$$

Notice that the impulse hold gives the variance $V_{PIH} = h/2$, while a zero order hold gives the variance $V_{PZOH} = h(3+\sqrt{3})/6$. The impulse hold is thus more effective than the zero order hold.

Event Based Control

For event based control we specify a region $-a < x < a$. No control action will be taken if the state is inside this region. Control actions are only taken at events t_k when $|x(t_k)| = a$. A simple strategy is to apply an impulse that drives the state to the origin, i.e. $x(t_k+0) = 0$. With this control law the closed loop system becomes a Markovian diffusion process of the type investigated in [18]. Let $T_{\pm d}$ denote the exit time i.e. the first time the process reaches the boundary $|x(t_k)| = a$ when it starts from the origin. The mean exit time can be computed from the fact that $t - x_t^2$ is a martingale between two impulses and thus

$$h_E := E(T_{\pm d}) = E(x^2_{T_{\pm d}}) = a^2.$$

The average sampling period thus equals $h_E = a^2$.

The probability distribution of x is given by the Kolmogorov forward equation for the Markov process

$$\frac{\partial f}{\partial t} = \frac{1}{2}\frac{\partial^2 f}{\partial x^2}(x) - \frac{1}{2}\frac{\partial f}{\partial x}(d)\delta_x + \frac{1}{2}\frac{\partial f}{\partial x}(-d)\delta_x,$$

with $f(-a) = f(a) = 0$, see [18]. This partial differential equation has the stationary solution

$$f(x) = (a - |x|)/a^2, \tag{9}$$

which can be verified by direct substitution. Notice that the distribution is not Gaussian but symmetric and triangular with the support $-a \le x \le a$. The steady state variance is

$$V_{EIH} = \frac{a^2}{6} = \frac{h_E}{6}. \tag{10}$$

Comparison

Summarizing the results we find that the loss functions are given by the Equations (4), (8) and (10) that

$$V_{PZOH} = \frac{3+\sqrt{3}}{6}h, \quad V_{PIH} = \frac{h}{2}, \quad V_{EIH} = \frac{a^2}{6} = \frac{h_E}{6}. \tag{11}$$

The variances are thus related as $4.7h{:}3h{:}a^2$. It follows that for periodic sampling a zero order hold increases the variance by 50% compared with impulse hold. To compare periodic sampling with event based sampling we will choose the parameter a so that the average sampling rates are the same. For event based sampling the average sampling period was $h_E = a^2$. Equating this with h gives $a^2 = h$. An event based controller thus gives a variance that is 3 times smaller than a controller with periodic sampling.

The reason for the differences is that the event based controller acts as soon as an error is detected. The reason why impulse holds give smaller variances than a zero order hold is because it is advantageous to act decisively as soon as a deviation is detected. In the particular case two thirds of the improvement is thus due to sampling and one third due to the impulse hold.

The behavior of the closed loop systems obtained with the different control strategies are illustrated by the simulation results shown in Figure 1. Simulation is performed by approximating continuous behavior by fast sampling. The same noise sequence is used in all simulations. The states are shown in

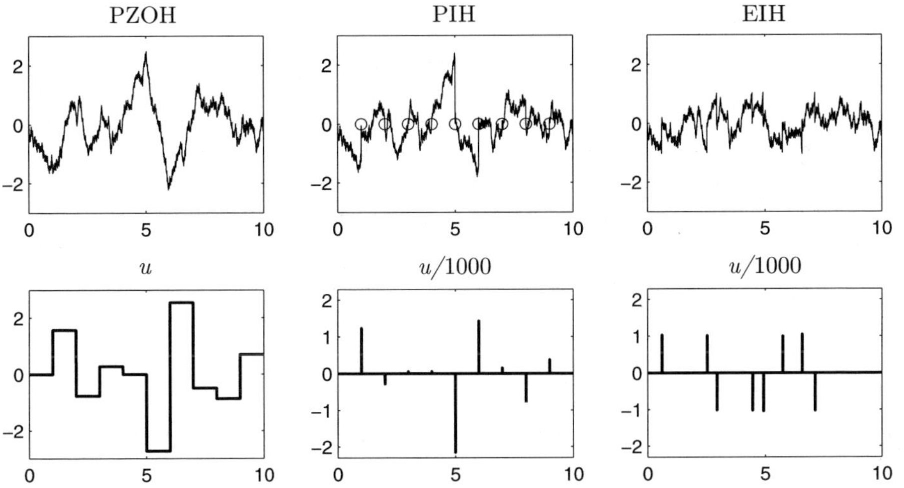

Fig. 1. Simulation of an integrator with periodic sampling and first order hold (*left*) periodic sampling and impulse hold (*center*) and event based control with impulse hold (*right*). The *top plots* shows the state and the control signals are shown in the *lower plots*. The number of samples are the same in all cases

the upper plots and the corresponding control signals in the lower plots. The left plots show results for periodic sampling with a zero order hold (ZOH), the center plots correspond to periodic sampling with impulse hold (IH) and the plots to the right show event based control with impulse hold EIH. The improved performance with event based control is clearly visible in the figure. Notice that the process state stays within the bounds all the time. In the plot of the process state for periodic sampling with impulse hold we have also marked the state at the sampling times with circles. With impulse sampling the state is reset instantaneously as is clearly seen in the figure. The advantage of the impulse hold is apparent by comparing the behavior of zero order hold and impulse sampling at time $t = 5$ where the state has a large positive value.

4 A General Structure

A block diagram of a system with event based control is shown in Figure 2. The system consists of the process, an event detector, an observer, and a control signal generator. The event detector generates a signal when an event occurs, typically when a signal passes a level, different events may be generated for up- and down-crossings. The observer updates the estimates when an event occurs and passes information to the control signal generator which generates the input signal to the process. The observer and the control signal generator run open loop between the events, the absence of an event is however information that can be used by the observer [26]. A simple special case is when the full state of the process is transmitted when an event occurs.

The control strategy is a combination of feedback and feedforward. Feedback actions occur only at the events. The actuator is driven by the control signal generator in open loop between the events. A consequence is that the behavior of the system is governed by the control signal generator. Design of the control signal generator is therefore a central issue.

It is interesting to compare with a conventional sampled data system where the events are generated by a clock and the behavior of the system is primarily determined by the control law. Such a system can also be represented by

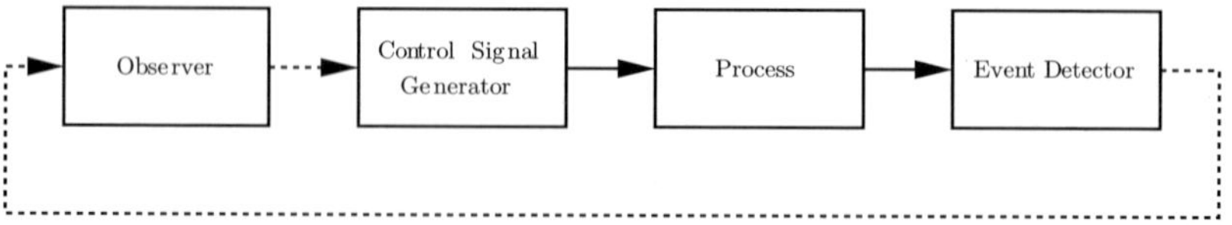

Fig. 2. Block diagram of a system with event based control. *Solid lines with filled arrows* denotes continuous signal transmission and the *dashed lines with open arrows* denotes event based signal transmission

Figure 2 but there is a block representing the control law inserted between the sampler and the control signal generator. For a conventional sampled system the behavior of the closed loop system is essentially determined by the control algorithm, but in an event based controller the behavior is instead determined by the control signal generator. Therefore it makes sense to use another name, even if the control signal generator can be regarded as a generalized hold. The different elements of the system in Figure 2 will now be discussed.

The Control Signal Generator

The design of hold circuits was discussed in the early literature on sampled-data systems, see [48] and [31]. When computer control became widely used the zero order hold became the standard solution. Linear holds were sometimes used when a smooth control signal was desired, [6]. There was a renewed interest of generalised data hold circuits around 1990 when it was discovered that the properties of a system can be changed significantly, [32]. The solutions proposed often led to irregular control signals which excited high frequency modes and gave poor robustness [19]. A properly designed control signal generator can however give improved performance as is shown in [45].

There has recently been a renewed interest in generalized hold in order to obtain systems that are insensitive to sampling jitter [44]. This is a central issue for event based systems where sampling can be very irregular. The difference between the different holds can be illustrated by a simple example. Consider the following first order system

$$\frac{dx}{dt} = x - u.$$

An unstable system is chosen because the differences will be more pronounced. Assume that an event based controller is designed based on an average sampling time t_0. We will compare an ordinary zero order hold and a hold with exponential decay. Consider the situation when $x(0) = 1$. A straightforward calculation shows that the control signals and the state behavior are given by

$$\begin{aligned} u_{PZOH} &= \frac{1}{1-e^{-t_0}}x(0) & \qquad x_{PZOH} &= \frac{1-e^{t-t_0}}{1-e^{-t_0}}x(0) \\ u_{EH} &= \frac{(a+1)e^{-t}}{1-e^{-(a+1)t_0}}x(0) & \qquad x_{EH} &= \frac{e^{-at}-e^{t-(a+1)t_0}}{1-e^{-(a+1)t_0}}x(0)\,. \end{aligned}$$

Figure 3 shows the state and the control signal for the different holds. The zero order hold gives a constant control signal but the exponential hold generates a control signal that is large initially and then decays. The behavior of the state is also of interest, all holds give the desired state $x = 0$ at the nominal time $t = t_0 = 2$ but the rate of change of the state at t_0 is quite different. The zero order hold gives the largest rate and the exponential hold with the fastest decay gives the smallest rate.

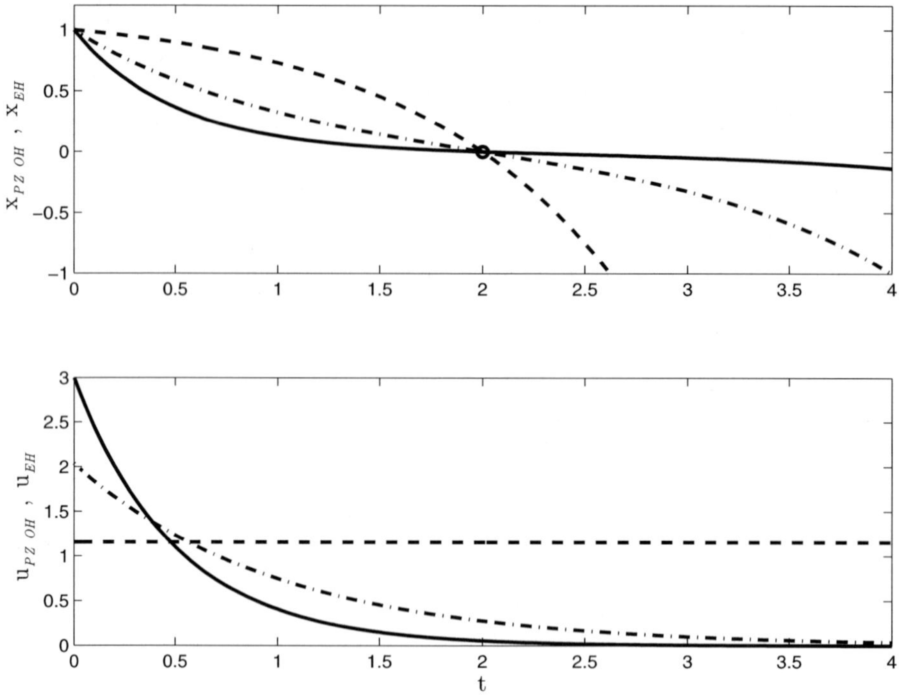

Fig. 3. Behavior of state and control variables after an event for systems with zero order hold (*dashed*) and exponential holds with $a = 1$ (*dash-dotted*) and $a = 2$ (*solid*). The signal is sampled at time $t = 0$ and the control signal is then applied in an open-loop manner. Notice that the states for both systems is the same at the nominal sampling time $t_0 = 2$, but that the rate of change of the state is much smaller with exponential hold.

The simple example shows that the hold circuit has important consequences. Data holds where the control signal decays after the event have the advantage that the behavior of the system is robust to changes in the times events occur. We can thus conclude that for event based control it is desirable to have holds that give control signals which are large initially and decay fast. The impulse hold where the control signal is an impulse is an extreme case. Holds with large control signals may however be undesirable for systems with poorly damped resonant modes. In the system discussed in Section 3 impulse sampling gave better performance than a zero order hold.

Since the hold circuit is important it is natural that it should be matched to the process and we will therefore briefly discuss methods for designing control signal generators circuits. There are many ways to do this, in fact many methods for control system design can be used. Consider for example the system described by

$$\frac{dx}{dt} = Ax + Bu, \tag{12}$$

with $x \in \mathbb{R}^n$, $u \in \mathbb{R}^p$. Design of the control signal generator is essentially the problem of finding an open loop control signal that drives the system from its state at the time of the event to a desired state. In regulation problems the desired state is typically a given constant for example $x = 0$. This can always be done in finite time if the system is reachable. There are many ways to generate the signal, we can for example use a dead-beat controller which drives the state to zero in finite time. Optimal control and model predictive control are other alternatives that are particularly useful when there are restriction on the control signal.

In this paper we will determine a state feedback $u = -Lx_c$ for the system (12) which drives the state to zero. Such a control signal can be generated from the dynamical system

$$\frac{dx_c}{dt} = (A - BL)x_c, \quad u = -Lx_c, \tag{13}$$

which is initialized with the process state at the event or the estimated process state. Notice that x_c is the controller state and that the control is applied in an open-loop manner like a feedforward signal.

Relations to Nonlinear Control

There is an advantage to generating the control signal from Equation 13. Assuming that there are no model errors and no disturbances, the controller state x_c is then equal to the process state x and the open loop control is identical to a closed loop control of the process with the control law $u = -Lx$. Since the control signal generator is a dynamical system the system can be analysed using nonlinear control theory which is highly attractive, see [28], [30]. An example where this idea is elaborated will be given in Section 5.

The Observer

When the state of the process is not measured directly it is suitable to reconstruct it using an observer. The observer problem for event based control is not a standard problem. Consider for example a system described by

$$dx = Axdt + Budt + dv, \quad dy = Cx\,dt + de, \tag{14}$$

where $x \in \mathbb{R}^n$, $u \in \mathbb{R}^p$ and $y \in \mathbb{R}$ and v and e are Wiener processes of appropriate dimensions with incremental covariances $R_1 dt$ and $R_2 dt$. Assume that an event is generated when the magnitude of the output y exceeds a. If no event is obtained we clearly have information that the measured signal is in the interval $-a < y < a$, when an event occurs we obtain a precise measurement of y. This is clearly a non-standard information pattern.

An ad hoc solution is to use the following approximate Kalman filter

$$\begin{aligned}\frac{d\hat{x}}{dt} &= A\hat{x} + Bu + K(t)(0 - C\hat{x}) \\ K(t) &= P(t)CR_2^{-1} \\ \frac{dP}{dt} &= AP + PA^T - PC^T R_2^{-1} CP + R_1,\end{aligned}$$

when no event occurs. A reasonable assumption for the measurement noise is $R_2 = a^2/(12t_s)$, where t_s is the average sampling rate, or the time elapsed since the last event. Recall that $a^2/12$ is the variance of a uniform distribution over the interval $(-a,\ a)$. When an event occurs the state is estimated by

$$\begin{aligned}\hat{x}^+ &= \hat{x} + PC(\bar{R}_2 + CPC^T)^{-1}(y_e - C\hat{x}) \\ P^+ &= P - PC(\bar{R}_2 + CPC^T)^{-1}C^T P,\end{aligned}$$

where $y_e = a$ or $-a$ depending on which boundary is crossed, and $\bar{R}_2$ is the variance of the detection error. The superscript + denotes the values immediately after the detection. A simple assumption is $\bar{R}_2 = 0$.

The filtering problem can be solved exactly by computing the conditional distribution of the state x of (14) given that $-a < y < a$. This problem is discussed in [26], where it is shown that at least for the accelerometer example in Section 5 the approximate Kalman filter discussed in Section 4 gives similar results. In [25] and [26] it is also shown that the exact solution to the filtering problem has interesting properties. For example it is shown that the conditional distribution is log-concave under quite general conditions.

5 An Example

As an illustration we will consider a simple model for a MEMS accelerometer, which consists of a mass supported a weak spring with a detector which detects small deviations of the mass from the reference position and a capacitive force actuator. In extreme cases the sensing is done by measuring tunnelling current via a very narrow tip, see [39], [35].

Neglecting the spring coefficient the system becomes a double integrator. Using normalized parameters the system can be described by the model

$$\frac{dx}{dt} = \begin{pmatrix} 0 & 1 \\ 0 & 0 \end{pmatrix} x + \begin{pmatrix} 0 \\ 1 \end{pmatrix} u \qquad (15)$$

$$y = \begin{pmatrix} 1 & 0 \end{pmatrix} x. \qquad (16)$$

It is assumed that the largest control signal is 1 and that an event is detected when $|y(t)| = a$. We also assume that the direction of the crossing is detected. In reality only position information is available, for simplicity we will here assume that the full state is available when an event occurs.

The state feedback approach is used to generate the control signal. Assuming a linear feedback $u = -Lx$ the closed loop system becomes

$$\begin{aligned}\frac{dx}{dt} &= \begin{pmatrix}0 & 1\\ 0 & 0\end{pmatrix} x + \begin{pmatrix}0\\ 1\end{pmatrix} u = \begin{pmatrix}0 & 1\\ -l_1 & -l_2\end{pmatrix} x\\ u &= -\begin{pmatrix}l_1 & l_2\end{pmatrix} x.\end{aligned} \tag{17}$$

The characteristic polynomial for the closed loop system is

$$s^2 + l_2 s + l_1.$$

Since the largest control signal is one it is natural to choose $l_1 = a^{-1}$. This means that full control action is used if the mass has zero velocity, and the output is at the boundary $|x_1| = a$ of the detection zone. Choosing $l_2 = 2/\sqrt{a}$ gives critical damping. The settling time after a disturbance is then of the order of $5\sqrt{a}$. With $a = 1$ we then get $l_1 = 1$ and $l_2 = 2$ and a settling time of 5.

When an event occurs the system (17) is initialized with the state equal to the state of the process and the control signal is then generated by running (17) in open loop. In this case the control signal generator is thus matched to the dynamics of the system. Since the actuator has limitations it is useful to limit the control signal, see [51]. The control signal then becomes

$$u = -\text{sat}(l_1\hat{x}_1 + l_2\hat{x}_2).$$

If the directions of the crossings are known it is possible to add a refinement by applying the full control power when the output leaves the band $-a < y < a$ and to apply the signal from the hold circuit when the signal enters the band again. This gives a relay action which ensures that the system quickly enters the detection zone. The complete nonlinear control law then becomes

$$u = \begin{cases} 1 & \textit{if} \quad x_1 \le -1\\ -\text{sat}(l_1\hat{x}_1 + l_2\hat{x}_2) & \textit{if} \quad -a < x_1 < -a\\ -1 & \textit{if} \quad x_1 \ge a\,. \end{cases} \tag{18}$$

The control signal generator is thus implemented by applying this nonlinear feedback law to the system (15) and running it in open loop. Another way to generate the hold signal is to observe that the pulse shape is given by

$$u = \begin{cases} 1 & \textit{if} \quad x_1 \le -1\\ -\text{sat}(\pm p_1(t) + p_2(t)v^*) & \textit{if} \quad -a < x_1 < -a\\ -1 & \textit{if} \quad x_1 \ge a\,, \end{cases}$$

where $p_1(t)$ and $p_2(t)$ are given functions of time which can be computed off line, v^* the velocity when x_1 enters the region $-a < x_1 < 1$, and the sign of the p_1-term depends on which side it enters from. A table look-up procedure can thus also be used to generate the control signal.

Equivalent Nonlinear System

If there are no disturbances and no modeling errors and if the control signal is generated by (13), the system with event based control behaves like the system (15) with the nonlinear feedback (18). Since this system is of second order it is straightforward to analyze its behavior. A phase plane of the system is shown in Figure 4 for $a = 1$. The phase plane is symmetric so it suffices to discuss half the region. It follows from (18) that $u = -1$ in the shaded region. The equations can then be integrated analytically in that region, and we find that the trajectories are given by the parabolic curves

$$x_1 = x_1(0) + \frac{1}{2}x_2^2(0) - \frac{1}{2}x_2.$$

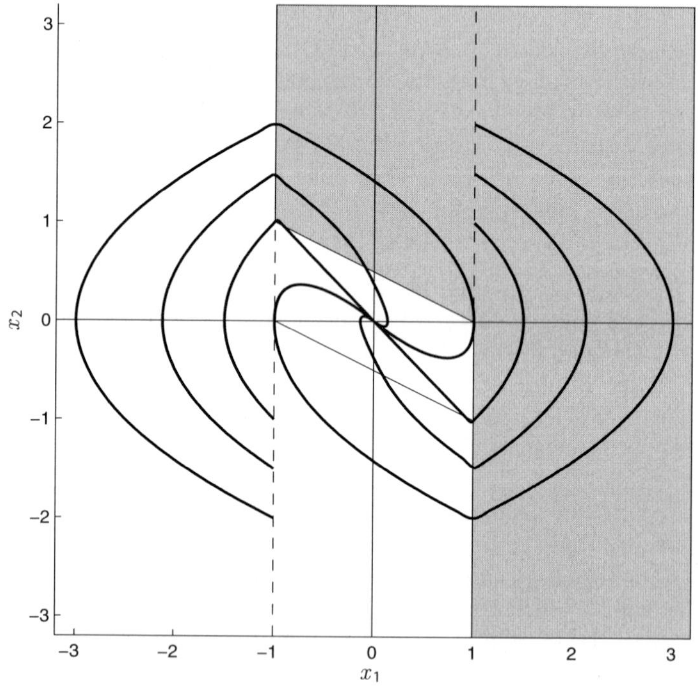

Fig. 4. Phase plane for the system when there are no modeling errors and no disturbances. The control signal is $u = -1$ in the *shaded zone*

A solution starting at $x_1(0) = -1$ with $x_2(0) \geq \sqrt(2)$ intersects the line $x_1 = 1$ at $x_2 = \pm\sqrt{x_2^2(0) - 2}$. The velocity x_2 decreases for each passage of the region $-1 < x_1 < 1$. Trajectories starting at $x_1(0) = -1$ with $1 \leq x_2(0) \leq \sqrt{2}$ enter the linear region, and trajectories starting at $x_1(0) = -1$ with $x_2(0) < 1$ remain in the rhomboid region. The origin is globally asymptotically stable and all trajectories entering the rhomboid region around the origin will remain in that region.

Response to Random Accelerations

The control law (18) implies that the system works like a system with relay feedback when the output is outside the detection region. This feedback attempts to force the output into the detection region. The action of the signal generator is to generate a signal that drives the state to the nominal zero position. This means that the signal generator provides a damping action.

The consequences of event based control is clearly seen in the response to a random acceleration. A simulation of such a case is shown in Figure 5. Notice

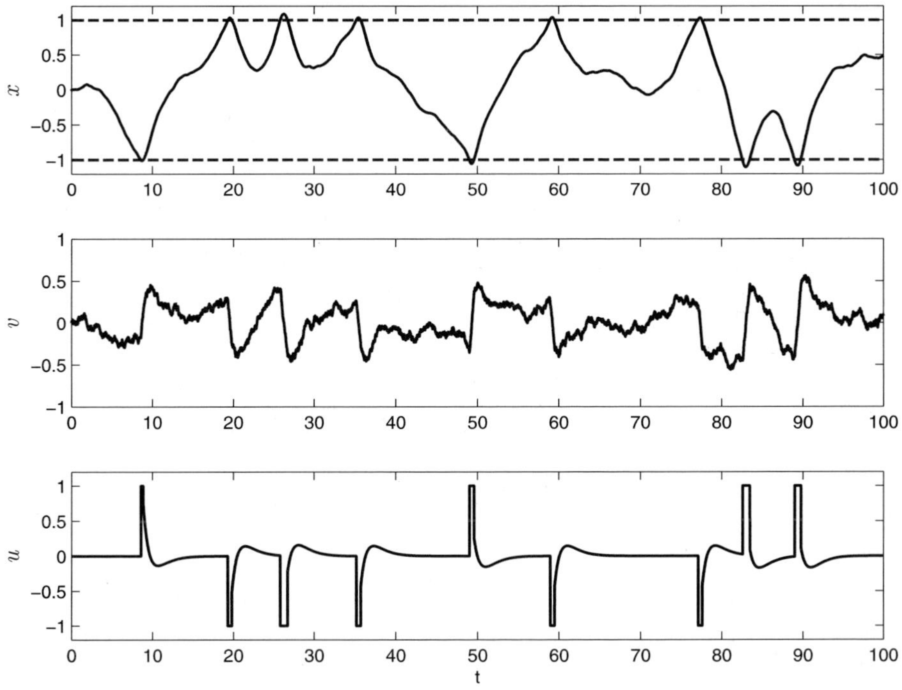

Fig. 5. Simulation of a MEMS accelerometer with event based feedback and random acceleration

that the deviation is kept between the limits even if the events are quite sparse. Also notice the shape of the pulses. They maintain the maximum values as long as the deviation is outside the detector limit, they jump at the events when the output enters the detection region and then they decay gracefully to zero. In this case it is essential to apply full restoring force when the output is outside the detection band. If this is not done the system may diverge after a large disturbance.

Response to a Constant Acceleration

The previous analysis is useful in order to understand the basic behavior of the system. It is, however, more relevant to explore the real purpose of the system by investigating how it responds to a constant acceleration Figure 6 shows another simulation when there is a constant acceleration. The top curve in the figure shows the position of the mass $y = x_1$, the center curve is the control signal u and the bottom curve is the integral of the acceleration error

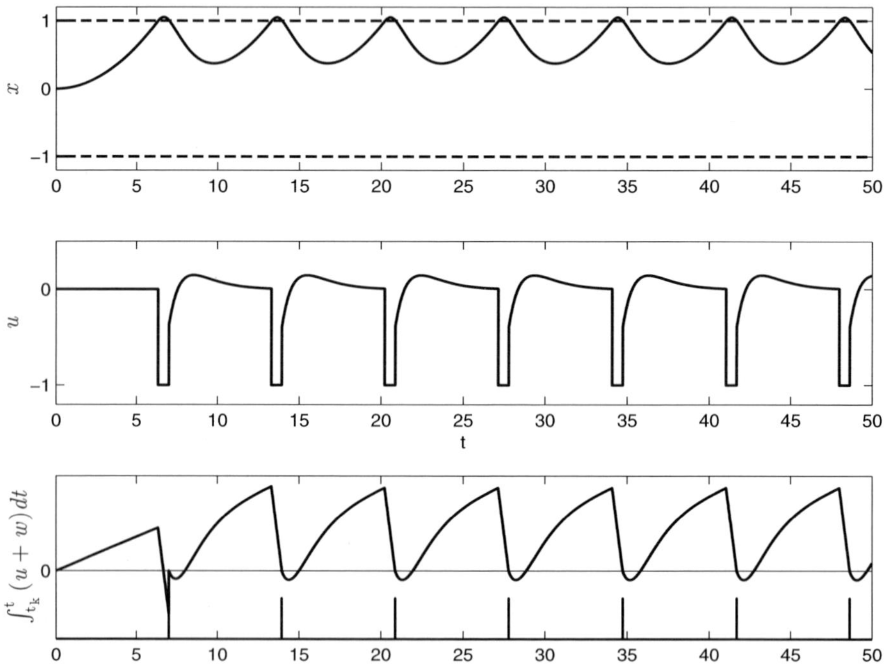

Fig. 6. Response to a constant acceleration $w = 0.05$. The *upper plot* shows position of the mass, the *middle plot* shows the control signal u and the *lower plot* shows the integral of the control error. The events are also indicated in the lower plot

e_a. The integral is reset to zero at the events formed by the up-crossings. The events are also marked in the figure.

Let the acceleration be w, and let the control signal be u. Assume that t_k and t_{k-1} are consecutive events where the same boundary is crossed. It then follows that

$$x(t_{k+1}) = x(t_k) + \int_{t_k}^{t_{k+1}} \int_{t_k}^{t} (w(\tau) + u(\tau)) d\tau = x(t_k) + \int_{t_k}^{t_{k+1}} (t_{k+1} - t)(w(t) + u(t)) dt.$$

Since $x(t_k) = x(t_{k-1})$ it follows that

$$\int_{t_k}^{t_{k+1}} \int_{t_k}^{t} u(t) dt = \int_{t_k}^{t_{k+1}} (t_{k+1} - t) u(t) dt = - \int_{t_k}^{t_{k+1}} (t_{k+1} - t) w(t) dt.$$

The double integral of the control signal between two consecutive events is thus a weighted average of the acceleration in the interval between the events.

A similar analysis shows that if the events represents crossings of different boundaries a correction term $4a/(t_{k+1} - t_k)^2$ should be added. The time resolution is given by the width of the detection interval. If there is a periodic solution we have $v(t_{k+1}) = v(t_k)$ and in this case we find that the integral of the control signal over an interval between two events is the average of the acceleration during that interval. This is illustrated in the lower plot in Figure 6 which shows the integral of the acceleration error. The integral is reset to zero at the events formed by the up-crossings. The events are also marked in the figure.

A phase plane is shown in Figure 7. The figure shows that the limit cycle is established quickly. Also notice the parabolic shape of the trajectories for

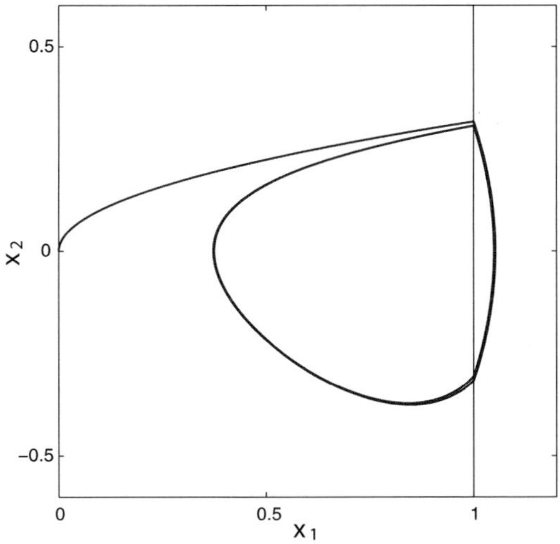

Fig. 7. Phase plane for the simulation in Figure 6

$x_1 > 1$. Exact conditions for limit cycles and their local stability can be computed using the method in [3].

6 Conclusions

Even if periodic sampling will remain the standard method to design computer controlled systems there are many problems where event based control is natural. Several examples of event based control were given in Section 2. In Section 3 we investigated a simple example where the event based and periodic control could be compared analytically. The example showed that the performance with event based control was superior to periodic sampling, and it also shown that that the control signal generation is important. A general architecture of an event based controller was discussed in Section 4. Key elements of the system are the event detector, the observer, and the control signal generator. Several ways to design the control signal generator was discussed. One method has the advantage that the event based system is equivalent to a nonlinear system. The results were illustrated with design of a controller for a simplified version of a MEMS accelerator. Even if event based systems have been used for a long time, the field is still at its infancy and there are many challenging theoretical problems that are not solved.

References

1. K.J. Åström. *On the choice of sampling rates in optimal linear systems.* Technical report rj-243, IBM Research, San José CA, 1963.
2. K.J. Åström. *Introduction to Stochastic Control Theory.* Academic Press, New York, 1970. Dover reprint published in 2006.
3. K.J. Åström. Oscillations in systems with relay feedback. In W. Miller, editor, *IMA Workshop on Adaptive Control.* Springer-Verlag, New York, 1993.
4. K.J. Åström and B. Bernhardsson. Comparison of periodic and event based sampling for first-order stochastic systems. *Proc. of the 14th IFAC World Congress,* pages 301–306, 1999.
5. K.J. Åström and B. Bernhardsson. Comparison of Riemann and Lebesque sampling for first order stochastic systems. *Proc. of the 41st IEEE Conf. on Decision and Contr.*, 9:2011–2016, 2002.
6. K.J. Åström and B. Wittenmark. *Computer-Controlled Systems.* Prentice Hall, 3rd edition, 1997. First published in 1984 edition.
7. J.P. Aubin. *Impulse differential inclusions and hybrid system: A viability approach.* Lecture notes, University of California, Berkeley, 1999.
8. A. Bensoussan and J.-L. Lions. *Impulse control and quasi-variational inequalities.* PhD thesis, Gauthier-Villars, Paris, 1984.
9. G. Bourdopoulos, A. Pnevmatikakis, V. Anastassopoulos, and T.L. Deliyannis. *Delta-Sigma Modulators: Modeling, Design and Applications.* Imperial College Press, London, 2003.

10. G.C. Buttazzo. *Hard Real-time Computing Systems: Predictable Scheduling Algorithms and Applications.* Kluwer, Amsterdam, 1997.
11. A. Cervin, D. Henriksson, B. Lincoln, J. Eker, and K.-E. Årzén. How does control timing affect performance? *IEEE Control Systems Magazine*, 23(3):16–30, 2003.
12. T. Chen and B.A. Francis. *Optimal Sampled-Data Control Systems.* Springer Verlag, New York, 1995.
13. C. Dase, J. Sullivan, and B. MacCleery. Motorcycle control prototyping using an FPGA-based embedded control system. *IEEE Control Systems Magazine*, 26(5):17–21, 2003.
14. S.S.P. DeWeerth, L. Nielsen, C.A. Mead, and K.J. Åström. A simple neuron servo. *IEEE Trans. Neural Networks*, 2(2):248–251, 1991.
15. M.D. DiBenedetto and A. Sangiovanni-Vincentelli. *Hybrid systems in computation and control.* Springer Verlag, New York, 2001.
16. S.J. Dodds. Adaptive, high precision, satellite attitude control for microprocessor implementation. *Automatica*, 17(4):563–573, 1981.
17. S.S. Draper, W. Wrigley, and J. Hovorka. *Inertial Guidance.* Pergamon Press, Oxford, 1960.
18. W. Feller. Diffusion processes in one dimension. *Trans. Am. Math. Soc.*, 55:1–31, 1954.
19. A. Feuer and G.C. Goodwin. Generalized sample hold functions – frequency domain analysis of robustness, sensitivity, and intersample behavior. *IEEE Trans. on Automat. Contr.*, 39(5):1042–1047, 1994.
20. I. Flügge-Lotz. *Discontinuous and Optimal Control.* McGraw-Hill, New York, 1968.
21. P.M. Frank. A continuous-time model for a PFM-controller. *IEEE Trans. on Automat. Contr.*, 25(5):782–784, 1979.
22. B. Friedland. Modeling systems for pulse-width modulated control. *IEEE Trans. on Automat. Contr.*, 21:739–746, 1976.
23. L. Guzzella and C. Onder. Past, present and future of automotive control. In M. C. Smith B. A. Francis and J. C. Willems, editors, *Control of Uncertain Systems: Modeling, Approximation and Design.* Springer-Verlag, Heidelberg, 2006.
24. E. Hendricks, M. Jensen, A. Chevalier, and T. Vesterholm. Problems in event based engine control. *Proc. of the 1994 Amer. Contr. Conf.*, 2:1585–1587, 1994.
25. T. Henningsson. *Logarithmic concave observers.* Master's thesis isrn lutfd2/tfrt--5747--se, Department of Automatic Control, Lund University, Sweden, 2005.
26. T. Henningsson and K.J. Åström. Log-concave observers. *Proc. of Mathematical Theory of Networks and Systems*, 2006.
27. R.K. Hobbie. *Intermediate Physics for Medicine and Biology.* Springer Verlag, 1997.
28. A. Isidori. *Nonlinear Control Systems.* Springer Verlag, New York, 3rd edition, 1995. First published 1985.
29. V Jacobson. Congestion avoidance and control. *Proc. of SIGCOMM '88*, pages 314–329, 1988.
30. M. Johansson. *Piecewise Linear Control Systems.* Springer Verlag, Heidelberg, 1999.
31. E.I. Jury. *Sampled-Data Control Systems.* John Wiley & Sons, New York, 1958.
32. P.T. Kabamba. Control of linear systems using generalized sampled-data hold functions. *IEEE Trans. on Automat. Contr.*, 32(9):772–783, 1987.

33. J. Keener and J. Sneyd. *Mathematical Physiology.* Springer Verlag, New York, 2001.
34. F.P. Kelley. Stochastic models of computer communication. *J. Royal Statistical Society*, B47(3):379–395, 1985.
35. M. Khammash, L. Oropeza-Ramos, and K.L. Turner. Robust feedback control design of an ultra-sensitive, high bandwidth tunneling accelerometer. *Proc. of the 2005 Amer. Contr. Conf.*, 6:4176–4180, 2005.
36. H. Kopetz. Should responsive systems be event triggered or time triggered? *IEICE Trans. on Information and Systems*, E76-D(10):1525–1532, 1993.
37. H. Kopetz. Time-triggered real-time systems. *Proc. of the 15th IFAC World Congress*, 2002.
38. W.H. Kwon, Y.H. Kim, S.J. Lee, and K.N. Paek. Event-based modeling and control for the burnthrough point in sintering control. *IEEE Trans. on Contr. Systems Tech.*, 7(1):31–41, 1999.
39. C.H. Liu and T.W. Kenny. A high-precision, wide-bandwidth micromachined tunneling accelerometer. *IEE/IEEE Journal of Microelectromechanical Systems*, 10(3):425–433, 2001.
40. C.A. Mead. *Analog VLSI and Neural Systems.* Addison-Wesley, Reading, Massachusetts, 1989.
41. J.E. Miller. The pulsed integrated pendulous accelerometer (pipa). In W. Miller, editor, *Space Navigation Guidance and Control.* Springer-Verlag, Technivision, Maidenhead, England, 1996. Agardograph 105.
42. S.R. Norsworthy, R. Schreier, and G. Temes. *Delta-Sigma Data Converters.* IEEE Press, New York, 1996.
43. M. Nougaret. A design method for first-order pulse-width-modulated systems. *Int. J. of Control*, 15(3):541–549, 1972.
44. K. Ohno, M. Hirata, and R. Horowitz. A comparative study of the use of the generalized hold function for HDDS. *Proc. of the 42nd IEEE Conf. on Decision and Contr.*, pages 5967–5972, 2003.
45. K. Ohno, M. Hirata, and R. Horowitz. A comparative study of the use of the generalized hold function for HDDS. *IEEE/ASME Transactions on Mechatronics*, 10(1):26–33, 2005.
46. B Pettersson. Production control of a complex integrated pulp and paper mill. *Tappi*, 52(11):2155–2159, 1969.
47. J. Pettersson, L. Ledung, and X. Zhang. Decision support for pulp mill operations based on large-scale optimization. *Control Systems*, 2006.
48. J.R. Ragazzini and G.F. Franklin. *Sampled-data Control Systems.* McGraw-Hill, New York, 1958.
49. H. Sira-Ramirez. A geometric approach to pulse-width modulated control in nonlinear dynamical systems. *IEEE Trans. on Automat. Contr.*, 34(2):184–187, 1989.
50. R.A. Skoog. On the stability of pulse-width-modulated feedback systems. *IEEE Trans. on Automat. Contr.*, 13(5):532–538, 1968.
51. A.R. Teel. A nonlinear small gain theorem for the analysis of control systems with saturation. *IEEE Trans. on Automat. Contr.*, 41(9):1256–1270, 1996.
52. Y.Z. Tsypkin. *Theory of Impulse Systems.* State Publisher for Physical and Mathematical Literature, Moscow, 1958.
53. Y.Z. Tsypkin. *Relay Control Systems.* Cambridge University Press, Cambridge, UK, 1984.

54. V.I. Utkin. Discontinuous control systems: State of the art in theory and applications. *Proc. of the 10th IFAC World Congress*, 1987.
55. H.R. Wilson. *Spikes, Decisions, and Actions: The Dynamical Foundations of Neuroscience.* Oxford University Press, 1999.
56. Y. Yamamoto. A retrospective view on sampled-data control systems. *CWI Quarterly*, 9(3):261–276, 1996.

Zero Dynamics and Tracking Performance Limits in Nonlinear Feedback Systems

A. Pedro Aguiar[1]*, João P. Hespanha[2], and Petar V. Kokotović[2]

[1] Dept. Electrical Eng. and Computers and the Institute for Systems and Robotics, Instituto Superior Técnico, Torre Norte 8, Av. Rovisco Pais, 1049-001 Lisbon, Portugal.

[2] Center for Control Engineering and Computation, University of California, Santa Barbara, CA 93106-9560, USA,

Summary. Among Alberto Isidori's many seminal contributions, his solution of the nonlinear tracking problem and the underlying concept of zero dynamics have had the widest and strongest impact. Here we use these results to investigate and quantify the limit to tracking performance posed by unstable zero dynamics. While some aspects of this limit are nonlinear analogs of Bode's T-integral formula, the dependence on the exosystem dynamics is an added complexity of nonlinear tracking.

1 Introduction

The concept of nonlinear *zero dynamics* is now firmly placed at the foundation of control theory. Its ability to reveal input-output properties and feedback limitations continues to stimulate many researchers to gain deeper insights and develop new design methods.

The foundational and pioneering role of Alberto Isidori in this area is well known. Some quarter of a century ago he captured the attention of most of active researchers in the field by the astonishing novelty of his ideas and brilliant clarity with which he presented them. Any one of his papers and talks was enough to convert his listeners to his way of thinking and motivate them to ask for his preprints and notes, which he most generously shared with his colleagues.

We use this opportunity to express our gratitude to Alberto Isidori and illustrate how his concepts of non-minimum phase nonlinear systems influenced our research. Isidori and coworkers introduced this concept in the 1980's, within the broader framework of input-output linearization theory [16, 11, 12, 15, 17, 13].

* The first author was supported by the FCT-ISR/IST pluriannual funding program through the POS_C Program that includes FEDER funds.

Difficulties with non-minimum phase linear systems have been known in classical feedback theory for many decades, especially in tracking and disturbance rejection problems. For linear transfer functions Bode has characterized the limitations of feedback connections via his seminal results on integral invariants in frequency domain. Bode showed that integrals of logarithmic sensitivities are constrained by unstable poles and zeros. Over the last several decades Bode-like results have been obtained for wider classes of linear time-invariant systems (see [21, 9, 22, 23, 4] and references therein).

The frequency domain form of Bode integrals makes it unclear whether such constraints apply to nonlinear systems. Occasionally one would even hear conjectures that, by introducing nonlinearities in the controller, Bode's constraints may be avoided.

A few years ago we approached the nonlinear feedback limitations problem using Isidori's work on nonlinear zero dynamics and normal forms. For this we first had to give a state space interpretation of Bode's integrals. We focused on the T-integral and followed a path which started with the 1972 "cheap control" result of Kwakernaak and Sivan [21]. This led us through the singular perturbation analysis [18, 26] to the explicit formulas of Qiu and Davison [23]. For a special case a nonlinear analog of the Bode T-integral was obtained by Seron et al. [24] while our general result is presented in [3]. The purpose of this text is to present a brief review on this line of research.

2 Bode T-Integral and Cheap Control

For single input-single output linear time invariant systems, the best attainable tracking performance is constrained by Bode's T-integral

$$\frac{1}{\pi}\int_0^{\infty} \log|T(j\omega)|\frac{d\omega}{\omega^2} + \frac{1}{2K_v} = \sum_{i=1}^{p}\frac{1}{\alpha_i}$$

with $T = GK(1+GK)^{-1}$ where G is the plant, K is a minimum phase controller, K_v is the velocity constant and $\alpha_1, \ldots, \alpha_p$ are the unstable zeros. Clearly, perfect tracking of a reference input that would result from $T(j\omega) = 1$ for all ω, is impossible in the presence of unstable zeros of plant G.

With a singular perturbation time scale decomposition of the cheap control tracking problem into the slow minimum energy stabilization of the zero dynamics and a rapid output regulation, Seron et al. [24] showed that the Bode T-invariant is, in fact, the minimum amount of output energy needed to stabilize the zero dynamics. This insight is gained from the linear normal form in which the output is the input into the zero dynamics subsystem and must be used for its stabilization. The amount of energy the output needs to stabilize the unstable zeros is therefore not available for tracking and appears as the energy of the tracking error which remains nonzero even when the gain is allowed to tend to infinity. In other words

$$\frac{1}{\pi}\int_0^\infty \log|T(j\omega)|\frac{d\omega}{\omega^2} + \frac{1}{2K_v} = \lim_{\epsilon\to 0}\frac{1}{2}\int_0^\infty e^2(t)\,dt.$$

Seron et al. have shown that this interpretation of Bode T-constraint applies to nonlinear systems having a normal form in which the system output is the sole input into the zero dynamics subsystem and plays the role of the stabilizing control in the corresponding nonlinear minimum energy problem.

The results and expressions summarized in the above discussion are for the tracking of a step input. To prepare for more general nonlinear results discussed in the next section, we briefly review the tracking problem for linear systems

$$\dot{x} = A\,x + Bu, \quad y = Cx + Du,$$

$x \in \mathbb{R}^n$, $u \in \mathbb{R}^m$, $y \in \mathbb{R}^q$, and reference signals $r(t) \in \mathbb{R}^q$ generated by a known *exosystem*

$$\dot{w} = Sw, \quad r = Qw.$$

Davison [7] and Francis [8] have shown that it is possible to design a feedback controller such that the closed-loop system is asymptotically stable and the output $y(t)$ converges to $r(t)$, if and only if (A, B) is stabilizable, (C, A) is detectable, the number of inputs is at least as large as the number of outputs $(m \geq q)$, and the zeros of (A, B, C, D) do not coincide with the eigenvalues of S. The *internal model approach*, [10, 8], designs the tracking controller

$$u(t) = Kx(t) + (\Gamma - K\Pi)w(t),$$

where $A + BK$ is Hurwitz, and Π and Γ satisfy

$$\begin{aligned} \Pi S &= A\Pi + B\Gamma, \\ 0 &= C\Pi + D\Gamma - Q. \end{aligned}$$

Then, the *tracking error* $e(t) := y(t) - r(t)$ converges to zero, and the transients

$$\tilde{x} := x - \Pi w, \quad \tilde{u} := u - \Gamma w$$

are governed by $\dot{\tilde{x}} = (A + BK)\tilde{x}$, $\tilde{u} = K\tilde{x}$.

Kwakernaak and Sivan [21] were the first to consider the cheap control problem

$$J_\epsilon := \min_{\tilde{u}} \int_0^\infty \left[\|y(t) - r(t)\|^2 + \epsilon^2\|\tilde{u}(t)\|^2\right] dt$$

and to demonstrate that in the presence of non-minimum phase zeros dynamics the limit $J_\epsilon \to J$ as $\epsilon \to 0$ is strictly positive.

Qiu and Davison [23] showed that for $r(t) = \eta_1 \sin \omega t + \eta_2 \cos \omega t$, $\eta = \mathrm{col}(\eta_1, \eta_2)$, the non-minimum phase zeros $z_1, z_2, \ldots, z_p$ determine the limit J as follows:

$$J = \eta' M \eta, \quad \operatorname{trace} M = \sum_{i=1}^{p} \left(\frac{1}{z_i - j\omega} + \frac{1}{z_i + j\omega} \right).$$

For more general reference signals, Su, Qiu, and Chen [25] give explicit formulas which show the dependence of J on the non-minimum phase zeros and their frequency-dependent directional information.

3 Performance Limits in Nonlinear Feedback Systems

The analogous nonlinear tracking problem

$$\dot{x} = f(x,u), \quad y = h(x,u), \tag{1}$$
$$\dot{w} = s(w), \qquad r = q(w), \tag{2}$$

where $f(0,0) = 0$, $s(0) = 0$, $h(0,0) = 0$, has been analyzed by Isidori and Byrnes [14]. They proved that this problem is solvable if and only if there exist smooth maps $\Pi(w)$ and $c(w)$, satisfying

$$\frac{\partial \Pi}{\partial w} s(w) = f(\Pi(w), c(w)), \quad \Pi(0) = 0, \tag{3a}$$
$$h(\Pi(w), c(w)) - q(w) = 0, \quad c(0) = 0. \tag{3b}$$

In [3] we consider the class of nonlinear systems which are locally diffeomorphic to systems in strict-feedback form (see for example [20, Appendix G])[3]:

$$\dot{z} = f_0(z) + g_0(z)\xi_1, \tag{4a}$$
$$\dot{\xi}_1 = f_1(z,\xi_1) + g_1(z,\xi_1)\xi_2,$$
$$\vdots$$
$$\dot{\xi}_{r_d} = f_{r_d}(z,\xi_1,\ldots,\xi_{r_d}) + g_{r_d}(z,\xi_1,\ldots,\xi_{r_d})u, \tag{4b}$$
$$y = \xi_1, \tag{4c}$$

where $z \in \mathbb{R}^{n_z}$, $\xi := \mathrm{col}(\xi_1, \ldots, \xi_{r_d})$, $\xi_i \in \mathbb{R}^m$, $\forall i \in \{1, \ldots, r_d\}$, $u \in \mathbb{R}^m$, and $y \in \mathbb{R}^m$. $f_i(\cdot)$ and $g_i(\cdot)$ are $\mathcal{C}^k$ functions of their arguments (for some large k), $f_i(0,\ldots,0) = 0$, and the matrices $g_i(\cdot)$, $\forall i \in \{1,\ldots,r_d\}$ are always nonsingular. We assume that initially the system is at rest, $(z,\xi) = (0,0)$.

[3] When convenient we use the compact form (1) for (4). In that case, $f(\cdot)$ denotes the vector field described by the right-hand-side of (4a)–(4b), $h(\cdot)$ the output map described by (4c), and $x = \mathrm{col}(z, \xi_1, \ldots, \xi_{r_d})$.

When the tracking problem is solvable, that is, when it is possible to design a continuous feedback law that drives the tracking error to zero, there exist maps $\Pi = \mathrm{col}(\Pi_0, \ldots, \Pi_{r_d})$, $\Pi_0 : \mathbb{R}^p \to \mathbb{R}^{n_z}$, $\Pi_i : \mathbb{R}^p \to \mathbb{R}^m$, $\forall i \in \{1, \ldots, r_d\}$, and $c : \mathbb{R}^p \to \mathbb{R}^m$ that satisfy (3). The locally diffeomorphic change of coordinates

$$\tilde{z} = z - \Pi_0(w), \tag{5a}$$

$$\tilde{\xi} := \mathrm{col}(\tilde{\xi}_1, \ldots, \tilde{\xi}_{r_d}), \tag{5b}$$

$$\tilde{\xi}_i = \xi_i - \Pi_i(w), \quad i = 1, \ldots, r_d \tag{5c}$$

$$\tilde{u} = u - c(w), \tag{5d}$$

transforms the system (4) into the *error system*

$$\begin{aligned}
\dot{\tilde{z}} &= \tilde{f}_0(\tilde{z}, w) + \tilde{g}_0(\tilde{z}, w)e, \\
\dot{\tilde{\xi}}_1 &= \tilde{f}_1(\tilde{z}, \tilde{\xi}_1, w) + \tilde{g}_1(\tilde{z}, \tilde{\xi}_1, w)\tilde{\xi}_2, \\
&\vdots \\
\dot{\tilde{\xi}}_{r_d} &= \tilde{f}_{r_d}(\tilde{z}, \tilde{\xi}_1, \ldots, \tilde{\xi}_{r_d}, w) + \tilde{g}_{r_d}(\tilde{z}, \tilde{\xi}_1, \ldots, \tilde{\xi}_{r_d}, w)\tilde{u}, \\
e &= \tilde{\xi}_1,
\end{aligned} \tag{6}$$

where

$$\begin{aligned}
\tilde{f}_0 &:= f_0(\tilde{z} + \Pi_0(w)) - f_0(\Pi_0(w)) + \Big[g_0(\tilde{z} + \Pi_0(w)) - g_0(\Pi_0(w))\Big]q(w), \\
\tilde{g}_0 &:= g_0(\tilde{z} + \Pi_0(w)),
\end{aligned}$$

$\tilde{f}_0(0, w) = 0$, $\tilde{g}_0(\tilde{z}, 0) = g_0(\tilde{z})$, and $\tilde{f}_i(\,\cdot\,)$, $\tilde{g}_i(\,\cdot\,)$, $\forall i \in \{1, \ldots, r_d\}$ are appropriately defined functions that satisfy $\tilde{f}_i(0, \ldots, 0, w) = 0$ and $\tilde{g}_i(\tilde{z}, \ldots, \tilde{\xi}_i, 0) = g_i(\tilde{z}, \ldots, \tilde{\xi}_i)$.

As in the work of Seron et al. [24], the singular perturbation separation of time scales gives rise to the following two optimal control problems.

Cheap control problem: For the system consisting of the error system (6) and the exosystem (2) with initial condition $\big(\tilde{z}(0), \tilde{\xi}(0), w(0)\big) = \big(\tilde{z}_0, \tilde{\xi}_0, w_0\big)$, find the optimal feedback law $\tilde{u} = \alpha^{cc}_{\delta,\epsilon}(\tilde{z}, \tilde{\xi}, w)$ that minimizes the cost functional

$$\frac{1}{2}\int_0^\infty \big(\|e(t)\|^2 + \delta\|\tilde{z}(t)\|^2 + \epsilon^{2r_d}\|\tilde{u}(t)\|^2\big)\, dt$$

for $\delta > 0$, $\epsilon > 0$. We denote by $J^{cc}_{\delta,\epsilon}(\tilde{z}_0, \tilde{\xi}_0, w_0)$ the corresponding optimal value. The best-attainable cheap control performance for tracking is then

$$J := \lim_{(\delta,\epsilon)\to 0} J^{cc}_{\delta,\epsilon}(\tilde{z}_0, \tilde{\xi}_0, w_0).$$

As shown by Krener [19], in some neighborhood of $(0, 0, 0)$ and for every $\delta > 0$, $\epsilon > 0$, the value $J^{cc}_{\delta,\epsilon}(\,\cdot\,, \cdot\,, \cdot\,)$ is $\mathcal{C}^{k-2}$ under the following assumption.

Assumption 1. *The linearization around* $(z, \xi) = (0, 0)$ *of system* (4) *is stabilizable and detectable, and the linearization around* $w = 0$ *of the exosystem* (2) *is stable.*

The fast part of the cheap control problem describes the rapid transient of $e(t)$ to its slow part represented by the minimum energy problem for the stabilization of zero dynamics.

Minimum-energy problem: For the system

$$\dot{\tilde{z}} = \tilde{f}_0(\tilde{z}, w) + \tilde{g}_0(\tilde{z}, w)e, \qquad \tilde{z}(0) = z_0, \tag{7a}$$

$$\dot{w} = s(w), \qquad w(0) = w_0, \tag{7b}$$

with e viewed as the input, find the optimal feedback law $e = \alpha_\delta^{me}(\tilde{z}, w)$ that minimizes the cost

$$\frac{1}{2}\int_0^\infty \left(\delta\|\tilde{z}(t)\|^2 + \|e(t)\|^2\right) dt,$$

for $\delta > 0$. We denote by $J_\delta^{me}(\tilde{z}_0, w_0)$ the corresponding optimal value. Under Assumption 1, $J_\delta^{me}(\,\cdot\,,\,\cdot\,)$ is $\mathcal{C}^{k-2}$ in some neighborhood of $(0, 0)$.

Our analysis reveals that the best-attainable cheap control performance J is equal to the least control effort (as $\delta \to 0$) needed to stabilize the corresponding zero dynamics system (7) driven by the tracking error e.

Theorem 1. *Suppose that Assumption 1 holds and that* (3) *has a solution in some neighborhood of* $w = 0$. *Then, for any* $(\tilde{z}(0), \tilde{\xi}(0), w(0)) = (\tilde{z}_0, \tilde{\xi}_0, w_0)$ *in some neighborhood of* $(0, 0, 0)$ *there exists a solution to the cheap control problem and the limit to tracking performance is*

$$J = \lim_{\delta \to 0} J_\delta^{me}.$$

A more detailed analysis leading to this theorem and its proof are soon to appear in [3].

For linear systems we obtain the following:

Corollary 1. *For linear systems with unstable zero-dynamics subsystem described by*

$$\dot{z} = F_0 z + G_0 y,$$

the limit to tracking performance is

$$J = \lim_{\delta \to 0} \frac{1}{2}\omega_0' \Pi_0' P_0(\delta) \Pi_0 \omega_0, \tag{8}$$

where $\omega_0 = \omega(0)$, *and* Π_0 *and* $P_0 > 0$ *satisfy*

$$\Pi_0 S = F_0 \Pi_0 + G_0 Q, \tag{9a}$$

$$F_0' P_0 + P_0 F_0 + \delta I = P_0 G_0 G_0' P_0. \tag{9b}$$

Formula (8) follows from the fact that the equations for the minimum-energy problem are

$$\dot{\tilde{z}} = F_0\tilde{z} + G_0 e,$$
$$\dot{w} = Sw$$

where $\tilde{z} = z - \Pi_0\omega$, and Π_0 is the solution of (9a). In this case the optimal feedback law for the minimum energy problem is $e = -G_0' P_0 \tilde{z}$ where $P_0 > 0$ is the solution of (9b), and $\frac{1}{2}\tilde{z}_0' P_0(\delta)\tilde{z}_0$ the corresponding optimal value. Note that $\tilde{z}_0 = z(0) - \Pi_0\omega(0) = -\Pi_0\omega_0$.

4 Illustrative Example

To illustrate the above results and show how the limits of tracking performance for nonlinear systems depend on the exosystem dynamics, we consider the following system

$$\dot{z} = -z + z^2 + \xi_1, \tag{10a}$$
$$\dot{\xi}_1 = \xi_2, \tag{10b}$$
$$\dot{\xi}_2 = u, \tag{10c}$$
$$y = \xi_1,$$

which is already in normal form. The zero-dynamics subsystem given by (10a) with $\xi_1 \equiv 0$ has an asymptotically stable equilibrium at $z = 0$. Suppose that the tracking task is to asymptotically track any reference $r(t)$ generated by the exosystem

$$\dot{\omega}_1 = a\omega_2, \tag{11a}$$
$$\dot{\omega}_2 = -a\omega_1, \tag{11b}$$
$$r(t) = q(\omega_1, \omega_2), \tag{11c}$$

where $a > 0$ and $q(\omega_1, \omega_2)$ is to be chosen later. When the maps $\Pi(w)$ and $c(w)$ satisfying (3) exist, we apply (5) and obtain the error system

$$\dot{\tilde{z}} = \big(2\Pi_0(\omega) - 1\big)\tilde{z} + \tilde{z}^2 + e, \tag{12a}$$
$$\dot{\tilde{\xi}}_1 = \tilde{\xi}_2, \tag{12b}$$
$$\dot{\tilde{\xi}}_2 = \tilde{u}, \tag{12c}$$
$$e = \tilde{\xi}_1.$$

The zero-dynamics of the error system are governed by (12a) with $e \equiv 0$ and (11a)–(11b). Clearly, the stability of the zero dynamics and hence

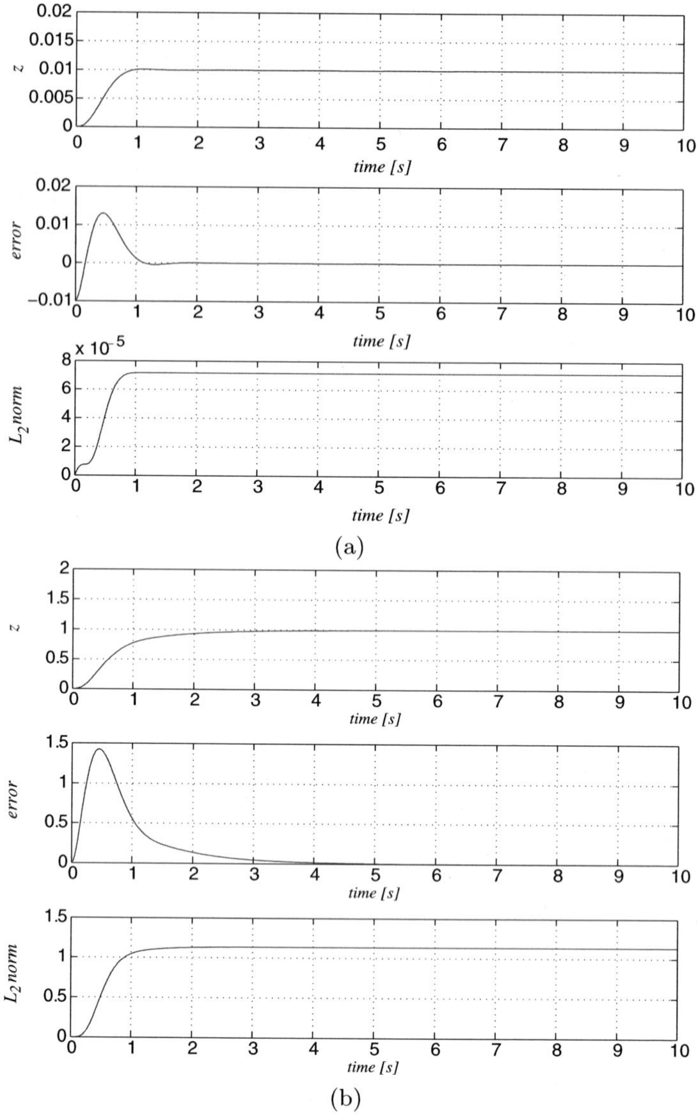

Fig. 1. State z, tracking error e, and $\int_0^t \|e(\tau)\|^2 \, d\tau$ for (a) $\omega(0) = (0.1, 0)'$ and (b) $\omega(0) = (1, 0)'$ (Note the 10^{-5} scale factor!)

the limits of tracking performance depend on the exosystem. In particular, the zero-dynamics are unstable for $2\Pi_0(\omega) > 1$. To illustrate this we let

$$\Pi_0(\omega_1, \omega_2) = \omega_1^2 + \omega_2^2$$

and then evaluate the corresponding $q(\omega_1, \omega_2)$ from

$$\frac{\partial \Pi_0}{\partial \omega_1} a\omega_2 - \frac{\partial \Pi_0}{\partial \omega_2} a\omega_1 = -\Pi_0(\omega_1, \omega_2) + \Pi_0(\omega_1, \omega_2)^2 + q(\omega_1, \omega_2)$$

as dictated by (3). We use this $q(\omega_1, \omega_2)$ in (11c) and perform a series of simulations. To compare the transient errors with different initial conditions $\omega(0) = \omega_0$, we define the normalized transient error $\bar{J} := \frac{J}{\|\omega_0\|^2}$.

Figure 1(a) displays the simulation results obtained with $a = 1\, rad/s$ and using a feedback law of the form $u = c(\omega) + k_0(z - \Pi_0(\omega)) + k_1(\xi_1 - \Pi_1(\omega)) + k_2(\xi_2 - \Pi_2(\omega))$. The initial conditions are $(z, \xi_1, \xi_2) = 0$, $\omega(0) = (0.1, 0)'$. In this case $\Pi_0(\omega_1, \omega_2) = 0.01$, so that the subsystem (12a) is locally input-to-state stable and the convergence to the desired reference signal is achieved with a negligibly small transient error $J \simeq 7.2 \times 10^{-5}$ and $\bar{J} \simeq 7.2 \times 10^{-3}$.

In contrast, Figure 1(b) shows the simulation results obtained with the same controller but with initial condition $\omega(0) = (1, 0)'$ which implies that $\Pi_0(\omega_1, \omega_2) = 1$ and, hence, the error zero-dynamics are not input-to-state stable. As it can be seen, the transient error and its normalized version have increased by several orders of magnitude to $J = \bar{J} \simeq 1.1$.

5 Concluding Remarks

We have shown that, analogous to the tracking problem for linear time-invariant non-minimum phase systems, the tracking performance for nonlinear non-minimum phase systems cannot be improved beyond a limit determined by the least amount of energy required to stabilize the zero dynamics of the tracking error system. In the nonlinear problem, these zero dynamics depend on the dynamics of the exosystem, which may destabilize them for some reference signals, as illustrated on an example.

Since non-minimum phase phenomena create a fundamental limit to the tracking performance that can not be removed by any controller redesign, a direction of practical interest would be to search for reformulations of the tracking problem that would be free of limitations, but still meaningful for applications. One such reformulation, pursued in our work [1, 6, 2, 5] is to replace the tracking problem by a less demanding path following problem, in which the speed along the prescribed geometric path is used as a free design parameter. As shown in [3], for a class of path following problems the limitations of the tracking problems can be avoided.

References

1. A.P. Aguiar, D.B. Dačić, J.P. Hespanha, and P.V. Kokotović. Path-following or reference-tracking? An answer relaxing the limits to performance. In *Proc. of IAV2004 - 5th IFAC/EURON Symp. on Intel. Auton. Vehicles*, Lisbon, Portugal, 2004.

2. A.P. Aguiar, J.P. Hespanha, and P.V. Kokotović. Path-following for nonminimum phase systems removes performance limitations. *IEEE Trans. on Automat. Contr.*, 50(2):234–239, 2005.
3. A.P. Aguiar, J.P. Hespanha, and P.V. Kokotović. Performance limitations in reference-tracking and path-following for nonlinear systems. *Automatica*, 2007. In press.
4. J. Chen, L. Qiu, and O. Toker. Limitations on maximal tracking accuracy. *IEEE Trans. on Automat. Contr.*, 45(2):326–331, 2000.
5. D.B. Dačić and P.V. Kokotović. Path-following for linear systems with unstable zero dynamics. *Automatica*, 42(10):1673–1683, 2006.
6. D.B. Dačić, M.V. Subbotin, and P.V. Kokotović. Path-following for a class of nonlinear systems with unstable zero dynamics. In *Proc. of the 43rd IEEE Conf. on Decision and Contr.*, Paradise Island, Bahamas, 2004.
7. E.J. Davison. The robust control of a servomechanism problem for linear time-invariant multivariable systems. *IEEE Trans. on Automat. Contr.*, 21(1):25–34, 1976.
8. B.A. Francis. The linear multivariable regulator problem. *SIAM J. Contr. Optimization*, 15(3):486–505, 1977.
9. B.A. Francis. The optimal linear-quadratic time-invariant regulator with cheap control. *IEEE Trans. on Automat. Contr.*, 24(4):616–621, 1979.
10. B.A. Francis and W.M. Wonham. The internal model principle of control theory. *Automatica*, 12(5):457–465, 1976.
11. A. Isidori. The matching of a prescribed linear input-output behavior in a nonlinear system. *IEEE Trans. on Automat. Contr.*, 30(3):258–265, 1985.
12. A. Isidori. *Nonlinear control systems: An introduction.* Springer-Verlag, New York, NY, 1985.
13. A. Isidori. *Nonlinear Control Systems.* Communications and Control Engineering Series. Springer-Verlag, Berlin, 3rd edition, 1995.
14. A. Isidori and C.I. Byrnes. Output regulation of nonlinear systems. *IEEE Trans. on Automat. Contr.*, 35(2):131–140, 1990.
15. A. Isidori and J.W. Grizzle. Fixed modes and nonlinear noninteracting control with stability. *IEEE Trans. on Automat. Contr.*, 33(10):907–914, 1988.
16. A. Isidori, A.J. Krener, C. Gori-Giorgi, and S. Monaco. Nonlinear decoupling via feedback: A differential geometric approach. *IEEE Trans. on Automat. Contr.*, 26(2):331–345, 1981.
17. A. Isidori and C. Moog. On the nonlinear equivalent of the notion of transmission zeros. In Byrnes C. and A. Kurzhanski, editors, *Modelling and Adaptive Control*, Lecture notes in information and control, pages 146–157. Springer-Verlag, Berlin, 1988.
18. A. Jameson and R.E. O'Malley. Cheap control of the time-invariant regulator. *Appl. Math. Optim.*, 1(4):337–354, 1975.
19. A.J. Krener. The local solvability of a Hamilton-Jacobi-Bellman PDE around a nonhyperbolic critical point. *SIAM J. Contr. Optimization*, 39(5):1461–1484, 2001.
20. M. Krstić, I. Kanellakopoulos, and P.V. Kokotović. *Nonlinear and Adaptive Control Design.* John Wiley & Sons, Inc., New York, USA, 1995.
21. H. Kwakernaak and R. Sivan. The maximal achievable accuracy of linear optimal regulators and linear optimal filters. *IEEE Trans. on Automat. Contr.*, 17(1):79–86, 1972.

22. R.H. Middleton. Trade-offs in linear control systems design. *Automatica*, 27(2):281–292, 1991.
23. L. Qiu and E.J. Davison. Performance limitations of nonminimum phase systems in the servomechanism problem. *Automatica*, 29:337–349, 1993.
24. M.M. Seron, J. H. Braslavsky, P.V. Kokotović, and D.Q. Maync. Fccdback limitations in nonlinear systems: From Bode integrals to cheap control. *IEEE Trans. on Automat. Contr.*, 44(4):829–833, 1999.
25. W. Su, L. Qiu, and J. Chen. Fundamental performance limitations in tracking sinusoidal signals. *IEEE Trans. on Automat. Contr.*, 48(8):1371–1380, 2003.
26. K. Young, P.V. Kokotović, and V. Utkin. A singular perturbation analysis of high-gain feedback systems. *IEEE Trans. on Automat. Contr.*, 22:931–938, 1977.

A Nonlinear Model for Combustion Instability: Analysis and Quenching of the Oscillations

Ioan D. Landau[1], Fethi Bouziani[1], and Robert R. Bitmead[2]

[1] Laboratoire d'Automatique de Grenoble, ENSIEG BP 46, 38402 Saint-Martin d'Hères, France
[2] Department of Mechanical & Aerospace Engineering, University of California, San Diego, La Jolla CA 92093-0411, USA

Summary. It is a great pleasure for us to contribute to this book dedicated to Alberto Isidori on the occasion of his sixty-fifth birthday. It is also, for the first author, the occasion to acknowledge a very long period of useful and pleasant exchange which started in 1973 (bilinear systems) and has continued through the years on various specific subjects.

The important contributions of Alberto Isidori to the control of nonlinear systems have had a tremendous impact in the control community. Feedback linearization was one of the important subjects developed by Alberto. While the oscillatory nonlinear system considered in this contribution requires specific techniques for its analysis, still a feedback linearization is used for quenching the oscillations.

1 Introduction

Combustion instability phenomena in gas-fueled turbomachinery are the focus of several investigations starting from more than 200 years ago [20]. Recent important research programs oriented toward this topic are currently conducted in some countries with industrial collaboration (*inter alia* USA, France and UK). The papers [16, 17, 8, 1] give an overview of this activity. These phenomena are extremely complex and hard to predict, but in most cases can be explained by an unsteady flame generating pressure waves which are reflected by physical boundaries back into the combustion process. In terms of system interpretation, this corresponds to a positive feedback coupling between the heat-release process and the acoustics of the combustion chamber, see Figure 1. The positive feedback coupling leads to the creation of nonlinear limit cycles. Generally, these phenomena occur only in confined environments. The Rijke tube [7] which consists of a flame placed inside an open-ended cylinder (Figure 2), is one simple demonstration that shows that interactions leading to instabilities can occur in even simple con-

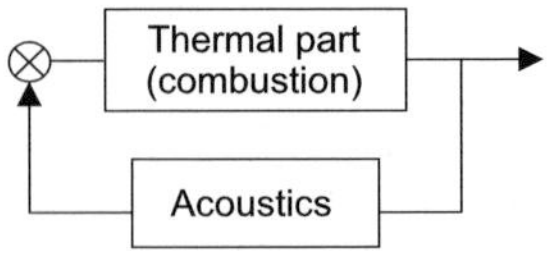

Fig. 1. Positive feedback coupling between the heat-release process and acoustics

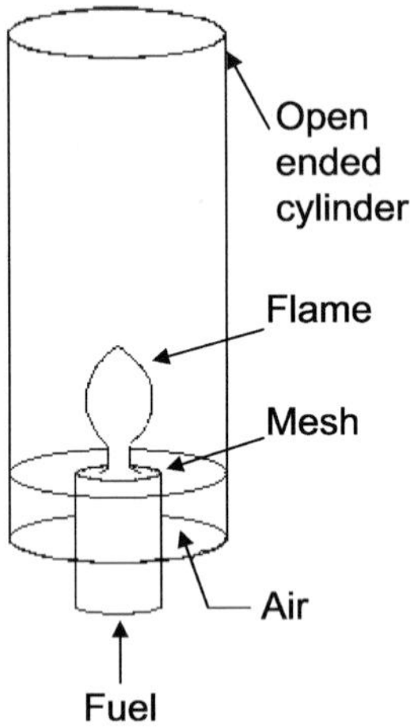

Fig. 2. Rijke tube

fined environments, where the combustion process and the acoustics interact.

The heart of the issue is that combustion at low equivalence ratio (fuel-to-air ratio) is desirable for reducing pollution. This occurs because low-equivalence-ratio combustion occurs at a low temperature, which in turn causes low levels of creation of Nitrogen-Oxygen compounds, NO_x. Unfortunately, as the equivalence ratio decreases the instability appears and manifests itself through strong self-sustained oscillations, which can be sometimes characterized by the coexistence of oscillations at several distinct non-harmonic frequencies. With the appearance of these oscillations, the flame front becomes spatially unstable and the combustion is alternatively hot and then cold; the pollution benefits disappear entirely.

From a practical perspective, the modulation of a fuel flow fraction into the combustor is possible as a control input to ameliorate the instability. This control action is of a multiplicative type. Such modulation has been widely tested in experiments as a candidate for active control for suppressing the combustion instability [19, 2]. In the literature, several active control methods have been proposed; an excellent overview of existing methods is given in [1]. Systematic design and implementation of active control requires a realistic low order model which exhibits the dominant dynamical effects. The parsimonious modelling of such nonlinear systems is an extremely difficult task, given that high-order, physics-based computational fluid dynamics codes can be unreliable in capturing the phenomenon well. One particular feature which any model should capture is the simultaneous coexistence of two non-harmonic oscillating modes in the uncontrolled signals [18, 9].

Establishing a low order model requires good knowledge of the physical phenomenon at the same time as the availability of mathematical tools allowing the analysis of nonlinear oscillating systems. The models presented in the literature are established according to various approaches. Some models are developed by a purely theoretical and phenomenological approach starting from physical equations of the combustion process, while others proceed by a purely experimental approach. However, there exist models which are established on a compromise between the theoretical approach and the experimental approach, so-called *gray-box* models. Among these models, that of Dunstan and Bitmead [10] provides a good base for understanding the instability mechanism. Unfortunately, in the absence of analytical tools this model loses much of its utility. For such a system, the first requirement of model validation can be seen as the association between this existing and appropriate powerful analysis methods. The Krylov-Bogoliubov (K-B) method (detailed in [3, 11, 13, 14, 15]) is such an analytical method capable of accommodating nonlinear oscillating systems. From our knowledge, this work is the first application of the K-B method to the analysis of combustion instabilities models. The effectiveness of this method combined with a relevant model may be considered as a major step forward in combustion instability modelling, since an approximate but physically reasonable model is attached to a cognate analysis approach.

The model presented in this chapter, inspired from [10], is an analytically tractable model for combustion instabilities. This model will be analyzed using K-B method and employed for studying the possibility of quenching of the oscillations in combustion instabilities by multiplicative control.

The chapter is organized as follows. In Section 2, a very brief summary of the first K-B approximation method for autonomous multi-resonator systems is presented. To illustrate the efficacy of the K-B approximation and of multiplicative control, a generalized van der Pol equation is developed as an oversimplified but candidate model for combustion instabilities in Section3. In Section 4, a multiple-resonator model with delay in the feedback path is presented and analyzed using the K-B method. This model is employed in Section 5 for studying the possibility of quenching the oscillations in combustion instabilities by multiplicative feedback control. Conclusions, remarks and directions for further work are given in Section 6.

2 First K-B Approximation for Autonomous Multi-Resonator Systems

Consider a system with n resonators described by differential equations of the form,

$$\frac{d^2x_j}{dt^2} + \omega_j^2 x_j = \epsilon f_j\left(x, \frac{dx}{dt}\right), \quad (j = 1, 2, \ldots, n), \tag{1}$$

where $x = \{x_1, \ldots, x_n\}$, $\frac{dx}{dt} = \{\frac{dx_1}{dt}, \ldots, \frac{dx_n}{dt}\}$ and ϵ is a small parameter. For the jth resonator, the first K-B approximation (for more details see Chapter 2 of [14]) proposes the solution

$$x_j = a_j \cos(\psi_j), \tag{2}$$

where $\psi_j = \omega_j t + \theta_j$, a_j and θ_j are slowly time-varying functions obeying the equations

$$\begin{cases} \frac{da_j}{dt} = -\frac{\epsilon}{2\omega_j} H_{jj}(a_1, \ldots, a_n, \theta_1, \ldots, \theta_n), \\ \frac{d\theta_j}{dt} = -\frac{\epsilon}{2\omega_j a_j} G_{jj}(a_1, \ldots, a_n, \theta_1, \ldots, \theta_n), \end{cases} \tag{3}$$

with H_{jj} and G_{jj} obtained from the function $f_j\left(x, \frac{dx}{dt}\right)$ by substituting

$$\begin{cases} x_k = a_k \cos(\omega_k t + \theta_k), \\ \frac{dx_k}{dt} = -a_k \omega_k \sin(\omega_k t + \theta_k), \end{cases} \quad (k = 1, 2, \ldots, n) \tag{4}$$

and by setting it in the form

$$\begin{aligned} & f_j\left(a_1 \cos(\omega_1 t + \theta_1), \ldots, a_n \cos(\omega_n t + \theta_n),\right. \\ & \quad \left. -a_1\omega_1 \sin(\omega_1 t + \theta_1), \ldots, -a_n\omega_n \sin(\omega_n t + \theta_n)\right) \\ & = H_{jj} \sin(\omega_j t + \theta_j) + G_{jj} \cos(\omega_j t + \theta_j) \\ & \quad + \sum_{\omega_j \not\approx \omega_\ell}^{r} \left(H_{\ell j} \sin(\omega_\ell t + \theta_\ell) + G_{\ell j} \cos(\omega_\ell t + \theta_\ell)\right), \end{aligned} \tag{5}$$

where ω_ℓ and θ_ℓ are integer linear combinations of $\omega_1, \ldots, \omega_n$ and $\theta_1, \ldots, \theta_n$, respectively, and r is the number of possible integer linear combinations of $\omega_1, \ldots, \omega_n$ different from ω_j. For x_j the coefficients of the fundamental term in (5) are used and the all other terms are eliminated.

3 Simple Case Study: Generalized Van der Pol Equation

In this section a reduced-order model for combustion instabilities with one single resonator (corresponding to a generalized van der Pol equation) will be considered to illustrate the potential effectiveness of the K-B method for analysis and of closed-loop multiplicative control for quenching the oscillations.

3.1 Model Analysis

Consider the van der Pol equation in generalized form (Figure 3).

$$\ddot{x} + \omega^2 x = \frac{d}{dt}\left\{\varphi_{v0} + \varphi_{v1}x - \frac{\varphi_{v3}}{3}x^3\right\}, \tag{6}$$

where ω is the natural frequency, φ_{v1} and φ_{v3} are arbitrary positive constants, φ_{v0} is an arbitrary constant, $p = x$ is the downstream pressure perturbation at the burning plane, and $q = \varphi_{v0} + \varphi_{v1}x - \frac{\varphi_{v3}}{3}x^3$ is the flame heat release rate.

Lemma 1. *For the van der Pol equation (6), the application of K-B approximation gives*

$$x = a\cos(\omega t + \theta) \tag{7}$$

with

$$\begin{cases} \dot{a} = \frac{\varphi_{v1}}{2}a\left(1 - \frac{\varphi_{v3}}{4\varphi_{v1}}a^2\right) \\ \dot{\theta} = 0\,. \end{cases} \tag{8}$$

Proof. Consider (6) and (1), in this case with $\epsilon = 1$,

$$f(x, \frac{dx}{dt}) = \varphi_{v1}\left(1 - \frac{\varphi_{v3}}{\varphi_{v1}}x^2\right)\frac{dx}{dt}. \tag{9}$$

Introducing $x = a\cos(\omega t + \theta)$ and $\dot{x} = -a\omega\sin(\omega t + \theta)$, $i = 1, 2$, in (9), yields

$$f(.,.) = -\omega\varphi_{v1}a\left(1 - \frac{\varphi_{v3}}{4\varphi_{v1}}a^2\right)\sin(\omega t + \theta) - \frac{\omega\varphi_{v3}a^3}{4}\sin\left(3(\omega t + \theta)\right). \tag{10}$$

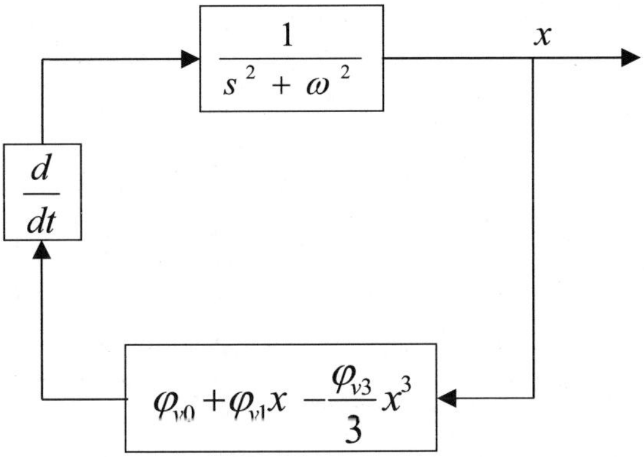

Fig. 3. A simplified (van der Pol) combustion instability model

Consequently, one can consider that (10) is in the form (5). Thus for x one obtains

$$\begin{cases} H_1 = -\omega\varphi_{v1}a\left(1 - \frac{\varphi_{v3}}{4\varphi_{v1}}a^2\right) \\ G_1 = 0 \end{cases} \tag{11}$$

By the application of rule (3) to (11), one deduces the result of Lemma 1. □

The system of equations (8) has two steady-state solutions. One steady-state solution corresponds to $a = 0$, and is unstable. The other steady-state solution corresponds to $a = 2\sqrt{\frac{\varphi_{v1}}{\varphi_{v3}}}$, and is locally stable. Therefore, we conclude for the uncontrolled generalized van der Pol equation, that the solution x is a self-sustained oscillation with steady frequency close to ω and with steady state amplitude close to $2\sqrt{\frac{\varphi_{v1}}{\varphi_{v3}}}$.

3.2 Quenching the Oscillations

Output-feedback control is introduced into the model by multiplying a function of the output x to capture the effect of fuel flow modulation on the heat release rate. The control strategy is characterized by the following differential equation.

$$\ddot{x} + \omega^2 x = \frac{d}{dt}\left\{(1 + \Phi(x))\left(\varphi_{v0} + \varphi_{v1}x - \frac{\varphi_{v3}}{3}x^3\right)\right\}, \tag{12}$$

where φ_{v0} is different from zero and the feedback law $\Phi(x)$ is a polynomial function of x. In order to quench self-sustained oscillation, the control law must force the system (12) to be asymptotically stable at the origin. Hence, one considers the following lemma.

Lemma 2. *For the following control low*

$$\Phi(x) = -Kx - \frac{1}{\varphi_{v0}}\left(\varphi_{v1}x - \frac{\varphi_{v3}}{3}x^3\right), \tag{13}$$

where K is a constant of the same sign as φ_{v0}, the system (12) is locally asymptotically stable at the origin.

Proof. Introducing (13) into (12), yields

$$\ddot{x}+\omega^2 x = -K\varphi_{v0}\dot{x} - 2\varphi_{v1}\left(K + \frac{\varphi_{v1}}{\varphi_{v0}}\right)x\dot{x} + \frac{4\varphi_{v3}}{3}\left(K + \frac{2\varphi_{v1}}{\varphi_{v0}}\right)x^3\dot{x} - 2\frac{\varphi_{v3}^2}{3\varphi_{v0}}x^5\dot{x}.$$

Expressing this equation in state equation form with $z_1 = x$ and $z_2 = \dot{x}$, one obtains

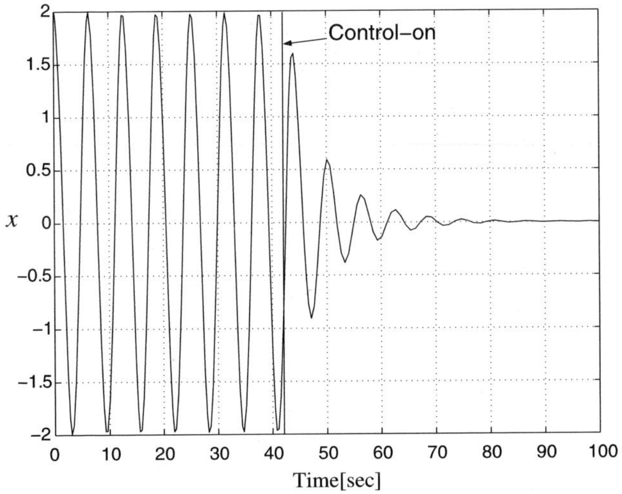

Fig. 4. Simulation test of multiplicative feedback quenching of oscillations in a model described by a generalized van der Pol equation, where $K = 1$, $\varphi_{v0} = 0.45$, $\varphi_{v1} = \varphi_{v1} = 0.1$ and $\omega = 1$

$$\begin{cases} \dot{z}_1 = z_2, \\ \dot{z}_2 = -\omega^2 z_1 - K\varphi_{v0} z_2 - 2\varphi_{v1}\left(K + \frac{\varphi_{v1}}{\varphi_{v0}}\right) z_1 z_2, \\ \qquad + \frac{4\varphi_{v3}}{3}\left(K + \frac{2\varphi_{v1}}{\varphi_{v0}}\right) z_1^3 z_2 - 2\frac{\varphi_{v3}^2}{3\varphi_{v0}} z_1^5 z_2. \end{cases}$$

Computation of the linearized system matrix around the origin gives

$$A_z = \begin{bmatrix} 0 & 1 \\ -\omega^2 & -K\varphi_{v0} \end{bmatrix}.$$

Since by assumption K and φ_{v0} have the same sign, the eigenvalues of matrix A_z will have negative real part. By using Lyapunov's indirect method, one can deduce that the system is locally asymptotically stable at the origin. □

The local asymptotic stability at the origin implies that quenching of the oscillation is possible and can occur around the origin in a local domain which can be estimated [12]. This quenching is illustrated by the simulation test presented in Figure 4.

4 Combustion Instability Model

The relationship between coupled van der Pol equations and the combustion instability model of Dunstan and Bitmead [10] has been discussed in [5].

A system of two coupled van der Pol equations has been considered as a basic model for combustion instabilities

$$\begin{cases} \frac{d^2x_1}{dt^2} + \omega_1^2 x_1 = \epsilon \frac{d}{dt}\left((x_1+x_2) - \frac{1}{3}(x_1+x_2)^3\right) \\ \frac{d^2x_2}{dt^2} + \omega_2^2 x_2 = \epsilon \frac{d}{dt}\left((x_1+x_2) - \frac{1}{3}(x_1+x_2)^3\right), \end{cases} \tag{14}$$

where ω_1 and ω_2 are the natural radian frequencies of the first and second resonators respectively and which can have arbitrary values with some modest provisions to be developed, and ε is a small positive quantity. The model has been successfully analyzed using the Krylov-Bogoliubov approach in [5]. However, this model does not include the cascade differentiator-plus-delay and the low pass filter (existing in Dunstan and Bitmead model [10]). While interesting results have been obtained concerning the existence or quenching of the oscillations in the system, the van der Pol model was not able to explain the simultaneous presence of two oscillating non-harmonic modes observed experimentally.

In order to make modelling more realistic in [6], the model based on two coupled van der Pol equations was further generalized by incorporating delay and filtering. The modifications lead to the model presented in Figure 5 and described by the following equations

$$\begin{cases} \ddot{x}_1 + \omega_1^2 x_1 = \frac{d}{dt} LPF\left\{\varphi_{v0} + \varphi_{v1}\dot{p}_\tau - \frac{\varphi_{v3}}{3}\dot{p}_\tau^3\right\} \\ \ddot{x}_2 + \omega_2^2 x_2 = \frac{d}{dt} LPF\left\{\varphi_{v0} + \varphi_{v1}\dot{p}_\tau - \frac{\varphi_{v3}}{3}\dot{p}_\tau^3\right\}, \end{cases} \tag{15}$$

where φ_{v0} is an arbitrary constant, φ_{v1} and φ_{v3} are arbitrary negative constants, τ is a transport time delay from nozzle to flame surface, LPF is the transfer operator of the low pass filter, $p = x_1 + x_2$ is the downstream pressure perturbation at the burning plane, $\dot{p}_\tau$ is the output of the delay-plus-differentiator block and q is the flame heat release rate.

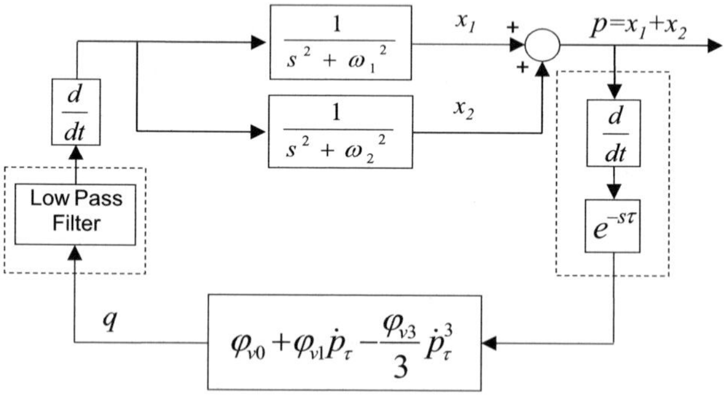

Fig. 5. Proposed combustion instability model

4.1 Model Analysis

Equation Development and Analysis

The first step of the analysis is to develop the left term of (15) in order to apply K-B approximation. The presence of delay and low pass filtering presents some difficulties in computing approximations, which will be treated by introducing some realistic assumptions which are effective in a stationary regime. Consider the system (15) and the form (1) (with $\epsilon = 1$), in this case

$$f_1 = f_2 = f = \frac{d}{dt} LPF \left\{ \varphi_{v0} + \varphi_{v1} \dot{p}_\tau - \frac{\varphi_{v3}}{3} \dot{p}_\tau^3 \right\}. \tag{16}$$

Replacing

$$\dot{x}_i = -a_i \omega_i \sin(\omega_i t + \theta_i), (i = 1, 2)$$

in $\dot{p}$, one obtains

$$\begin{aligned} &\dot{p} = \dot{x_1} + \dot{x_2} \\ &\Rightarrow \dot{p} = -\omega_1 a_1 \sin(\omega_1 t + \theta_1) - \omega_2 a_2 \sin(\omega_2 t + \theta_2) \\ &\quad\;\; = \omega_1 a_1 \cos(\omega_1 t + \theta_1 + \tfrac{\pi}{2}) + \omega_2 a_2 \cos(\omega_2 t + \theta_2 + \tfrac{\pi}{2}), \end{aligned}$$

which after adding the delay, takes the form

$$\dot{p}_\tau = \omega_1 a_{1\tau} \cos(\omega_1 t + \theta_{1\tau} + \tfrac{\pi}{2} - \omega_1 \tau) + \omega_2 a_{2\tau} \cos(\omega_2 t + \theta_{2\tau} + \tfrac{\pi}{2} - \omega_2 \tau) \tag{17}$$

where $a_{1\tau}$, $a_{2\tau}$, $\theta_{1\tau}$ and $\theta_{2\tau}$ are amplitudes and phases after the delay block, respectively. Since, for K-B approximation the amplitudes a_i and the phases θ_i $(i = 1, 2)$ are slowly time-varying functions, and in order to approximate the time delay block, we propose the following assumption.

Assumption 1. *For small time delay τ, the quantities $\mid a_i - a_{i\tau} \mid$ and $\mid \theta_i - \theta_{i\tau} \mid$ $(i = 1, 2)$ can be neglected.*

Assumption 1 allows the following approximations

$$\begin{cases} a_{i\tau} = a_i - (a_i - a_{i\tau}) \approx a_i, \\ \theta_{i\tau} = \theta_i - (\theta_i - \theta_{i\tau}) \approx \theta_i. \end{cases} (i = 1, 2) \tag{18}$$

To get an expression in the form of (5), the low pass filter block must also be approximated. Therefore, we consider a second assumption.

Assumption 2. *The low pass filter is linear and its dynamics are much faster than the evolution of amplitudes and phases in the K-B approximation.*

The utility of Assumption 2 is that, for an input given as the sum of sinusoidal terms (such as (5)), the output will be equal to the sum of the outputs of each term, and for sinusoidal inputs with slowly time-varying amplitudes and phases the rise time will be neglected, and the amplitudes and phases will be considered as constant parameters. Therefore, Assumption 2 leads to the following approximation

$$LPF\,(a\cos(\omega t+\theta)) \approx G(\omega)a\cos(\omega t+\theta-\phi(\omega)) \tag{19}$$

where a, ω and θ are the amplitude, the frequency and the phase of the input, respectively, $G(\omega)$ and $\phi(\omega)$ are the gain and the phase of the filter at frequency ω. We use the following notation in the remaining of paper.

$$\psi_{k_1k_2} = (k_1\omega_1+k_2\omega_2)t+(k_1\theta_1+k_2\theta_2), \tag{20}$$

$$A_{k_1k_2} = \omega_1^{|k_1|}\omega_2^{|k_2|}G(k_1\omega_1+k_2\omega_2), \tag{21}$$

$$\chi_{k_1k_2} = \frac{(k_1+k_2)\pi}{2}-(k_1\omega_1+k_2\omega_2)\tau-\phi(k_1\omega_1+k_2\omega_2). \tag{22}$$

The analysis to come will show that this notation has some significance. Expression (20) corresponds to the linear combinations of frequencies ω_1 and ω_2 present in the development of f, and the expressions (21) and (22) are the corresponding gains and phases introduced by the delay plus differentiator block and by filtering.

Lemma 3. *Consider the expression (17) and Assumptions 1 and 2. Then one obtains the following development*

$$\begin{aligned}
f \approx & -\varphi_{v1}\Bigg\{\omega_1A_{10}a_1\Big(1-\frac{\varphi_{v3}}{\varphi_{v1}}\Big(\frac{(\omega_1a_1)^2}{4}+\frac{(\omega_2a_2)^2}{2}\Big)\Big)\sin(\psi_{10}+\chi_{10})\\
&+\omega_2A_{01}a_2\Big(1-\frac{\varphi_{v3}}{\varphi_{v1}}\Big(\frac{(\omega_2a_2)^2}{4}+\frac{(\omega_1a_1)^2}{2}\Big)\Big)\sin(\psi_{01}+\chi_{01})\Bigg\}\\
&+\varphi_{v3}\Bigg\{\frac{\omega_1A_{30}a_1^3}{4}\sin(\psi_{30}+\chi_{30})+\frac{(2\omega_1-\omega_2)A_{2-1}a_1^2a_2}{4}\sin(\psi_{2-1}+\chi_{2-1})\\
&+\frac{(2\omega_1+\omega_2)A_{21}a_1^2a_2}{4}\sin(\psi_{21}+\chi_{21})+\frac{(\omega_1+2\omega_2)A_{12}a_1a_2^2}{4}\sin(\psi_{12}+\chi_{12})\\
&+\frac{\omega_2A_{03}a_2^3}{4}\sin(\psi_{03}+\chi_{03})+\frac{(2\omega_2-\omega_1)A_{-12}a_2^2a_1}{4}\sin(\psi_{-12}+\chi_{-12})\Bigg\}.
\end{aligned} \tag{23}$$

Proof. Substituting approximations (18) into (17), yields

$$\dot{p}_\tau \approx \omega_1a_1\cos(\omega_1t+\theta_1+\tfrac{\pi}{2}-\omega_1\tau)+\omega_2a_2\cos(\omega_2t+\theta_2+\tfrac{\pi}{2}-\omega_2\tau). \tag{24}$$

Whence, using (24), (20) and trigonometrical simplifications,

$$
\begin{aligned}
\varphi_{v0}+&\varphi_{v1}\dot{p}_{-\tau}-\frac{\varphi_{v3}}{3}\dot{p}^3_{-\tau}\\
&\approx \varphi_{v0}+\varphi_{v1}\omega_1 a_1\left(1-\frac{\varphi_{v3}}{\varphi_{v1}}\left(\frac{(\omega_1 a_1)^2}{4}+\frac{(\omega_2 a_2)^2}{2}\right)\right)\cos(\psi_{10}+\frac{\pi}{2}-\omega_1\tau)\\
&+\varphi_{v1}\omega_2 a_2\left(1-\frac{\varphi_{v3}}{\varphi_{v1}}\left(\frac{(\omega_2 a_2)^2}{4}+\frac{(\omega_1 a_1)^2}{2}\right)\right)\cos(\psi_{01}+\frac{\pi}{2}-\omega_2\tau)\\
&-\varphi_{v3}\left\{\frac{(\omega_1 a_1)^3}{12}\cos\left(\psi_{30}+\frac{3\pi}{2}-3\omega_1\tau\right)+\right.\\
&+\frac{\omega_1^2\omega_2 a_1^2 a_2}{4}\cos\left(\psi_{2-1}+\frac{\pi}{2}-(2\omega_1-\omega_2)\tau\right)\\
&+\frac{\omega_1^2\omega_2 a_1^2 a_2}{4}\cos\left(\psi_{21}+\frac{3\pi}{2}-(2\omega_1+\omega_2)\tau\right)\\
&+\frac{\omega_1\omega_2^2 a_1 a_2^2}{4}\cos\left(\psi_{12}+\frac{3\pi}{2}-(2\omega_2+\omega_1)\tau\right)\\
&+\frac{(\omega_2 a_2)^3}{12}\cos\left(\psi_{03}+\frac{3\pi}{2}-3\omega_2\tau\right)\\
&\left.+\frac{\omega_1\omega_2^2 a_2^2 a_1}{4}\cos\left(\psi_{-12}+\frac{\pi}{2}-(2\omega_2-\omega_1)\tau\right)\right\} \qquad (25)
\end{aligned}
$$

Using (25), (19) and (21), one arrives at (23). □

Result (23) yields the frequency set

$$W=\{\omega_1,\omega_2,3\omega_1,3\omega_2,2\omega_1+\omega_2,\omega_1+2\omega_2,2\omega_1-\omega_2,2\omega_2-\omega_1\}\,, \qquad (26)$$

which will be very important for identifying the different situations depending on the proximity of frequencies in W. Consequently, one has the following three situations:

1) $\omega_1 \not\approx \{\omega_2, 3\omega_2, \frac{\omega_2}{3}\}$,
2) $\omega_1 \approx \omega_2$: mutual synchronization with close frequencies
3) $\omega_1 \approx 3\omega_2$ (respectively $\omega_2 \approx 3\omega_1$): Mutual synchronization with multiple frequencies

Cases 2 and 3 were considered in [5]. However, experimental results [10] reveal that Case 1 occurs in practice and that both modes can oscillate freely without synchronization. Hence, we limit our study here to Case 1. This is both new and more practically significant than the results in [5].

K-B Approximation of the Model

The second analytical task is to find an expression for the model output using the result (23) and to analyze the evolution of this output in different situations.

Lemma 4. *Consider the condition* $\omega_1 \not\approx \{\omega_2, 3\omega_2, \frac{\omega_2}{3}\}$ *and the result (23). The application of K-B approximation gives*

$$x_i = a_i \cos(\omega_i t + \theta_i), \; (i = 1, 2) \tag{27}$$

with

$$\begin{cases} \dot{a}_1 = \frac{\varphi_{v1} A_{10} \cos(\chi_{10})}{2} a_1 \left(1 - \frac{\varphi_{v3}}{\varphi_{v1}} \left(\frac{(\omega_1 a_1)^2}{4} + \frac{(\omega_2 a_2)^2}{2}\right)\right), \\ \dot{a}_2 = \frac{\varphi_{v1} A_{01} \cos(\chi_{01})}{2} a_2 \left(1 - \frac{\varphi_{v3}}{\varphi_{v1}} \left(\frac{(\omega_2 a_2)^2}{4} + \frac{(\omega_1 a_1)^2}{2}\right)\right), \\ \dot{\theta}_1 = \frac{\varphi_{v1} A_{10} \sin(\chi_{10})}{2} \left(1 - \frac{\varphi_{v3}}{\varphi_{v1}} \left(\frac{(\omega_1 a_1)^2}{4} + \frac{(\omega_2 a_2)^2}{2}\right)\right), \\ \dot{\theta}_2 = \frac{\varphi_{v1} A_{01} \sin(\chi_{01})}{2} \left(1 - \frac{\varphi_{v3}}{\varphi_{v1}} \left(\frac{(\omega_2 a_2)^2}{4} + \frac{(\omega_1 a_1)^2}{2}\right)\right), \end{cases} \tag{28}$$

Proof. By application of rules (2) and (3) to the expression (23), one finds the result. □

From (28), one can see that the coupled parameters are a_1 and a_2. Therefore, the system dynamics depend essentially on the evolution of amplitudes a_1 and a_2. The analytical determination of equilibrium points gives

$$a_1 = 0 \text{ and } a_2 = 0, \tag{29}$$

$$a_1 = \frac{2}{\omega_1} \sqrt{\frac{\varphi_{v1}}{\varphi_{v3}}} \text{ and } a_2 = 0, \tag{30}$$

$$a_1 = 0 \text{ and } a_2 = \frac{2}{\omega_2} \sqrt{\frac{\varphi_{v1}}{\varphi_{v3}}}, \tag{31}$$

$$a_1 = \frac{2}{\omega_1} \sqrt{\frac{\varphi_{v1}}{3\varphi_{v3}}} \text{ and } a_2 = \frac{2}{\omega_2} \sqrt{\frac{\varphi_{v1}}{3\varphi_{v3}}}. \tag{32}$$

The stability of each equilibrium point leads to a particular regime. Consequently, one distinguishes four operation regimes, which will be elaborated and explained shortly.

1) Asymptotically stable system.
2) Two generators with competitive quenching.
3) Simultaneous self-sustained oscillations.
4) Total instability.

For the stability study one can apply Lyapunov's indirect method, which uses the stability properties of the linearized system around the equilibrium point. The computation of the characteristic polynomial leads to the following result.

$$P(\lambda) = \lambda^2 - \frac{\varphi_{v1}}{2} \Big\{ A_{10} \cos(\chi_{10}) \Big(1 - \frac{\varphi_{v3}}{\varphi_{v1}} \Big(\frac{3(\omega_1 a_1)^2}{4} + \frac{(\omega_2 a_2)^2}{2}\Big)\Big)$$
$$+ A_{01} \cos(\chi_{01}) \Big(1 - \frac{\varphi_{v3}}{\varphi_{v1}} \Big(\frac{3(\omega_2 a_2)^2}{4} + \frac{(\omega_1 a_1)^2}{2}\Big)\Big)\Big\} \lambda$$

$$+\frac{A_{10}A_{01}\cos(\chi_{10})\cos(\chi_{01})}{4}\Big\{\varphi_{v1}^2\Big(1-\frac{\varphi_{v3}}{\varphi_{v1}}\Big(\frac{3(\omega_1 a_1)^2}{4}+$$
$$+\frac{(\omega_2 a_2)^2}{2}\Big)\Big)\Big(1-\frac{\varphi_{v3}}{\varphi_{v1}}\Big(\frac{3(\omega_2 a_2)^2}{4}+\frac{(\omega_1 a_1)^2}{2}\Big)\Big)-\varphi_{v3}^2(\omega_1\omega_2 a_1 a_2)^2\Big\}. \quad (33)$$

The characteristic polynomial obtained is second order. Therefore, the stability can be verified by testing the signs of polynomial coefficients.

Asymptotically stable system

The system is asymptotically stable around the origin, if and only if the equilibrium point (29) is asymptotically stable. Introducing (29) in the general characteristic polynomial (33), one obtains the following.

$$P(\lambda)=\lambda^2-\frac{\varphi_{v1}}{2}\Big\{A_{10}\cos(\chi_{10})+A_{01}\cos(\chi_{01})\Big\}\lambda$$
$$+\frac{A_{10}A_{01}\cos(\chi_{10})\cos(\chi_{01})\varphi_{v1}^2}{4},$$

which has two stable zeros if the following conditions are respected

$$\begin{cases}-\varphi_{v1}\Big\{A_{10}\cos(\chi_{10})+A_{01}\cos(\chi_{01})\Big\}>0\\ A_{10}A_{01}\cos(\chi_{10})\cos(\chi_{01})\varphi_{v1}^2>0\end{cases}\Longleftrightarrow\begin{cases}\varphi_{v1}\cos(\chi_{10})<0\\ \varphi_{v1}\cos(\chi_{01})<0\,.\end{cases} \quad (34)$$

Provided (34) is satisfied and the initial states are close to the origin, the amplitudes of both oscillations converge to the equilibrium point (29). Figure 6 shows an example of a simulation test, where the conditions (34) are satisfied under realistic parameters values.

Two generators with competitive quenching

The two generators with competitive quenching regime occurs when both equilibrium points (30) and (31) are locally stable. Substituting (30) into (33), one gets

$$P(\lambda)=\lambda^2+\varphi_{v1}\Big\{A_{10}\cos(\chi_{10})+\frac{1}{2}A_{01}\cos(\chi_{01})\Big\}\lambda$$
$$+\frac{A_{10}A_{01}\cos(\chi_{10})\cos(\chi_{01})\varphi_{v1}^2}{2}.$$

The local stability of (30) is satisfied if and only if

$$\begin{cases}\varphi_{v1}\Big\{A_{10}\cos(\chi_{10})+\frac{1}{2}A_{01}\cos(\chi_{01})\Big\}>0\\ A_{10}A_{01}\cos(\chi_{10})\cos(\chi_{01})\varphi_{v1}^2>0\end{cases}\Longleftrightarrow\begin{cases}\varphi_{v1}\cos(\chi_{10})>0\\ \varphi_{v1}\cos(\chi_{01})>0\,.\end{cases} \quad (35)$$

By symmetry, for the equilibrium point (31) one finds the same conditions.

Provided that the conditions (35) are satisfied, the amplitudes of x_1 and x_2 converge to one of both possible equilibrium points (30) and (31). Depending on the initial states, one of the generators is excited, while the oscillations of the other generator are entirely quenched. Figure 7 shows an example of a simulation test in this competitive quenching regime.

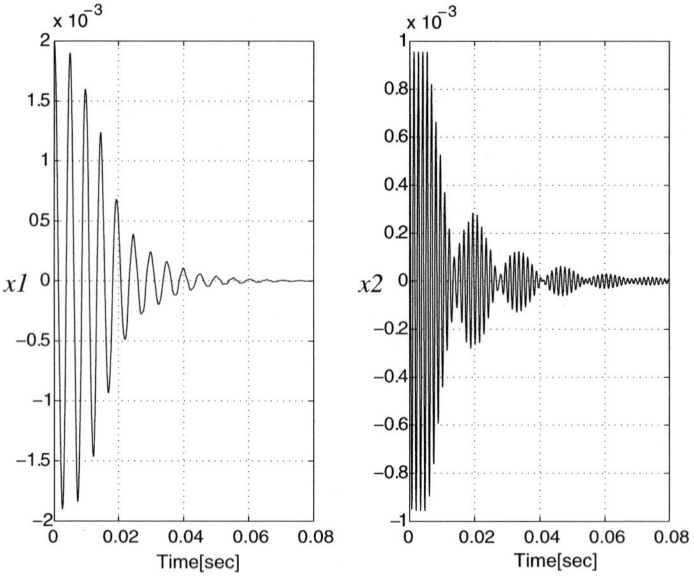

Fig. 6. Simulation test for $\omega_1 = 2\pi \times 210$, $\omega_2 = 2\pi \times 740$, $\varphi_{v0} = 0.45$, $\varphi_{v1} = -0.135$, $\varphi_{v3} = -5.4 \times 10^{-3}$, $LPF = \frac{2\pi \times 500}{s+2\pi \times 500}$ and $\tau = 5.5 \times 10^{-3}$

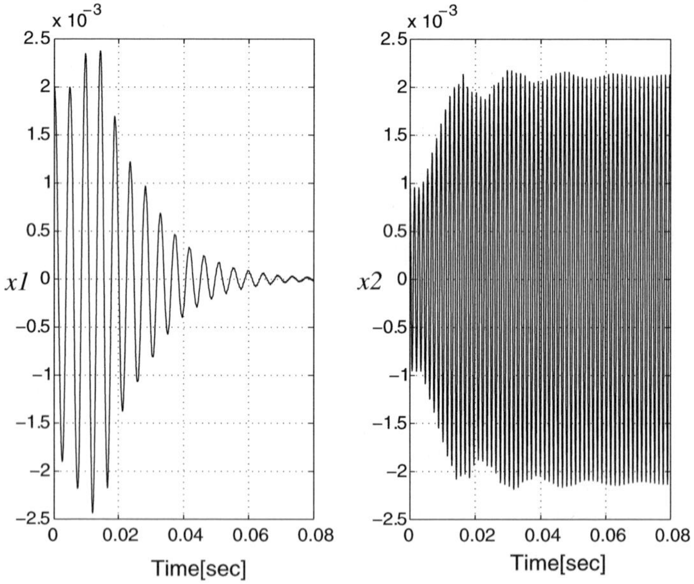

Fig. 7. Simulation test for $\omega_1 = 2\pi \times 210$, $\omega_2 = 2\pi \times 740$, $\varphi_{v0} = 0.45$, $\varphi_{v1} = -0.135$, $\varphi_{v3} = -5.4 \times 10^{-3}$, $LPF = \frac{2\pi \times 500}{s+2\pi \times 500}$ and $\tau = 3.5 \times 10^{-3}$

Simultaneous self-sustained oscillations

Simultaneous and persistent oscillation of both resonators occurs when the equilibrium point (32) is stable. Introducing (32) in (33), one obtains

$$P(\lambda) = \lambda^2 + \tfrac{\varphi_{v1}}{3}\Big\{A_{10}\cos(\chi_{10}) + A_{01}\cos(\chi_{01})\Big\}\lambda - \tfrac{A_{10}A_{01}\cos(\chi_{10})\cos(\chi_{01})\varphi_{v1}^2}{3}\,.$$

The equilibrium point (32) is locally stable if and only if

$$\begin{cases} \varphi_{v1}\Big\{A_{10}\cos(\chi_{10}) + A_{01}\cos(\chi_{01})\Big\} > 0 \\ -A_{10}A_{01}\cos(\chi_{10})\cos(\chi_{01})\varphi_{v1}^2 > 0 \end{cases}$$
$$\Longleftrightarrow \begin{cases} \varphi_{v1}\big(A_{10}\cos(\chi_{10}) + A_{01}\cos(\chi_{01})\big) > 0 \\ \cos(\chi_{10})\cos(\chi_{01}) < 0\,. \end{cases} \tag{36}$$

By satisfying (36), it is possible to have simultaneous self-sustained oscillations, the amplitudes of x_1 and x_2 converge to the equilibrium point (32). By self-sustained oscillations it is meant that both oscillators are excited without synchronization. Figure 8 shows the stationary part of a simulation test example in the simultaneous self-sustained oscillations regime.

Total instability

When the conditions (34), (35) and (36) are not satisfied, there does not exist a stable equilibrium point. Therefore, there is no stable limit cycle and the

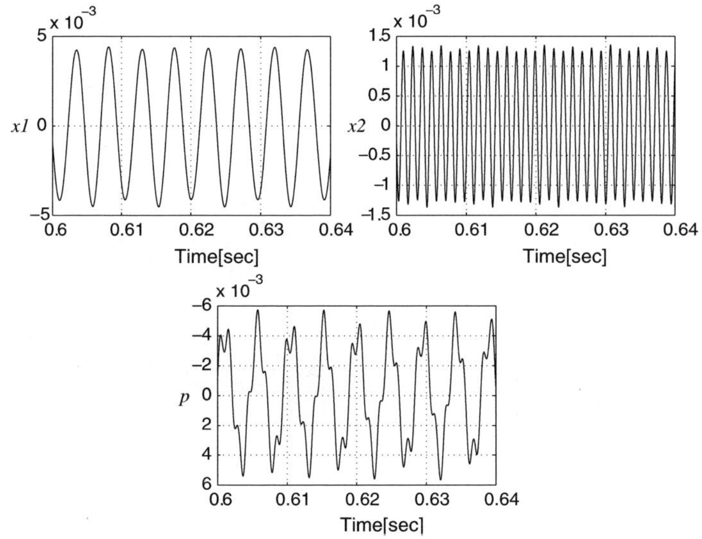

Fig. 8. Simulation test for $\omega_1 = 2\pi \times 210$, $\omega_2 = 2\pi \times 740$, $\varphi_{v0} = 0.45$, $\varphi_{v1} = -0.135$, $\varphi_{v3} = -5.4 \times 10^{-3}$, $LPF = \frac{2\pi \times 500}{s + 2\pi \times 500}$ and $\tau = 4.8 \times 10^{-3}$

amplitudes of both oscillations diverge. By total instability it is meant that for any non-equilibrium initial state, the state of the system diverges.

Analysis of Results

The results demonstrate the existence of four operation regimes when the ratio of natural frequencies is different from $1, 3$ and $\frac{1}{3}$. The occurrence of each operation regime depends on the satisfaction of certain conditions. The simulations confirm the quality of estimated amplitudes using the K-B approximation. The conditions for each regime contain essentially the phases χ_{01} and χ_{10} introduced by the delay and low pass filtering respectively. The phase domain conditions (34), (35) and (36) are independent. So with fixed delay and low-pass filter it is not possible to have operation in more than one regime. The results suggest that it is perhaps possible in certain cases to estimate the delay from the measurement of the oscillations (number, frequency, amplitude). To illustrate the importance of delay, Figure 9 depicts the operation regimes for several values of τ and for other parameters fixed near to the practical values. One observes the occurrence of various regimes as a function of the delay.

From a practical point of view, the interesting situation is the simultaneous self-sustained oscillatory regime which is represented in Figure 9. The fact that, the amplitudes of harmonics ω_1 and ω_2 take values $\frac{2}{\omega_1}\sqrt{\frac{\varphi_{v1}}{3\varphi_{v3}}}$ and $\frac{2}{\omega_2}\sqrt{\frac{\varphi_{v1}}{3\varphi_{v3}}}$ respectively, shows clearly that they depend inversely on the values of frequencies (a phenomenon which has been observed in practice). The results of Figure 9 have a very important practical implication: the design of the combustion system influences the type of combustion instability which may occur.

5 Quenching the Oscillations

For the development of an effective control method for quenching both oscillation modes present in combustion instabilities, the model (14) is taken

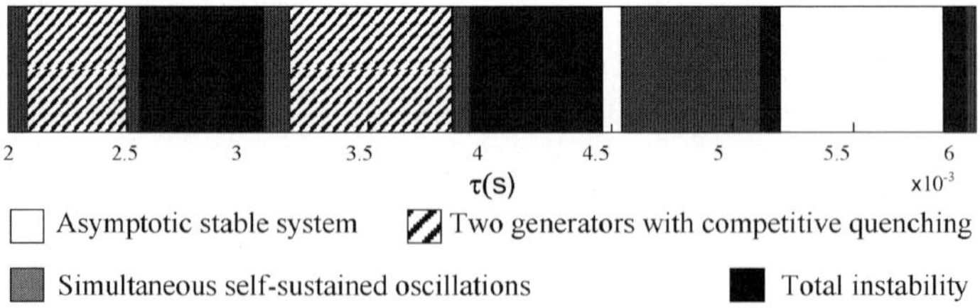

Fig. 9. The operation regimes as a function of τ, $\omega_1 = 2\pi \times 210$, $\omega_2 = 2\pi \times 740$, $\varphi_{v0} = 0.45$, $\varphi_{v1} = -0.135$, $\varphi_{v3} = -5.4 \times 10^{-3}$ and $LPF = \frac{2\pi \times 500}{s+2\pi \times 500}$

with the addition of a multiplicative control action (modulation of the fuel flow). The multiplicative effect of the control action must be included in the representation to capture the modulation of a fraction of the fuel flow u into the combustion chamber and its consequent effect on the heat release rate [10, 9]. This leads to the following differential equations

$$\begin{cases} \ddot{x}_1 + \omega_1^2 x_1 = \frac{d}{dt} LPF \left\{ (1+u) \left(\varphi_{v0} + \varphi_{v1} \dot{p}_\tau - \frac{\varphi_{v3}}{3} \dot{p}_\tau^3 \right) \right\} \\ \ddot{x}_2 + \omega_2^2 x_2 = \frac{d}{dt} LPF \left\{ (1+u) \left(\varphi_{v0} + \varphi_{v1} \dot{p}_\tau - \frac{\varphi_{v3}}{3} \dot{p}_\tau^3 \right) \right\} . \end{cases} \quad (37)$$

To deal with combustion instabilities, different control approaches have been considered in practice [1]. One can use a feedback control which is based on the pressure measurement alone. Two types of control law can be considered, one linear the other nonlinear. The linear law requires the availability of x_1 and x_2 which can be obtained by using appropriate band pass filters [4]. Here we are interested in nonlinear feedback which uses the pressure measurement ($p = x_1 + x_2$), since it gives effective results and is subject to fewer constraints. To study the effects of such control we continue to use the K-B method. The following assumption is proposed concerning the validity of the K-B approximation.

Assumption 3. *Let a_1 and a_2 be the amplitudes of the oscillations x_1 and x_2 obtained from the K-B approximation. If a_1 and a_2 are asymptotically locally (globally) stable at the origin, then the original system is asymptotically locally stable at the origin.*

For nonlinear feedback, the pressure measurement p is differentiated and delayed with a delay τ to obtain $\dot{p}_\tau$, which is introduced into a nonlinear function Φ, to obtain a control law $u = \Phi(p, \dot{p}_\tau)$. This control strategy is explained in the block diagram shown in Figure 10. This control law will be used to add damping and to compensate the physical feedback caused by the coupling between the thermal heat-release process and the acoustics of the combustion chamber. If such a control is designed, the system will be asymptotically stable at the origin and the quenching of oscillations will occur. This can be interpreted also as a feedback linearization which in addition stabilizes the system. The results are summarized in the following lemma

Lemma 5. *For the control law*

$$\Phi\left(p, \dot{p}_\tau\right) = -Kp - \frac{1}{\varphi_{v0}} \left(\varphi_{v1} \dot{p}_\tau - \frac{\varphi_{v3}}{3} \dot{p}_\tau^3 \right), \quad (38)$$

where K is a constant satisfying the conditions

$$\begin{cases} K \varphi_{v0} \cos(\phi(\omega_1)) > 0 \\ K \varphi_{v0} \cos(\phi(\omega_2)) > 0, \end{cases} \quad (39)$$

the system is locally asymptotically stable at the origin.

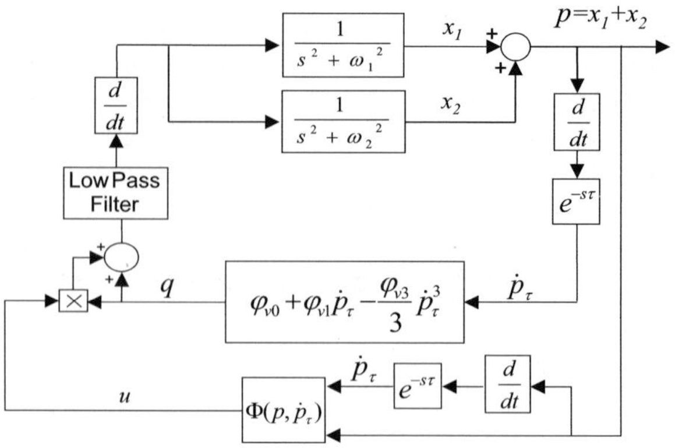

Fig. 10. Block diagram of nonlinear feedback

Proof: Replacing the function (38) in (37) yields,

$$\begin{cases} \ddot{x}_1+\omega_1^2 x_1 = \frac{d}{dt}LPF\{\varphi_{v0}-K\varphi_{v0}p-K\varphi_{v1}p\dot{p}_\tau+K\frac{\varphi_{v3}}{3}p\dot{p}_\tau^3 \\ \qquad -\frac{\varphi_{v1}^2}{\varphi_{v0}}\dot{p}_\tau^2+\frac{2\varphi_{v1}\varphi_{v3}}{3\varphi_{v0}}\dot{p}_\tau^4-\frac{\varphi_{v3}^2}{9\varphi_{v0}}\dot{p}_\tau^6\} \\ \ddot{x}_2+\omega_2^2 x_2 = \frac{d}{dt}LPF\{\varphi_{v0}-K\varphi_{v0}p-K\varphi_{v1}p\dot{p}_\tau+K\frac{\varphi_{v3}}{3}p\dot{p}_\tau^3 \\ \qquad -\frac{\varphi_{v1}^2}{\varphi_{v0}}\dot{p}_\tau^2+\frac{2\varphi_{v1}\varphi_{v3}}{3\varphi_{v0}}\dot{p}_\tau^4-\frac{\varphi_{v3}^2}{9\varphi_{v0}}\dot{p}_\tau^6\}. \end{cases}$$

Therefore, for the K-B approximation (1) with $\epsilon=1$ one may consider the following choice

$$\begin{aligned} f_1=f_2=f=\frac{d}{dt}LPF\{&\varphi_{v0}-K\varphi_{v0}p-K\varphi_{v1}p\dot{p}_\tau+K\frac{\varphi_{v3}}{3}p\dot{p}_\tau^3 \\ &-\frac{\varphi_{v1}^2}{\varphi_{v0}}\dot{p}_\tau^2+\frac{2\varphi_{v1}\varphi_{v3}}{3\varphi_{v0}}\dot{p}_\tau^4-\frac{\varphi_{v3}^2}{9\varphi_{v0}}\dot{p}_\tau^6\}. \end{aligned} \tag{40}$$

Introducing $x_i=a_i\cos(\omega_i t+\theta_i)$ and $\dot{x}_i=-a_i\omega_i\sin(\omega_i t+\theta_i)$, $i=1,2$, in (40), using the approximation (24) and after trigonometric simplifications and using Assumption 2, one obtains the expression

$$\begin{aligned} f\approx\ & K\varphi_{v0}\omega_1 G(\omega_1)a_1\Big[\cos(\phi(\omega_1))\sin(\omega_1 t+\theta_1)-\sin(\phi(\omega_1))\cos(\omega_1 t+\theta_1)\Big] \\ &+K\varphi_{v0}\omega_2 G(\omega_2)a_2\Big[\cos(\phi(\omega_2))\sin(\omega_2 t+\theta_2)-\sin(\phi(\omega_2))\cos(\omega_2 t+\theta_2)\Big] \\ &+\sum_{\omega_\ell\not\approx\omega_1\wedge\omega_2}^{r}\left(H_\ell(a_1,a_2,\theta_1,\theta_2)\sin(\omega_\ell t+\theta_\ell)+G_\ell(a_1,a_2,\theta_1,\theta_2)\cos(\omega_\ell t+\theta_\ell)\right). \end{aligned}$$

Applying the rule (3) for the amplitudes leads to the following approximations

$$\begin{cases} \frac{da_1}{dt}=-\frac{1}{2}G(\omega_1)K\varphi_{v0}\cos(\phi(\omega_1))a_1 \\ \frac{da_2}{dt}=-\frac{1}{2}G(\omega_2)K\varphi_{v0}\cos(\phi(\omega_2))a_2. \end{cases} \tag{41}$$

These are linear differential equations for the uncoupled amplitudes a_1 and a_2. The amplitudes are globally asymptotically stable at the origin if the

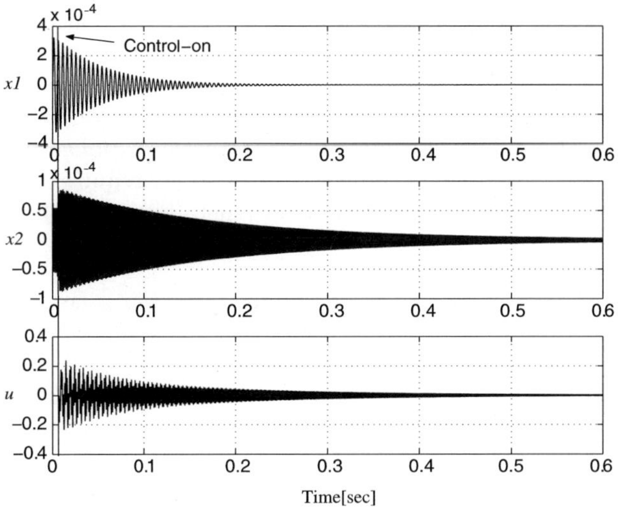

Fig. 11. Linear feedback control applied at the first appearance of oscillations

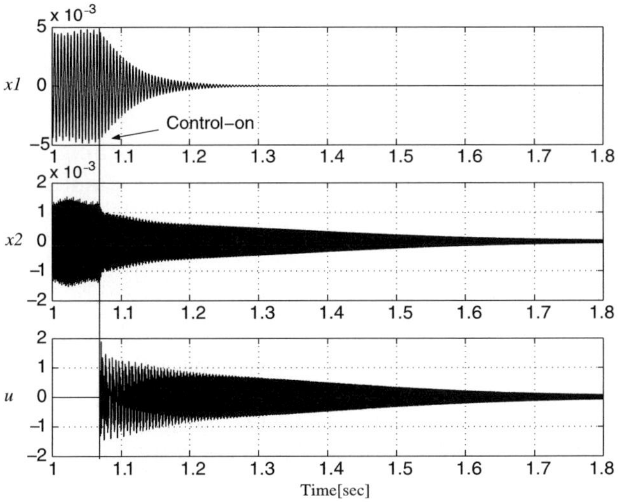

Fig. 12. Nonlinear feedback control applied after the oscillation has reached steady-state

conditions (39) are satisfied. Appealing to Assumption 3, one deduces the result.

This control has been tested in simulation, with the same model parameters used in simultaneous self-sustained oscillations and with gain $K = 100$ for the two possible scenarios. In the first scenario, the control is applied at the appearance of oscillations in x_1 or x_2. This is presented in Figure 11. In the second scenario, the control is applied after the oscillations in x_1 and x_2 have already reached steady-state operation. This is presented in Figure 12.

The control law yields asymptotic local stability at the origin. The domain of amplitudes a_1 and a_2 where quenching oscillations is guaranteed, is delimited by the boundary of validity the K-B method applied to (37). It is important to note however, that such a control strategy requires good knowledge of the parameters values in the combustion instability model, particularly of the value of the delay τ.

6 Conclusion

This paper presents a model for combustion instabilities which is analytically tractable. Its analysis is carried out using the Krylov-Bogoliubov (K-B) method. The analysis has shown that the model can explain the coexistence of several distinct non-harmonic frequencies observed in practice. The chapter has also shown and quantified the effect of the delay for the occurrence of the various phenomena observed.

Once this method of analysis has been made available and taking advantage of the mutiplicative control which can be implemented by amplitude modulation of the fuel flow, a feedback control methodology for quenching oscillations in combustion instabilities has been developed. The possibility of quenching oscillations has been analyzed also by the K-B method.

Further work will focus on robustness of these quenching approaches with respect to model parameters uncertainties as well as a quantitative evaluation of the stability domains.

References

1. A.M. Annaswamy and A.F. Ghoniem. Active control of combustion instability: Theory and practice. *IEEE Control Systems Magazine*, 22(6):37–54, Dec 2002.
2. A. Banaszuk, K.B. Ariyur, M. Krstic, and C. Jacobson. An adaptive algorithm for control of combustion instability. *Automatica*, 40:2155–2162, 2004.
3. N.N. Bogoliubov and Y.A. Mitropolski. *Asymptotic Methods in the Theory of Nonlinear Oscillations.* Hindustan Publishing Corp, Delhi, and Gordon and Breach, New York, 1961.
4. F. Bouziani, I.D. Landau, and R.R. Bitmead. Quenching oscillations in combustion instabilities using model-based closed-loop multiplicative control. Technical report, Laboratoire d'Automatique de Grenoble, ENSIEG BP 46, 38402 Saint-Martin d'Hères, France, November 2006.
5. F. Bouziani, I.D. Landau, R.R. Bitmead, and A. Voda-Besançon. An analytically tractable model for combustion instability. *Proc. of the 44rd IEEE Conf. on Decision and Contr.*, 2005.
6. F. Bouziani, I.D. Landau, R.R. Bitmead, and A. Voda-Besançon. Analysis of a tractable model for combustion instability: the effect of delay and low pass filtering. *Proc. of the 45rd IEEE Conf. on Decision and Contr.*, 2006.

7. P. Chatterjee, U. Vandsburger, W.R. Saunders, V.W. Khanna, and W.T. Baumann. On the spectral characteristics of a self-excited Rijke tube combustor: numerical simulation and experimental measurements. *JSV*, 283:573–588, 2005.
8. A.P. Dowling. The calculation of thermoacoustic oscillations. *Journal Sound & Vibration*, 180(4):557–581, 1995.
9. W. J. Dunstan, R. R. Bitmead, and S. M. Savaresi. Fitting nonlinear low-order models for combustion instability control. *Control Engineering Practice*, pages 1301–1317, 2001.
10. W.J. Dunstan. *System Identification of Nonlinear Resonant Systems.* PhD thesis, University of California, San Diego, 2003.
11. C. Hayashi. *Nonlinear Oscillations in Physical Systems.* McGraw-Hill Book Co, New York, 1964. Reprinted by Princeton University Press, 1985.
12. H.K. Khalil. *Nonlinear Systems.* MacMillan, New York, 1992.
13. P.S. Landa. *Nonlinear Oscillations and Waves in Dynamical Systems.* Kluwer, 1996.
14. P.S. Landa. *Regular and Chaotic Oscillations.* Springer, New York, 2000.
15. I. D. Landau and R.R. Bitmead. On the method of Krylov and Bogoliubov for the analysis of nonlinear oscillations. Technical report, Mechanical and Aerospace Engineering Department, University of California, San Diego, 9500 Gilman Drive, La Jolla CA 92093-0411, USA, Jan 2004.
16. K. McManus, F. Han, W. Dunstan, C. Barbu, and M. Shah. Modeling and control of combustion dynamics in industrial gas turbines. *Proc. ASME Turbo-Expo*, pages 567–575, 2004.
17. M. Mettenleiter, E. Haile, and S. Candel. Adaptive control of aeroacoustic instabilities. *Journal Sound & Vibration*, 230(4):761–789, 2000.
18. R. M. Murray, C. A. Jacobson, R. Casas, A. I. Khibnik, C. R. Johnson Jr, R. R. Bitmead, A. A. Peracchio, and W. M. Proscia. System identification for limit cycling systems: a case study for combustion instabilities. *Proc. of the 1998 Amer. Contr. Conf.*, pages 2004–2008, 1998.
19. G.D. Roy. *Advances in Chemical Propulsion: Science to Technology.* CRC – Taylor & Francis, Boc Raton, Fl USA, 2001.
20. J. Tyndall. *Sound.* D. Appleton & Company, New York, 1867.

Convexification of the Range-Only Station Keeping Problem

Ming Cao and A. Stephen Morse

Department of Electrical Engineering, Yale University, New Haven, CT 06520, USA

This paper is dedicated to Alberto Isidori on the occasion of his 65th birthday.

Summary. Using concepts from switched adaptive control theory plus a special parameterization of the class of 2×2 nonsingular matrices, a tractable and provably correct solution is given to the three landmark station keeping problem in the plane in which range measurements are the only sensed signals upon which station keeping is to be based. The performance of the overall system degrades gracefully in the face of increasing measurement and miss-alignment errors, provided the measurement errors are not too large.

1 Introduction

"Station keeping" is a term from orbital mechanics which refers to the "practice of maintaining the orbital position of satellites in geostationary orbit" {Wikipedia}. In this paper as in [3], we take station keeping to mean the practice of keeping a mobile autonomous agent in a position in the plane which is determined by prescribed distances from two or more landmarks. We refer to these landmarks as neighboring agents because we envision solutions to the station keeping problem as potential solutions to multi-agent formation maintenance problems. We are particularly interested in solutions to the station keeping problem in which the only signals available to the agent whose position is to be maintained, are noisy range measurements from its neighbors[1]. Our approach to station keeping builds on the work initiated in [3] where we treated station keeping as a problem in switched adaptive control. We continue with the same approach in this paper but now deal directly with an important computational issue which was not addressed in [3]. In particular, the control system considered in [3] requires an algorithm capable

[1] We are indebted to B. D. O. Anderson for making us aware of this problem.

of minimizing with respect to the four entries in a 2×2 nonsingular matrix P, a cost function of the form $M(X, P) = \text{trace}\{[I\ P]\, X\, [I\ P]'\}$ where X is a 4×4 positive semi-definite matrix. What makes the problem difficult is the constraint that P must be non-singular, since this leads to an non-convex optimization problem. The main contribution of this paper is to explain how to avoid this difficulty by utilizing the fact that any 2×2 non-singular matrix B can be written as $B = U(I + L)S$ where U is a specially structured matrix from a finite set, L is strictly lower triangular and S is symmetric and positive definite [12]. This fact enables us to modify the optimization problem just described, so that instead of having a non-convex problem to solve, one has a finite set of convex problems instead. Not only does the modification lead to convex programming problems, but also programming problems which can each be solved efficiently using semi-definite programming methods [22].

Work on the range-only station keeping problem already exists [9, 20, 2] and related work on range-only source localization can be found in [5, 4]. The station keeping problem is closely related to the Simultaneous Localization and Mapping {SLAM} problem [11, 16, 6, 23], which is also called the Concurrent Mapping and Localization problem [7, 21]. SLAM is the process of building a map of an unknown environment by using mobile robots' sensed information and simultaneously estimating those robots' locations by using this map. The station keeping problem with one autonomous agent and multiple landmarks can be cast as a SLAM problem in which the map describes the positions of the landmarks and the autonomous agent is the robot to be localized. There are several approaches to the SLAM problem, such as those based on Kalman filters [19, 1] and those using sequential Monte Carlo techniques [8, 24]. Kalman filtering based methods apply to linearized observation models and assume that the measurement errors are Gaussian. Since most of the sensory data from the range-only measurements are nonlinear and with non-Gaussian errors, the limitation of the Kalman filter method in this context is obvious. Sequential Monte Carlo based methods use nonlinear observation models and do not require suitable probabilistic models for measurement noises, but do require large numbers of samples; typically such methods are computationally difficult to implement. There are also several interesting and new set-based techniques addressed to the range-only SLAM problem [9, 20], but these have not been validated mathematically.

Several features of the station keeping method proposed here distinguish it from SLAM-based methods. First, SLAM algorithms seek to localize and map whereas the approach here focuses sharply and exclusively on the ultimate goal of moving an agent to its assigned position; no attempt is made to localize the assigned position and because of this, the approach taken here is fundamentally different than the more indirect SLAM approach. Second, the present method uses a provably correct switched adaptive control algorithm, whereas the SLAM-based methods do not.

In Section 2 we formulate the station keeping problem of interest. Error models appropriate to the solution to the problem are developed in Section 3. Some of the error equations developed have appeared previously in [9, 20, 18] and elsewhere. In Section 4 we present a switched adaptive control system which solves the three neighbor station keeping problem for a point modelled agent. The control system consists of a "multi-estimator" $\mathbb{E}$, a "multi-controller" $\mathbb{C}$, a "monitor" $\mathbb{M}$ and a "dwell-time switching logic" $\mathbb{S}$. These terms and definitions have been discussed before in [14, 15] and elsewhere. In Section 4.3, the output of the monitor is defined to be a parameter-dependent, scalar-valued signal of the form $M(W, P) = \text{trace}\{\begin{bmatrix} I & P \end{bmatrix} W \begin{bmatrix} I & P \end{bmatrix}'\}$ where W is a 4×4 positive semi-definite signal generated by the monitor and P is a 2×2 non-singular matrix of parameters taking values in a compact but non-convex parameter space $\mathcal{P}$. Although this particular definition is intuitive and natural for the adaptive solution to the station keeping problem, as we've already noted, the definition leads to non-convex optimization problem. To avoid this, use is made of the previously mentioned fact that any 2×2 non-singular matrix B can be written as $B = U(I + L)S$ where U is a specially structured matrix from a finite set, L is strictly lower triangular and S is symmetric and positive definite [12]. In Section 4.3 $M(\cdot)$ is redefined as $M(W, U, L, S) = \text{trace}\{\begin{bmatrix}(I - L)U' & S\end{bmatrix} X \begin{bmatrix}(I - L)U' & S\end{bmatrix}'\}$ where L and S take values in compact convex sets $\mathcal{L}$ and $\mathcal{S}$ respectively. More detailed descriptions of these sets are derived in Section 6. In Section 7 it is then explained how to re-formulate the resulting problem of minimizing $M(W, U, L, S)$ over $\mathcal{L} \times \mathcal{S}$ for fixed W and U, as a semi-definite programming problem.

Because of the re-parameterization just outlined, the resulting switched adaptive control is completely tractable and easy to implement. In addition, it has especially desirable properties. For example, in the absence of errors the control causes agent positioning to occur exponentially fast; moreover it guarantees that performance will degrade gracefully in the face of increasing measurement and miss-alignment errors, provided the measurement errors are not too large. In Section 5 we sketch the ideas upon which these claims are based.

2 Formulation

Let $n > 1$ be an integer. The system of interest consists of $n + 1$ points in the plane labelled $0, 1, 2, \ldots, n$ which will be referred to as agents. Let $x_0, x_1, \ldots, x_n$ denote the coordinate vector of current positions of agents $0, 1, 2, \ldots n$ respectively with respect to a common frame of reference. Assume that the formation is suppose to come to rest and moreover that agents $1, 2, \ldots, n$ are already at their proper positions in the formation and are at rest. Thus

$$\dot{x}_i = 0, \quad i \in \{1, 2, 3, \ldots, n\}. \tag{1}$$

We further assume that the nominal model for how agent 0 moves is a kinematic point model of the form

$$\dot{x}_0 = u \tag{2}$$

where u is an open loop control taking values in $\mathbb{R}^2$.

Suppose that agent 0 can sense its distances $y_1, y_2, \ldots, y_n$ from neighboring agents $1, 2, \ldots, n$ with uniformly bounded, additive errors $\epsilon_1, \epsilon_2, \ldots, \epsilon_n$ respectively. Thus

$$y_i = ||x_i - x_0|| + \epsilon_i, \quad i \in \{1, 2, \ldots, n\} \tag{3}$$

where $||\cdot||$ denotes the Euclidean 2-norm. Suppose in addition that agent 0 is given a set of non-negative numbers $d_1, d_2, \ldots, d_n$, where d_i represents a desired distance from agent 0 to agent i. The problem is to devise a control law depending on the d_i and the y_i which, were the ϵ_i all zero, would cause agent 0 to move to a position in the formation which, for $i \in \{1, 2, \ldots, n\}$, is d_i units from agent i. We call this the *n neighbor station keeping problem.* We shall also require the controllers we devise to guarantee that errors between the y_i and their desired values eventually become small if the measurement errors are all small.

Let x^* denote the target position to which agent 0 would have to move were the station keeping problem solvable. Then x^* would have to satisfy

$$d_i = ||x_i - x^*||, \quad i \in \{1, 2, \ldots, n\}. \tag{4}$$

There are two cases to consider.

1) If $n = 2$, there will be two solutions x^* to (4) if $|d_1 - d_2| < ||x_1 - x_2|| < d_1 + d_2$ and no solutions if either $|d_1 - d_2| > ||x_1 - x_2||$ or $||x_1 - x_2|| > d_1 + d_2$. We will assume that two solutions exist and that the target position is the one closest to the initial position of agent zero.
2) If $n \geq 3$, there will exist a solution x^* to (4) only if agents 1 through n are aligned in such a way so that the circles centered at the x_i of radii d_i all intersect at least one point. If the x_i are so aligned and at least three x_i are not co-linear, then x^* is even unique. Such alignments are of course exceptional. To account for the more realistic situation when points are out of alignment, we will assume instead of (4), that there is a value of x^* for which
$$d_i = ||x^* - x_i|| + \bar{\epsilon}_i, \quad i \in \{1, 2, \ldots, n\} \tag{5}$$
where each $\bar{\epsilon}_i$ is a small miss-alignment error.

Our specific control objective can now be stated. Devise a feedback control for agent 0, using the d_i and measurements y_i, which bounds the induced $\mathcal{L}^2$ gains from each ϵ_i and each $\bar{\epsilon}_i$ to each of the errors

$$e_i = y_i^2 - d_i^2, \quad i \in \{1, 2, 3, \ldots, n\}. \tag{6}$$

We will address this problem using well known concepts and constructions from adaptive control.

3 Error Models

The controllers which we propose to study will all be based on suitably defined error models. We now proceed to develop these models.

3.1 Error Equations

To begin, we want to derive a useful expression for each e_i. In view of (3)

$$y_i^2 = ||x_i - x_0||^2 + 2\epsilon_i||x_i - x_0|| + \epsilon_i^2.$$

But

$$||x_i - x_0||^2 = ||x_i - x^*||^2 + 2(x^* - x_i)'\bar{x}_0 + ||\bar{x}_0||^2$$

where

$$\bar{x}_0 = x_0 - x^*. \tag{7}$$

Moreover from (5)

$$d_i^2 = ||x_i - x^*||^2 + 2\bar{\epsilon}_i||x_i - x^*|| + \bar{\epsilon}_i^2.$$

From these expressions and the definition of e_i in (6) it follows that

$$e_i = 2(x^* - x_i)'\bar{x}_0 + ||\bar{x}_0||^2 + 2\epsilon_i||\bar{x}_0|| + \eta_i \tag{8}$$

where

$$\eta_i = 2\epsilon_i||x_i - x_0|| + \epsilon_i^2 - 2\bar{\epsilon}_i||x_i - x^*|| - \bar{\epsilon}_i^2 - 2\epsilon_i||\bar{x}_0||.$$

Note that $|||x_i - x_0|| - ||\bar{x}_0||| \leq ||x_i - x^*||$ because of the triangle inequality and the definition of $\bar{x}_0$ in (7). From this and (5) it is easy to see that

$$|\eta_i| \leq (|\epsilon_i| + |\bar{\epsilon}_i|)\gamma_i \tag{9}$$

where $\gamma_i = 2d_i + |\epsilon_i - \bar{\epsilon}_i|$.

3.2 Station Keeping with $n = 3$ Neighbors

In this section we consider the case when $n = 3$. We shall assume that x_1, x_2, and x_3 are not co-linear. Note first that we can write

$$\dot{\bar{x}}_0 = u \tag{10}$$

because of (2) and the fact that $\bar{x}_0 = x_0 - x^*$. Let

$$e = \begin{bmatrix} e_1 - e_3 \\ e_2 - e_3 \end{bmatrix}$$

and define $q = B\bar{x}_0$, where

$$B = 2\left[x_3 - x_1 \; x_3 - x_2\right]'. \tag{11}$$

The error model for this case is then

$$e = q + \epsilon||B^{-1}q|| + \eta \tag{12}$$
$$\dot{q} = Bu \tag{13}$$

where

$$\epsilon = 2\begin{bmatrix}\epsilon_1 - \epsilon_3\\ \epsilon_2 - \epsilon_3\end{bmatrix} \qquad \eta = \begin{bmatrix}\eta_1 - \eta_3\\ \eta_2 - \eta_3\end{bmatrix}.$$

Our assumption that the x_i are not co-linear implies that B is non-singular. Note that since B is nonsingular, $x_0 = x^*$ whenever $q = 0$. This in turn will be the case when $e = 0$ provided $\epsilon = 0$ and $\eta = 0$. The term $||B^{-1}q||\epsilon$ can be regarded as a perturbation and can be dealt with using standard small gain arguments. Essentially linear error models like (12), (13) can also be derived for any $n > 3$.

3.3 Station Keeping with $n = 2$ Neighbors

In the two-neighbor case we've assumed that $|d_1 - d_2| < ||x_1 - x_2|| < d_1 + d_2$ and thus that two solutions x^* to (4) exist. We will assume that $\bar{x}_0$ has been defined so that $||\bar{x}_0(0)||$ is the smaller of the two possibilities. As before, and for the same reason, (10) holds. For this version of the problem we define

$$e = \begin{bmatrix}e_1\\ e_2\end{bmatrix}.$$

Let $q = B\bar{x}_0$, where now

$$B = 2\left[x^* - x_1 \; x^* - x_2\right]'. \tag{14}$$

The error model for this case is then

$$e = q + \epsilon||B^{-1}q|| + ||B^{-1}q||^2\mathbf{1} + \eta \tag{15}$$
$$\dot{q} = Bu \tag{16}$$

where

$$\mathbf{1} = \begin{bmatrix}1\\ 1\end{bmatrix} \qquad \epsilon = 2\begin{bmatrix}\epsilon_1\\ \epsilon_2\end{bmatrix} \qquad \eta = \begin{bmatrix}\eta_1\\ \eta_2\end{bmatrix}.$$

Note that our assumption that $|d_1 - d_2| < ||x_1 - x_2|| < d_1 + d_2$ implies that x_1, x_2, x^* are not co-linear. This in turn implies that B is non-singular. The essential difference between this error model and the error model for the three neighbor case is that the two-neighbor agent model has a quadratic function of state in its readout equation whereas the three-neighbor error model does not.

4 Station Keeping Supervisory Controller

In this section we will develop a set of controller equations aimed at solving the station keeping problem with three neighbors. Because of its properties, the controller we propose can also be used for the two neighbor version of the problem; however in this case meaningful results can only be claimed if agent 0 starts out at a position which is sufficiently close to its target x^*. For ease of reference, we repeat the error equations of interest.

$$e = q + \epsilon||B^{-1}q|| + \eta \tag{17}$$

$$\dot{q} = Bu. \tag{18}$$

In the sequel we will assume that $||\epsilon|| \leq \epsilon^*, \quad t \geq 0$ where ϵ^* is a positive constant which satisfies the constraint

$$\epsilon^* < \frac{1}{||B^{-1}||}. \tag{19}$$

Note that this constraint says that the allowable measurement error bound will decrease as agents 1, 2, and 3 are positioned closer and closer to colinear and/or further and further away from agent 0. While we are unable to fully justify this assumption at this time, we suspect that it is intrinsic and is not specific to the particular approach to station keeping which we are following. Our suspicion is prompted in part by the observation that the map $q \longmapsto q + \epsilon||B^{-1}q||$ will be invertible for all $||\epsilon|| \leq \epsilon^*$ if and only if (19) holds.

The type of control system we intend to develop assumes that B is unknown, but requires one to define at the outset a closed bounded subset of 2×2 non-singular matrices $\mathcal{P} \subset \mathbb{R}^{2\times 2}$ which is big enough so that it can be assumed that $B \in \mathcal{P}$. It is clear that because of the non-singularity requirement, just about any reasonably defined parameter space $\mathcal{P}$ which satisfies these conditions would not be convex, or even the union of a finite number of convex sets. This has important practical implications which we will elaborate on later.

The supervisory control system to be considered consists of a "multi-estimator" $\mathbb{E}$, a "multi-controller" $\mathbb{C}$, a "monitor" $\mathbb{M}$ and a "dwell-time switching logic" $\mathbb{S}$. These terms and definitions have been discussed before in [14, 15] and elsewhere. They are fairly general concepts, have specific meanings, and apply to a broad range of problems. Although there is considerable flexibility in how one might define these component subsystems, in this paper we shall be quite specific. The numbered equations which follow, are the equations which define the supervisory controller we will consider.

4.1 Multi-Estimator $\mathbb{E}$

For the problem of interest, the multi-estimator $\mathbb{E}$ is defined by the two equations

$$\dot{z}_1 = -\lambda z_1 + \lambda e \tag{20}$$

$$\dot{z}_2 = -\lambda z_2 + u \tag{21}$$

where λ is a design constant which must be positive but is otherwise unconstrained.

Note that the signal $\rho = z_1 + Bz_2 - q$ satisfies

$$\dot{\rho} = -\lambda\rho + \lambda(\epsilon||B^{-1}q|| + \eta).$$

For $P \in \mathcal{P}$, let $\bar{e}_P$ denote the Pth *output estimation error*

$$\bar{e}_P = z_1 + Pz_2 - e.$$

The relevant relationships between these signals when $P = B$ can be conveniently described by the block diagram in Figure 1. The diagram describes

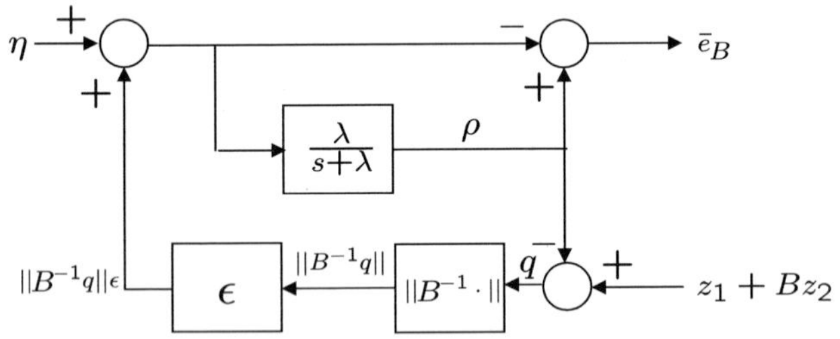

Fig. 1. Subsystem

a nonlinear dynamical system with inputs η and $z_1 + Bz_2$ and outputs $\bar{e}_B$. It is easy to verify that this system is globally exponentially stable with stability margin no smaller than $\lambda(1-\epsilon^*||B^{-1}||)$ because of the measurement constraint (19) discussed earlier. The diagram clearly implies that if ϵ and η were 0, $\bar{e}_B$ would tend to 0; in this case $z_1 + Bz_2$ would therefore be an asymptotically correct estimate of $e = q$. We exploit these observations below.

4.2 Multi-Controller $\mathbb{C}$

The multi-controller $\mathbb{C}$ we propose to study is simply

$$u = -\lambda\widehat{B}^{-1}e \tag{22}$$

where $\widehat{B}$ is a suitably defined piecewise constant switching signal taking values in $\mathcal{P}$. The definition of u has been crafted so that the "closed-loop parameterized system" matrix $-\lambda PP^{-1}$ is stable with "stability margin" λ for all $P \in \mathcal{P}$. Other controllers which accomplish this could also be used {e.g., $u = -\lambda\widehat{B}^{-1}(z_1 + \widehat{B}z_2)$}. The consequence of this definition of u is predicted by the certainty equivalence stabilization theorem [10] and is as follows. Let $\bar{e}_{\widehat{B}} = z_1 + \widehat{B}z_2 - e$ and define the so-called *injected sub-system* to be the system with input $\bar{e}_{\widehat{B}}$ and output $z_1 + Bz_2$ which results when $z_1 + Bz_2 - \bar{e}_{\widehat{B}}$ is substituted for e in the closed loop system determined by (20), (21) and (22). Thus

$$
\begin{aligned}
\dot{z}_1 &= \lambda\widehat{B}z_2 - \lambda\bar{e}_{\widehat{B}} \\
\dot{z}_2 &= -\lambda\widehat{B}^{-1}z_1 - 2\lambda z_2 + \lambda\widehat{B}^{-1}\bar{e}_{\widehat{B}}.
\end{aligned}
$$

Certainty equivalence implies that this system, viewed as a dynamical system with input $\bar{e}_{\widehat{B}}$, is also stable with stability margin λ for each fixed $\widehat{B} \in \mathcal{P}$. In this special case one can deduce this directly using the state transformation $\{z_1, z_2\} \longmapsto \{z_1, z_1 + \widehat{B}z_2\}$. For this system to have stability margin λ means that for any positive number $\lambda_0 < \lambda$ the matrix $\lambda_0 I + A(\widehat{B})$ is exponentially stable for all constant $\widehat{B} \in \mathcal{P}$. Here

$$
A(\widehat{B}) = \begin{bmatrix} 0 & \lambda\widehat{B} \\ -\lambda\widehat{B}^{-1} & -2\lambda I \end{bmatrix}
$$

which is the state coefficient matrix of the injected system.

In the sequel, we fix λ_0 at any positive value such that $\lambda_0 < \lambda(1 - \epsilon^*)||B||^{-1}$. This number turns out to be a lower bound on the convergence rate for the entire closed-loop control system.

We need to pick one more positive design parameter, called a *dwell time* τ_D. This number has to be chosen large enough so that the injected linear system defined above is exponentially stable with stability margin λ for every "admissible" piecewise constant switching signal $\widehat{B} : [0, \infty) \rightarrow \mathcal{P}$, where by *admissible* we mean a piecewise constant signal whose switching instants are separated by at least τ_D time units. This is easily accomplished because each $\lambda_0 I + A(P)$, $P \in \mathcal{P}$ is a stability matrix. All that's required then is to pick τ_D large enough so that the induced norm {any matrix norm} of each matrix $e^{\{\lambda_0 I + A(P)\}t}$, $P \in \mathcal{P}$, is less than 1.

It is useful for analysis to add to Figure 1, two copies of the injected system just defined, one $\{\Sigma_1\}$ with output $e = z_1 + Bz_2 - \bar{e}_{\widehat{B}}$ and the other $\{\Sigma_2\}$ with output $z_1 + Bz_2$. The multiple copies are valid because the injected system is an exponentially stable linear system. The resulting system is shown in Figure 2.

Note that if there were a gain between $\bar{e}_B$ and $\bar{e}_{\widehat{B}}$, and if ϵ were small enough, the overall system shown in Figure 2 would be exponentially stable

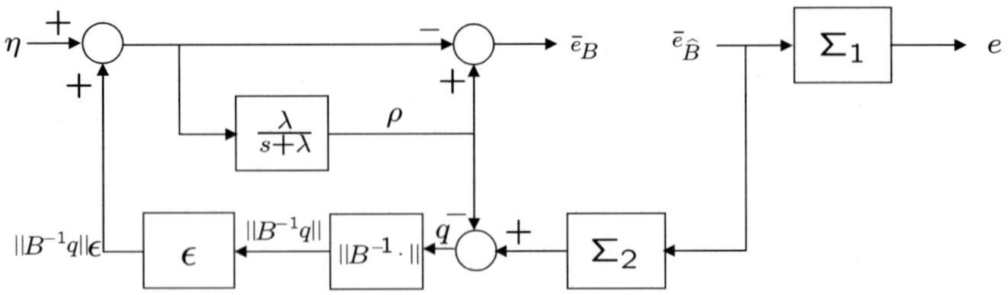

Fig. 2. Subsystem for analysis

and bounded η would produce bounded e. We return to this observation later.

4.3 Monitor $\mathbb{M}$

The state dynamic of monitor $\mathbb{M}$ is defined by the equation

$$\dot{W} = -2\lambda_0 W + \begin{bmatrix} z_1 - e \\ z_2 \end{bmatrix} \begin{bmatrix} z_1 - e \\ z_2 \end{bmatrix}' \tag{23}$$

where W is a "weighting matrix" which takes values in the linear space $\mathcal{X}$ of 4×4 symmetric matrices; although not crucial, for simplicity we will require $\mathbb{M}$ to be initialized at zero; thus $W(0) = 0$. This clearly implies that $W(t)$ is positive semi-definite for all $t \geq 0$. Note that it takes only 10 differential equations rather than 16 to generate W because of symmetry.

The output of $\mathbb{M}$ - First Pass

The output of $\mathbb{M}$ is a parameter dependent "monitoring signal" which for the moment we define to be $\mu_P = M(W, P)$ where $M : \mathcal{X} \times \mathcal{P} \to \mathbb{R}$ is the scalar-valued function

$$M(X, P) = \text{trace}\{[I \ P]\, X\, [I \ P]'\}.$$

The μ_P are helpful in motivating the definition of $\mathbb{M}$ and the switching logic $\mathbb{S}$ which follows; however, they are actually not used anywhere in the implemented system. It is obvious that they could not be because there are infinitely many of them.

Note that for any $P \in \mathcal{P}$,

$$\dot{\mu}_P = -2\lambda_0 \mu_P + \text{trace}(\{z_1 - e + Pz_2\}\{z_1 - e + Pz_2\}')$$

so

$$\dot{\mu}_P = -2\lambda_0 \mu_P + ||z_1 - e + Pz_2||^2.$$

But $\bar{e}_P = z_1 - e + Pz_2$. Therefore

$$\dot{\mu}_P = -2\lambda_0\mu_P + ||\bar{e}_P||^2$$

and

$$M(W,P) = \int_0^t e^{-2\lambda_0(t-s)}||\bar{e}_P||^2 ds.$$

Thus if we introduce the exponentially weighted 2-norm

$$||\omega||_t = \sqrt{\int_0^t \{e^{\lambda_0 s}||\omega(s)||\}^2 ds}$$

where ω is a piecewise continuous signal, then

$$M(W(t),P) = e^{-2\lambda_0 t}||\bar{e}_P||_t^2, \quad t \geq 0.$$

Minimizing $M(W(t),P)$ with respect to P and setting $\widehat{B}(t)$ to the resulting minimizing value, would then yield an inequality of the form

$$||\bar{e}_{\widehat{B}}||_t \leq ||e_B||_t.$$

Were it possible to accomplish this at every instant of time and were $\widehat{B}$ changing slowly enough so that all of the time-varying subsystems in Figure 2 were exponentially stable, then one could conclude that for ϵ^* sufficiently small, the resulting overall system with input η and output e would be stable with respect to the exponentially weighted norm we've been discussing. It is of course not possible to carry out these steps instantly and even if it were, $\widehat{B}$ would likely be changing too fast for the time-varying subsystems in Figure 2 to be exponentially stable. Were we to continue with this definition of μ_P, we would nonetheless, want to minimize $M(W(t),P)$ from time to time and in doing so would end up with an input-output stable system. In fact the implementation of dwell time switching proposed in [3] requires such minimizations to be carried out. But were we to proceed with this approach, we'd run head on into an important practical problem which we want to address.

A Non-Convex Parameter Space

Note that even though $M(X,P)$ is a quadratic positive semi-definite function of the elements of P, the problem of minimizing $M(X,P)$ over $\mathcal{P}$ is still very complex because $\mathcal{P}$ is not typically convex or even a finite union of convex sets. Thus if we were to use such a parameter space and proceed as we've just outlined, we'd be faced with an intractable non-convex optimization problem. The root of the problem stems from the requirement that the algebraic curve

$$\mathcal{C} = \{P : p_{11}p_{22} - p_{12}p_{21} = 0\}$$

in $\mathbb{R}^{2\times 2}$ on which P is singular cannot intersect $\mathcal{P}$. One way to deal with this difficulty relies on an idea called "cyclic switching" which was specifically devised to deal with this type of problem [17, 13]. Cyclic switching is roughly as follows. First $\mathcal{P}$ is allowed to contain singular matrices, in which case it is reasonable to assume that it is a finite union of compact convex sets. Minimization over $\mathcal{P}$ thus becomes a finite number of standard convex programming problems. For minimizing values of $\widehat{B}$ which turn out to be close to or on $\mathcal{C}$, one uses a specially structured switching controller in place of (22) – one which does not require $\widehat{B}$ to be nonsingular. This controller is used for a specific length of time over which a "switching cycle" takes place. At the end of the cycle, minimization of $M(W, \widehat{B})$ is again carried out; if $\widehat{B}$ is again close to $\mathcal{C}$, another switching cycle is executed. On the other hand, if $\widehat{B}$ is not close to $\mathcal{C}$, the standard certainty equivalence control (22) is used. Cyclic switching is completely systematic and can be shown to solve the singularity problem of interest here. The main disadvantage of cyclic switching is that it introduces additional complexity.

There is another possible way to deal with the singularity problem. What we'd really like is to construct a parameter space $\mathcal{P}$ which is a finite union of convex sets, defined so that every matrix in $\mathcal{P}$ is nonsingular and, in addition, the matrices in $\mathcal{P}$ correspond to a "large" class of possible positions of agents $1, 2, 3$. Keep in mind that the convex subsets whose union defines such a $\mathcal{P}$, *can* overlap. This suggests the following problem.

Problem 1 (Convex Covering Problem). Suppose that we are given a compact subset $\mathcal{P}_0$ of a finite dimensional space which is disjoint from a second closed subset $\mathcal{C}$ {typically an algebraic curve}. Define a *convex cover* of $\mathcal{P}_0$ to mean a finite set of possibly overlapping convex subsets $\mathcal{E}_i$ such that the union of the $\mathcal{E}_i$ contains $\mathcal{P}_0$ but is disjoint from $\mathcal{C}$. One could then define $\mathcal{P}$ to be the union of the $\mathcal{E}_i$.

The existence of such a convex cover can be established as follows[2]. Let d denote the shortest distance between $\mathcal{P}_0$ and $\mathcal{C}$; thus $d = \min\{||p - s|| : p \in \mathcal{P}_0,\ s \in \mathcal{C}\}$. Since $\mathcal{P}_0$ and $\mathcal{C}$ are disjoint, $d > 0$. Let r be any positive number less than d and for each $p \in \mathcal{P}_0$ let $\mathcal{B}(p) = \{q : ||q - p|| < r,\ q \in \mathcal{P}_0\}$. Then for each $p \in \mathcal{P}_0$, the closure of $\mathcal{B}(p)$ and $\mathcal{C}$ are disjoint. Moreover the set of all $\mathcal{B}(p)$ is an open cover of $\mathcal{P}_0$. Thus by the Heine-Borel Theorem, there is a finite subset of the $\mathcal{B}(p)$, say $\mathcal{B}_1, \mathcal{B}_2, \ldots, \mathcal{B}_m$ which covers $\mathcal{P}_0$. Setting $\mathcal{E}_i$ equal to the closure of $\mathcal{B}_i$, $i \in \{1, 2, \ldots, m\}$ thus provides a convex cover of $\mathcal{P}_0$ whose union is disjoint from $\mathcal{C}$. Of course this construction would typically produce a cover containing many more convex subsets than might be needed. The question then is how might one going about constructing a convex cover consisting of the smallest number of subsets possible? This unfortunately appears to be a very difficult problem. Nonetheless, its solution could provide an attractive

[2] We thank Ji Liu for pointing this out to us.

alternative to the approach to station keeping which we've outlined in this paper.

There is a third way to avoid the tractability problem which is the approach which we will take here. The key idea is to use a different parameterization which we describe next.

Re-parameterization

Let $\mathcal{U}$ denote the set of all 2×2 matrices U, where each U is a matrix of 0's, 1's and -1's having exactly one nonzero entry in each row and column; there are exactly eight such matrices. It is known [12] that any 2×2 nonsingular matrix M can be written as $M = U(I+L)S$ for some $U \in \mathcal{U}$, some strictly lower triangular matrix L and some symmetric positive definite matrix S. This suggests that we consider a parameter space

$$\mathcal{P} = \{U(I+L)S : \{U, L, S\} \in \mathcal{U} \times \mathcal{L} \times \mathcal{S}\}$$

where $\mathcal{L}$ is a compact, convex subset of the linear space of strictly lower triangular 2×2 matrices and $\mathcal{S}$ a compact, convex subset of the convex set of all 2×2 positive definite matrices. Notice that this definition of $\mathcal{P}$ satisfies both the compactness requirement and the requirement that its elements are all non-singular matrices. Of course one needs to also make sure that $\mathcal{L}$ and $\mathcal{S}$ are large enough so that $B \in \mathcal{P}$. We will say more about how to do this later. For the present we will assume that $B \in \mathcal{P}$ and thus that there are matrices $U_B \in \mathcal{U}$, $L_B \in \mathcal{L}$ and $S_B \in \mathcal{S}$ such that

$$B = U_B(I + L_B)S_B.$$

In the sequel we will show that it is possible to meaningfully redefine the type of optimization referred to above as the problem of minimizing a function $J(U, L, S)$ over the set $\mathcal{U} \times \mathcal{L} \times \mathcal{S}$. While this set is not convex, $\mathcal{L} \times \mathcal{S}$ is. Moreover, as we shall see, for each fixed $U \in \mathcal{U}$, $J(U, L, S)$ is a convex, quadratic function of the entries in L and S. Because of this, the minimization of $J(U, L, S)$ over $\mathcal{U} \times \mathcal{L} \times \mathcal{S}$ boils down to solving eight convex programming problems, one for each $U \in \mathcal{U}$.

The Output of $\mathbb{M}$ – Second Pass

In the light of the preceding discussion we now re-define $\mathbb{M}$'s output to be $\mu_{\{U,L,S\}} = M(W, U, L, S)$ where now $M : \mathcal{X} \times \mathcal{U} \times \mathcal{L} \times \mathcal{S} \rightarrow \mathbb{R}$ is

$$M(X, U, L, S) = \text{trace}\{\left[(I-L)U'\ S\right] X \left[(I-L)U'\ S\right]'\}. \tag{24}$$

In this case it is easy to see that

$$M(W(t), U, L, S) = e^{-2\lambda_0 t}||(I-L)U'\bar{e}_P||_t^2, \quad t \geq 0$$

where $P = U(I + L)S$. In deriving this expression for M we've made use of the easily verified formulas $U' = U^{-1}$, $U \in \mathcal{U}$ and $(I + L)^{-1} = I - L$, $L \in \mathcal{L}$.

The matrix $\widehat{B}$ used in the definition of u in (22) is now defined by the formula

$$\widehat{B} = \widehat{U}(I + \widehat{L})\widehat{S} \tag{25}$$

where $\{\widehat{U}, \widehat{L}, \widehat{S}\}$ is a piecewise constant switching signal taking values in $\mathcal{U} \times \mathcal{L} \times \mathcal{S}$. This signal will be generated by a "dwell-time switching logic" which will be described next.

4.4 Dwell-Time Switching Logic $\mathbb{S}$

For our purposes a *dwell-time switching logic* $\mathbb{S}$, is a hybrid dynamical system whose input and output are W and $\widehat{B}$ respectively, and whose state is the ordered triple $\{X, \tau, \{\widehat{U}, \widehat{L}, \widehat{S}\}\}$. Here X is a discrete-time matrix which takes on sampled values of W, and τ is a continuous-time variable called a *timing*

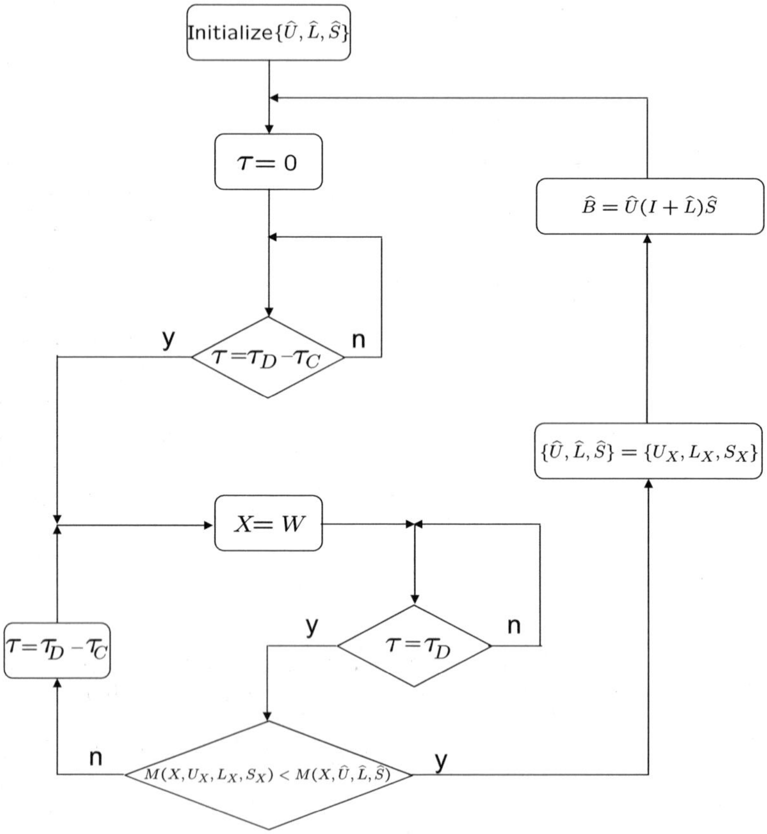

Fig. 3. Dwell-time switching logic $\mathbb{S}$

signal. τ takes values in the closed interval $[0, \tau_D]$. Also assumed pre-specified is a *computation time* $\tau_C \leq \tau_D$ which bounds from above for any $X \in \mathcal{W}$, the time it would take to compute a value $\{U, L, S\} \in \mathcal{U} \times \mathcal{L} \times \mathcal{S}$ which minimizes $M(X, U, L, S)$. Between "event times," τ is generated by a reset integrator according to the rule $\dot{\tau} = 1$. Event times occur when the value of τ reaches either $\tau_D - \tau_C$ or τ_D; at such times τ is reset to either 0 or $\tau_D - \tau_C$ depending on the value of $\mathbb{S}$'s state. $\mathbb{S}$'s internal logic is defined by the flow diagram shown in Figure 3 where $\{U_X, L_X, S_X\}$ denotes a value of $\{U, L, S\} \in \mathcal{U} \times \mathcal{L} \times \mathcal{S}$ which minimizes $M(X, U, L, S)$.

The definition of $\mathbb{S}$ clearly implies that its output $\widehat{B}$ is an admissible switching signal. This means that switching cannot occur infinitely fast and thus that existence and uniqueness of solutions to the differential equations involved is not an issue.

Note that implementation of the switching logic just described requires an algorithm capable of minimizing trace$\{M(X, U, L, S)\}$ over $\mathcal{U} \times \mathcal{L} \times \mathcal{S}$ for various values of $X \in \mathcal{X}$. As we've already explained, for each fixed $U \in \mathcal{U}$, and $X \in \mathcal{X}$, minimization of trace$\{M(X, U, L, S)\}$ reduces to a convex programming problem. Thus for each $X \in \mathcal{X}$, it is enough to solve eight convex programming problems, one for each value of $U \in \mathcal{U}$; the results of these eight computations can then be compared to find the values of U, L and S which attain a global minimum of trace$\{M(X, U, L, S)\}$ over $\mathcal{U} \times \mathcal{L} \times \mathcal{S}$. In other words, by making use of the parameterization we've been discussing, we've been able to reformulate the overall adaptive algorithm in such a way that at each event time all that is necessary is to solve eight, independent quadratic programming problems, one for each $U \in \mathcal{U}$. Of course each of these eight problems may still be challenging. In Section 7 we will explain how each can be reformulated as a semi-definite programming problem.

5 Results

The results which follow rely heavily on the following proposition which characterizes the effect of the monitor-dwell time switching logic subsystem.

Proposition 1. *Suppose that* $W(0) = 0$, *that* $\widehat{B} = \widehat{U}(I + \widehat{L})\widehat{S}$ *is the response of the monitor-switching logic subsystem* $\{\mathbb{M}, \mathbb{S}\}$ *to any continuous input signals* e, z_1, *and* z_2 *taking values in* $\mathbb{R}^2$, *and that for* $\{U, L, S\} \in \mathcal{U} \times \mathcal{L} \times \mathcal{S}$, $\bar{e}_P = (z_1 - e) + P z_2$ *where* $P = U(I + L)S$. *For each real number* $\gamma > 0$ *and each fixed time* $T > 0$, *there exists piecewise-constant signals* $H : [0, \infty) \to \mathbb{R}^{2\times 4}$ *and* $\psi : [0, \infty) \to \{0, 1\}$ *such that*

$$|H(t)| \leq \gamma, \quad t \geq 0 \tag{26}$$

$$\int_0^\infty \psi(t)dt \leq 4(\tau_D + \tau_C) \tag{27}$$

and

$$||(1-\psi)(\bar{e}_{\widehat{B}} - Hz) + \psi\bar{e}_B||_T \leq \delta||\bar{e}_B||_T \tag{28}$$

where $z = \begin{bmatrix} z_1' & z_2' \end{bmatrix}'$,

$$\delta = 1 + 8\alpha^2 \left(\frac{1 + \text{diameter}\{\mathcal{P}\}}{\gamma}\right)^4,$$

and

$$\alpha = \max_{L \in \mathcal{L}} ||I + L||.$$

This proposition is a minor modification of a similar proposition proved in [14, 15]. The proposition summarizes the key consequences of dwell time switching which are needed to analyze the system under consideration. While the inequality in (28) is more involved than the inequality $||\bar{e}_{\widehat{B}}||_t \leq ||\bar{e}_B||_t$ mentioned earlier, the former is provably correct whereas the latter is not. Despite its complexity, (28) can be used to establish input-output stability with respect to the exponentially weighted norm $||\cdot||_t$. The idea is roughly as follows. Fix $T > 0$ and pick γ small enough so that $\lambda_0 I + A(\widehat{B}) + (1-\psi)D(\widehat{B})H$ is exponentially stable where $A(\widehat{B})$ is the state evolution matrix of the injected system defined at the beginning of Section 4.2 and $D(\widehat{B}) = \begin{bmatrix} -\lambda I' & \lambda(\widehat{B}^{-1})' \end{bmatrix}'$. The fact that ψ has a finite $\mathcal{L}^1$ norm {cf. (27)}, implies that $\lambda_0 I + A(\widehat{B}) + (1-\psi)D(\widehat{B})H + \psi \begin{bmatrix} 0 & \widehat{B} - B \end{bmatrix}$ is exponentially stable as well. Next define

$$\bar{e} = (1-\psi)(\bar{e}_{\widehat{B}} - Hz) + \psi\bar{e}_B.$$

Then

$$||\bar{e}||_T \leq \delta||\bar{e}_B||_T \tag{29}$$

because of (28). The definition of $\bar{e}$ implies that

$$\bar{e}_{\widehat{B}} = \bar{e} + (1-\psi)Hz + \psi \begin{bmatrix} 0 & \widehat{B} - B \end{bmatrix} z.$$

Substitution into the injected system defined earlier yields the exponentially stable system

$$\dot{z} = \{A(\widehat{B}) + (1-\psi)D(\widehat{B})H + \psi \begin{bmatrix} 0 & \widehat{B} - B \end{bmatrix}\}z + D(\widehat{B})\bar{e}$$

with input $\bar{e}$. Now add to Figure 1, two copies of the system just defined, one $\{\bar{\Sigma}_1\}$ with output $e = \begin{bmatrix} I & \widehat{B} \end{bmatrix} z - \{\bar{e} + (1-\psi)Hz + \psi \begin{bmatrix} 0 & \widehat{B} - B \end{bmatrix} z\}$ and the other $\{\bar{\Sigma}_2\}$ with output $z_1 + Bz_2 = \begin{bmatrix} I & B \end{bmatrix} z$. Like before, the multiple copies are valid because the matrix $A(\widehat{B}) + (1-\psi)D(\widehat{B})H + \psi \begin{bmatrix} 0 & \widehat{B} - B \end{bmatrix}$ is exponentially stable. The resulting overall system is shown in Figure 4.

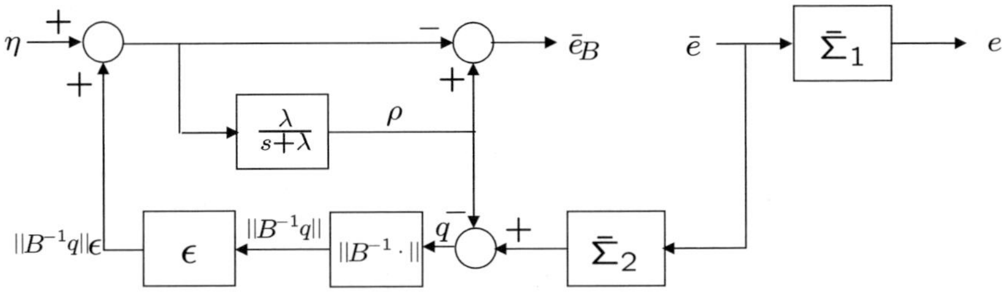

Fig. 4. Snapshot at time T of the overall subsystem for analysis

In the light of (29) it is easy to see that if the bound ϵ^* on ϵ is sufficiently small, the induced gain of this system from η to e with respect to $||\cdot||_T$ is bounded by a finite constant g_T. It can be shown that g_T in turn, is bounded above by a constant g not depending on T [15]. Since this is true for all T, it must be true that g bounds the induced gain from η to e with respect to $||\cdot||_\infty$.

The following results are fairly straightforward consequences of these ideas. Detailed proofs, specific to the problem at hand, can be found in the full-length version of this paper. The results are as follows:

1) If all measurement errors ϵ_i and all miss-alignment errors $\bar{e}_i$ are zero, then, no matter what its initial value, $x_0(t)$ tends to the unique solution x^* to (4) as fast as $e^{-\lambda_0 t}$.
2) If the measurement errors ϵ_i and the miss-alignment errors $\bar{e}_i$ are not all zero, and the ϵ_i sufficiently small, then no matter what its initial value, $x_0(t)$ tends to a value for which the norm of the error e is bounded by a constant times the sum of the norms of the ϵ_i and the $\bar{\epsilon}_i$.

Before leaving this section, it should be mentioned that success with the new parameterization we've proposed, of course comes with a price. Note that the gain δ which appears in the statement of Proposition 1 is an increasing function of α, and moreover $\alpha > 1$. Thus the effect of re-parameterization is, in essence, to increase the "gain" around the loop containing $\bar{\Sigma}_2$ in Figure 4. This in turn, reduces the stability margin associated with ϵ and also increases overall induced gain from η to e.

6 Definitions for $\mathcal{L}$ and $\mathcal{S}$

So far we have assumed that $\mathcal{L}$ is a compact, convex subset of the linear space of strictly lower triangular 2×2 matrices and that $\mathcal{S}$ is a compact, convex subset of the set of positive definite 2×2 matrices. The assumptions are sufficient to ensure that any matrix in

$$\mathcal{P} = \{U(I + L)S : (U, L, S) \in \mathcal{U} \times \mathcal{L} \times \mathcal{S}\}$$

is invertible and also that the minimization of

$$M(X,U,L,S) = \text{trace}\{\left[(I-L)U'\ S\right] X \left[(I-L)U'\ S\right]'\}$$

over $\mathcal{L} \times \mathcal{S}$ for any fixed $U \in \mathcal{U}$ and any fixed positive semi-definite 2×2 matrix X, is a convex programming problem. But we've not yet explained how to explicitly define $\mathcal{L}$ and $\mathcal{S}$. To do this, it makes sense to first define bounds for B which are meaningful for the problem at hand. Towards this end, suppose that agent 0 has a limited sensing radius ρ. Since we've assumed that agent 0 can sense the distances to agents $1, 2$, and 3, it must be true that $||x_3 - x_1|| \leq 2\rho$ and $||x_3 - x_1|| \leq 2\rho$. But $B = 2\left[x_3 - x_1\ x_3 - x_2\right]'$. Prompted by this we will assume that $\sqrt{B'B} \leq \beta_2 I$ where $\beta_2 = 4\rho$.

We've also assumed that agents $1, 2$ and 3 are not positioned along a line; this is equivalent to B being nonsingular. One measure of B's non-singularity, is its smallest singular value. Prompted by this, we will assume that there is a positive number β_1 such that $\sqrt{B'B} \geq \beta_1 I$; β_1 might be chosen empirically to reflect the degree to which the three leader agents are non co-linear in a given formation. We shall assume that such a number has been chosen and moreover that $\beta_1 < \beta_2$. In summary we suppose that bounds β_1 and β_2 have been derived such that

$$\beta_1 I \leq \sqrt{B'B} \leq \beta_2 I \tag{30}$$

where β_1 and β_2 are distinct positive numbers. It is obvious that the set of matrices B satisfying these inequalities is not convex.

Our next objective is to define $\mathcal{L}$ and $\mathcal{S}$ so that any matrix B satisfying (30) is in $\mathcal{P}$. Let $\mathcal{L}$ be the set of all strictly lower triangular 2×2 matrices $L = \left[l_{ij}\right]$ for which

$$|l_{21}| \leq 1 + \sqrt{2}\frac{\beta_2}{\beta_1}. \tag{31}$$

In addition, let $\mathcal{S}$ be the set of all 2×2, symmetric matrices satisfying

$$\sigma_1 I \leq S \leq \sigma_2 I \tag{32}$$

where

$$\sigma_1 = \frac{1}{\left(2\sqrt{1+\left(\frac{\beta_2}{\beta_1}\right)^2}\right)}\beta_1$$

$$\sigma_2 = \left(2\sqrt{1+\left(\frac{\beta_2}{\beta_1}\right)^2}\right)\beta_2. \tag{33}$$

It is now shown that any matrix B satisfying (30) is in $\mathcal{P}$.

As a first step, let us note that b_{11} and b_{21} cannot both be zero because B is nonsingular. If $|b_{11}| \geq |b_{21}|$, let

$$U = \begin{bmatrix} \text{sign}\{b_{11}\} & 0 \\ 0 & \text{sign}\{b_{11}d\} \end{bmatrix}$$

$$L = \begin{bmatrix} 0 & 0 \\ \frac{u_{22}b_{21} - u_{11}b_{12}}{|b_{11}|} & 0 \end{bmatrix}$$

$$S = \begin{bmatrix} |b_{11}| & u_{11}b_{12} \\ u_{11}b_{12} & \frac{b_{12}^2 + |d|}{|b_{11}|} \end{bmatrix}. \tag{34}$$

On the other hand, if $|b_{21}| > |b_{12}|$, let

$$U = \begin{bmatrix} 0 & -\text{sign}\{b_{21}d\} \\ \text{sign}\{b_{21}\} & 0 \end{bmatrix}$$

$$L = \begin{bmatrix} 0 & 0 \\ \frac{u_{12}b_{11} - u_{21}b_{22}}{|b_{21}|} & 0 \end{bmatrix}$$

$$S = \begin{bmatrix} |b_{21}| & u_{21}b_{22} \\ u_{21}b_{22} & \frac{b_{22}^2 + |d|}{|b_{21}|} \end{bmatrix}. \tag{35}$$

In either case it is easy to verify that $B = U(I + L)S$. It is also clear that in either case $U \in \mathcal{U}$, that L is strictly lower triangular and that S is symmetric. Thus to prove that $B \in \mathcal{P}$, it is sufficient to show that in either of the two cases, L and S satisfy (31) and (32) respectively. We will do this only for the case $|b_{11}| \geq |b_{21}|$ as similar reasoning applies to the case $|b_{21}| < |b_{11}|$.

Let us note from (34) that $|l_{21}| \leq \left|\frac{b_{21}}{b_{11}}\right| + \left|\frac{b_{12}}{b_{11}}\right|$. By assumption $|b_{11}| \geq |b_{21}|$; this implies that $\left|\frac{b_{21}}{b_{11}}\right| \leq 1$ so $|l_{21}| \leq 1 + \left|\frac{b_{12}}{b_{11}}\right|$. Now from (30), $\beta_1 \leq \sqrt{b_{11}^2 + b_{21}^2}$, so $\beta_1 \leq \sqrt{2b_{11}^2} = \sqrt{2}|b_{11}|$; also from (30), $|b_{12}| \leq \beta_2$. Therefore $\left|\frac{b_{12}}{b_{11}}\right| \leq \sqrt{2}\frac{\beta_2}{\beta_1}$. It follows that l_{21} satisfies (31).

Next observe that $B'B = S(I+L)'U'U(I+L)S = S(I+L)'(I+L)S$. Now $(I+L)'(I+L) \leq (2+|l_{12}|^2)I$. Therefore $B'B \leq (2+|l_{12}|^2)S^2$. From this and (30), it follows that $S^2 \geq \frac{\beta_1^2}{2+|l_{12}|^2}I$. From (31),

$$l_{21}^2 \leq 2\left(1 + 2\frac{\beta_2^2}{\beta_1^2}\right). \tag{36}$$

Therefore $S^2 \geq \frac{\beta_1^4}{4(\beta_1^2+\beta_2^2)}I = \sigma_1^2 I$.

Finally observe that $S = (I - L)U'B$ and thus that $S^2 = B'U(I - L)'(I - L)U'B$. But $(I-L)'(I-L) \leq (2+|l_{12}|^2)I$. Therefore $S^2 \leq (2+|l_{12}|^2)B'UU'B =$

$(2+|l_{12}|^2)B'B$. From this (30), and (36) it follows that $S^2 \leq 4(1+\frac{\beta_2^2}{\beta_1^2})\beta_2^2 I$. Therefore S satisfies both inequalities in (32). This means that $B \in \mathcal{P}$.

7 Semi-Definite Programming Formulation

Fix $U \in \mathcal{U}$, and let $X \in \mathcal{X}$ be a given positive semi-definite matrix. To implement the dwell time switching logic defined in Section 4.4, it is necessary to make use of an algorithm capable of minimizing over $\mathcal{L} \times \mathcal{S}$, a cost function of the form

$$N(L,S) = \text{trace}\{\left[(I-L)U'\ S\right] X \left[(I-L)U'\ S\right]'\}. \tag{37}$$

Our aim is to explain how to reformulate this convex optimization problem as a convex semi-definite programming problem over the space $\mathcal{Y} \times \mathcal{L} \times \mathcal{Y}$ where $\mathcal{Y}$ is the linear space of 2×2 symmetric matrices[3]. As a first step towards this end, we exploit two easily proved facts. First, if (L_1, S_1) minimizes $N(L,S)$ over $\mathcal{L} \times \mathcal{S}$, then $(\{\left[(I-L_1)U_1'\ S_1\right] X \left[(I-L_1)U_1'\ S_1\right]'\}, L_1, S_1)$ minimizes

$$\bar{N}(Y,L,S) = \text{trace}\{Y\}$$

over $\mathcal{Y} \times \mathcal{L} \times \mathcal{S}$ subject to the constraint that $Y - [(I-L_1)U_1'\ S_1]X[(I-L_1)U_1'\ S_1]'$ is positive semi-definite. Second, if (Y_2, L_2, S_2) minimizes $\bar{N}(Y,L,S)$ over $\mathcal{Y} \times \mathcal{L} \times \mathcal{S}$ subject to the constraint that $Y - [(I-L_1)U_1'\ S_1]X[(I-L_1)U_1'\ S_1]'$ is positive semi-definite, then (L_2, S_2) minimizes $N(L,S)$ over $\mathcal{L} \times \mathcal{S}$. In other words, the optimization problem of interest is equivalent to minimizing the cost $\bar{N}(Y,L,S)$ over $\mathcal{Y} \times \mathcal{L} \times \mathcal{S}$ subject to the constraint

$$Y - \left[(I-L)U'\ S\right] X \left[(I-L)U'\ S\right]' \geq 0. \tag{38}$$

To proceed, let us next observe that the matrix to the left in the above inequality, is the Schur complement of the matrix

$$Q = \begin{bmatrix} I & R'\left[(I-L)U'\ S\right]' \\ \left[(I-L)U'\ S\right]R & Y \end{bmatrix}$$

where R is any matrix such that $X = RR'$. Thus the matrix inequality in (38) is equivalent to the matrix inequality

$$Q \geq 0. \tag{39}$$

Moreover the constraint that $S \in \mathcal{S}$ is equivalent to $S \in \mathcal{Y}$ and the pair of linear matrix inequality constraints $\sigma_2 I - S \geq 0$ and $S - \sigma_1 I \geq 0$. These constraints can be combined with (39) to give finally the constraint

[3] We are indebted to Ali Jadbabai for making us aware of this simplification.

$$\begin{bmatrix} Q & 0 & 0 \\ 0 & \sigma_2 I - S & 0 \\ 0 & 0 & S - \sigma_1 I \end{bmatrix} \geq 0. \tag{40}$$

Thus we've reduced the optimization problem of interest to minimizing $\bar{N}(Y, L, S)$ over $\mathcal{Y} \times \mathcal{L} \times \mathcal{Y}$ subject to (40). Since (31) is equivalent to two linear inequality constraints, the problem to which we've been led is a conventional convex, semi-definite programming problem [22]. Of course to carry out this optimization, one needs also an standard algorithm to factor a positive semi-definite matrix X as $X = RR'$.

8 Concluding Remarks

In this paper we have devised a tractable solution to the three neighbor station keeping problem in which range measurements are the only sensed signals upon which station keeping is to be based. The solution is the same as that in [3] except that here a special parameterization is used to avoid the non-convex optimization problem which must be solved in order to implement the algorithm in [3]. The solution in this paper is provably correct and the performance of the resulting system degrades gracefully in the face of increasing measurement and miss-alignment errors, provided the measurement errors are not too large. We have used standard constructions from adaptive control to accomplish this. Because of the exponential stability of the overall system, the same control algorithm will solve the two agent station keeping problem provided the agent is initially not too far from its target position.

Acknowledgments

This research was supported by the National Science Foundation, the US Army Research Office, and by a gift from the Xerox Corporation.

References

1. L. Armesto and J. Tornero. SLAM based on Kalman filter for multi-rate fusion of laser and encoder measurements. In *Proceedings of the 2004 IEEE/RSJ International Conference on Intelligent Robots and Systems*, pages 1860–1865, 2004.
2. M. Cao and A.S. Morse. Mainaining an autonomous agent's position in a moving formation with range-only measurements. In *Proc. of the 2007 European Control Conference*, 2007. to be presented.
3. M. Cao and A.S. Morse. Station-keeping in the plane with range-only measurements. In *Proc. of the 2007 Amer. Contr. Conf.*, 2007. to be presented.

4. S. H. Dandach, B. Fidan, S. Dasgupta, and B.D.O. Anderson. Adaptive source localization by mobile agents. In *Proc. of the 45rd IEEE Conf. on Decision and Contr.*, pages 2045–2050, 2006.
5. S. H. Dandach, B. Fidan, S. Dasgupta, and B.D.O. Anderson. A continuous time linear adaptive source localization algorithm robust to persistent drift. *Systems & Control Letters*, 2007. To appear.
6. M. W. Dissanayake, P. Newman, H. F. Durrant-Whyte, S. Clark, and M. Csorba. An experimental and thoretical investigation into simultaneous localisation and map building. *Experimental Robotics*, IV:265–274, 2000.
7. H.J.S. Feder, J.J. Leonard, and C.M. Smith. Adaptive mobile robot navigation and mapping. *International Journal of Robotics Research*, 18:650–668, 1999.
8. D. Fox, W. Burgard, F. Dellaert, and S. Thrun. Monte Carlo localization: Efficient position estimation for mobile robots. In *Proceedings of the National Conference on Artificial Intelligence (AAAI)*, 1999.
9. B. Grocholsky, E. Stump, P. Shiroma, and V. Kumar. Control for localization of targets using range-only sensors. In *Proceedings of the International Symposium on Experimental Robotics*, 2006.
10. J.P. Hespanha and A.S. Morse. Certainty equivalence implies detectability. *Systems & Control Letters*, pages 1–13, 1999.
11. J.J. Leonard and H.F. Durrant-Whyte. Simultaneous map building and localization for an autonomous mobile robot. In *Proceedings of the 1991 IEEE/RSJ International Workshop on Intelligent Robots and Systems*, pages 1442–1447, 1991.
12. A.S. Morse. A gain matrix decomposition and some of its applications. *SCL*, 21:1–10, 1993.
13. A.S. Morse. Cyclic switching and supervisory control. In *Proceedings of the 1995 IFAC Symposium on Nonlinear Control Systems Design*, pages 499–502, 1995.
14. A.S. Morse. Analysis of a supervised set-point control system containing a compact continuum of finite dimensional linear controllers. In *Proc. of Mathematical Theory of Networks and Systems*, 2004.
15. A.S. Morse. Logically switched dynamical systems. In *Nonlinear and Optimal Control Theory*, pages 1–84. Springer-Verlag, 2005. To appear.
16. P. Newman. *On the Structure and Solution of the Simultaneous Localisation and Map Building Problem.* Ph.D dissertation, University of Sydney, Australian Centre for Field Robotics, 1999.
17. F.M. Pait and A.S. Morse. A cyclic switching strategy for parameter-adaptive control. *IEEE Trans. on Automat. Contr.*, 39(6):1172–1183, 1994.
18. A.H. Sayed and A. Tarighat. Network-based wireless location. *IEEE Signal Processing Magazine*, pages 24–40, July 2005.
19. R. Smith, M. Self, and P. Cheeseman. Estimating uncertain spatial relationships in robotics. In I. Cox and G. Wilfong, editors, *Autonomous Robot Vehicles.* Springer-Verlag, 1990.
20. E. Stump, B. Grocholsky, and V. Kumar. Extensive representaions and algorithms for nonlinear filtering and estimation. In *Proceedings of the 7th Annual Workshop on the Algorithmic Foundations of Robotics*, 2006.
21. S. Thrun, D. Fox, and W. Burgard. A probabilistic approach to concurrent mapping and localization for mobile robots. *Machine Learning*, 31:29–53, 1998.
22. L. Vandenberghe and S. Boyd. Semidefinite programming. *SIAM Review*, 38:49–95, 1996.

23. S.B. Williams, M.W.. Dissanayake, and H.F. Durrant-Whyte. Towards multi-vehicle simultaneous localisation and mapping. In *Proceedings of the 2002 IEEE International Conference on Robotics and Automation*, pages 2743–2748, 2002.
24. D.C.K. Yuen and B.A. MacDonald. An evaluation of the Sequential Monte Carlo technique for simultaneous localisation and map-building. In *Proceedings of the 2003 IEEE International Conference on Robotics and Automation*, pages 1564–1569, 2003.

Control of Hydraulic Devices, an Internal Model Approach

Kurt Schlacher and Kurt Zehetleitner

Institute of Automatic Control and Control Systems Technology, Johannes Kepler University, Linz, Austria

Summary. Hydraulic devices are used, where high forces must be generated by small devices. A disadvantage of these devices is the nonlinear behavior, especially of big devices, which are used e.g. in steel industries. Since one cannot measure the whole state in these applications, an output feedback controller is presented, which is based on a reduced observer. In addition disturbance forces, which can be described by exosystems, are taken into account. Their effect is eliminated in the steady state by the internal model approach. The presented design is applied to the hydraulic gap control of mill stands in rolling mills, such that unknown slowly varying disturbance forces and forces caused by eccentricities of the rolls are rejected.

1 Introduction

Hydraulic devices are often used, where very high forces must be generated by small devices provided there is enough space for the supply tanks. If one takes the elasticity of the oil into account, which is necessary for fast or big devices, then the intrinsic non linear behavior of these devices cannot be neglected any more. Often these hydraulic devices drive non linear mechanisms with one degree of freedom (1-DOF) such that the position of a point of the mechanism or the generated force at a point is controlled. Such a configuration is input to state linearizable, see e.g. [5], therefore, the controller design problem looks quite simple. But one cannot measure the velocity of the mechanism and its value is not derivable from the position signal because of the noise. Therefore, two approaches are possible; the first one avoids the use of the velocity signal in the control law, see e.g. [7], the second applies observers, see e.g. [4]. In addition several disturbance forces act on such a system. Typical representatives are slowly varying forces and sinusoidal forces caused by imperfections of the mechanism like eccentricities. The design goal is to eliminate the effect of the disturbances at least in the steady state. This demand brings the internal model approach into the play, see e.g. [1], [6], which is applied to the hydraulic gap control problem of stands in steel rolling mills.

According to the program presented above this contribution is organized as follows. In Section 2 the mathematical models of the mechanical and hydraulic part are derived. The design goals and the methods to achieve them are presented in Section 3, where the force and position control problem are formulated. In Section 4 the previous developed control loop design is applied to hydraulic gap control of four high mill stands equipped with a single acting hydraulic piston. Simulations show both, the applicability of the methods to industrial problems, as well as high achievable performance. Finally, this contribution finishes with some conclusions.

2 The Mechanics and Hydraulics

The plants under consideration are simple mechanical systems with one degree of freedom, which are driven by hydraulic pistons. E.g. Figure 1 shows a looper, where a hydraulic piston moves an arm with a roll. Such devices are used to control the tension of steel strips, to balance different rolling velocities, etc. We assume, that the mechanical part is described by the equations

$$\begin{aligned} \dot{x} &= \partial_p H \\ \dot{p} &= -\partial_x H - d\left(\tfrac{p}{m}\right) \end{aligned} \tag{1}$$

with Hamiltonian

$$H\left(x, p, F_h, F_1, \ldots, F_s\right) = \frac{1}{2m\left(x\right)} p^2 + V_p\left(x\right) - xF_h + \sum_{l=1}^{s} h_i F_i \ . \tag{2}$$

Here and in the following x, p denote the position and the momentum of the piston, $m\left(x\right)$ the mass, $V_p\left(x\right)$ the potential energy, d the coefficient of viscous

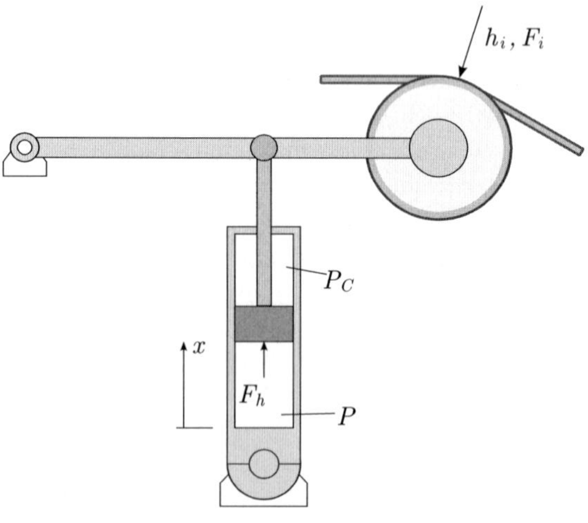

Fig. 1. A Looper, a hydromechanic device

friction. The hydraulic force F_h is the control input of the mechanical part, whereas the forces F_i describe disturbances, discussed later. The C^1 functions $h_i(x)$ are related to the points of attack x_i of the forces F_i.

We restrict the description of the hydraulic part to the quasi static case. Therefore, any solution must pull back the closed form

$$\mathrm{d}U - T\mathrm{d}S + P\mathrm{d}V$$

with the internal energy U, the temperature T, the entropy S, the pressure P, the volume V to zero. This relation is valid for a thermodynamically closed system. Introducing specific quantities $U = uM$, $S = sM$, $\rho V = M$ with the mass M and the density ρ of the hydraulic fluid, we get

$$\begin{aligned} &\mathrm{d}(Mu) - T\mathrm{d}(Ms) + P\mathrm{d}\frac{M}{\rho} \\ &= M\underbrace{\left(\mathrm{d}u - T\mathrm{d}s + P\mathrm{d}\left(\frac{1}{\rho}\right)\right)}_{=0} + \left(u - Ts + \frac{P}{\rho}\right)\mathrm{d}M \, . \end{aligned}$$

In a thermodynamically open system the underbraced term vanishes. Since the process under consideration is isoentropic, this means $\mathrm{d}s = 0$, we end up with

$$\underbrace{\mathrm{d}(Mu)}_{U} + P\mathrm{d}V - \underbrace{\left(u + \frac{P}{\rho}\right)}_{=h}\mathrm{d}M \, , \tag{3}$$

where h denotes the specific enthalpy. To complete the description, a constitutive relation is required. The choice

$$\partial_\rho P = \frac{E}{\rho}$$

with the bulk modulus $E \in \mathbb{R}^+$ leads to

$$\begin{aligned} P - P_0 &= E\int_{\rho_0}^{\rho} \frac{\mathrm{d}\tau}{\tau} = E\ln\left(\frac{\rho}{\rho_0}\right) \\ &= E\ln\left(\frac{M}{M_0}\frac{V_0}{V}\right) \, , \end{aligned} \tag{4}$$

where the index zero refers to a fixed thermodynamic reference state. The energy U stored in the liquid in the volume of the piston follows from (3), (4) as

$$\begin{aligned} U - U_0 &= -\int_{V_0}^{V}\left(E\ln\left(\frac{M}{M_0}\frac{V_0}{\tau}\right) + P_0\right)\mathrm{d}\tau + \int_{M_0}^{M} h\mathrm{d}M \\ &= \left(E\ln\left(\frac{M}{M_0}\right) + P_0 + E\right)V_0 - \left(E\ln\left(\frac{M}{M_0}\frac{V_0}{V}\right) + P_0 + E\right)V \\ &\quad + \int_{t_0}^{t} \dot{M}\mathrm{d}M \end{aligned}$$

with the mass flow $\dot{M}$ into the piston. Obviously, the function H_P,

$$H_P = AEV \ln\left(\frac{M}{M_0}\frac{V_0}{V}\right) + A\left(P_0 + E\right)\left(V - V_0\right) \tag{5}$$

with the effective cross section A of the piston is a potential of the hydraulic force $F_h = AP$.

To finalize the derivation of the equations of motion, we assume that x_0, $p_0 = 0$, ρ_0, M_0, $F_i = 0$ is an equilibrium of the system. Let P denote the pressure of chamber one and let the pressure $P_C = P_0$ of the second chamber with the same effective cross section A like chamber one be constant. Introducing displacement coordinates $\Delta x = x - x_0$, $\Delta M = M - M_0$, we rewrite (1) as

$$\begin{aligned} \Delta\dot{x} &= \partial_p \Delta H \\ \dot{p} &= -\partial_{\Delta x}\Delta H - d\left(\tfrac{p}{m}\right) \end{aligned} \tag{6}$$

with Hamiltonian

$$\begin{aligned} &\Delta H\left(\Delta x, p, \Delta M, F_1, \ldots, F_s\right) \\ &\quad = H\left(x_0 + \Delta x, p, 0, F_1, \ldots, F_s\right) - H\left(x_0, 0, 0, 0, \ldots, 0\right) \\ &\quad\quad + H_P\left(M_0 + \Delta M, V\right) - AP_0 \,, \end{aligned} \tag{7}$$

see (5). The dynamics of the hydraulic part takes the simple form

$$\Delta\dot{M} = \frac{M}{V} Q_c \,, \tag{8}$$

where the control input Q_c is the volumetric flow into chamber one of the piston. Here, leakages are neglected.

In industrial applications the measured quantities are the piston position x and the piston pressure P. Therefore, we choose the output y,

$$\begin{aligned} y_1 &= \Delta x \\ y_2 &= \Delta P = E\ln\left(1 + \tfrac{\Delta M}{M_0}\right) - E\ln\left(1 + \tfrac{\Delta V}{V_0}\right) \,, \end{aligned} \tag{9}$$

where $V(x) = V(x_0) + A\Delta x$ is the volume of the piston filled with hydraulic fluid. It is also worth mentioning that the function ΔH, see (7) is a Liapunov function of the coupled system (6), (8), (9) for the case $Q_c = 0$. Roughly speaking the piston behaves like a non linear spring in this case with potential ΔH_P of (5).

Often the input Q_c is generated by a servo valve, here a critically centered three land four way spool valve with input x_v is used, see Figure 2. We neglect its dynamics and use the static relations

$$Q_c = \begin{cases} g\left(x_v\right)\sqrt{P_S - P} \,, & x_v \geq 0 \\ -g\left(x_v\right)\sqrt{P - P_T} \,, & x_v \leq 0 \,, \end{cases} \tag{10}$$

where P_S, P_T are the supply and tank pressures. The function g is monotonously increasing and meets $g(0) = 0$. Since (10) is invertible, one can take Q_c as control input. Often one says, that the spool valve is compensated.

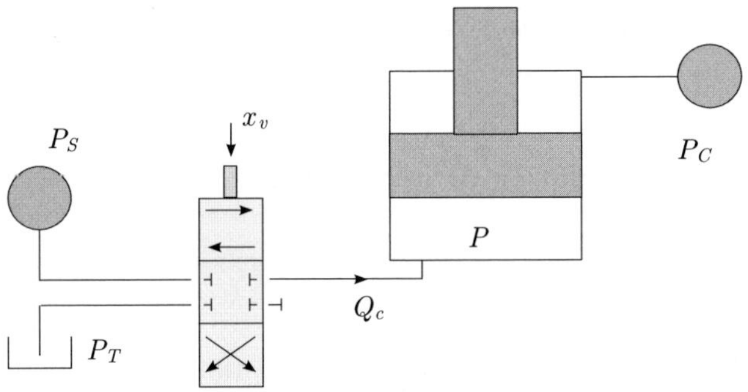

Fig. 2. A spool valve

Sometimes it is advantageous to use the coordinates (x, v, P), $p = vm$. With the Lagrangian ΔL,

$$\Delta L\left(\Delta x, v, \Delta M, F_1, \ldots, F_s\right) = \left(pv - \Delta H\right)_{p=vm}$$

one derives the relation

$$\begin{aligned} \Delta\dot{x} &= v \\ m\dot{v} + \tfrac{v^2}{2}\partial_{\Delta x} m &= \partial_{\Delta x} V_p - dv + A\Delta P - \textstyle\sum_{l=1}^{s} \partial_{\Delta x} h_i F_i \end{aligned} \tag{11}$$

for the mechanical part, see (1), (2), and gets

$$V\Delta\dot{P} = E\left(-Av + Q_c\right) \tag{12}$$

for the hydraulic part, see (8), (9).

3 Design Goals and Methods

A hydraulic actuator is the heart of many industrial applications, because it can produce very high forces in some milliseconds. The main disadvantage is the non linear behavior, especially for applications with high dynamics. Here we discuss two problems. The first one is to keep a point x_p of the mechanism as close as possible to a desired position. The second problem, the force control problem, is to keep a generated force F_p at a point x_p as close as possible to its desired value. In the two cases disturbances F_i are acting on the mechanical part. We distinguish two disturbances, internal and external ones. Here, the external disturbance is a constant force $F_1 = F_d$ modeled by the exosystem

$$\dot{F}_d = 0\ . \tag{13}$$

The internal disturbance is a force caused by imperfections of the mechanical part. A typical representative is a force caused by the eccentricity of rotating bodies. Since the angular velocity $\omega \in \mathbb{R}^+$ of a rotating body is often known or can easily be measured, an appropriate model for the eccentricity force F_2 is given by

$$\begin{aligned}\dot{F}_e &= \underbrace{\begin{bmatrix} 0 & -\omega \\ \omega & 0 \end{bmatrix}}_{\Omega} F_e \\ F_2 &= \underbrace{\begin{bmatrix} c_1 & c_2 \end{bmatrix}}_{c} F_e\end{aligned} \tag{14}$$

with $c_1^2 + c_2^2 = 1$. Therefore, the models of Section 2 must be completed by the models (13), (14) to take these disturbances into account.

One of the design goals could be that neither F_1 nor F_2 have an effect on the position x_p. Since x_p is not measured one has to reach this goal indirectly. Often the effect of F_1, F_2 on x_p is known in the static case. Then one can compensate this effect by a suitable choice of the piston position x at least for the static behavior. This is industrial standard for the case $F_2 = 0$, where known elastic deformations are eliminated. In the case of force control it also often desirable that the effect of F_1, F_2 on the piston pressure is eliminated.

A simple check shows, that the models (6), (8) or (11), (12) are input to state linearizable. The main problem is that the piston velocity v is not measurable in most of the industrial applications. Also high gain methods will fail because the measurements are noisy. Therefore, two approaches are possible in principle. The first one avoids the use of v for feedback. Several successful applications of this strategy exist, see e.g. [7], [8], [9], [10]. The second approach is based on the design of observers for the velocity v and for the disturbances, see e.g. [4]. Here we follow the second strategy.

The models developed in the previous section show the dependency of m on x. Since we deal with 1-DOF models, we can get rid of this disadvantage by the transformation

$$\begin{aligned}\Delta\bar{x} &= \varphi(\Delta x) = \int_{x_0}^{x_0+\Delta x} \sqrt{\frac{m(\tau)}{m_0}} \mathrm{d}\tau \\ \bar{v} &= \sqrt{\frac{m}{m_0}} v \\ \bar{p} &= \sqrt{\frac{m_0}{m}} p \,.\end{aligned}$$

The equations (11), (12) take the form

$$\begin{aligned}\Delta\dot{\bar{x}} &= \bar{v} \\ m_0 \dot{\bar{v}} &= \overline{V}_p' - d\sqrt{\frac{m_0}{m}} \bar{v} + A \Delta P - \bar{h}_1' F_1 - \bar{h}_2' F_2 \\ \frac{\Delta\dot{P}}{E} &= -\frac{A}{\overline{V}} \sqrt{\frac{m_0}{m}} \bar{v} + \frac{1}{\overline{V}} Q_c\end{aligned} \tag{15}$$

in the new coordinates with the measured quantities Δx, ΔP with $\bar{V} = V \circ \varphi^{-1}$, $\overline{V}_p = V_p \circ \varphi^{-1}$. Here and in the following we use the shortcut $'$ for $\partial_{\Delta \bar{x}}$.

3.1 The Reduced Observer

It is a matter of fact, that a model of the exosystems must be included into the controller to reach the design goals, see [1], [2], [6]. Our choice is the design of a reduced observer for the linear and time variant subsystem

$$\begin{bmatrix} \dot{\bar{v}} \\ \dot{F}_d \\ \dot{F}_{e,1} \\ \dot{F}_{e,2} \end{bmatrix} = \begin{bmatrix} \frac{-d}{\sqrt{m_0 m}} & -\frac{\bar{h}_1'}{m_0} & -\frac{\bar{h}_2' c_1}{m_0} & -\frac{\bar{h}_2' c_2}{m_0} \\ 0 & 0 & 0 & 0 \\ 0 & 0 & 0 & -\omega \\ 0 & 0 & \omega & 0 \end{bmatrix} \begin{bmatrix} \bar{v} \\ F_d \\ F_{e,1} \\ F_{e,2} \end{bmatrix} + \begin{bmatrix} \frac{\bar{V}_p' + AP}{m_0} \\ 0 \\ 0 \\ 0 \end{bmatrix} \tag{16}$$

based on the measurement of $\bar{x}$, P. Since equation (15) shows that both $\Delta\dot{P}$ and $\Delta\dot{\bar{x}}$ depend on $\bar{v}$, we use $\Delta\bar{x}$ to stabilize the observer. With the transformation

$$\begin{aligned} w_1 &= \bar{v} + k_1 \Delta\bar{x} + d\chi \\ w_2 &= F_d + k_2 \Delta\bar{x} \\ w_3 &= F_{e,1} + k_3 \Delta\bar{x} \\ w_4 &= F_{e,2} + k_4 \Delta\bar{x} \\ \chi &= \int_{m_0}^{m} \frac{1}{\sqrt{m_0 \tau}} \frac{d\tau}{\tau'} \end{aligned} \tag{17}$$

and $\tau(x) = m(x)$ one gets the system

$$\dot{w} = Aw + r \tag{18}$$

with driving term r and

$$\begin{aligned} w^T &= \begin{bmatrix} w_1 \; w_2 \; w_3 \; w_4 \end{bmatrix} \\ r^T &= \begin{bmatrix} r_1 \; r_2 \; r_3 \; r_4 \end{bmatrix} \\ A &= \begin{bmatrix} k_1 & -\frac{\bar{h}_1'}{m_0} & -\frac{\bar{h}_2' c_1}{m_0} & -\frac{\bar{h}_2' c_2}{m_0} \\ k_2 & 0 & 0 & 0 \\ k_3 & 0 & 0 & -\omega \\ k_4 & 0 & \omega & 0 \end{bmatrix} \\ r_1 &= -k_1 (k_1 \Delta\bar{x} + \chi) + \frac{\bar{h}_1'}{m_0} k_2 + \frac{\bar{h}_2' c_1}{m_0} k_3 + \frac{\bar{h}_2' c_2}{m_0} k_4 \\ r_2 &= -k_2 (k_1 \Delta\bar{x} + \chi) \\ r_3 &= -k_3 (k_1 \Delta\bar{x} + \chi) + \omega k_4 \\ r_4 &= -k_4 (k_1 \Delta\bar{x} + \chi) - \omega k_3 \; . \end{aligned}$$

Because of constructional reasons the relations

$$\bar{h}_1'(\Delta \underline{x}) = \bar{h}_{1,0}' + \Delta \bar{h}_1'(\Delta \bar{x}) \,, \quad \left|\bar{h}_{1,0}'\right| > \left|\Delta \bar{h}_1'\right|$$
$$\bar{h}_2'(\Delta \bar{x}) = \bar{h}_{2,0}' + \Delta \bar{h}_2'(\Delta \bar{x}) \,, \quad \left|\bar{h}_{2,0}'\right| > \left|\Delta \bar{h}_2'\right|$$

with $\bar{h}_{1,0}', \bar{h}_{2,0}' \in \mathbb{R}$ are met in the operating range. In this special case one can always find numbers k_i, $i = 1, \ldots, 4$ such that

$$A_0 = \begin{bmatrix} k_1 & -\frac{\bar{h}_{1,0}'}{m_0} & -\frac{c_1 \bar{h}_{2,0}'}{m_0} & -\frac{c_2 \bar{h}_{2,0}'}{m_0} \\ k_2 & 0 & 0 & 0 \\ k_3 & 0 & 0 & -\omega \\ k_4 & 0 & \omega & 0 \end{bmatrix}$$

is a Hurwitz matrix and such that the function V

$$V = \frac{1}{2} w^T P w$$
$$0 = A_0^T P + P A_0 + Q$$

with $Q > 0$ is a time invariant Liapunov function of the time variant system (18) for $r = 0$. Following the standard approach, one chooses a trivial observer

$$\dot{\hat{w}} = A \hat{w} + r$$

for (18), where $\hat{}$ indicates the estimated value, and derives for the error $e^T = \left[e_1 \; e_2 \; e_3 \; e_4 \right]$,

$$\begin{aligned} e_1 &= \widehat{\bar{v}} - \hat{v} &&= -\hat{w}_1 + k_1 \Delta \bar{x} + \chi \\ e_2 &= F_d - \hat{F}_d &&= -\hat{w}_2 + k_2 \Delta \bar{x} \\ e_3 &= F_{e,1} - \hat{F}_{e,1} &&= -\hat{w}_3 + k_3 \Delta \bar{x} \\ e_4 &= F_{e,2} - \hat{F}_{e,2} &&= -\hat{w}_4 + k_4 \Delta \bar{x} \end{aligned} \tag{19}$$

the asymptotic stable dynamics

$$\dot{e} = A e \,. \tag{20}$$

3.2 Force Control

In the case of pure force control, one tries to keep a force F_p at a point x_p on the mechanics, as close as possible to its desired value $F_{p,\text{des}}$. We assume, there exists a function $F_p - F_{p,\text{des}} = f(\Delta \bar{x}, A \Delta P, F_1, F_2)$, $0 = f(0,0,0,0)$, which connects F_p with the other forces in the steady state. The control problem is to keep the controlled output y_c,

$$y_c = f(\Delta \bar{x}, A \Delta P, F_1, F_2) \tag{21}$$

as close as possible to zero. The choice of the ideal dynamics

$$\dot{y}_c = -\alpha y_c, \qquad \alpha \in \mathbb{R}^+ , \tag{22}$$

leads to a control law of the type

$$Q_c = Q_c\left(\Delta\bar{x}, \bar{v}, \Delta P, F_1, F_2\right) \tag{23}$$

because $\dot{y}_c$ depends already on Q_c, see (15). Obviously, one has to replace $\bar{v}$, F_1, F_2 by their estimates $\hat{\bar{v}}$, $\hat{F}_1$, $\hat{F}_2$ for the implementation. The arising stability problem will be discussed only shortly for the application.

3.3 Position Control

The goal of position control is to keep the position $\bar{x}_p$ of a point of the mechanism as close as possible to its desired value $\bar{x}_{p,\mathrm{des}}$. Since x_p is not measurable, we assume there exists a static compensation $\Delta\bar{x}_{\mathrm{des}} = \bar{x}_{\mathrm{des}} - x_0 = f\left(\Delta x_p, \Delta\bar{x}, F_1, F_2\right)$, $0 = f\left(0,0,0,0\right)$, $\Delta x_p = x_p - x_{p,\mathrm{des}}$ with the corrected piston position $\bar{x}_{\mathrm{des}}$ in the transformed coordinates. Therefore, the controller must keep the controlled output y_c,

$$y_c = \Delta\bar{x} - f\left(0, \Delta\bar{x}, F_1, F_2\right) \tag{24}$$

as close as possible to zero. The choice of the ideal dynamics

$$\begin{aligned} &\dddot{y} + \alpha_2\ddot{y} + \alpha_1\dot{y} + \alpha_0 y = 0 \\ &s^3 + \alpha_2 s^2 + \alpha_1 s + \alpha_0 = \left(s^2 + 2\zeta s\omega_0 + \omega_0^2\right)\left(s + \omega_1\right) , \end{aligned} \tag{25}$$

with $\zeta, \omega_0, \omega_1 \in \mathbb{R}^+$ leads again to a control law of the type

$$Q_c = Q_c\left(\Delta\bar{x}, \bar{v}, \Delta P, F_1, F_2\right) \tag{26}$$

because $\Delta\dddot{\bar{x}}$ depends on Q_c, see (15). Again, one has to replace $\bar{v}$, F_1, F_2 by their estimates $\hat{\bar{v}}$, $\hat{F}_1$, $\hat{F}_2$ for the implementation of (26). Like above the stability problem will be discussed for the application, only.

4 Gap Control of a Four High Stand

The approach presented in the previous sections is now applied to the hydraulic gap control problem in steel rolling mills. Figure 3 presents a sketch of a four high mill stand, consisting of a hydraulic actuator acting on the upper backup and work roll system. The steel strip is located between the upper and the lower work roll. There are two problems in hydraulic gap control. The first one is to control the thickness of the strip. Since the gauge meter for the thickness measurement is mounted after the stand, one controls the position of the upper work roll. But the applied forces cause elastic deformations of the stand, as well as of the rolls. These deformations must be compensated at least statically. The second problem is to control the roll force, which acts on the strip.

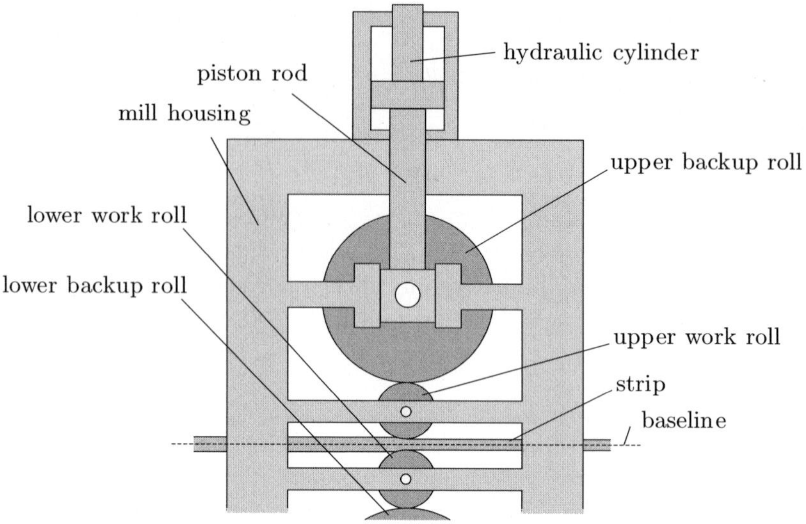

Fig. 3. Sketch of a four high mill stand

Again the roll force is not measurable. Therefore, one controls the pressure of the piston. Depending on the construction also elastic deformations have to be taken into account. In the following we consider a simplified model of the setup, and neglect the dynamics of the mill housing and, moreover, consider the piston to be rigidly connected to work and backup roll system. There exist more complex models, which also describe the plastic deformation of the strip, see e.g. [3]. Even for this simple one dimensional model one has to deal with implicit nonlinear equations for the deformation. For the following we approximate this model by a simple spring, mass, damper system with mass m, constant c of the linear spring, constant d of the viscose damper and an unknown, but slowly varying disturbance force F_d. Figure 4 depicts this simplified one mass model, which is a standard one in steel industries, because it reflects the reality quite well. Further we assume that one chamber of the single acting piston is connected to the spool valve, which again is connected to the supply tanks, and that the other one with a tank with constant pressure P_C. We assume that the spool valve is compensated, see (10), and choose the volumetric flow Q_c as the plant input. After some operation time the rolls are deformed and must be exchanged, which is quite costly. Usually the deformation can be approximated by a sinusoidal eccentricity. Here we take only the eccentricity of the work rolls into account, because they get more damaged than the backup rolls. But it is no problem to extend the approach by the eccentricity of the later ones. In addition the roll force model used online to determine the required roll force, see e.g. [3], is only an approximative one. Its deviation from the reality is described by a slowly varying force. Therefore, we complete our model with the forces F_1, F_2 caused by imperfections of the roll

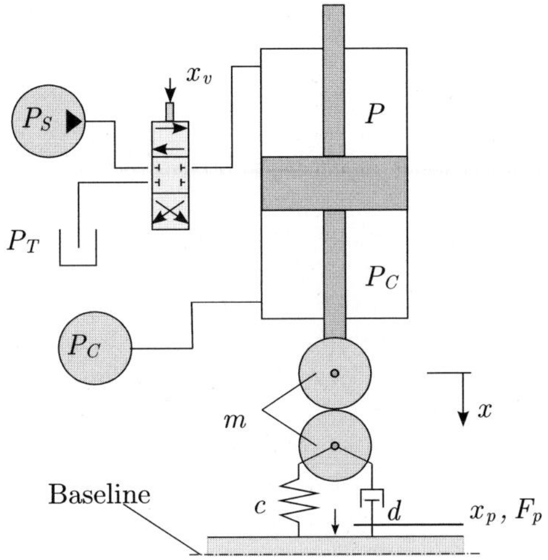

Fig. 4. 1-DOF model of a four high mill stand with hydraulic actuator

force model and by the eccentricity of the work rolls and derive the equations

$$\begin{aligned} \Delta\dot{x} &= v \\ m\dot{v} &= -c\Delta x - dv + A\Delta P - F_1 + F_2, \\ (V_0 + A\Delta x)\,\Delta\dot{P} &= E\,(Q_c - Av)\,, \end{aligned} \tag{27}$$

with the hydraulic force $\Delta F_h = A\Delta P$, see also (15). The effect of gravity is taken into account by the steady state conditions. The equations (13), (14) with $c_1 = 1$, $c_2 = 0$ are used to model F_1, F_2. Thus we end up with a model of the form (11), (12) with $m = m_0$. It is worth mentioning that the angular velocity ω, see (14), of the work rolls must be controlled precisely. Therefore, measurements of ω are always available.

4.1 Reduced Observer

The observer for the non measurable quantities v, F_1, F_2 is designed according to Section 3.1. Since the mass $m = m_0$ is constant and the functions h_1, h_2 meet $h_1 = h_2 = x$, the dynamic matrices of (16), (18) and therefore, the error dynamics (20) of the reduced observer are time invariant. Here A and w, see (17), (19) take the values

$$A = \begin{bmatrix} k_1 & -\frac{1}{m_p} & \frac{1}{m_p} & 0 \\ k_2 & 0 & 0 & 0 \\ k_3 & 0 & 0 & -\omega \\ k_4 & 0 & \omega & 0 \end{bmatrix}, \quad w = \begin{bmatrix} v_p + k_1 x_p \\ F_d + k_2 x_p \\ F_{e,1} + k_3 x_p \\ F_{e,2} + k_4 x_p \end{bmatrix}. \tag{28}$$

It is easy to see that one can render the observer asymptotically stable by a suitable choice of the parameters k_i, $i = 1, \ldots, 4$.

4.2 Force Control

The force controller is designed according to Subsection 3.2, and we try to keep the roll force F_p, which acts at the position x_p, see Figure 4, as close as possible to its desired value. The static relation

$$F_p = \Delta F_h - F_1$$

implies the choice of the controlled output $y_c = F_p - F_{p,\mathrm{des}}$, see (21). The control law (23) can easily be calculated from (22). It is worth mentioning that it is independent of the internal force F_2. Nevertheless F_e must be estimated, since it enters the observer, see (28). Another approach for this eccentricity problem is worth to be mentioned. If one eliminates the effect of F_d on the piston pressure ΔP, then one eliminates its effect on F_p, see [8], [9].

It remains to show stability of the whole configuration. It is easy to see that the operating point, the equilibrium $F_1 = F_2 = 0$, is locally asymptotically stable by Liapunov's indirect method, which allows us to find a Liapunov function. Because of the special structure of the models (13), (14) one can construct an ISS-Liapunov function, see [6], locally for at least sufficient small values of $|F_d|$, $\|F_e\|$.

4.3 Position Control

The position control problem is to keep the position of the point x_p, where the roll is in contact with the strip, see Figure 4, as close as possible to its desired value. Since we cannot measure x_p, we control the piston position x such that elastic deformations are eliminated at least in the static case, see Subsection 3.3. Therefore, we move the piston in such a manner that the eccentricity gets eliminated. Following this idea we derive the controlled output y_c, see (24),

$$y_c = \Delta x + \frac{F_2}{c}\,, \tag{29}$$

which is also a measure of the strip exit thickness. The control law (26) follows from (25) in a straightforward manner. We have to estimate the non measurable signals v, F_d, F_e, and use the reduced observer presented in the Subsection 4.1. Since the stability investigations for the whole system are totally analogous to the force control problem of Subsection 4.2, they are omitted here.

4.4 Simulation Results

The simulations are based on the relations (27) with disturbances generated by the exosystems (13) and (14) with $\omega = 15\,\frac{rad}{s}$. The model parameters are presented in Table 1. The results for the force controller of Subsections 3.2, 4.2 with $\alpha = -10$, see (22) are presented in Figure 5. The choice for the

Table 1. Simulation parameters for the 1-DOF model

Variable	Value	Unit
m	50000	$[kg]$
d	10^7	$\left[\frac{\mathrm{Ns}}{\mathrm{m}}\right]$
c	2×10^9	$\left[\frac{\mathrm{N}}{\mathrm{m}}\right]$
V_0	0.0715	$[\mathrm{m}^3]$
A	0.6779	$[\mathrm{m}^2]$

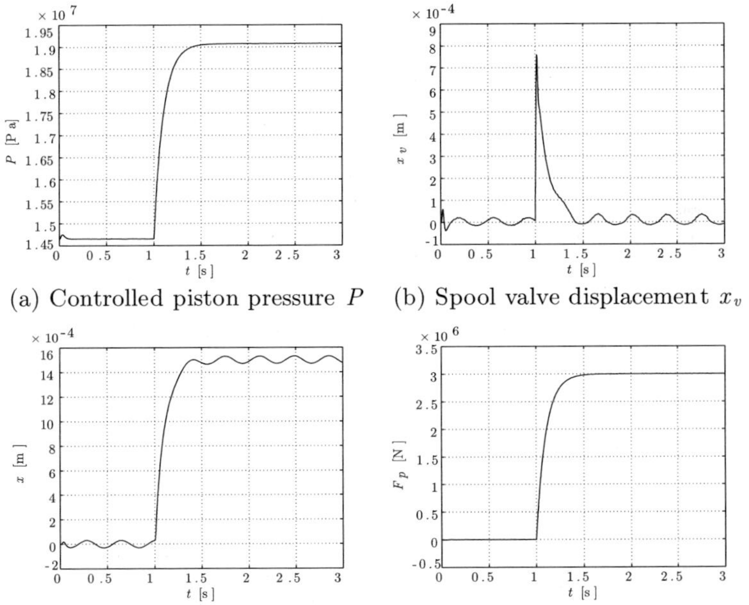

(a) Controlled piston pressure P (b) Spool valve displacement x_v

(c) Piston position x (d) Controlled force F_p

Fig. 5. Simulation results: force control

eigenvalues λ_i, $i = 1, \ldots, 4$ of the observer, see (28), is $\lambda_i = -120$. Here and later small leakages and noise of industrial sensors are taken into account. The setup values are $F_d = 5\mathrm{MN}$, and $P_0 = 146 \cdot 10^5\,\mathrm{P}a$, $P_C = 80 \cdot 10^4\,\mathrm{P}a$, which imply $F_p = A\Delta P - F_d = 0$. We choose $\hat{v} = \hat{F}_d = \hat{F}_e = 0$ for the observer. At $t = 1\,\mathrm{s}$ the reference value for $F_{p,ref}$ is changed to 8 MN. The closed loop shows excellent tracking and disturbance rejection. E.g. the effect of the eccentricity force F_e on the controlled output vanishes almost.
Figure 6 shows the estimated values of disturbance forces F_d, F_e.

The controller derived in Subsection 4.3 is applied to the position control problem. Figure 7 presents the simulation results for two scenarios. In the first scenario, dashed lines, no eccentricity compensation has been carried

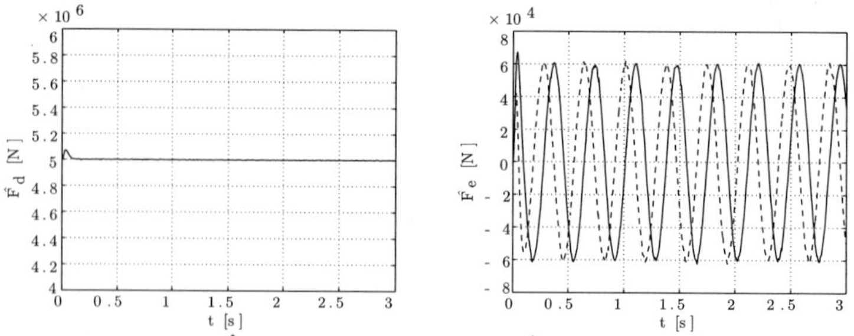

(a) $\hat{F}_d$ (b) $\hat{F}_{e,1}$ (cont.), $\hat{F}_{e,2}$ (dashed)

Fig. 6. Observer signals: force control

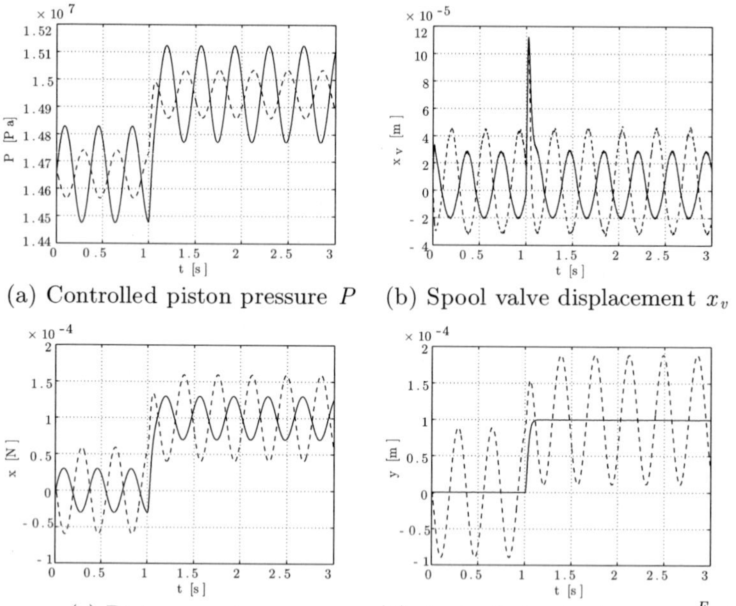

(a) Controlled piston pressure P (b) Spool valve displacement x_v

(c) Piston position x (d) Controlled output $y = x + \frac{F_{e,1}}{c}$

Fig. 7. Simulation results: position control

out, i.e., the estimated values for F_e are unused. The second one, lines are continuous, shows the eccentricity compensation according to Subsection 4.3. The values of the parameter (25) of the control law (26) are $\omega_0 = 70$, $\zeta = 1$, $\omega_1 = 70$ in both cases. In addition the desired controlled output (29) makes a step of height 0.1mm at $t = 1$. The simulations show that the influence of the eccentricity can be almost eliminated. Here, the observer is initialized correctly such that no initial error is visible.

Finally it is worth mentioning that for both the force and position control problem the design parameters of the controllers are chosen quite conservatively to make the result better comparable with the ones published in literature. Therefore, there exists a significant potential for the further improvement of such hydraulic devices.

5 Conclusions

This contribution shows, how nonlinear output feedback with internal models can significantly improve the performance of industrial control systems, with the help of hydraulic devices. Such devices are of interest, where high forces are supposed to be generated by small devices. One of the limiting factors is the non linear behavior of such devices, since high dynamical behavior is not reachable by classical engineering methods, but can be achieved by the presented approach. Here, the authors want to thank Professor Alberto Isidori for his excellent course about "Robust Autonomous Control: an Internal Model Approach", that he gave at the Johannes Kepler University, Linz, Austria in the year 2005. This course gave the initial ideas for the application of output feedback with internal models to problems in steel industries.

References

1. C.I. Byrnes and A. Isidori. Limit sets, zero dynamics and internal models in the problem of nonlinear output regulation. *IEEE Trans. on Automat. Contr.*, 48:1712–1723, 2003.
2. C.I. Byrnes and A. Isidori. Nonlinear internal models for output regulation. *IEEE Trans. on Automat. Contr.*, 49:2244–2247, 2004.
3. S. Fuchshumer. A non-circular arc roll gap model for control applications in steel rolling mills. In *CD Proceedings of the 2005 IEEE International Conference on Control Applications*, Toronto, Canada, 8 2005.
4. G. Grabmair and K. Schlacher. Energy based nonlinear control of hydraulically actuated mechanical systems. In *Proc. of the 44rd IEEE Conf. on Decision and Contr.*, Sevilla, Spain, 12 2005.
5. A. Isidori. *Nonlinear Control Systems.* Springer, London, UK, 3rd edition, 1995.
6. A. Isidori, L. Marconi, and A. Serrani. *Robust Autonomous Guidance: An internal Model Approach.* Springer, London, UK, 2003.
7. A. Kugi. *Non-linear Control Based on Physical Models.* Springer, London, UK, 2003.
8. A. Kugi, W. Haas, K. Schlacher, K. Aistleitner, H. Frank, and G. Rigler. Active compensation of roll eccentricity in rolling mills. *IEEE Transactions on Industry Applications*, 36(2):625–632, 2000.
9. A. Kugi, K. Schlacher, and G. Keintzel. Position control and active eccentricity compensation in rolling mills. *Automatisierungstechnik at*, 47(8):342–349, 1999.
10. K. Schlacher, S. Fuchshumer, G. Grabmair, J. Holl, and G. Keintzel. Active vibration rejection in steel rolling mills. In *Proc. of the 16th IFAC World Congress*, Prague, Czech Republic, 2005.

Hybrid Zero Dynamics of Planar Bipedal Walking

Jessy W. Grizzle[1] and Eric R. Westervelt[2]

[1] Control Systems Laboratory, Electrical Engineering and Computer Science Department, University of Michigan, 1301 Beal Avenue, Ann Arbor, MI 48109-2122 USA,

[2] Department of Mechanical Engineering, The Ohio State University, 201 West 19th Avenue, Columbus, OH 43210-1142 USA,

Summary. Models of bipedal robots in motion are fundamentally hybrid due to the presence of continuous phases, discrete transitions, and unilateral constraints arising from the contact forces between the robot and the ground. A major challenge in the control of bipedal robots has been to create a feedback theory that provides systematic synthesis methods, provable correctness and computational tools for designing asymptotically stable, periodic walking motions, especially walking motions that are dynamic unlike the quasi-static, flat-footed gaits that are prevalent in today's machines. This chapter highlights the fundamental role of *zero dynamics* in obtaining truly dynamic walking gaits that include underactuated phases. The theoretical analysis is verified with experimental work.

1 Introduction

Feedback control is an integral part of any biped, whether biological or mechanical. With the exception of "passive" robots that exhibit a very limited range of stable walking on an inclined plane, without any sensing and control [5], bipeds are dynamically unstable. Said another way, without a properly functioning control system, a biped stumbles and falls.

Models of bipedal robots are quite complex. They are hybrid, nonlinear, and typically, high dimensional. In addition, as will be explained later, the continuous portion of the dynamics is effectively underactuated. A further complication is that a steady walking cycle is a non-trivial periodic motion. This means that standard stability tools for static equilibria do not apply. Instead, one must use tools appropriate for the study of periodic orbits, such as Poincaré return maps. It is of course well known how to use numerical methods to compute a Poincaré return map and to find fixed points of it [12]. The drawback in such a direct approach is that *it does not yield sufficient insight for feedback design and synthesis.* An extension of the notion of the zero dynamics to the hybrid models arising in bipedal locomotion leads to a feedback design process in which Poincaré stability analysis can be directly and insightfully incorporated into feedback synthesis.

Early definitions of the zero dynamics of a time-invariant nonlinear control system were proposed by Krener and Isidori in 1980 (using controlled-invariant distributions), by Byrnes and Isidori in 1984, and Marino in 1985 (using inverse systems) as a tool for feedback design and stability analysis. An important refinement of the concept was achieved by Isidori and Moog in 1988 [10], where three equivalent state-space characterizations of the zero dynamics of a linear time-invariant system were evaluated and compared for nonlinear systems, including the now-well-known definition of the zero dynamics as the restriction dynamics to the largest controlled-invariant manifold contained in the zero set of the output. Which of the definitions to adopt in nonlinear control was not settled until the hugely-influential 1991 paper [3] by Byrnes and Isidori that treated stabilization of equilibrium points on the basis of the zero dynamics. The notion of a hybrid zero dynamics builds on this fundamental work.

For the hybrid closed-loop system consisting of a biped robot, its environment, and a given feedback controller, the objective during the analysis phase is to be able to determine if periodic orbits exist and, if they exist, whether they are asymptotically stable. In the ensuing feedback synthesis phase, the objective is to optimize over a class of stabilizing feedback controllers in order to achieve performance objectives, such as minimal peak actuator torques and walking with a given average speed.

2 Why Study Underactuation?

An important source of complexity in a bipedal robot is the degree of actuation of the model, or more precisely, the degree of *underactuation*. It is assumed here that the robot's legs are terminated in points, and consequently, no actuation is possible at the end of the stance leg. It follows that the mechanical model is underactuated during single support, as opposed to fully actuated (a control at each joint and at the contact point with the ground). One could be concerned that "real robots have feet", and thus, while the analysis of point-feet models may be of interest mathematically, it is "misguided for practical robotics". Focusing on underactuation is important for at least two reasons.

On the one hand, it is interesting to prove, both theoretically and experimentally, that elegant walking and running motions are possible with a mechanically simple robot (no feet). On the other hand, if human walking is taken as the defacto standard against which mechanical bipedal walking is to be compared, then the flat-footed walking achieved by current robots needs to be improved. In particular, toe roll toward the end of the single support phase needs to be allowed as part of the gait design. Currently, this is not allowed specifically because it leads to underactuation[3], which cannot be treated with a control design philosophy based on trajectory tracking and the quasi-static stability criterion, known as the Zero Moment Point (ZMP) [14], as is currently practiced widely in the bipedal robotics community.

3 Hybrid Model of a Bipedal Walker

This section introduces a hybrid dynamic model for walking motions of a planar bipedal robot with point feet. The robot is assumed to consist of $N \geq 2$ rigid links with mass connected via rigid, frictionless revolute joints to form a single open kinematic chain lying in a plane. It is further assumed that there are two identical sub-chains called the legs, connected at a common point called the hip, and, optionally, additional sub-chains that may be identified as a torso, arms, a tail, etc. Since each leg end is terminated in a point, either the robot does not have feet, or it is walking tip-toe. A typical allowed robot is depicted in Fig. 1, which is intentionally suggestive of a human form. All motions will be assumed to take place in the sagittal plane and consist of successive phases of *single support* (stance leg on the ground and swing leg in the air) and *double support* (both legs on the ground). Conditions that guarantee the leg ends alternate in ground contact – while other links such as the torso or arms remain free – will be imposed during control design. A rigid impact is used to model the contact of the swing leg with the ground. Further details on the model are given in [17, Sec. II], along with assumptions on the walking gait (symmetric, motion from left to right, instantaneous double support phase, no slipping or rebound at impact).

The distinct phases of walking naturally lead to mathematical models that are comprised of two parts: the differential equations describing the dynamics during the swing phase and a model that describes the dynamics when a leg end impacts the ground. For the models developed here, the ground – also called a walking surface – is assumed to be smooth and perpendicular to the gravitational field, that is, the ground is assumed to be flat as opposed to sloped or terraced.

3.1 Lagrangian Swing Phase Model

The swing phase model corresponds to a pinned open kinematic chain. It is assumed that only symmetric gaits are of interest, and hence it does not matter which leg end is pinned. The swapping of the roles of the legs will be accounted for in the impact model.

Let $\mathcal{Q}$ be the N-dimensional configuration manifold of the robot when the stance leg end is acting as a pivot and let $q := (q_1; \cdots; q_N) \in \mathcal{Q}$ be

[3] When the foot is rotating about the toe, one effectively has a point contact with no actuation.

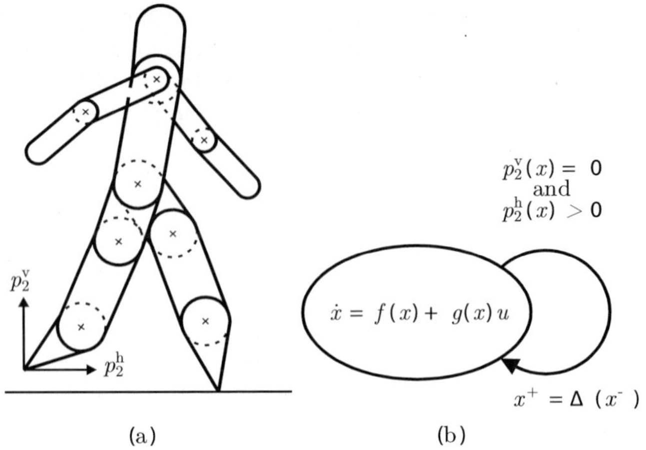

Fig. 1. (**a**) A typical planar robot model analyzed here. For later use, Cartesian coordinates are indicated at the swing leg end. (**b**) Hybrid model of walking with point feet. Key elements are the continuous dynamics of the single support phase, written in state space form as $\dot{x} = g(x) + g(x)u$, the switching or impact condition, $p_2^v(q) = 0$ and $p_2^h(q) > 0$, which detects when the height of the swing leg above the walking surface is zero and the swing leg is in front of the stance leg, and the re-initialization rule coming from the impact map, Δ

a set of generalized coordinates and denote the potential and kinetic energies by $V(q)$ and $K(q,\dot{q}) = \frac{1}{2}\dot{q}'D(q)\dot{q}$, respectively, where the inertial matrix D is positive definite on $\mathcal{Q}$. The dynamic model is easily obtained with the method of Lagrange, yielding the mechanical model

$$D(q)\ddot{q} + C(q,\dot{q})\dot{q} + G(q) = B(q)u, \tag{1}$$

where $u = (u_1; \cdots; u_{N-1}) \in \mathbb{R}^{(N-1)}$, where u_i is the torque applied between the two links connected by joint-i, *and there is no torque applied between the stance leg and ground.* The model is written in state space form by defining

$$\dot{x} = \begin{bmatrix} \dot{q} \\ D^{-1}(q)\left[-C(q,\dot{q})\dot{q} - G(q) + B(q)u\right] \end{bmatrix} \tag{2}$$

$$=: f(x) + g(x)u \tag{3}$$

where $x := (q;\dot{q})$. The state space of the model is $\mathcal{X} = T\mathcal{Q}$. For each $x \in \mathcal{X}$, $g(x)$ is a $2N \times (N-1)$ matrix; its i-th column is denoted by g_i. In natural coordinates $(q;\dot{q})$ for $T\mathcal{Q}$, g is independent of $\dot{q}$.

3.2 Impact Model

The impact of the swing leg with the ground at the end of a step is represented with the rigid (i.e., perfectly inelastic) contact model of [6, 13]. This model effectively collapses the impact phase to an instant in time. The impact forces are consequently represented by impulses, and a discontinuity or jump is allowed in the velocity component of the robot's state, with the configuration variables remaining continuous or constant during the impact. Since we are assuming a symmetric walking gait, we can avoid having to use two swing phase models – one for each leg playing the role of the stance leg – by relabeling the robot's coordinates at impact. The coordinates must be relabeled because the roles of the legs must be swapped. Immediately after swapping, the former swing leg is in contact with the ground and is poised to take on the role of the stance leg.

The *relabeling* of the generalized coordinates is given by a matrix, R, acting on q with the property that $RR = I$, i.e., R is a circular matrix. The result of the impact and the relabeling of the states provides an expression

$$x^+ = \Delta(x^-) \tag{4}$$

where $x^+ := (q^+; \dot{q}^+)$ (resp. $x^- := (q^-; \dot{q}^-)$) is the state value just after (resp. just before) impact and

$$\Delta(x^-) := \begin{bmatrix} \Delta_q \, q^- \\ \Delta_{\dot{q}}(q^-) \, \dot{q}^- \end{bmatrix}. \tag{5}$$

The impact map is linear in the generalized velocities. Further details are given in [7, 17].

3.3 Overall Hybrid Model

A hybrid model of walking is obtained by combining the swing phase model and the impact model to form a system with impulse effects. Assume that the trajectories of the swing phase model possess finite left and right limits, and denote them by $x^-(t) := \lim_{\tau \nearrow t} x(\tau)$ and $x^+(t) := \lim_{\tau \searrow t} x(\tau)$, respectively. The model is then

$$\Sigma : \begin{cases} \dot{x} = f(x) + g(x)u, & x^- \notin \mathcal{S} \\ x^+ = \Delta(x^-), & x^- \in \mathcal{S}, \end{cases} \tag{6}$$

where the switching set is chosen to be

$$\mathcal{S} := \{(q, \dot{q}) \in T\mathcal{Q} \mid p_2^{\mathrm{v}}(q) = 0,\ p_2^{\mathrm{h}}(q) > 0\}. \tag{7}$$

In words, a trajectory of the hybrid model is specified by the swing phase model until an impact occurs. An impact occurs when the state "attains" the set $\mathcal{S}$, which represents the walking surface. At this point, the impact of the swing leg with the walking surface results in a very rapid change in the velocity components of the state vector. The impulse model of the impact compresses the impact event into an instantaneous moment in time, resulting in a discontinuity in the velocities. The ultimate result of the impact model is a new initial condition from which the swing phase model evolves until the next impact. In order for the state not to be obliged to take on two values at the "impact time", the impact event is, roughly speaking, described in terms of the values of the state "just prior to impact" at time "t^-" and "just after impact" at time "t^+". These values are represented by the left and right limits, x^- and x^+, respectively. Solutions are taken to be right continuous and must have finite left and right limits at each impact event. Figure 1 gives a graphical representation of this discrete-event system.

A *step* of the robot is a solution of (6) that starts with the robot in double support, ends in double support with the configurations of the legs swapped, and contains only one impact event. *Walking* is a sequence of steps.

4 Feedback Design via Posture Control

Any attempt to describe walking, even something as simple as the difference between human-like walking (knees bent forward) and bird-like walking (knees bent backward), inevitably leads to a description of the *posture* or *shape* of the robot throughout a step. In other words, a description of walking involves at least a partial specification of the path followed in the configuration space of the biped; see Fig. 2. The following is one possible way to express this mathematically: let $q_{\mathrm{b}} = (q_1; \cdots; q_{N-1})$ be a set of *body* coordinates for the robot and let θ be the angle of some point of the robot with respect to an inertial frame, and assume moreover that θ has been chosen so that it is strictly monotonic throughout the step. Then the path of the robot in the configuration space can be expressed as

$$0 = q_{\mathrm{b}} - h_d(\theta). \tag{8}$$

A natural objective is therefore: determine a feedback controller that drives asymptotically to zero the output function

$$y = h(q) := q_{\mathrm{b}} - h_d(\theta). \tag{9}$$

This leads to two questions.

1) **An analysis question:** when will a given choice of $h_d(\theta)$ lead to an asymptotically stable periodic orbit (i.e., a stable walking motion)?
2) **A synthesis question:** how to make a choice of $h_d(\theta)$ that will yield an asymptotically stable periodic orbit meeting physically motivated requirements such as: energy efficiency; the robot walks at a desired speed; and the reaction forces at the leg end are such that the contact between the stance and the ground behaves as a pivot?

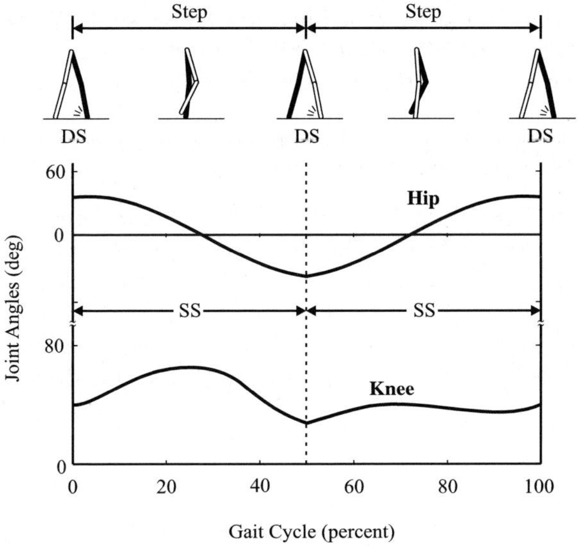

Fig. 2. Joint angles for a four-link walker over a complete gait cycle, that is, two steps. The gait cycle consists of two phases each of double support (DS) and single support (SS). Depicted are the relative hip angle and the knee angle of the leg drawn in white. The first single support phase can be thought of as a graph of (8) for the relative angles of the hip and knee during the swing phase, and the second single support phase is a graph of (8) for the relative angles of the hip and knee during the stance phase. The angle θ in (8) can be taken as the angle of the hip with respect to the ground contact point of the stance leg

Addressing the first question leads to the notion of the hybrid zero dynamics, the focus of this chapter. A finite parametrization of possible paths $h_d(\theta, \alpha)$ via Bézier polynomials and the use of parameter optimization have been employed to address the second question.

5 The Zero Dynamics of Walking

The zero dynamics of the hybrid model (6) with output (9) are developed in a two-step process. First, the zero dynamics of the (non-hybrid) nonlinear model consisting of the swing phase dynamics (3) and the output (9) are characterized, and then, second, an impact invariance condition is imposed on the swing-phase zero dynamics manifold in order to obtain the hybrid zero dynamics. For general hybrid systems and output functions, this approach to forming a hybrid zero dynamics is less general than directly applying the output zeroing definition of Isidori et al. [3, 9, 10]. From results in [17, 18], however, it can be deduced that for an N-degree of freedom biped model with one degree of underactuation and an $N-1$ dimensional output vector of the form (9), if there exists at least one periodic solution of the hybrid model (6) that zeros the output and is transversal to $\mathcal{S}$, then the approach followed here and the definition used in [9] are equivalent.

5.1 The Swing Phase Zero Dynamics

The objective is to characterize the zero dynamics manifold and associated restriction dynamics for the swing-phase model (3) and output (9). The zero dynamics, by definition, is independent of the choice of coordinates and the application of regular state variable feedback [9, pp. 228]. Express the mechanical model (1) in the generalized coordinates $q = (q_{\mathrm{b}}; \theta)$. It is proven in [8, pp. 562] that the model (3) is globally feedback equivalent to

$$\ddot{q}_{\mathrm{b}} = v \tag{10a}$$

$$\dot{\theta} = \frac{\bar{\sigma}_N}{d_{N,N}(q_{\mathrm{b}})} - J^{\mathrm{norm}}(q_{\mathrm{b}})\dot{q}_{\mathrm{b}} \tag{10b}$$

$$\dot{\bar{\sigma}}_N = -\frac{\partial V}{\partial \theta}(q_{\mathrm{b}}, \theta), \tag{10c}$$

where

$$J^{\mathrm{norm}}(q_{\mathrm{b}}) = \frac{1}{d_{N,N}(q_{\mathrm{b}})}\left[d_{N,1}(q_{\mathrm{b}}), \cdots, d_{N,N-1}(q_{\mathrm{b}})\right], \tag{11}$$

$d_{j,k}$ is the j-k-element of D, $\bar{\sigma}_N$ is the generalized momentum conjugate to $q_N = \theta$, and v is the new input coming from the feedback transformation. Taking $\tilde{x} := (q_{\mathrm{b}}; \theta; \dot{q}_{\mathrm{b}}; \bar{\sigma}_N)$, the swing-phase model after feedback is expressed in state variable form as

$$\dot{\tilde{x}} = \begin{bmatrix} \dot{q}_{\mathrm{b}} \\ \frac{\bar{\sigma}_N}{d_{N,N}(q_{\mathrm{b}})} - J^{\mathrm{norm}}(q_{\mathrm{b}})\dot{q}_{\mathrm{b}} \\ v \\ -\frac{\partial V}{\partial \theta}(q_{\mathrm{b}}, \theta) \end{bmatrix} =: \tilde{f}(\tilde{x}) + \tilde{g}(\tilde{x})v. \tag{12}$$

Decoupling Matrix

Differentiating (9) twice along the dynamics (12) gives

$$\ddot{y} = L_{\tilde{f}}^2 h(\tilde{x}) + L_{\tilde{g}} L_{\tilde{f}} h(q) v. \tag{13}$$

A simple calculation gives that the decoupling matrix is

$$L_{\tilde{g}} L_{\tilde{f}} h(\tilde{q}) = \mathrm{I}_{(N-1)\times(N-1)} + \underbrace{\frac{\partial h_d(\theta)}{\partial \theta}}_{(N-1)\times 1} \underbrace{J^{\mathrm{norm}}(q_{\mathrm{b}})}_{1\times(N-1)}. \tag{14}$$

It follows that [18]

$$\det(L_{\tilde{g}}L_{\tilde{f}}h)(q) = 1 + J^{\mathrm{norm}}(q_{\mathrm{b}})\frac{\partial h_d(\theta)}{\partial \theta} \tag{15}$$

and is nonzero if, and only if,

$$d_{N,N}(q_{\mathrm{b}}) + \left[d_{N,1}(q_{\mathrm{b}}), \cdots, d_{N,(N-1)}(q_{\mathrm{b}})\right]\frac{\partial h_d(\theta)}{\partial \theta} \neq 0. \tag{16}$$

Moreover, on the open set $T\tilde{\mathcal{Q}} \subset T\mathcal{Q}$ where the determinant of the decoupling matrix is nonzero, the inverse of the decoupling matrix is

$$\left[L_{\tilde{g}}L_{\tilde{f}}h(q)\right]^{-1} = \mathrm{I}_{(N-1)\times(N-1)} + \frac{1}{\det(L_{\tilde{g}}L_{\tilde{f}}h)(q)}\frac{\partial h_d(\theta)}{\partial \theta}J^{\mathrm{norm}}(q_{\mathrm{b}}). \tag{17}$$

The swing phase zero dynamics manifold is then[4]

$$\mathcal{Z} := \{x \in T\tilde{\mathcal{Q}} \mid h(x) = 0, L_f h(x) = 0\}; \tag{18}$$

it is a smooth two-dimensional embedded submanifold of $T\mathcal{Q}$. The feedback control

$$\tilde{u}^*(\tilde{x}) = -\left[L_{\tilde{g}}L_{\tilde{f}}h(q)\right]^{-1} L^2_{\tilde{f}}h(\tilde{x}) \tag{19}$$

renders $\mathcal{Z}$ invariant under the closed-loop vector field $\tilde{f} + \tilde{g}\tilde{u}^*$. The zero dynamics vector field is the restriction

$$f_{\mathrm{zero}} := \tilde{f} + \tilde{g}\tilde{u}^*\Big|_{\mathcal{Z}}. \tag{20}$$

The zero dynamics is given by

$$\dot{z} = f_{\mathrm{zero}}(z), \tag{21}$$

for $z \in \mathcal{Z}$.

Computing Terms in the Zero Dynamics

In the coordinates $(q_{\mathrm{b}}; \theta; \dot{q}_{\mathrm{b}}; \dot{\theta})$, the zero dynamics manifold can be written as

$$\mathcal{Z} = \left\{(q_{\mathrm{b}}; \theta; \dot{q}_{\mathrm{b}}; \dot{\theta}) \;\middle|\; q_{\mathrm{b}} = h_d(\theta),\ \dot{q}_{\mathrm{b}} = \frac{\partial h_d(\theta)}{\partial \theta}\dot{\theta}\right\}. \tag{22}$$

On $\mathcal{Z}$, the generalized momentum conjugate to θ becomes

$$\bar{\sigma}_N = I(\theta)\dot{\theta}, \tag{23}$$

[4] By [9, pp. 230], $L_{\tilde{f}}h = L_f h$; moreover, (12) and (3) have the same zero dynamics because they are related by a regular state variable feedback [9].

where the *virtual inertia* $I(\theta)$ is given by

$$I(\theta) := \left[d_{N,N}(q_b) + \left[d_{N,1}(q_b), \cdots, d_{N,(N-1)}(q_b) \right] \frac{\partial h_d(\theta)}{\partial \theta} \right] \Bigg|_{q_b = h_d(\theta)} . \tag{24}$$

On $\mathcal{Z}$, the invertibility of the decoupling matrix establishes a bijective relationship between $\bar{\sigma}_N$ and $\dot{\theta}$,

$$\dot{\theta} = \frac{\bar{\sigma}_N}{I(\theta)}. \tag{25}$$

Restricting (10c) to $\mathcal{Z}$,

$$\dot{\bar{\sigma}}_N = -\frac{\partial V}{\partial \theta}(q_b, \theta) \Bigg|_{q_b = h_d(\theta)} . \tag{26}$$

Defining $\xi_1 := \theta$ and $\xi_2 := \bar{\sigma}_N$, it follows that the swing phase zero dynamics is

$$\dot{\xi}_1 = \kappa_1(\xi_1)\xi_2 \tag{27a}$$
$$\dot{\xi}_2 = \kappa_2(\xi_1), \tag{27b}$$

where

$$\kappa_1(\xi_1) = \frac{1}{I(\xi_1)} \tag{28a}$$
$$\kappa_2(\xi_1) = -\frac{\partial V}{\partial \theta} \Bigg|_{q_b = h_d(\theta),\ \theta = \xi_1} . \tag{28b}$$

It is emphasized that these terms can be determined directly from the Lagrangian of the swing-phase model and the term h_d of the output (9). In particular, there is no need to invert the inertia matrix, as would be required if the zero dynamics were computed directly from (3).

5.2 The Hybrid Zero Dynamics

To obtain the hybrid zero dynamics, the zero dynamics manifold must be *invariant under the impact map*, that is

$$\Delta(\mathcal{S} \cap \mathcal{Z}) \subset \mathcal{Z}. \tag{29}$$

If $\mathcal{S} \cap \mathcal{Z}$ is nonempty, then, due to the form of the output (9), $\mathcal{S} \cap \mathcal{Z}$ is a smooth one-dimensional embedded submanifold of $T\tilde{\mathcal{Q}}$ if, and only, if $p_2^{\mathrm{v}}(h_d(\theta), \theta)$ has constant rank on its zero set. Furthermore, when the decoupling matrix is invertible, the following statements are equivalent [17]:

(a) (29) holds;

(b) $h \circ \Delta|_{(\mathcal{S}\cap\mathcal{Z})} = 0$ and $L_f h \circ \Delta|_{(\mathcal{S}\cap\mathcal{Z})} = 0$; and
(c) there exists at least one point $(q_0^-; \dot{q}_0^-) \in \mathcal{S} \cap \mathcal{Z}$ such that $\bar{\sigma}_N \neq 0$, $h \circ \Delta_q(q_0^-) = 0$, and $L_f h \circ \Delta(q_0^-, \dot{q}_0^-) = 0$.

Definition 1 (Hybrid zero dynamics [17]). *Consider the hybrid model (6) and output (9). Suppose that the decoupling matrix (14) is invertible and let $\mathcal{Z}$ and $\dot{z} = f_{\text{zero}}(z)$ be the associated zero dynamics manifold and zero dynamics of the swing phase model. Suppose that $\mathcal{S} \cap \mathcal{Z}$ is a smooth, one-dimensional, embedded submanifold of $T\mathcal{Q}$. Suppose further that $\Delta(\mathcal{S} \cap \mathcal{Z}) \subset \mathcal{Z}$. Then the nonlinear system with impulse effects,*

$$\Sigma_{\text{zero}} : \begin{cases} \dot{z} = f_{\text{zero}}(z), & z^- \notin \mathcal{S} \cap \mathcal{Z} \\ z^+ = \Delta(z^-), & z^- \in \mathcal{S} \cap \mathcal{Z}, \end{cases} \tag{30}$$

with state manifold $\mathcal{Z}$, is the hybrid zero dynamics.

In the local coordinates $(\xi_1; \xi_2)$, $\mathcal{S} \cap \mathcal{Z}$ and $\Delta : (\xi_1^-; \xi_2^-) \to (\xi_1^+; \xi_2^+)$ simplify to

$$\mathcal{S} \cap \mathcal{Z} = \left\{ (\xi_1^-; \xi_2^-) \mid \xi_1^- = \theta^-, \xi_2^- \in \mathbb{R} \right\} \tag{31a}$$
$$\xi_1^+ = \theta^+ \tag{31b}$$
$$\xi_2^+ = \delta_{\text{zero}}\, \xi_2^-, \tag{31c}$$

where δ_{zero} is a constant that may be readily computed using (4) and (9) and where θ^- and θ^+ satisfy

$$p_2^{\text{v}}(h_d(\theta^-), \theta^-)) = 0, \qquad p_2^{\text{h}}(h_d(\theta^-), \theta^-) > 0, \tag{32a}$$
$$p_2^{\text{v}}(h_d(\theta^+), \theta^+)) = 0, \qquad p_2^{\text{h}}(h_d(\theta^+), \theta^+) < 0. \tag{32b}$$

The hybrid zero dynamics is thus given by (27) during the swing phase, and at impact with $\mathcal{S} \cap \mathcal{Z}$, the re-initialization rules (31b) and (31c) are applied.

For $\theta^+ \leq \xi_1 \leq \theta^-$, define

$$V_{\text{zero}}(\xi_1) := -\int_{\theta^+}^{\xi_1} \frac{\kappa_2(\xi)}{\kappa_1(\xi)}\, d\xi. \tag{33}$$

A straightforward computation shows that $\mathcal{L}_{\text{zero}} := K_{\text{zero}} - V_{\text{zero}}$ [17], where

$$K_{\text{zero}} = \frac{1}{2} \left(\frac{\dot{\xi}_1}{\kappa_1(\xi_1)} \right)^2, \tag{34}$$

is a Lagrangian of the swing-phase zero dynamics (27). This implies, in particular, that the total energy $\mathcal{H}_{\text{zero}} := K_{\text{zero}} + V_{\text{zero}}$ is constant along solutions of the swing-phase zero dynamics.

5.3 Existence and Stability of Periodic Orbits

The analysis of periodic orbits of the hybrid zero dynamics forms the basis for proposing feedback laws that induce exponentially stable walking motions in the full-dimensional hybrid model. Take the Poincaré section to be $\mathcal{S} \cap \mathcal{Z}$ and let

$$\rho : \mathcal{S} \cap \mathcal{Z} \to \mathcal{S} \cap \mathcal{Z} \tag{35}$$

denote the Poincaré (first return) map[5] of the hybrid zero dynamics. Using the fact that the total energy $\mathcal{H}_{\text{zero}}$ is constant along solutions of the continuous portion of the dynamics, the Poincaré map may be shown to be

$$\rho(\zeta_2^-) = \delta_{\text{zero}}^2 \, \zeta_2^- - V_{\text{zero}}(\theta^-), \tag{36}$$

where $\zeta_2^- := \frac{1}{2}(\xi_2^-)^2$, and its domain of definition is

$$\mathcal{D}_{\text{zero}} = \left\{ \zeta_2^- > 0 \mid \delta_{\text{zero}}^2 \, \zeta_2^- - V_{\text{zero}}^{\max} > 0 \right\}, \tag{37}$$

where

$$V_{\text{zero}}^{\max} := \max_{\theta^+ \le \xi_1 \le \theta^-} V_{\text{zero}}(\xi_1). \tag{38}$$

The domain $\mathcal{D}_{\text{zero}}$ is non-empty if, and only if, $\delta_{\text{zero}}^2 > 0$. Whenever $\delta_{\text{zero}}^2 < 1$, the fixed point of (36),

$$\zeta_2^* := -\frac{V_{\text{zero}}(\theta^-)}{1 - \delta_{\text{zero}}^2}, \tag{39}$$

will be exponentially stable as long as it belongs to $\mathcal{D}_{\text{zero}}$. The conditions for there to exist an exponentially stable periodic orbit of (30) are thus

$$\frac{\delta_{\text{zero}}^2}{1 - \delta_{\text{zero}}^2} V_{\text{zero}}(\theta^-) + V_{\text{zero}}^{\max} < 0 \tag{40a}$$

$$0 < \delta_{\text{zero}}^2 < 1. \tag{40b}$$

Periodic orbits of the hybrid zero dynamics are periodic orbits of the full-dimensional model. Two different feedback controllers are provided in [11, 17] for exponentially stabilizing these orbits in the full-dimensional model, (3).

6 Experimental Data

The hybrid zero dynamics has played an integral role in the design of walking gaits for a prototype bipedal robot called RABBIT [4]; see Fig. 3. The analytical results of Section 5.2 are rendered useful in feedback synthesis by introducing a finite parametrization of the output (9). In particular, the function h_d is constructed from Bézier polynomials [1], which in turn introduces free parameters α into the hybrid zero dynamics (30),

[5] This is in general a partial map.

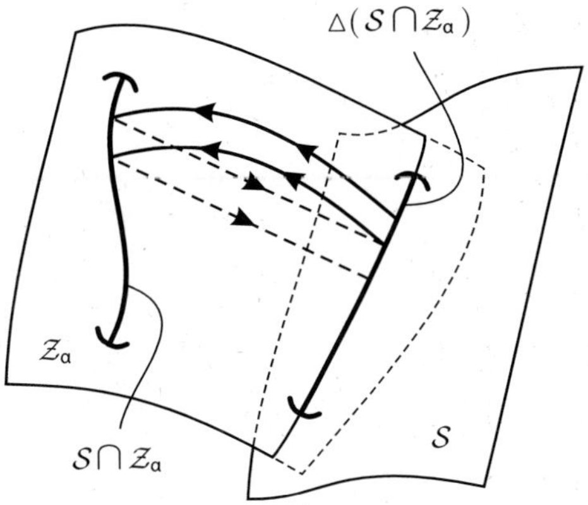

(a)

Fig. 3. In (**a**) geometry of the closed-loop system. In (**b**), the prototype RABBIT, which was developed as a French National Project by the CNRS [4]; the robot is housed in LAG, the Automatic Control Laboratory of Grenoble

$$\Sigma_{\text{zero},\alpha} : \begin{cases} \dot{z} = f_{\text{zero},\alpha}(z), & z^- \notin \mathcal{S} \cap \mathcal{Z}_\alpha \\ z^+ = \Delta(z^-), & z^- \in \mathcal{S} \cap \mathcal{Z}_\alpha, \end{cases} \tag{41}$$

through

$$h_\alpha(q) := q_{\text{b}} - h_d(q, \alpha). \tag{42}$$

Moreover, the hybrid invariance condition (29) can be imposed analytically; see [17, Thm. 4].

A minimum-energy cost criterion

$$J(\alpha) = \frac{1}{\text{step length}} \int_0^{\text{step time}} ||u_\alpha^*(t)||_2^2 dt \tag{43}$$

is posed, where

$$u_\alpha^* := -\left[L_g L_f h_\alpha\right]^{-1} L_f^2 h_\alpha \Big|_{\mathcal{Z}} \tag{44}$$

is the (unique) input to the model (3) realizing the periodic orbit of the hybrid zero dynamics. Constraints based on (40) are easily imposed to guarantee that periodic orbits exist and are exponentially stable. Additional constraints are used to prescribe walking at a desired average speed, with the (unilateral) forces on the support leg lying in the allowed friction cone. Parameter optimization is then used to minimize the cost $J(\alpha)$. Whenever a solution exists, the result is a provably stable, closed-loop system with satisfied design constraints. Typical solutions times for computing the optimal parameter value are thirty seconds to one minute in MATLAB. For further details on the procedure, see [17].

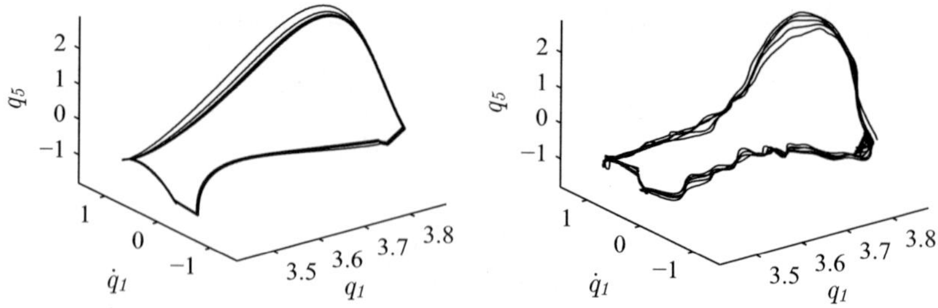

(a) Simulation (b) Experiment

Fig. 4. Limit cycles of the closed-loop hybrid systems corresponding to asymptotically stable walking

Feedback control designs based on the hybrid zero dynamics have been extensively evaluated on RABBIT. As reported in [16], natural walking motions were obtained with remarkably little trial and error. Figure 4 compares a limit cycle obtained on the robot with a limit cycle obtained with the same controller on a simulation model. For videos of RABBIT walking and running, see [2, 15].

7 Conclusions

The notion of zero dynamics has become ubiquitous in nonlinear control theory and practice. This chapter has reviewed an extension of the zero dynamics to a class of hybrid models relevant for the analysis of walking gaits in bipedal robots. The hybrid zero dynamics provides an effective tool for the analysis and synthesis of feedback controllers that induce exponentially stable, periodic walking motions in bipedal robots. Its utility has been confirmed experimentally.

8 Acknowledgments

The work of J.W. Grizzle and E.R. Westervelt was supported by National Science Foundation grants ECS-0600869 and CMS-0408348, respectively.

References

1. P. Bézier. *Numerical Control: Mathematics and Applications.* John Wiley & Sons, New York, 1972.
2. G. Buche. ROBEA Home Page. `http://robot-rabbit.lag.ensieg.inpg.fr/English/`, 2007.

3. C. Byrnes and A. Isidori. Asymptotic stabilization of nonlinear minimum phase systems. *IEEE Trans. on Automat. Contr.*, 376:1122–1137, 1991.
4. C. Chevallereau, G. Abba, Y. Aoustin, F. Plestan, E. R. Westervelt, C. Canudas de Wit, and J. W. Grizzle. RABBIT: a testbed for advanced control theory. *IEEE Control Systems Magazine*, 23(5):57–79, 2003.
5. S. H. Collins, A. Ruina, R. Tedrake, and M. Wisse. Efficient bipedal robots based on passive-dynamic walkers. *Science*, 307:1082–1085, 2005.
6. B. Espiau and A. Goswani. Compass gait revisited. In *Proc. of the IFAC Symposium on Robot Control, Capri, Italy*, pages 839–846, September 1994.
7. J. W. Grizzle, G. Abba, and F. Plestan. Asymptotically stable walking for biped robots: Analysis via systems with impulse effects. *IEEE Trans. on Automat. Contr.*, 46:51–64, January 2001.
8. J. W. Grizzle, C. H. Moog, and C. Chevallereau. Nonlinear control of mechanical systems with an unactuated cyclic variable. *IEEE Trans. on Automat. Contr.*, 30(5):559–576, May 2005.
9. A. Isidori. *Nonlinear Control Systems.* Springer-Verlag, Berlin, 3rd edition, 1995.
10. A. Isidori and C. H. Moog. On the nonlinear equivalent of the notion of transmission zeros. In C. Byrnes and A. Kurzhanski, editors, *Proc. of the IIASA Conference: Modeling and Adaptive Control*, pages 146–157, Berlin, 1988. Springer-Verlag.
11. B. Morris and J. W. Grizzle. A restricted Poincaré map for determining exponentially stable periodic orbits in systems with impulse effects: Application to bipedal robots. In *Proc. of the 44rd IEEE Conf. on Decision and Contr.*, pages 4199–4206, 2005.
12. T. S. Parker and L. O. Chua. *Practical Numerical Algorithms for Chaotic Systems.* Springer-Verlag, New York, 1989.
13. C.-L. Shih and W. A. Gruver. Control of a biped robot in the double-support phase. *IEEE Trans. on Systems, Man and Cybernetics*, 22(4):729–735, 1992.
14. M. Vukobratović, B. Borovac, D. Surla, and D. Stokic. *Biped Locomotion.* Springer-Verlag, Berlin, 1990.
15. E. R. Westervelt. Eric Westervelt's publications. `http://www.mecheng.osu.edu/~westerve/publications/`, 2007.
16. E. R. Westervelt, G. Buche, and J. W. Grizzle. Experimental validation of a framework for the design of controllers that induce stable walking in planar bipeds. *International Journal of Robotics Research*, 23(6):559–582, 2004.
17. E. R. Westervelt, J. W. Grizzle, and D. E. Koditschek. Hybrid zero dynamics of planar biped walkers. *IEEE Trans. on Automat. Contr.*, 48(1):42–56, 2003.
18. E. R. Westervelt, B. Morris, and K. D. Farrell. Analysis results and tools for the control of planar bipedal gaits using hybrid zero dynamics. *Autonomous Robots*, 2007. In press.

Part IV

Regulation

Hybrid Systems: Limit Sets and Zero Dynamics with a View Toward Output Regulation

Chaohong Cai[1], Rafal Goebel[2], Ricardo G. Sanfelice[1], and Andrew R. Teel[1]

[1] Center for Control, Dynamical Systems, and Computation; Department of Electrical and Computer Engineering; University of California, Santa Barbara, CA 93106-9560, USA.

[2] Address: 3518 NE 42 St., Seattle, WA 98105, USA.

Summary. We present results on omega-limit sets and minimum phase zero dynamics for hybrid dynamical systems. Moreover, we give pointers to how these results may be useful in the future for solving the output regulation problem for hybrid systems. We highlight the main attributes of omega-limit sets and we show, under mild conditions, that they are asymptotically stable. We define a minimum phase notion in terms of omega-limit sets and establish an equivalent Lyapunov characterization. Then we study the feedback stabilization problem for a class of minimum phase, relative degree one hybrid systems. Finally, we discuss output regulation for this class of hybrid systems. We illustrate the concepts with examples throughout the paper.

1 Introduction

This paper is written as a tribute to Professor Alberto Isidori for all of the important concepts and results he has introduced in the nonlinear control systems area over his illustrious career. Following the adage that imitation is the highest form of flattery, we have chosen for this tribute to emulate some of Professor Isidori's recent results on limit sets, zero dynamics, and output regulation [4, 7, 5, 3, 6].The novelty of our results comes from the setting that we consider: hybrid dynamical systems. These systems contain state variables that are capable of evolving continuously (flowing) and/or evolving discontinuously (jumping). In particular, systems with logical modes that interact with continuous states can be modeled in this framework. Hybrid systems have been studied in the literature for multiple decades (early notable references include [31, 28]), with the majority of progress having occurred since the early 1990s as codified, for example, in the books [30, 23, 19]. Recently, we have established mild sufficient conditions for robustness in hybrid dynamical systems [13, 14]. Along the way, these conditions have led to a generalization of results on ω-limit sets of trajectories and of LaSalle's invariance principle [26], and to general results on the existence of smooth Lyapunov functions (converse theorems) for hybrid systems [9, 10]. These results come together in the present paper, where we take inspiration from Isidori and Byrnes to establish results on Ω-limit sets (limit sets of sets of initial conditions) for hybrid systems, to show under mild conditions that these sets are asymptotically stable, to show how this notion can lead to a non-equilibrium characterization of asymptotically stable zero dynamics for hybrid systems, including converse Lyapunov theorems for a "minimum phase" property, and to give a stabilization result, related to nonlinear output regulation, for a class of minimum phase, relative degree one hybrid systems. Perhaps eventually, following the trail blazed by Professor Isidori, these results will be used for a more general theory of output regulation for hybrid systems and/or output regulation using hybrid controllers. We conclude this short introduction by noting that hybrid controllers have already appeared in the context of output regulation; see, for example, [27].

2 Hybrid Dynamical Systems

For the purposes of this paper, a hybrid system $\mathcal{H}$ is specified by the data (F, G, C, D) and a state space $O \subset \mathbb{R}^n$ where F is a set-valued mapping from O to $\mathbb{R}^n$ called the "flow map", G is a set-valued mapping from O to $\mathbb{R}^n$ called the "jump map", $C \subset O$ is called the "flow set" and indicates where in the state space flows may occur, $D \subset O$ is called the "jump set" and indicates from where in the state space jumps may occur.

We denote by x the state of the hybrid system $\mathcal{H}$ which can include both the so-called "continuous variables" and the so-called "discrete variables", or modes. A hybrid system $\mathcal{H}$ can be expressed as

$$\mathcal{H}\begin{cases} \dot{x} \in F(x) & x \in C \\ x^+ \in G(x) & x \in D, \end{cases}$$

which is suggestive of the meaning of solution to $\mathcal{H}$. Following [13, 14] and also [11] (cf. [1], and [21]), a solution to a hybrid system is a function defined on a hybrid time domain satisfying certain conditions. Let $\mathbb{R}_{\geq 0} := [0, +\infty)$ and $\mathbb{N} := \{0, 1, 2, \ldots\}$. A set $S \subset \mathbb{R}_{\geq 0} \times \mathbb{N}$ is a *compact hybrid time domain* if

$$S = \bigcup_{j=0}^{J-1} ([t_j, t_{j+1}], j)$$

for some finite sequence of times $0 = t_0 \leq t_1 \leq t_2 \ldots \leq t_J$. The set S is a *hybrid time domain* if for all $(T, J) \in S$,

$$S \cap ([0, T] \times \{0, 1, \ldots J\})$$

is a compact hybrid domain. By a *hybrid arc* we understand a pair consisting of a hybrid time domain dom ϕ and a function ϕ : dom $\phi \to \mathbb{R}^n$ such that

$t \mapsto \phi(t,j)$ is locally absolutely continuous for fixed j and $(t,j) \in \text{dom } \phi$. We will not mention dom ϕ explicitly, but always assume that given a hybrid arc ϕ, the set dom ϕ is exactly the set on which ϕ is defined.

A hybrid arc $\phi : \text{dom } \phi \to O$ is a *solution to* $\mathcal{H}$ if $\phi(0,0) \in C \cup D$ and:

(S1) for all $j \in \mathbb{N}$ and almost all t such that $(t,j) \in \text{dom } \phi$,

$$\phi(t,j) \in C, \quad \dot{\phi}(t,j) \in F(\phi(t,j));$$

(S2) for all $(t,j) \in \text{dom } \phi$ such that $(t,j+1) \in \text{dom } \phi$,

$$\phi(t,j) \in D, \quad \phi(t,j+1) \in G(\phi(t,j)).$$

A solution is called *nontrivial* if dom ϕ contains at least one point different from $(0,0)$, *complete* if dom ϕ is unbounded, *Zeno* if it is complete but the projection of dom ϕ onto $\mathbb{R}_{\geq 0}$ is bounded, and *maximal* if it is not a truncation of another solution ϕ' to some proper subset of dom ϕ'. The notation $\mathcal{S}_{\mathcal{H}}(\mathcal{X})$ indicates the set of maximal solutions to $\mathcal{H}$ from the set of initial conditions $\mathcal{X}$. Note that when $x^0 \notin C \cup D$, $\mathcal{S}_{\mathcal{H}}(x^0) = \emptyset$.

Standing Assumption 1 (Hybrid Basic Conditions) *State space $O \subset \mathbb{R}^n$ is open; sets C and D are closed relative*[3] *to O; mappings F and G are outer semicontinuous and locally bounded*[4] *on O; $F(x)$ is nonempty and convex for all $x \in C$; $G(x)$ is nonempty and contained in O for all $x \in D$.*

These (mild) assumptions on the data of $\mathcal{H}$ are needed to guarantee that, among other properties, the sets of solutions to $\mathcal{H}$ have good sequential compactness properties.

Theorem 1. *(sequential compactness, [14, Theorem 4.4]) Let $\phi_i : \text{dom } \phi_i \to \mathbb{R}^n$, $i = 1, 2, \ldots$, be a locally eventually bounded with respect to O sequence of solutions*[5] *to $\mathcal{H}$. Then there exists a subsequence of ϕ_i's graphically converging to a solution of $\mathcal{H}$. Such a limiting solution is complete if each ϕ_i is complete, or more generally, if no subsequence of ϕ_i's has uniformly bounded domains (i.e. for any $m > 0$, there exists $i_m \in \mathbb{N}$ such that for all $i > i_m$, there exists $(t,j) \in \text{dom } \phi_i$ with $t + j > m$).*

We refer the reader to [14] (see also [13]) for more details on and consequences of Standing Assumption 1.

---

[3] A set C is closed relative to O if $C = O \cap \overline{C}$.

[4] A set-valued mapping G defined on an open set O is *outer semicontinuous* if for each sequence $x_i \in O$ converging to a point $x \in O$ and each sequence $y_i \in G(x_i)$ converging to a point y, it holds that $y \in G(x)$. It is *locally bounded* if, for each compact set $K \subset O$ there exists $\mu > 0$ such that $G(K) := \cup_{x \in K} G(x) \subset \mu\mathbb{B}$, where $\mathbb{B}$ is the open unit ball in $\mathbb{R}^n$. For more details, see [25, Chapter 5].

[5] A sequence $\{\phi_i\}_{i=1}^{\infty}$ of hybrid trajectories is *locally eventually bounded* with respect to an open set O if for any $m > 0$, there exists $i_0 > 0$ and a compact set $K \subset O$ such that for all $i > i_0$, all $(t,j) \in \text{dom } \phi_i$ with $t + j < m$, $\phi_i(t,j) \in K$.

A more general approach to the study of hybrid systems is to consider abstract hybrid systems given by a collection of hybrid arcs satisfying certain properties but not associated to any particular data. These abstract hybrid systems have been introduced in [26] and are called *sets of hybrid trajectories*. Sets of hybrid trajectories parallel the concept of generalized semiflows, but with elements, given by hybrid arcs (or equivalently, following [26], given by *hybrid trajectories*), that can flow and/or jump. When they satisfy the sequential compactness property stated in Theorem 1, convergence results for sets of hybrid trajectories have been presented in [26]. For the sake of simplicity, in this paper we present results for hybrid systems $\mathcal{H}$ with data (F, G, C, D) and state space O, but extensions to sets of hybrid trajectories are possible.

Regarding existence of solutions to $\mathcal{H}$, conditions were given in [14] (see also [1]) for the existence of nontrivial solutions from $C \cup D$ that are either complete or "blow up". In words, these conditions are that at every point in $C \setminus D$ flowing should be possible and at every point in D, the map G maps to $C \cup D$. These conditions are automatically satisfied when $C \cup D = O$.

In what follows, we do not necessarily assume that solutions are either complete or blow up. Moreover, given a hybrid system $\mathcal{H}$ and a set $\mathcal{Y} \subset O$ that is closed relative to O, we denote the restriction of $\mathcal{H}$ to $\mathcal{Y}$ by the hybrid system $\mathcal{H}|_{\mathcal{Y}}$ which has data $(F, G, C \cap \mathcal{Y}, D \cap \mathcal{Y})$ and state space O. Note that $\mathcal{H}|_{\mathcal{Y}}$ still satisfies the hybrid basic conditions.

3 Ω-Limit Sets

The results in this section pertain to Ω-limit sets of sets of initial conditions for hybrid dynamical systems satisfying Standing Assumption 1. They extend to these systems some of the results in [16] as specialized to finite-dimensional systems. Since the solutions to hybrid systems are often not unique, the results here resemble those for generalized semiflows in [2] and [22], where nonuniqueness of solutions to continuous-time systems is permitted.

Consider a hybrid system $\mathcal{H}$ with state space O and data (F, G, C, D) satisfying Standing Assumption 1. For a given set $\mathcal{X} \subset O$, we define the Ω-limit set of $\mathcal{X}$ for $\mathcal{H}$ as:

$$\begin{aligned}\Omega_{\mathcal{H}}(\mathcal{X}) := \{ & y \in \mathbb{R}^n : \\ & y = \lim_{i \to \infty} \phi_i(t_i, j_i),\ \phi_i \in \mathcal{S}_{\mathcal{H}}(\mathcal{X}),\ (t_i, j_i) \in \operatorname{dom} \phi_i\ ,\ t_i + j_i \to \infty \Big\}.\end{aligned}$$

Clearly, there are connections between Ω-limit sets of sets of initial conditions for hybrid systems and ω-limit sets of solutions to hybrid systems, as pursued together with various hybrid invariance principles in [26]. We do not pursue such connections here other than to observe that, letting $\omega(\phi)$ denote the

ω-limit set of the solution ϕ to the hybrid system $\mathcal{H}$, we have

$$\bigcup_{x\in\mathcal{X},\phi\in\mathcal{S}_{\mathcal{H}}(x)} \omega(\phi) \subset \Omega_{\mathcal{H}}(\mathcal{X})$$

but that the opposite set containment does not necessarily hold.

We also define, for each $i \in \mathbb{N}$,

$$\mathcal{R}^i_{\mathcal{H}}(\mathcal{X}) := \{y \in O : y = \phi(t,j),\ \phi \in \mathcal{S}_{\mathcal{H}}(\mathcal{X}),\ (t,j) \in \operatorname{dom}\ \phi\ ,\ t+j \geq i\}\ .$$

We note that if $i' > i$ then $\mathcal{R}^{i'}_{\mathcal{H}}(\mathcal{X}) \subset \mathcal{R}^i_{\mathcal{H}}(\mathcal{X})$. Because of this, we say that the sequence of sets $\mathcal{R}^i_{\mathcal{H}}(\mathcal{X})$ is *nested.* Below, $\mathbb{B}$ denotes the open unit ball in $\mathbb{R}^n$.

Lemma 1. *Let $\mathcal{X} \subset O$. Then*[6]

$$\Omega_{\mathcal{H}}(\mathcal{X}) = \lim_{i\to\infty} \mathcal{R}^i_{\mathcal{H}}(\mathcal{X}) = \bigcap_i \overline{\mathcal{R}^i_{\mathcal{H}}(\mathcal{X})}\ . \tag{1}$$

Equivalently, for each $\varepsilon > 0$ and $\rho > 0$ there exists i^* such that for all $i \geq i^*$

(a) $\Omega_{\mathcal{H}}(\mathcal{X}) \cap \rho\overline{\mathbb{B}} \subset \mathcal{R}^i_{\mathcal{H}}(\mathcal{X}) + \varepsilon\overline{\mathbb{B}}$
(b) $\mathcal{R}^i_{\mathcal{H}}(\mathcal{X}) \cap \rho\overline{\mathbb{B}} \subset \Omega_{\mathcal{H}}(\mathcal{X}) + \varepsilon\overline{\mathbb{B}}$.

In what follows, we aim to clarify various attributes of the set $\Omega_{\mathcal{H}}(\mathcal{X})$. All of the subsequent attributes will be established under the assumption that the sets $\mathcal{R}^i_{\mathcal{H}}(\mathcal{X})$ are uniformly bounded with respect to O for large i:

Assumption 1. *The set $\mathcal{X} \subset O$ is such that the hybrid system $\mathcal{H}$ is* eventually uniformly bounded *from $\mathcal{X}$, i.e., there exist a compact set $K \subset O$ and a nonnegative integer i^* such that $\mathcal{R}^i_{\mathcal{H}}(\mathcal{X}) \subset K$ for all $i \geq i^*$.*

Remark 1. The notion of eventual uniform boundedness agrees with the property defined in [16, p. 8] of a compact set (contained in O) attracting $\mathcal{X}$ under the solutions of the system $\mathcal{H}$. The papers [4, 5, 3] use the term "uniformly attracts". ◁

Remark 2. Assumption 1 does not necessarily imply that $\mathcal{R}^i_{\mathcal{H}}(\mathcal{X})$ is nonempty for all i. Under Assumption 1, $\Omega_{\mathcal{H}}(\mathcal{X})$ is nonempty if and only if $\mathcal{R}^i_{\mathcal{H}}(\mathcal{X})$ is nonempty for all i. ◁

Since the sequence of sets $\mathcal{R}^i_{\mathcal{H}}(\mathcal{X})$ is nested, it is enough to verify that $\mathcal{R}^{i^*}_{\mathcal{H}}(\mathcal{X}) \subset K$ for some nonnegative integer i^* in order to establish Assumption 1. In particular, if $\mathcal{R}^0_{\mathcal{H}}(\mathcal{X})$, i.e., the reachable set from $\mathcal{X}$, is contained in a compact subset of O then $\mathcal{H}$ is eventually uniformly bounded from $\mathcal{X}$. The following examples show that it is possible for Assumption 1 to hold without $\mathcal{R}^0_{\mathcal{H}}(\mathcal{X})$ being bounded.

---

[6] A sequence of sets $S_i \subset \mathbb{R}^n$ converges to $S \subset \mathbb{R}^n$ (i.e. $\lim_{i\to\infty} S_i = S$) if for all $x \in S$ there exists a convergent sequence of $x_i \in S_i$ such that $\lim_{i\to\infty} x_i = x$ and, for any sequence of $x_i \in S_i$ and any convergent subsequence x_{i_k}, we have $\lim_{k\to\infty} x_{i_k} \in S$. For more details, see [25, Chapter 4].

Example 1. Consider the (hybrid) system with data $F(x) = -x^3$ and $\mathcal{X} = C := \mathbb{R}$ (and $D := \emptyset$). The solutions from $\mathcal{X}$ are unique, with $|x(t)| = \frac{|x(0)|}{\sqrt{1+2x(0)^2 t}}$. Thus, $\mathcal{R}^1_{\mathcal{H}}(\mathcal{X}) \subset \left[-\frac{1}{\sqrt{2}}, \frac{1}{\sqrt{2}}\right]$. It follows that $\mathcal{H}$ is eventually uniformly bounded from $\mathcal{X}$. △

Example 2. In the preceding example, the set $\mathcal{X}$ was not bounded whereas in this example it is. In [29] an example was given of a scalar, time-varying, locally Lipschitz, differential equation $\dot{x} = f(t,x)$, with $|f(t,x)| \le c|x|^3$ for some real number $c > 0$, where the origin is uniformly globally attractive (meaning that for each $R > 0$ and $\varepsilon > 0$ there exists $T > 0$ such that $|x(t_\circ)| \le R$ and $t \ge t_\circ + T$ implies $|x(t)| \le \varepsilon$) but not uniformly globally stable. In particular, the overshoots from the set $|x(t_\circ)| = 1$ grow to infinity with $t_\circ$. Define the set-valued map $F : \mathbb{R}^2 \rightrightarrows \mathbb{R}^2$ by

$$F(\xi) := \begin{bmatrix} -\xi_1^2 \\ f(\xi_1^{-1}, \xi_2) \end{bmatrix} \quad \forall \xi_1 \neq 0 \, , \qquad F(0,\xi_2) := \begin{bmatrix} 0 \\ [-c,c]\,\xi_2^3 \end{bmatrix}$$

and set $C = \mathbb{R}^2$, $D = \emptyset$, so that the hybrid basic conditions are satisfied. Let $\mathcal{X} = (0,1] \times [-1,1]$. Note that $\frac{d}{dt}\left(\xi_1^{-1}(t)\right) = 1$ and thus the behavior of ξ_2 matches that of the system $\dot{x} = f(t,x)$ with $t_\circ = \xi_1(0)^{-1}$. Due to the results in [29], $\mathcal{R}^0_{\mathcal{H}}(\mathcal{X})$ is not contained in a compact subset of $\mathbb{R}^2$. Now, since $\xi_1(0) \in (0,1]$ and $\dot{\xi}_1(t) = -\xi_1^2$, we have $\xi_1(t) \in (0,1]$ for all $t \ge 0$. Using uniform global attractivity, there exists an integer i^* such that $|\xi_2(t)| \le 1$ for all $t \ge i^*$ and all $\xi(0) \in \mathcal{X}$. Thus, $\mathcal{R}^{i^*}_{\mathcal{H}}(\mathcal{X}) \subset \mathcal{X}$, the latter being contained in the compact set $[0,1] \times [-1,1]$. It follows that $\mathcal{H}$ is eventually uniformly bounded from $\mathcal{X}$. △

Example 3. In the preceding example, the set $\mathcal{X}$ was not compact and the system did not exhibit finite escape times from $\mathcal{X}$. Consider the hybrid system $F(x) = x^3$, $C = \{0\} \cup [1/2, \infty)$, $G(1/8) = \{0,1\}$, $G(x) = 0$ otherwise, $D = (-\infty, 1/2]$ and take $\mathcal{X} = [-1/4, 1/4]$. The solutions first jump either to zero or, if initialized at 1/8, possibly to one. From $x = 0$, the solution remains at zero for all hybrid time, either flowing or jumping. From $x = 1$, the solutions escape to infinity in one unit of time. It follows that $\mathcal{R}^2_{\mathcal{H}}(\mathcal{X}) = \{0\}$ and thus $\mathcal{H}$ is eventually uniformly bounded from $\mathcal{X}$. △

The following proposition gives a realistic scenario in which Assumption 1 is equivalent to the assumption that $\mathcal{R}^0_{\mathcal{H}}(\mathcal{X})$ is contained in a compact subset of O.

Proposition 1. *Under Assumption 1, if $\overline{\mathcal{X}}$ is a compact subset of O and every maximal solution starting in $\overline{\mathcal{X}}$ either has an unbounded hybrid time domain or is bounded with respect to O then $\mathcal{R}^0_{\mathcal{H}}(\mathcal{X})$ is contained in a compact subset of O.*

We will now focus on invariance properties for $\Omega_{\mathcal{H}}(\mathcal{X})$. We say that a set $O_1 \subset O$ is *weakly backward invariant* if for each $q \in O_1$, $N > 0$, there exist $x^0 \in O_1$

and at least one $\phi \in \mathcal{S}_{\mathcal{H}}(x^0)$ such that for some $(t^*, j^*) \in \operatorname{dom} \phi$, $t^* + j^* \geq N$, we have $\phi(t^*, j^*) = q$ and $\phi(t, j) \in O_1$ for all $(t, j) \preceq (t^*, j^*)$, $(t, j) \in \operatorname{dom} \phi$. This definition was used to characterize invariance properties for the ω-limit set of a hybrid trajectory in [26]. A similar property, but for continuous-time systems, is called "negative semi-invariance" in [22, Definition 5].

We say that a set $O_1 \subset O$ is *strongly pre-forward invariant* if, for each $x^0 \in O_1$ and each $\phi \in \mathcal{S}_{\mathcal{H}}(x^0)$, we have $\phi(t, j) \in O_1$ for all $(t, j) \in \operatorname{dom} \phi$. The prefix "pre" is used here since we do not assume that maximal solutions starting in O_1 have an unbounded hybrid time domain.

The next theorem asserts that, under Assumption 1, the properties of weak backward invariance and uniform attractivity from $\mathcal{X}$ are generic for $\Omega_{\mathcal{H}}(\mathcal{X})$. These results parallel some of the results in [22, Theorem 1] for continuous-time, generalized semiflows. The result below also gives a condition for strong pre-forward invariance, which parallels a part of [2, Lemma 3.4] for continuous-time, generalized semiflows.

Theorem 2. *Under Assumption 1, the set $\Omega_{\mathcal{H}}(\mathcal{X})$ is contained in O, compact, weakly backward invariant, and for each $\varepsilon > 0$ there exists i^* such that, for all $i \geq i^*$, $\Omega_{\mathcal{H}}(\mathcal{X}) \subset \mathcal{R}^i_{\mathcal{H}}(\mathcal{X}) + \varepsilon\mathbb{B}$ and $\mathcal{R}^i_{\mathcal{H}}(\mathcal{X}) \subset \Omega_{\mathcal{H}}(\mathcal{X}) + \varepsilon\mathbb{B}$. If, in addition, $\Omega_{\mathcal{H}}(\mathcal{X}) \subset \mathcal{R}^0_{\mathcal{H}}(\mathcal{X}) \cup \mathcal{X}$ then $\Omega_{\mathcal{H}}(\mathcal{X})$ is strongly pre-forward invariant.*

Remark 3. If $\mathcal{X} \subset C \cup D$ then $\mathcal{X} \subset \mathcal{R}^0_{\mathcal{H}}(\mathcal{X})$. Otherwise, neither the containment $\mathcal{R}^0_{\mathcal{H}}(\mathcal{X}) \subset \mathcal{X}$ nor the containment $\mathcal{X} \subset \mathcal{R}^0_{\mathcal{H}}(\mathcal{X})$ necessarily holds. ◁

The next examples show that if the extra condition for strong pre-forward invariance, $\Omega_{\mathcal{H}}(\mathcal{X}) \subset \mathcal{R}^0_{\mathcal{H}}(\mathcal{X}) \cup \mathcal{X}$, is removed, then strong pre-forward invariance may fail.

Example 4. Consider the hybrid system with data $F(x) = -x$, $C = \mathbb{R}$, $G(x) = 1$, $D = \{0\}$ and take $\mathcal{X} = \{-1\}$. It is not difficult to verify that $\Omega_{\mathcal{H}}(\mathcal{X}) = \{0\}$. However, there is a solution starting at the origin that jumps to the value one, thus leaving $\Omega_{\mathcal{H}}(\mathcal{X})$, before flowing back toward the origin. Thus, $\Omega_{\mathcal{H}}(\mathcal{X})$ is not strongly (pre-)forward invariant. △

Example 5. This example is a purely continuous-time system. Consider the (hybrid) system with data $F(x) = -x$ for $x < 0$ and $F(x) = x^{1/3}$ for $x \geq 0$, $C = \mathbb{R}$ and $D = \emptyset$. With $\mathcal{X} = \{-1\}$, we again have $\Omega_{\mathcal{H}}(\mathcal{X}) = \{0\}$. However, there is a solution starting at the origin satisfying $\phi(t, 0) = (2t/3)^{3/2}$ for all $t \geq 0$. Thus, $\Omega_{\mathcal{H}}(\mathcal{X})$ is not strongly (pre-)forward invariant. △

The preceding examples motivate considering a weaker notion of forward invariance, as an alternative to strong (pre-)forward invariance. It would make sense to define weak *pre*-forward invariance, i.e., to require the existence of a solution remaining in the set that is nontrivial but not necessarily complete, at least at points where nontrivial solutions exist. However, the condition on completeness of solutions we give below actually implies more than what a weak *pre*-forward invariance notion would; it actually guarantees the existence of a complete solution remaining in $\Omega_{\mathcal{H}}(\mathcal{X})$ (this follows by weak backward invariance of $\Omega_{\mathcal{H}}(\mathcal{X})$). Therefore, our definition will actually insist on the existence of one *complete* solution remaining in the set. In this way, following [26], we say that a set $O_1 \subset O$ is *weakly forward invariant* if for each $x \in O_1$, there exists at least one complete solution $\phi \in \mathcal{S}_{\mathcal{H}}(x)$ with $\phi(t,j) \in O_1$ for all $(t,j) \in \text{dom } \phi$. We note that, in the context of continuous-time, generalized semiflows, the reference [2] combines weak forward invariance and weak backward invariance into a single property called quasi-invariance.

The next examples show that weak forward invariance can fail without extra assumptions, beyond Assumption 1.

Example 6. This example shows that there may not be any nontrivial solutions from points in $\Omega_{\mathcal{H}}(\mathcal{X})$. Consider the system with data $F(x) = x-1$, $C = [1,2]$, $D = \emptyset$ and take $\mathcal{X} = C$. Then it is not difficult to verify that $\Omega_{\mathcal{H}}(\mathcal{X}) = \mathcal{X}$ but that from the point $x = 2$, which belongs to $\Omega_{\mathcal{H}}(\mathcal{X})$, there are no nontrivial solutions. △

Example 7. This example shows that it is possible to have the existence of nontrivial solutions but not one that remains in $\Omega_{\mathcal{H}}(\mathcal{X})$. Consider the hybrid system with data $F(x) = -x$, $C = [-1,1]$, $G(x) = 10 + x$, $D = [-1,1] \cup \{10\}$ and $\mathcal{X} = \{1\}$. It is not difficult to verify that $\Omega_{\mathcal{H}}(\mathcal{X}) = \{0, 10\}$ but from the point $x = 10$ there is only one solution and it jumps to the value 20, i.e., leaves $\Omega_{\mathcal{H}}(\mathcal{X})$, which doesn't belong to $C \cup D$. △

Example 8. This example shows that weak forward invariance can fail even when the system is a purely continuous-time system with constraints. Consider the system with data $F(x) = \begin{bmatrix} 0 & x_2 - 1 \end{bmatrix}^T$, $C = \left\{x \in \mathbb{R}^2 : x_1 \geq 0 \text{ or } x_2 \geq 0\right\}$, $D = \emptyset$ and take $\mathcal{X} = \{x \in C : x_1 < 0\}$. It is not difficult to verify that $\Omega_{\mathcal{H}}(\mathcal{X}) = \{x \in C : x_1 \leq 0\,, x_2 \geq 0\}$. Thus, the origin belongs to $\Omega_{\mathcal{H}}(\mathcal{X})$. There is only one solution starting at the origin and it immediately leaves $\Omega_{\mathcal{H}}(\mathcal{X})$ by virtue of the x_2 component of the solution becoming negative. △

In order to guarantee weak forward invariance of $\Omega_{\mathcal{H}}(\mathcal{X})$, we will assume that the hybrid system $\mathcal{H}$ is *eventually complete from* $\mathcal{X}$, i.e., there exists a nonnegative integer i^* such that, for all $i \geq i^*$, every maximal solution starting in $\mathcal{R}^i_{\mathcal{H}}(\mathcal{X})$ has an unbounded hybrid time domain. (Note: this still doesn't guarantee that $\mathcal{R}^i_{\mathcal{H}}(\mathcal{X})$ is nonempty for all i and thus still doesn't guarantee that $\Omega_{\mathcal{H}}(\mathcal{X})$ is nonempty.) Since the sequence of sets $\mathcal{R}^i_{\mathcal{H}}(\mathcal{X})$ is nested, it is enough to verify this property for solutions starting in $\mathcal{R}^{i^*}_{\mathcal{H}}(\mathcal{X})$ for some nonnegative integer i^*. Example 3 has already shown that it is possible for $\mathcal{H}$ to be eventually complete from $\mathcal{X}$ without being complete from $\mathcal{X}$.

We will see that eventual completeness combined with the previous condition for strong pre-forward invariance will guarantee strong pre-forward invariance with complete solutions. We say that a set $O_1 \subset O$ is *strongly forward invariant* if it is strongly pre-forward invariant and each maximal solution starting in O_1 is complete, i.e., has an unbounded hybrid time domain.

The next theorem establishes weak forward invariance under Assumption 1 and the assumption that the system $\mathcal{H}$ is eventually complete from $\mathcal{X}$. This parallels a part of [22, Lemma 3.4] on continuous-time, generalized semiflows.

Theorem 3. *Under Assumption 1, if the hybrid system $\mathcal{H}$ is eventually complete from $\mathcal{X}$ then $\Omega_{\mathcal{H}}(\mathcal{X})$ is weakly forward invariant. If, in addition, $\Omega_{\mathcal{H}}(\mathcal{X}) \subset \mathcal{R}^0_{\mathcal{H}}(\mathcal{X}) \cup \mathcal{X}$ then $\Omega_{\mathcal{H}}(\mathcal{X})$ is strongly forward invariant.*

Below we state results on the Ω-limit sets of the restriction of $\mathcal{H}$ to some subset in the state space O.

Proposition 2. *If the set $\mathcal{Y} \subset O$ is closed relative to O and strongly pre-forward invariant for $\mathcal{H}$, then $\Omega_{\mathcal{H}|_{\mathcal{Y}}}(\mathcal{X}) = \Omega_{\mathcal{H}}(\mathcal{X} \cap \mathcal{Y})$.*

Theorem 4. *Suppose Assumption 1 holds. Define $M := \Omega_{\mathcal{H}}(\mathcal{X})$. Then $M \subset \Omega_{\mathcal{H}|_M}(M)$.*

Remark 4. The opposite containment, $\Omega_{\mathcal{H}|_M}(M) \subset M$, does not necessarily hold as demonstrated by Example 4 or Example 7. Clearly, the only way this containment can fail is if M is not forward invariant for the system $\mathcal{H}|_M$. This requires jumps from M that leave M, as in the referenced examples. ◁

The following corollaries of Theorem 4 are related to [4, Lemma 4.1] and the reduction principle for Ω-limit sets given in [3, Lemma 5.2].

Corollary 1. *Suppose Assumption 1 holds, $\mathcal{Z} \subset O$, and that $\mathcal{Y} \subset O$ is closed relative to O. If $\Omega_{\mathcal{H}}(\mathcal{X}) \subset \mathcal{Y} \cap \mathcal{Z}$ then $\Omega_{\mathcal{H}}(\mathcal{X}) \subset \Omega_{\mathcal{H}|_{\mathcal{Y}}}(\mathcal{Z})$.*

Corollary 2. *Suppose Assumption 1 holds, $\mathcal{Z} \subset O$ is compact, and that $\mathcal{Y} \subset O$ is closed relative to O. Suppose, for the system $\mathcal{H}$ that, for each $\varepsilon > 0$ there exists $T > 0$ such that for each $x \in \mathcal{X}$, each $\phi \in \mathcal{S}_{\mathcal{H}}(x)$, and each $(t, j) \in \operatorname{dom} \phi$ with $t + j \geq T$, we have $|\phi(t, j)|_{\mathcal{Y}\cap\mathcal{Z}} \leq \varepsilon$ [7]. Then $\Omega_{\mathcal{H}}(\mathcal{X}) \subset \mathcal{Y} \cap \mathcal{Z}$. In particular, we have $\Omega_{\mathcal{H}}(\mathcal{X}) \subset \Omega_{\mathcal{H}|_{\mathcal{Y}}}(\mathcal{Z})$.*

Remark 5. Corollary 2 is a useful tool for stability analysis in cascade-connected systems (for example, recovering Corollaries 10.3.2 and 10.3.3 in [18]). ◁

The properties established in Theorems 2, 3 and 4 above parallel analogous results for continuous-time dynamical systems, as summarized in [16, Chapter 2], that have been fundamental to the work in [4, 7, 5, 3]. In subsequent sections, we will use these properties in a manner that parallels how results for continuous-time systems were used in these latter references. In particular, we will show how these results impact robust stability and control results for hybrid systems. The next section addresses the notion of asymptotic stability that we use.

[7] In what follows, given $x \in \mathbb{R}^n$ and $S \subset \mathbb{R}^n$, $|x|_S := \inf\{|x - s| : s \in S\}$.

4 Pre-Asymptotically Stable Compact Sets

Pre-asymptotic stability (pre-AS) is a generalization of standard asymptotic stability to the setting where completeness or even existence of solutions is not required. Pre-AS was introduced in [10] as an equivalent characterization of the existence of a smooth Lyapunov function for a hybrid system. It is a natural stability notion for hybrid systems, since often the set $C \cup D$ does not cover the state space O and because local existence of solutions is sometimes not guaranteed. As we will see subsequently, not insisting on local existence of solutions can make it easier to characterize certain dynamic properties, such as the minimum phase property, and to give stronger converse Lyapunov theorems for such properties.

Consider the hybrid system $\mathcal{H}$. Let $\mathcal{A} \subset O$ be compact. We say that

- $\mathcal{A}$ is *pre-stable* for $\mathcal{H}$ if for each $\varepsilon > 0$ there exists $\delta > 0$ such that any solution to $\mathcal{H}$ with $|\phi(0,0)|_{\mathcal{A}} \leq \delta$ satisfies $|\phi(t,j)|_{\mathcal{A}} \leq \varepsilon$ for all $(t,j) \in \operatorname{dom} \phi$;
- $\mathcal{A}$ is *pre-attractive* for $\mathcal{H}$ if there exists $\delta > 0$ such that any solution ϕ to $\mathcal{H}$ with $|\phi(0,0)|_{\mathcal{A}} \leq \delta$ is bounded with respect to O and if it is complete then $\phi(t,j) \to \mathcal{A}$ as $t + j \to \infty$;
- $\mathcal{A}$ is *uniformly pre-attractive* if there exists $\delta > 0$ and for each $\varepsilon > 0$ there exists $T > 0$ such that any solution ϕ to $\mathcal{H}$ with $|\phi(0,0)|_{\mathcal{A}} \leq \delta$ is bounded with respect to O and $|\phi(t,j)|_{\mathcal{A}} \leq \varepsilon$ for all $(t,j) \in \operatorname{dom} \phi$ satisfying $t + j \geq T$;
- $\mathcal{A}$ is *pre-asymptotically stable* if it is both pre-stable and pre-attractive;
- ($\mathcal{A}$ is *asymptotically stable* if it is pre-asymptotically stable and there exists $\delta > 0$ such that any maximal solution ϕ to $\mathcal{H}$ with $|\phi(0,0)|_{\mathcal{A}} \leq \delta$ is complete.)

The set of all $x \in C \cup D$ from which all solutions are bounded with respect to O and the complete ones converge to $\mathcal{A}$ is called the *pre-basin of attraction* of $\mathcal{A}$.

Clearly, these stability definitions cover classical stability notions. They also cover some unexpected situations, such as in the following example.

Example 9. Consider the (hybrid) system $\dot{x} = Ax, x \in C$ where $A \in \mathbb{R}^{2\times 2}$ has complex eigenvalues with positive real part and $C := \{x \in \mathbb{R}^2 : x_1 x_2 \leq 0\}$ (and $D := \emptyset$). Because of the structure of the matrix A, there is a number $T > 0$ such that solutions to $\dot{x} = Ax$ starting on the unit circle in the set C can flow for no more than T units of time before leaving C. It follows from homogeneity that no solutions are complete and thus the origin is pre-attractive, in fact, uniformly pre-attractive. Moreover, defining $c := \exp(AT)$, we have $|\phi(t,0)| \leq c|\phi(0,0)|$. Thus the origin is pre-stable. In summary, the origin is pre-asymptotically stable with pre-basin of attraction given as C. △

The following results come from [10] and are used to establish many of the subsequent statements in this paper.

Lemma 2. *For system $\mathcal{H}$, if the compact set $\mathcal{A} \subset O$ is strongly pre-forward invariant and uniformly pre-attractive, then $\mathcal{A}$ is pre-asymptotically stable.*

Lemma 3. *Let the set $O_1 \subset O$ be open, and let the set $\mathcal{A} \subset O_1$ be nonempty and compact. For system $\mathcal{H}$, the following statements are equivalent:*

- *The set $\mathcal{A}$ is pre-asymptotically stable with pre-basin of attraction containing $O_1 \cap (C \cup D)$, and O_1 is strongly pre-forward invariant;*
- *For each function $\omega : O_1 \to \mathbb{R}_{\geq 0}$ that is a proper indicator*[8] *for $\mathcal{A}$ on O_1, there exists a smooth Lyapunov function $V : O_1 \to \mathbb{R}_{\geq 0}$ for $(O_1, F, G, C, D, \omega)$ on O_1, that is, there exist class-$\mathcal{K}_\infty$ functions α_1, α_2 such that*

$$\begin{aligned} \alpha_1(\omega(x)) \leq V(x) \leq \alpha_2(\omega(x)) \qquad & \forall x \in O_1 \ , \\ \max_{f \in F(x)} \langle \nabla V(x), f \rangle \leq -V(x) \qquad & \forall x \in O_1 \cap C \ , \\ \max_{g \in G(x)} V(g) \leq e^{-1} V(x) \qquad & \forall x \in O_1 \cap D \ . \end{aligned}$$

Lemma 4. *For system $\mathcal{H}$, if the compact set $\mathcal{A} \subset O$ is pre-asymptotically stable, then its pre-basin of attraction is open relatively to $C \cup D$, and there exists an open set $O_1 \subset O$ that is strongly pre-forward invariant and equals to $O_1 \cap (C \cup D)$.*

The combination of Lemmas 3 and 4 not only provides Lyapunov characterizations of pre-asymptotic stability (and even a strong result on converse Lyapunov theorems for pre-asymptotic stability), but also allows us to establish an equivalent Lyapunov characterization of hybrid systems with pre-asymptotically stable zero-dynamics in the next section.

In the following result we give sufficient conditions for $\Omega_{\mathcal{H}}(\mathcal{X})$ to be pre-asymptotically stable for hybrid systems.

Theorem 5. *Suppose Assumption 1 holds. If the set $\mathcal{X} \subset O$ is such that each solution starting in $\mathcal{X}$ is bounded with respect to O and $\Omega_{\mathcal{H}}(\mathcal{X}) \subset \mathrm{int}(\mathcal{X})$ then $\Omega_{\mathcal{H}}(\mathcal{X})$ is a compact pre-asymptotically stable set with pre-basin of attraction containing $\mathcal{X} \cap (C \cup D)$.*

Corollary 3. *Suppose for the system $\mathcal{H}$ that there exist $T > 0$ and compact sets $\mathcal{X} \subset O$ and $\mathcal{X}_o \subset O$ such that $\mathcal{X}_o \subset \mathrm{int}(\mathcal{X})$ and, each solution ϕ starting in $\mathcal{X}$ is bounded with respect to O and $\phi(t,j) \in \mathcal{X}_o$ for all $(t,j) \in \mathrm{dom}\,\phi$ with $t+j \geq T$. Then Assumption 1 holds and $\Omega_{\mathcal{H}}(\mathcal{X}) \subset \mathcal{X}_o \subset \mathrm{int}(\mathcal{X})$. In particular, $\Omega_{\mathcal{H}}(\mathcal{X})$ is a compact pre-asymptotically stable set with pre-basin of attraction containing $\mathcal{X} \cap (C \cup D)$.*

---

[8] Given an open set O_1 containing a compact set $\mathcal{A}$, a continuous function $\omega : O_1 \to \mathbb{R}_{\geq 0}$ is *proper* on O_1 if $\omega(x_i) \to \infty$ when x_i converge to the boundary of O_1 or $|x_i| \to \infty$, and is a *proper indicator for $\mathcal{A}$ on O_1* if it is proper on O_1 and satisfies $\{x \in O_1 : \omega(x) = 0\} = \mathcal{A}$.

Theorem 5 parallels the stability result for omega limit sets of sets in [16, Lemma 2.0.1] for continuous-time nonlinear systems (see also [4, Lemma 2.1]).

It is obvious that the weaker condition $\Omega_{\mathcal{H}}(\mathcal{X}) \subset \mathcal{X}$ does not imply pre-asymptotic stability for $\Omega_{\mathcal{H}}(\mathcal{X})$. For example, consider any Lipschitz differential equation where the origin is an unstable equilibrium point and take $\mathcal{X} = \{0\}$. It is even possible to have $\Omega_{\mathcal{H}}(\mathcal{X}) \subset \mathcal{X}$ and have $\Omega_{\mathcal{H}}(\mathcal{X})$ globally attractive without having $\Omega_{\mathcal{H}}(\mathcal{X})$ pre-asymptotically stable. For example, consider the system in [15, pp. 191-194] where the origin is globally attractive but not stable.

The following result gives sufficient conditions for pre-asymptotic stability of $\Omega_{\mathcal{H}_{|\mathcal{Y}}}(\mathcal{X})$ (recall that $\mathcal{H}_{|\mathcal{Y}} = (F, G, C \cap \mathcal{Y}, D \cap \mathcal{Y})$).

Proposition 3. *Suppose Assumption 1 holds and let $\mathcal{Y}$ be closed relative to O. Suppose, for the hybrid system $\mathcal{H}$, each solution starting in $\mathcal{X} \cap \mathcal{Y}$ is bounded with respect to O, and $\Omega_{\mathcal{H}}(\mathcal{X} \cap \mathcal{Y}) \subset \operatorname{int}(\mathcal{X})$. Then the hybrid system $\mathcal{H}_{|\mathcal{Y}}$ is eventually uniformly bounded from $\mathcal{X}$, each solution of $\mathcal{H}_{|\mathcal{Y}}$ starting in $\mathcal{X}$ is bounded with respect to O and $\Omega_{\mathcal{H}_{|\mathcal{Y}}}(\mathcal{X}) \subset \operatorname{int}(\mathcal{X})$. In particular, for the system $\mathcal{H}_{|\mathcal{Y}}$, $\Omega_{\mathcal{H}_{|\mathcal{Y}}}(\mathcal{X})$ is a compact pre-asymptotically stable set with pre-basin of attraction containing $\mathcal{X} \cap \mathcal{Y} \cap (C \cup D)$* [9].

We emphasize that none of the assumptions in the results above have guaranteed that $\Omega_{\mathcal{H}}(\mathcal{X})$ is nonempty. One may ask, in the case when $\Omega_{\mathcal{H}}(\mathcal{X})$ is empty, when one can still guarantee the existence of a compact, pre-asymptotically stable set contained in the interior of $\mathcal{X}$ with pre-basin of attraction containing $\mathcal{X}$. Such a characterization is given next.

Proposition 4. *Let Assumption 1 hold. Suppose the set $\mathcal{X} \subset O$ is such that each solution starting in $\mathcal{X}$ is bounded with respect to O and $\Omega_{\mathcal{H}}(\mathcal{X}) \subset \operatorname{int}(\mathcal{X})$. There exists a nonempty, compact, pre-asymptotically stable set $\mathcal{A} \subset \operatorname{int}(\mathcal{X})$ with pre-basin of attraction containing $\mathcal{X} \cap (C \cup D)$ if and only if there exists a point $x \in \operatorname{int}(\mathcal{X})$ such that either $x \notin C \cup D$ or $\mathcal{R}^0_{\mathcal{H}}(x) \subset \operatorname{int}(\mathcal{X})$.*

5 Minimum Phase Zero Dynamics

In this section, we address the concept of zero dynamics and the minimum phase property for hybrid systems with inputs. For an introduction to these concepts for non-hybrid nonlinear control systems, see [17, Chapter 6].

[9] The assumptions used in Proposition 3 are related to [4, Assumption 1]. In particular, the set $\mathcal{Y}$ in the proposition below should be associated with the set $\{(z, w) : w \in W\}$ where W is characterized in [4, Assumption 0] and $\mathcal{X}$ should be associated with the set $\{(z, w) : z \in Z\}$ where Z is a set of initial conditions given in [4]. Furthermore, if $\mathcal{Y}$ is strongly (pre-)forward invariant (cf. [4, Assumption 0]), then Proposition 2 says that $\Omega_{\mathcal{H}}(\mathcal{X} \cap \mathcal{Y}) = \Omega_{\mathcal{H}_{|\mathcal{Y}}}(\mathcal{X})$.

Consider the control-hybrid system

$$\mathcal{H}^u \begin{cases} \dot{x} = f(x,u) & (x,u) \in C \\ x^+ = g(x,u) & (x,u) \in D \end{cases} \tag{2}$$

with state space $O \subset \mathbb{R}^n$, where $f : C \to \mathbb{R}^n$ and $g : D \to O$ are continuous, and $C, D \subset O \times \mathbb{R}^m$ are closed relative to $O \times \mathbb{R}^m$. Solutions of $\mathcal{H}^u$ are defined in a manner that is analogous to the definition of solutions for $\mathcal{H}$ in Section 2. The signal u is a hybrid control signal, i.e., like a hybrid arc but instead of being locally absolutely continuous in t, it only needs to be locally bounded and measurable. A solution is a pair (x, u) consisting of a hybrid arc and a hybrid control signal that share the same hybrid time domain. In particular, it is not possible to pick the domain of the hybrid control signal independently from the domain of the state trajectory.

Associate to (2) an additional constraint $(x,u) \in \mathcal{Y}$, i.e., consider the control-hybrid system $\mathcal{H}^u_{|\mathcal{Y}}$. The "zero dynamics" (of $\mathcal{H}^u$ relative to $\mathcal{Y}$) is given by the hybrid system $\mathcal{H}^u_{|\mathcal{Y}}$. Let $\mathcal{N}$ denote the class of functions from $\mathbb{R}_{\geq 0}$ to $\mathbb{R}_{\geq 0}$ that are continuous and nondecreasing. Given $\gamma \in \mathcal{N}$, we use $\mathcal{H}^{u \to \gamma}_{|\mathcal{Y}}$ to denote the hybrid system with the data

$$\begin{aligned} F_{\gamma,\mathcal{Y}}(x) &:= \overline{\text{co}}\{z \in \mathbb{R}^n : z = f(x,u),\ (x,u) \in C \cap \mathcal{Y},\ |u| \leq \gamma(|x|)\}\ , \\ G_{\gamma,\mathcal{Y}}(x) &:= \{z \in \mathbb{R}^n : z = g(x,u),\ (x,u) \in D \cap \mathcal{Y},\ |u| \leq \gamma(|x|)\}\ , \\ C_{\gamma,\mathcal{Y}} &:= \{x \in \mathbb{R}^n : \exists u \in \mathbb{R}^m \text{ such that } (x,u) \in C \cap \mathcal{Y},\ |u| \leq \gamma(|x|)\}\ , \\ D_{\gamma,\mathcal{Y}} &:= \{x \in \mathbb{R}^n : \exists u \in \mathbb{R}^m \text{ such that } (x,u) \in D \cap \mathcal{Y},\ |u| \leq \gamma(|x|)\}\ . \end{aligned} \tag{3}$$

Let $\mathcal{Y}_* := \{x \in \mathbb{R}^n : \exists u \in \mathbb{R}^m \text{ such that } (x,u) \in \mathcal{Y},\ |u| \leq \gamma(|x|)\}$. The zero dynamics of $\mathcal{H}^u$ relative to $\mathcal{Y}$ is said to be *(robustly) pre-asymptotically stable from the compact set* $\mathcal{X} \subset O$ if, for each $\gamma \in \mathcal{N}$, the system $\mathcal{H}^{u \to \gamma}_{|\mathcal{Y}}$ is such that each solution starting in $\mathcal{X}$ is bounded with respect to O and $\Omega_{\mathcal{H}^{u \to \gamma}_{|\mathcal{Y}}}(\mathcal{X}) \subset \text{int}(\mathcal{X}) \cap \mathcal{Y}_*$; moreover, if $\Omega_{\mathcal{H}^{u \to \gamma}_{|\mathcal{Y}}}(\mathcal{X})$ is empty then there exists $x \in \text{int}(\mathcal{X}) \cap \mathcal{Y}_*$ such that either $x \notin C_{\gamma,\mathcal{Y}} \cup D_{\gamma,\mathcal{Y}}$ or $\mathcal{R}^0_{\mathcal{H}^{u \to \gamma}_{|\mathcal{Y}}}(x) \subset \text{int}(\mathcal{X}) \cap \mathcal{Y}_*$. When the zero dynamics of $\mathcal{H}^u$ relative to $\mathcal{Y}$ is (robustly) pre-asymptotically stable from the compact set $\mathcal{X} \subset O$, we will say that $\mathcal{H}^u_{|\mathcal{Y}}$ is *strongly minimum phase* relative to $\mathcal{X}$.

Example 10. Consider the nonlinear control system defined on $\mathbb{R}^3 \times \mathbb{R}$

$$\begin{aligned} \dot{x}_1 &- x_1^3 + x_1 u + x_2^2 u^2 \\ \dot{x}_2 &= x_3 \\ \dot{x}_3 &= q(x_1, x_2, x_3) + u \\ y &= x_2 \end{aligned} \tag{4}$$

where $q : \mathbb{R}^3 \to \mathbb{R}$ is continuous such that

$$|x_1| > 1 \qquad \Longrightarrow \qquad x_1 \left(x_1^3 - x_1 q(x_1, 0, 0)\right) < 0 \,. \tag{5}$$

To check the minimum phase property, we must consider, for each function $\gamma \in \mathcal{N}$, the behavior of the (hybrid) system

$$\begin{bmatrix} \dot{x}_1 \\ \dot{x}_2 \\ \dot{x}_3 \end{bmatrix} \in \left\{ \begin{bmatrix} x_1^3 + x_1 u + x_2^2 u^2 \\ x_3 \\ q(x_1, x_2, x_3) + u \end{bmatrix} , \ |u| \leq \gamma(|x|) \right\} \qquad (x, u) \in \mathcal{Y} \tag{6}$$

where $\mathcal{Y} := \left\{(x, u) \in \mathbb{R}^3 \times \mathbb{R} : x_2 = 0\right\}$. We check the minimum phase property relative to a compact set $\mathcal{X}$ containing the set $[-1, 1] \times \{0\} \times \{0\}$ in its interior. We note that, regardless of the function γ, in order to flow in the set $\mathcal{Y}$ we must have $x_2(t) = x_3(t) = 0$ for all t in the maximal interval of definition and $\dot{x}_3(t) = 0$ for almost all such t. From this it follows that flowing is only possible from points $(x_1, 0, 0)$ such that $|q(x_1, 0, 0)| \leq \gamma(|x_1|)$. Whenever flowing is possible, it must be the case that, for almost all t,

$$\dot{x}_1(t) = x_1(t)^3 - x_1(t) q(x_1(t), 0, 0) \,.$$

Using (5), the set $[-1, 1] \times \{0\} \times \{0\}$ is strongly forward invariant for (6). It also follows from (5) that, when the Ω-limit set of (6) is nonempty, it is contained in the set $[-1, 1] \times \{0\} \times \{0\}$ which, by assumption, is contained in $\operatorname{int}(\mathcal{X}) \cap \mathcal{Y}_*$. Then, the system (4) is minimum phase relative to $\mathcal{X}$. When the Ω-limit set is empty, which is the case for some functions $\gamma \in \mathcal{N}$ and q satisfying (5) (for example, consider $q(x_1, 0, 0) = 1 + 2x_1^2$ and $\gamma \equiv 0$), there are no complete solutions to (6). In turns out that, due to (5), we can take any point x the set $[-1, 1] \times \{0\} \times \{0\}$ and get that $\mathcal{R}^0_{\mathcal{H}^{u \to \gamma}_{|\mathcal{Y}}}(x) \in \operatorname{int}(\mathcal{X}) \cap \mathcal{Y}_*$. This shows that the system (4) is minimum phase relative to $\mathcal{X}$. △

Following the ideas in the example above, one can compare the zero dynamics notion given above to the description used in [4]. In the latter case, one identifies a subset of $\mathcal{Y}$, called the zero dynamics kernel, that is viable at every point and a (unique) feedback control selection that makes the zero dynamics kernel viable. In contrast, we work with the dynamics on all of $\mathcal{Y}$ and do not insist on viability. A possible advantage of the latter approach is that it leads to an equivalent Lyapunov characterization of the minimum phase property where the Lyapunov function is shown to be decreasing on all of $\mathcal{Y}$, not just on the zero dynamics kernel and not just for certain control values. This result is provided next and is a consequence of Lyapunov characterizations of pre-asymptotic stability in the last section.

Theorem 6. *Let $\mathcal{X} \subset O$ be compact. For system (2), the following statements are equivalent.*

(a) $\mathcal{H}^u_{|\mathcal{Y}}$ is strongly minimum phase relative to $\mathcal{X}$;

(b) For each $\gamma \in \mathcal{N}$ there exists a nonempty open set $O_1 \subset O$ containing $\mathcal{X}$, and a nonempty compact set $\mathcal{A} \subset \mathrm{int}(\mathcal{X})$ such that for each proper indicator $\omega : O_1 \to \mathbb{R}_{\geq 0}$ for $\mathcal{A}$ on O_1 there exists a smooth function $V : O_1 \to \mathbb{R}_{\geq 0}$ and class-$\mathcal{K}_\infty$ functions α_1 and α_2 such that

$$\begin{aligned} \alpha_1(\omega(x)) \leq V(x) \leq \alpha_2(\omega(x)) \quad & \forall x \in O_1 \ , \\ \langle \nabla V(x), f(x,u) \rangle \leq -V(x) \quad & \forall (x,u) \in C \cap \mathcal{Y}, \ |u| \leq \gamma(|x|) \ , \\ V(g(x,u)) \leq e^{-1} V(x) \quad & \forall (x,u) \in D \cap \mathcal{Y}, \ |u| \leq \gamma(|x|) \ . \end{aligned}$$

There have been several alternative characterizations of minimum phase zero dynamics that have appeared in the literature. In [24], the authors provide a notion of minimum phase (relative to an equilibrium point) that again asks for viability of a zero dynamics kernel and the existence of a stabilizing control selection, but allows for other control selections that are destabilizing. The system in [24, Example 1] is minimum phase in the sense of [24] but it is not strongly minimum phase in the sense of the current paper. Compared to what we have proposed, one could call the notion in [24] a *weak* minimum phase property (like the distinction between weak and strong invariance.) It is easy to define a weak minimum phase property in the context of Ω-limit sets for hybrid systems by replacing u by a stabilizing, locally bounded feedback, forming the convex hull and considering Ω-limit sets. Lyapunov characterizations of this property would also be straightforward, with the Lyapunov function decreasing everywhere that the output is zero but only for control values close to those of the stabilizing feedback.

The minimum phase property is also addressed in [20, Definition 3] where more general notions, output-input stability [20, Definition 1] and weak uniform 0-detectability, are introduced. In output-input stability, the state and the input should be bounded by the output and its derivatives plus a function of the norm of the state that decays with time. When evolving in the set where the output is zero, so that the derivatives are also zero, this asks for convergence of the input and state to zero, which is a property that is similar to our strong minimum phase property in the case where the Ω-limit set is the origin and the functions γ are required to be zero at zero. Many interesting phenomena appear by considering dynamics outside of the output zeroing set, and this is in large part the focus of the paper [20]. Included in this work is a Lyapunov characterization of weak uniform 0-detectability, which is like output-input stability but without imposing a bound on the input. The authors of [20] also provide an example that partially motivates bounding the inputs by some function of the state in the Lyapunov characterization of the minimum phase property.

The work [12] also considers a strong minimum phase property, much like the one we have presented but for equilibria, and discusses its Lyapunov characterization. In [12], the decrease condition for the Lyapunov function is in the set where the output *and* all of its derivatives are zero, a set related to the zero dynamics kernel mentioned above. This is in contrast to our result when the Lyapunov function decreases everywhere in the set where the output is zero. Like the example mentioned in [20], [12, Example 2] again motivates restricting the size of the input as a function of the size of the state in order to get a converse Lyapunov theorem.

6 Feedback Stabilization for a Class of Strongly Minimum Phase, Relative Degree One Hybrid Systems

Consider the control-hybrid system

$$\mathcal{H}^u \begin{cases} \left.\begin{array}{l} \dot{z} = \hat{f}(z,\zeta) \\ \dot{\zeta} = q(z,\zeta) + u \end{array}\right\} & (z,\zeta) \in \hat{C} \\ \\ \left.\begin{array}{l} z^+ = \hat{g}(z,\zeta) \\ \zeta^+ = r(z,\zeta) \end{array}\right\} & (z,\zeta) \in \hat{D} \end{cases}$$

where $z \in \mathbb{R}^{n_1}$; $\zeta, u \in \mathbb{R}^{n_2}$; $O = \mathbb{R}^{n_1+n_2}$ is the state space; $\hat{f}, \hat{g} : \mathbb{R}^{n_1} \times \mathbb{R}^{n_2} \to \mathbb{R}^{n_1}$ and $q, r : \mathbb{R}^{n_1} \times \mathbb{R}^{n_2} \to \mathbb{R}^{n_2}$ are continuous functions and the sets $\hat{C}$ and $\hat{D}$ are closed relatively to O (per the Standing Assumption 1). We investigate the effect of the feedback control algorithm $u = -k\zeta$ with $k > 0$ to be specified. We make suitable assumptions, made explicit below, that guarantee this feedback steers ζ to zero while keeping the entire state bounded. The zero dynamics corresponding to the output $y = \zeta$ play a crucial role.

To match the notation of the previous section, we define $f = [\hat{f}\ q]^T$, $g = [\hat{g}\ r]^T$, $C := \hat{C} \times \mathbb{R}^m$ and $D := \hat{D} \times \mathbb{R}^m$. We also define $\hat{\mathcal{Y}} := \{(z,\zeta) : \zeta = 0\}$ and $\mathcal{Y} := \hat{\mathcal{Y}} \times \mathbb{R}^m$. We note that $\mathcal{Y}_\star = \hat{\mathcal{Y}}$.

Assumption 2. *Let $\mathcal{X} \subset O$ be compact. The system $\mathcal{H}^u_{|\mathcal{Y}}$ is strongly minimum phase relative to $\mathcal{X}$.*

Regardless of $\gamma \in \mathcal{N}$, as long as the solutions of $\mathcal{H}^{u\to\gamma}_{|\mathcal{Y}}$ exist, they are solutions of the hybrid system

$$\mathcal{H}_\circ \begin{cases} \left.\begin{array}{l} \dot{z} = \hat{f}(z,0) \\ \dot{\zeta} = 0 \end{array}\right\} & (z,\zeta) \in \hat{C} \cap \hat{\mathcal{Y}} \\ \\ \left.\begin{array}{l} z^+ = \hat{g}(z,0) \\ \zeta^+ = r(z,0) \end{array}\right\} & (z,\zeta) \in \hat{D} \cap \hat{\mathcal{Y}}\ . \end{cases} \tag{7}$$

Thus, to check the conditions for the strong minimum phase property, it is enough to check them for the system (7). It is worth noting that, depending on γ, the system (7) may have more solutions than the zero dynamics. This is because, for a given $\gamma \in \mathcal{N}$ and a certain z, the zero value may not belong to the set $\{q(z,0)\} + \gamma(|z|)\overline{\mathbb{B}}$. However, when $\gamma \in \mathcal{N}$ is such that $|q(z,0)| \leq \gamma(|z|)$ for all z then the solutions of the zero dynamics agree with the solutions of (7).

Define $\mathcal{A}_\circ := \Omega_{\mathcal{H}_\circ}(\mathcal{X})$, where $\mathcal{X}$ is given in Assumption 2, for the case where $\Omega_{\mathcal{H}_\circ}(\mathcal{X})$ is nonempty. Otherwise, one can take $\mathcal{A}_\circ$ to be either the point in $\text{int}(\mathcal{X}) \cap \hat{\mathcal{Y}}$ that is not in $\hat{C} \cup \hat{D}$ or else the reachable set for $\mathcal{H}_\circ$ from the point in $\text{int}(\mathcal{X}) \cap \hat{\mathcal{Y}}$ having the property that this reachable set is contained in $\text{int}(\mathcal{X}) \cap \hat{\mathcal{Y}}$. Necessarily the set $\mathcal{A}_\circ$ is pre-asymptotically stable for $\mathcal{H}_\circ$. Let O_1 be the largest open set such that, for the system $\mathcal{H}_\circ$, the pre-basin of attraction for $\mathcal{A}_\circ$ is $O_1 \cap (\hat{C} \cup \hat{D}) \cap \hat{\mathcal{Y}}$. We note that $(z,\zeta) \in O_1$ does not put a restriction on ζ.

In addition to the strong minimum phase assumption, we make some simplifying assumptions on the functions q, r, $\hat{f}$ and $\hat{g}$ in order to give the flavor for the kinds of results that are possible. With a good knowledge of the nonlinear control literature, the reader may be able to see the directions in which these assumptions can be relaxed, especially in light of the converse Lyapunov theorem for the strong minimum phase property, as given in Theorem 6. (Also see the discussion in Section 7.)

We let $K \subset O_1$ denote a compact set over which we expect the closed-loop system to operate. It can, and should, be chosen to contain a neighborhood of $\mathcal{A}_\circ$. In order to state the assumptions succinctly, we make the definitions

$$F(z,\zeta,u) := \begin{bmatrix} \hat{f}(z,\zeta) \\ q(z,\zeta) + u \end{bmatrix}, \qquad G(z,\zeta) := \begin{bmatrix} \hat{g}(z,\zeta) \\ r(z,\zeta) \end{bmatrix}.$$

Assumption 3. *There exist $c > 0$ and $\delta > 0$ such that*

(a) a) $(z,\zeta) \in K \cap \hat{D} \implies |r(z,\zeta)| \le c|\zeta|$;
b) $(z,\zeta) \in \left(K \cap \hat{C}\right) + \delta\mathbb{B} \implies |q(z,\zeta)| \le c|\zeta|$;

(b) There exists a closed set $\hat{D}_e \subset O$ such that $\hat{D} \subset \hat{D}_e$ and $G(\hat{D}) \cap \hat{D}_e = \emptyset$;

(c) for almost all $(z,\zeta) \in \left(K \cap \hat{C}\right) + \delta\mathbb{B}$ and all u such that $\langle \zeta, u\rangle \le 0$,

$$-\langle \nabla |(z,\zeta)|_{\hat{D}_e}, F(z,\zeta,u)\rangle \le c\,. \tag{8}$$

Remark 6. The condition (8) is certainly satisfied when $|(z,\zeta)|_{\hat{D}_e}$ is independent of ζ, i.e., when the jump condition depends only on z. The condition in the third item guarantees that the flow for ζ, which can be controlled, is given enough time to dominate the jump behavior of ζ. In particular, it rules out Zeno solutions for the closed-loop control system. ◁

Theorem 7. *Under Assumptions 2-3, there exists $k^* \ge 0$ such that for each $k \ge k^*$, using the feedback control law $u = -k\zeta$ in the control system $\mathcal{H}^u$ results in the following property: The set $\mathcal{A}_\circ$ is pre-asymptotically stable with pre-basin of attraction containing the set of all initial conditions having the property that the ensuing solutions remain in the set K.*

In the proof of this result, we use the positive semidefinite Lyapunov function $V(z,\zeta) := \rho(|(z,\zeta)|_{\hat{D}_e})|\zeta|^2$ to establish, under the stated assumptions, that there exist $k^* \ge 0$ and an uniform bound on the ζ component of all solutions to $\mathcal{H}^u$ remaining in K when using the control law $u = -k\zeta$, $k \geq k^*$ and that trajectories remaining in K converge uniformly to $\hat{\mathcal{Y}}$. Then Corollary 2 and Theorem 5 are used to draw the stated conclusion.

Example 11. Consider the hybrid system given by

$$\mathcal{H}^u \begin{cases} \left.\begin{matrix} \begin{bmatrix}\dot{z}_1\\ \dot{z}_2\end{bmatrix} = \begin{bmatrix}z_2\\ -g\end{bmatrix} \\ \begin{bmatrix}\dot{\zeta}_1\\ \dot{\zeta}_2\end{bmatrix} = \begin{bmatrix}u\\ u\end{bmatrix}\end{matrix}\right\} & (z,\zeta) \in \hat{C} := \{(z,\zeta) : z_1 \geq 0, \zeta_1 = \zeta_2\} \\ \\ \left.\begin{matrix} \begin{bmatrix}z_1\\ z_2\end{bmatrix}^+ = \begin{bmatrix}a\\ 0\end{bmatrix} \\ \begin{bmatrix}\zeta_1\\ \zeta_2\end{bmatrix}^+ = \begin{bmatrix}\zeta_1 + \eta(\zeta_1)\\ \zeta_2 + \eta(\zeta_2)\end{bmatrix}\end{matrix}\right\} & (z,\zeta) \in \hat{D} := \{(z,\zeta) : z_1 = 0, z_2 \leq 0\} \ , \end{cases}$$

where $z, \zeta \in \mathbb{R}^2$, $u \in \mathbb{R}$, $\eta : \mathbb{R} \to \mathbb{R}$ is a locally Lipschitz continuous function satisfying $\eta(0) = 0$, $a > 0, g > 0$, and the state space is given by $O := \mathbb{R}^4$. Let the control law be given by $u = -k\zeta$. The hybrid system $\mathcal{H}_u$ can be interpreted as a simplified model of an actuated particle, with horizontal position given by ζ_1, moving on a concave-shaped surface and experiencing impacts with a free-falling particle, with height z_1, vertical velocity z_2, and horizontal position ζ_2. In this setting, the goal is to stabilize the horizontal position of the actuated particle to $\zeta_1 = 0$ under the effect of the impacts with the free-falling particle, which occur when $z_1 = 0$ and $z_2 \leq 0$, and affect the position of the actuated particle by $\eta(\zeta_1)$. The horizontal position of the free-falling particle (ζ_2) tracks the position of the actuated particle (ζ_1) to guarantee the collision. At impacts, the free-falling particle is repositioned to the height given by a with zero vertical velocity. In the simplified model given above, the actuated particle moves only horizontally but the effect of the free-falling particle impacting with the actuated particle on a concave-shaped surface are captured in the function η. In this particular physical situation, the function η will be such that it has the same sign as its argument. Note that we do not need to assume this as our result hold for more general functions η.

The solutions to the zero dynamics of $\mathcal{H}^u$, denoted by $\mathcal{H}^u_{|\mathcal{Y}}$ with $\mathcal{Y} = \hat{\mathcal{Y}} \times \mathbb{R}$, $\hat{\mathcal{Y}} = \{(z,\zeta) : \zeta = 0\}$, are such that $\zeta = 0$ and the z-component of the solutions is reset to $[a\ 0]^T$, then flows until the jump set is reached, and then it is reset to $[a\ 0]^T$ from where this evolution is repeated. (For the illustration given by the physical system above, the solutions to $\mathcal{H}^u_{|\mathcal{Y}}$ are such that the actuated particle stays at $\zeta_1 = 0$ and the free-falling particle, with $\zeta_2 = 0$, falls from $z_1 = a$ with zero velocity, then impacts with the actuated particle, and then is reset to $z_1 = a$, $z_2 = 0$ again for another free fall.) We now check that Assumption 2 holds. For any compact set $\mathcal{X} = \mathcal{X}_1 \times \mathcal{X}_2 \subset O$, $\mathcal{X}_1, \mathcal{X}_2 \subset \mathbb{R}^2$, such that $\mathcal{X}_1$ contains a neighborhood of $[0,a] \times [-\sqrt{2a\,g}, 0]$ and $\mathcal{X}_2$ contains a neighborhood of $\{0\} \subset \mathbb{R}^2$, the zero dynamics of the system

$\mathcal{H}^u$ relative to $\mathcal{Y}$ is (robustly) pre-asymptotically stable (in fact, the omega limit set $\Omega_{\mathcal{H}^{u\to\gamma}_{|\mathcal{Y}}}(\mathcal{X})$ can be explicitly computed to check that it is nonempty and satisfies $\Omega_{\mathcal{H}^{u\to\gamma}_{|\mathcal{Y}}}(\mathcal{X}) \subset \mathrm{int}(\mathcal{X}) \cap \hat{\mathcal{Y}})$. Let $K \subset O$ be compact. We now check Assumption 3. Items 1.a and 1.b hold by inspection. Item 2 holds with $\hat{D}_e = \hat{D}$ since after every jump we have $z_1 = a > 0$. Item 3 automatically holds since $|(z,\zeta)|_{\hat{D}_e}$ is independently of ζ. Note that Assumption 3 holds for every set $K \subset O$. It follows by Theorem 7 that there exists $k^* > 0$ such that the set $\mathcal{A}_\circ = \Omega_{\mathcal{H}^{u\to\gamma}_{|\mathcal{Y}}}(\mathcal{X})$ is pre-asymptotically stable for the hybrid system $\mathcal{H}^u$ with $u = -k\zeta$, $k \geq k^*$. △

7 Comments on Output Regulation and Conclusions

We conclude this paper by comparing the assumptions of the previous section to the assumptions that are in place in the (non-hybrid) output regulation [4] problem after a preliminary compensator is introduced to cancel the term $q(z,0)$, found in the $\dot{\zeta}$ equation, as the term evolves along solutions of the zero dynamics. (See the immersion assumptions in [4, 5] and the relaxation in [6]; we acknowledge that we have not given any thought to accomplishing this preliminary step in the context of hybrid systems).

First we note that, even in the presence of a Poisson stable exosystem, our strong minimum phase assumption holds under [4, Assumption 1]. This can be achieved by restricting the flow (and jump) sets to the forward invariant set W of [4, Assumption 0] and recognizing that our strong minimum phase property is expressed in terms of *pre*-asymptotic stability, so that there is nothing to check for solutions that start outside of W. Moreover, with the forward invariance assumption on W and the other assumptions in [4] the Ω-limit set for $\mathcal{H}_\circ$ (see (7)) is non-empty the dynamics restricted to the zero dynamics kernel is complete and bounded.

Thus, the main extra condition we are assuming is that

$$|q(z,\zeta)| = 0 \qquad \forall (z,\zeta) \in C \cap \mathcal{Y}$$

whereas the preliminary steps in output regulation only provide that

$$|q(z,\zeta)| = 0 \qquad \forall (z,\zeta) \in C \cap \mathcal{Y} \cap \mathcal{A}_\circ \, .$$

(See, for example, the assumptions in [7, Proposition 4.1].) With such a relaxed assumption, and using the control $u = -k\zeta$, the interconnection of z and ζ will not behave like a cascade of systems, like it did in the previous section. Like in the continuous-time case, the analysis for the full interconnection would then require either a small gain argument or a full-state Lyapunov argument. We do not pursue such an approach here for hybrid systems, but we do mention that such arguments for general hybrid systems are in preparation. We also add here that, unlike in the non-hybrid case (see [7]), exponential stability for hybrid systems with Lipschitz data does not necessarily imply local input-to-state stability with finite gain (see the counterexample in [8, Example 1]). Thus, exponential stability for the zero dynamics will not guarantee that one can achieve asymptotic stability for $\mathcal{A}_\circ$ using a feedback of the form $u = -k\zeta$. Nevertheless, even without exponential stability for the zero dynamics, a nonlinear feedback of ζ should be able to achieve the goals of output regulation: driving ζ to zero while keeping the full state bounded.

We conjecture that the preliminary steps of output regulation can be solved for a class of minimum phase, relative degree one hybrid systems, like those considered in the previous section, and that emerging tools for the analysis of interconnected hybrid systems will permit concluding output regulation results that parallel what is known in the continuous-time case.

The present paper should at least put into place the pieces related to the characterization of Ω-limit sets that are required to start tackling output regulation for hybrid systems.

Acknowledgments

This work was supported by the Air Force Office of Scientific Research under grant number F9550-06-1-0134 and the National Science Foundation under grant numbers ECS-0622253 & CCR-0311084.

References

1. J.-P. Aubin, J. Lygeros, M. Quincampoix, S.S. Sastry, and N. Seube. Impulse differential inclusions: a viability approach to hybrid systems. *IEEE Trans. on Automat. Contr.*, 47(1):2–20, 2002.
2. J.M. Ball. Continuity properties and global attractors of generalized semiflows and the Navier-Stokes equations. *Journal of Nonlinear Science*, 7(5):475–502, 1997.
3. C.I. Byrnes, F. Celani, and A. Isidori. Omega-limit sets of a class of nonlinear systems that are semiglobally practically stabilized. *Int. J. of Robust and Nonlinear Control*, 15:315 – 333, 2005.
4. C.I. Byrnes and A. Isidori. Limit sets, zero dynamics, and internal models in the problem of nonlinear output regulation. *IEEE Trans. on Automat. Contr.*, 48(10):1712– 1723, 2003.
5. C.I. Byrnes and A. Isidori. Nonlinear internal models for output regulation. *IEEE Trans. on Automat. Contr.*, 49(12):2244– 2247, 2004.
6. C.I. Byrnes, A. Isidori, L. Marconi, and L. Praly. Nonlinear output regulation without immersion. In *Proc. of the 44rd IEEE Conf. on Decision and Contr.*, pages 3315 – 3320, 2005.
7. C.I. Byrnes, A. Isidori, and L. Praly. On the asymptotic properties of a system arising in non-equilibrium theory of output regulation. *Mittag Leffler Institute, Stockholm, Sweden*, 2003.
8. C. Cai and A.R. Teel. Results on input-to-state stability for hybrid systems. In *Proc. of the 44rd IEEE Conf. on Decision and Contr.*, pages 5403–5408, 2005.

9. C Cai, A.R. Teel, and R. Goebel. Converse Lyapunov theorems and robust asymptotic stability for hybrid systems. In *Proc. of the 2005 Amer. Contr. Conf.*, pages 12–17, 2005.
10. C. Cai, A.R. Teel, and R. Goebel. Results on existence of smooth Lyapunov functions for asymptotically stable hybrid systems with nonopen basin of attraction. In *Proc. of the 2007 Amer. Contr. Conf.*, 2007.
11. P. Collins. A trajectory-space approach to hybrid systems. In *Proc. of Mathematical Theory of Networks and Systems*, 2004.
12. C. Ebenbauer and F. Allgöwer. Minimun-phase property of nonlinear systems in terms of a dissipation inequality. In *Proc. of the 2004 Amer. Contr. Conf.*, pages 1737 – 1742, 2004.
13. R. Goebel, J.P. Hespanha, A.R. Teel, C. Cai, and R.G. Sanfelice. Hybrid systems: Generalized solutions and robust stability. In *Proc. of the 6th IFAC Symposium in Nonlinear Control Systems*, pages 1–12, 2004.
14. R. Goebel and A.R. Teel. Solutions to hybrid inclusions via set and graphical convergence with stability theory applications. *Automatica*, 42(4):573–587, 2006.
15. W. Hahn. *Stability of Motion.* Springer Verlag, 1967.
16. J.H. Hale, L.T. Magalhaes, and W.M. Oliva. *Dynamics in finite dimensions.* Springer Verlag, 2002.
17. A. Isidori. *Nonlinear Control Systems.* Springer, 1995.
18. A. Isidori. *Nonlinear Control Systems II.* Springer, 1999.
19. D. Liberzon. *Switching in Systems and Control.* Birkhauser, 2003.
20. D. Liberzon, A.S. Morse, and E.D. Sontag. Output-input stability and minimum-phase nonlinear systems. *IEEE Trans. on Automat. Contr.*, 47(3):422–436, 2002.
21. J. Lygeros, K.H. Johansson, S.N. Simić, J. Zhang, and S.S. Sastry. Dynamical properties of hybrid automata. *IEEE Trans. on Automat. Contr.*, 48(1):2–17, 2003.
22. V.S. Melnik and J. Valero. On attractors of multivalued semi-flows and differential inclusions. *Set-Valued Analysis*, 6(1):83–111, 2004.
23. A. N. Michel, L. Wang, and B. Hu. *Qualitative Theory of Dynamical Systems.* Pure and Applied Mathematics, Marcel Dekker, 2001.
24. D. Nešić, E. Skafidas, I.M.Y. Mareels, and R.J. Evans. Minimum phase properties for input non-affine nonlinear systems. *IEEE Trans. on Automat. Contr.*, 44(4):868–872, 1999.
25. R.T. Rockafellar and R.J-B Wets. *Variational Analysis.* Springer Verlag, 1998.
26. R.G. Sanfelice, R. Goebel, and A.R. Teel. Results on convergence in hybrid systems via detectability and an invariance principle. In *Proc. of the 2005 Amer. Contr. Conf.*, pages 551–556, 2005.
27. A. Serrani. Rejection of harmonic disturbances at the controller input via hybrid adaptive external models. *Automatica*, 42(11):1977–1985, 2006.
28. L. Tavernini. Differential automata and their discrete simulators. *Nonlinear Analysis, Theory, Methods & Applications*, 11(6):665–683, 1987.
29. A.R. Teel and L. Zaccarian. On *uniformity* in definitions of global asymptotic stability for time-varying nonlinear systems. *IEEE Trans. on Automat. Contr.*, 42(12):2219–2222, 2006.
30. A. van der Schaft and H. Schumacher. *An Introduction to Hybrid Dynamical Systems.* Lecture Notes in Control and Information Sciences. Springer Verlag, 2000.
31. H.S. Witsenhausen. A class of hybrid-state continuous-time dynamic systems. *IEEE Trans. on Automat. Contr.*, 11(2):161–167, 1966.

Essential and Redundant Internal Models in Nonlinear Output Regulation

Lorenzo Marconi[1] and Laurent Praly[2]

[1] CASY-DEIS, University of Bologna, Italy.
[2] CAS, École des Mines de Paris, Fontainebleau, France.

Summary. This paper is focused on the problem of output regulation for nonlinear systems within the main framework developed in [23]. The main goal is to complement that theory with some new results showing how the dimension of the internal model-based regulator can be reduced by preserving the so-called internal model property. It is shown how the problem of reducing the regulator dimension can be approached by identifying "observability" parts of the so-called steady-state input generator system. A local analysis based on canonical geometric tools and local observability decomposition is also presented to identify lower bounds on the regulator dimension. Possible benefits in designing redundant internal models are also discussed.

This work is dedicated to Prof. Alberto Isidori
on the occasion of his 65th birthday.

1 Introduction

One of the main issue in control theory is in the ability to capture information about the plant to be supervised and the environment in which it operates and to employ such a knowledge in the design of the controller in order to achieve prescribed performances. A well-known control framework where such an issue is particularly emphasized, is the one of output regulation (see, besides others, [4], [19]) in which the problem is to design a regulator able to asymptotically offset the effect, on a controlled system, of persistent exogenous signals which are thought as generated by an autonomous system (the so-called exosystem) of known structure but unknown initial condition. Indeed, as pioneered in a linear setting in [12] and in a nonlinear setting in [18], the controller, to succeed in enforcing the desired asymptotic properties, is necessarily required to be designed by employing the a-priori knowledge of the environment in which the plant operates provided, in the classical framework, by the structure of the exo-system. This, in turn, has led to the fundamental concept of *internal model* and to the identification of design procedures for *internal model-based regulators*. To this respect the crucial property required to any regulator solving the problem is to be able to generate all possible *steady state*"feed-forward" control inputs needed to enforce an identically zero regulation error, namely the control inputs able to render invariant the so-called zero error manifold. This is what, in the important work [4], has been referred to as *internal model property*.

The design of regulators with the internal model property in a nonlinear context necessarily requires the ability to address two major points. The first regards the extension of the notion of steady state for nonlinear systems which, clearly, is instrumental to properly formulate the internal model property. The number of attempts along this direction which appeared in the related literature started with the work [18], in which the steady state has been characterized in terms of the solution of the celebrated *regulator equations* (somewhere also referred to as Francis-Isidori-Byrnes equations), and culminated with the notion recently given in [4]. In this work the authors showed how the right mathematical tool to look for is the omega limit set of a set and, upon this tool, they built up a non-equilibrium framework of output regulation.

The second critical point to be addressed consists of identifying methodologies to design regulators which on one hand posses the internal model property, and, on the other hand, enforce in the closed-loop system a steady state with zero regulated error. This double requirement justifies the usual regulator structure constituted by a first dynamical unit (the internal model), designed to provide the needed steady-state control action, and a second dynamical unit (the stabilizer), whose role is to effectively steer the closed-loop trajectories towards the desired steady-state. Of course the design of the two units are strongly interlaced in the sense that the ability of designing a stabilizer is affected by the specific structure of the internal model which, as a consequence, has to be identified with an eye to the available stabilization tools. The need of satisfying simultaneously the previous two properties motivated the requirement, characterizing all the frameworks appeared in literature, that the dynamical system defining all possible "feed-forward" inputs which force an identically zero regulation error be "immersed" into a system exhibiting certain structural properties. This requirement is what, in literature, is referred to as "immersion assumption". This is the side where, in the literature of the last fifteen years or so, the research attempts have mostly concentrated by attempting to weaken even more the immersion assumption. At the beginning, the system in question was assumed to be immersed into a linear *known* observable system (see [15], [21], [3], [24]). This assumption has been then weakened, in the framework of adaptive nonlinear regulation (see [25]), by asking immersion into a linear *un-known* (but linearly parameterized) system. Subsequent extensions have been presented in [6] (where immersion into a linear system having a nonlinear output map is assumed) and in [7] (where immersion into a nonlinear system linearizable by output injection is assumed). Finally the recent works in [5] and [8] (see also [9]) have definitely focused the attention on the design of nonlinear internal models requiring immersion into *nonlinear* systems described, respectively, in a canonical observability form and in a nonlinear adaptive observability form.

As clearly pointed out in [9], the inspiring idea in all the previous works was to adopt methodologies for the design of the internal model inherited by the design of observers. This perspective, along with the new theory to design nonlinear observers proposed in [20] and developed in [1], played a crucial role to completely drop the immersion assumption in the work [23]. In plain words the main achievement in [23] has been to show that the steady state input rendering invariant a compact attractor to be stabilized by output feedback can be dynamically generated, in a robust framework, by an appropriately designed regulator without any specific condition on this input (required, on the contrary, in the past through the immersion assumption).

This paper aims to extend [23] by exploring conditions under which the dimension of the controller can be decreased while preserving the internal model property and, on the other side, to show potential advantage in the regulator design resulting from a redundant implementation of the internal model. The major achievement in the reduction results is to show that the identification of an "essential" internal model is intimately related to the identification of "observability" parts of the so-called steady-state input generator system. Motivated by this result we show how a local analysis based on canonical geometric tools and local observability decomposition is useful to identify lower bounds on the regulator dimension. On the other side, it is presented a result showing that implementing a not essential internal model, in the sense better specified in the paper, leads to a simplification in the structure of the stabilizer which can be taken linear. Basically, the results presented in the paper reveal a trade-off between the redundancy of the internal model and the simplicity of the stabilizer.

The paper is organized as follows. In the next section we briefly review the framework of output regulation and the solution given in [23]. Section 3, articulated in two subsections, present the new results regarding essential regulators and the potential advantage coming from redundant internal models. Finally, Section 4 presents some concluding remarks.

2 The Framework of Output Regulation

2.1 The Class of Systems and the Problem

The typical setting where the problem of nonlinear output regulation is formulated is the one in which it is given a smooth nonlinear system described in thc form[3]

[3] The form (1) is easily recognized to be the well-known *normal form* with relative degree 1 and unitary high-frequency gain (see [17]). As discussed in [23],

$$\begin{aligned} \dot{z} &= f(w,z,y) \\ \dot{y} &= q(w,z,y) + u\,, \end{aligned} \tag{1}$$

with state $(z,y) \in \mathbb{R}^n \times \mathbb{R}$ and *control* input $u \in \mathbb{R}$ and measurable output y, influenced by an exogenous input $w \in \mathbb{R}^s$ which is supposed to be generated by the smooth *exosystem*

$$\dot{w} = s(w) \tag{2}$$

whose initial state $w(0)$ is supposed to range on an *invariant* compact set $W \subset \mathbb{R}^s$. Depending on the control scenario, the variable w may assume different meanings. It may represent exogenous disturbances to be rejected and/or references to be tracked. It may also contain a set of (constant or time-varying) uncertain parameters affecting the controlled plant. Associated with (1) there is a *regulated* error $e \in \mathbb{R}$ expressed as

$$e = h(w,z,y) \tag{3}$$

in which $h : \mathbb{R}^s \times \mathbb{R}^n \times \mathbb{R} \to \mathbb{R}$ is a smooth function.

For system (1)–(2)–(3) *the problem of semiglobal output regulation* is defined as follows. Given arbitrary compact sets $Z \subset \mathbb{R}^n$ and $Y \subset \mathbb{R}$ find, if possible, an output feedback controller of the form

$$\begin{aligned} \dot{\eta} &= \varphi(\eta, y) \\ u &= \varrho(\eta, y) \end{aligned} \tag{4}$$

with state $\eta \in \mathbb{R}^\nu$ and a compact set $M \subset \mathbb{R}^\nu$ such that, in the associated closed-loop system (1), (2), (4) the positive orbit of $W \times Z \times Y \times M$ is bounded and, for each $w(0), z(0), y(0), \eta(0) \in W \times Z \times Y \times M$, $\lim_{t\to\infty} e(t) = 0$ uniformly in $w(0), z(0), y(0), \eta(0)$.

As in [23], we approach the solution of the problem at issue under the following assumption formulated on the zero dynamics (with respect to the input u and output y) of system (1), namely on the system

$$\begin{aligned} \dot{w} &= s(w) \\ \dot{z} &= f(w,z,0)\,. \end{aligned} \tag{5}$$

Note that, as a consequence of the fact that W is an *invariant* set for $\dot{w} = s(w)$, the closed cylinder $\mathcal{C} := W \times \mathbb{R}^n$ is locally invariant for (2)–(5) and thus it is natural to regard this system on $\mathcal{C}$ and endow the latter with the subset topology. This, indeed, is done in all the forthcoming analysis and, in particular, in the next assumption.

Assumption 1. *There exists a compact set $\mathcal{A} \subset \mathbb{R}^{s+n}$ which is locally asymptotically stable for (5) with a domain of attraction which contains the set of initial conditions $W \times Z$. Furthermore, $h(w,z,0) = 0$ for all $(w,z) \in \mathcal{A}$.*

Section 2.2, (see also [8]) the more general case (higher relative degree and not unitary high frequency gain) can be dealt with with simple modifications which, for sake of compactness, are not repeated here.

Following [4] this assumption can be regarded as a "weak" minimum phase assumption, with the adjective weak to highlight the fact that the "forced" zero dynamics of the plant $\dot{z} = f(w, z, 0)$ is not required to posses input-to-state stability (with respect to the input w) properties nor that the "unforced" $\dot{z} = f(0, z)$ dynamics exhibit equilibrium points with prescribed stability properties.

2.2 The Asymptotic Regulator in [23]

The regulator proposed in [23] to solve the problem at hand is a system of the form

$$\begin{aligned} \dot{\eta} &= F\eta + Gu \qquad \eta \in \mathbb{R}^m \\ u &= \gamma(\eta) + v \\ v &= -\kappa(y) \,, \end{aligned} \tag{6}$$

in which $m > 0$, $(F, G) \in \mathbb{R}^{m \times m} \times \mathbb{R}^m$ is a controllable pair with F Hurwitz and $\gamma : \mathbb{R}^m \to \mathbb{R}$ and $\kappa : \mathbb{R} \to \mathbb{R}$ are suitable continuous maps. The initial condition of (6) is supposed to be in an arbitrary compact set $M \subset \mathbb{R}^m$.

The key result proved in [23] is that, under the only assumption stated in Section 2.1, there exist a lower bound for m, a choice of the pair (F, G) and of the maps γ and κ such that the regulator (6) succeeds in solving the problem at hand. In this subsection we run very briefly over the key steps and ideas followed in [23] to prove this, which are instrumental for the forthcoming analysis in Section 3.

First of all, for sake of compactness, define $\mathbf{z} := \mathrm{col}(w, z)$ and rewrite system (5) as $\dot{\mathbf{z}} = \mathbf{f}_0(\mathbf{z})$ where

$$\mathbf{f}_0(\mathbf{z}) := \mathrm{col}(s(w), f(w, z, 0)) \,. \tag{7}$$

Consistently set $\mathbf{q}_0(\mathbf{z}) := q(w, z, 0)$. A key role in the regulator (6) is played by the function $\gamma(\,\cdot\,)$ which is supposed to be an at least continuous function satisfying the design formula

$$\mathbf{q}_0(\mathbf{z}) + \gamma \circ \tau(\mathbf{z}) = 0 \qquad \forall \mathbf{z} \in \mathcal{A} \tag{8}$$

with the function $\tau : \mathcal{A} \to \mathbb{R}^m$ a continuous function satisfying

$$L_{\mathbf{f}_0} \tau(\mathbf{z}) = F\tau(\mathbf{z}) - G\mathbf{q}_0(\mathbf{z}) \qquad \forall \mathbf{z} \in \mathcal{A} \tag{9}$$

where $L_{\mathbf{f}_0}$ denotes the Lie derivative along $\mathbf{f}_0$.

In order to motivate the design formulas (8)–(9), consider the closed-loop system (1), (2), (6) given by

$$\begin{aligned} \dot{w} &= s(w) \\ \dot{z} &= f(w, z, y) \\ \dot{\eta} &= F\eta + G\gamma(\eta) + v \\ \dot{y} &= q(w, z, y) + v \,. \end{aligned} \tag{10}$$

The crucial property exhibited by this system is that, by the fact that the set $\mathcal{A}$ is forward invariant for (5) (as a consequence of the fact that $\mathcal{A}$ is locally asymptotically stable for (5)) and by (8), (9), the set

$$\text{graph}(\tau) \times \{0\} = \{(\mathbf{z}, \eta, y) \in \mathcal{A} \times \mathbb{R}^m \times \mathbb{R} \ : \ \eta = \tau(\mathbf{z}), y = 0\} \qquad (11)$$

is a forward invariant set for (10) (with $v \equiv 0$) on which, by assumption, the regulation error e is identically zero. This, in turn, makes it possible to consider the problem of output regulation as a *set stabilization problem* in which the issue is to design the function κ so that the set (11) is locally asymptotically stable for (10) with $v = \kappa(y)$ with a domain of attraction containing the set of initial conditions. Both the existence of a γ (and of the pair (F, G)) satisfying (8), (9) and the existence of κ so that the set (11) is locally asymptotically stable for (10) with $v = \kappa(y)$ are issues which have been investigated in [23] and [22]. In the remaining part of the section we present the main result along this direction. We start with a proposition presenting the main result as far as the existence of γ is concerned (see Propositions 2 and 3 in [23]).

Proposition 1. *Set*

$$m \geq 2(s+n) + 2 .$$

There exist an $\ell > 0$ and a set $\mathcal{S} \subset \mathbb{C}$ of zero Lebesgue measure such that if $\sigma(F) \subset \{\lambda \in \mathbb{C} \ : \ Re\lambda < -\ell\} \setminus \mathcal{S}$, then there exists a function $\tau : \mathcal{A} \to \mathbb{R}^m$ solution of (9) which satisfies the partial injectivity condition

$$|\mathbf{q}_0(\mathbf{z}_1) - \mathbf{q}_0(\mathbf{z}_2)| \leq \varrho(|\tau(\mathbf{z}_1) - \tau(\mathbf{z}_2)|) \qquad \textit{for all } \mathbf{z}_1, \mathbf{z}_2 \in \mathcal{A} \qquad (12)$$

where ϱ is a class-$\mathcal{K}$ function. As a consequence of (12) there exists a continuous function γ satisfying (8).

On the other hand the problem of designing the function κ so that (11) is locally asymptotically stable for (10) with $v = \kappa(y)$ can be successfully handled by means of a generalization of the tools proposed in [27] (see also [17]) for stabilization of minimum-phase systems via high-gain output feedback. Here, in particular, is where the "weak" minimum phase assumption presented in Section 2.1 plays a role. The main result in this direction is presented in the following proposition collecting the main achievements of Theorems 1 and 2 and Proposition 1 in [23].

Proposition 2. *Let the pair (F, G) and the function γ be fixed according to Proposition 1. There exists a continuous κ such that the set* graph$(\tau) \times \{0\}$ *is asymptotically stable for (10) with $v = -\kappa(y)$ with a domain of attraction containing $W \times Z \times X \times Y$.*
Furthermore, if γ is also locally Lipschitz and $\mathcal{A}$ is locally exponentially stable for (5) then there exists a $k^\star > 0$ such that for all $k \geq k^\star$, the set (11) is locally asymptotically stable for (10) with $v = -ky$.

The issue of providing an explicit expression of γ, whose existence is guaranteed by Proposition 1, has been dealt with, in an exact and approximated way, in the work [22]. For compactness we present only one of the two expressions of γ given in [22] to which the interested reader is referred for further details. In formulating the expression of γ it is argued that the class-$\mathcal{K}$ function ϱ in (12) satisfies

$$\varrho(|x_3 - x_1|) \le \varrho(|x_3 - x_2|) + \varrho(|x_1 - x_2|) \qquad \forall\,(x_1, x_2, x_3) \in \mathbb{R}^{3m}\,. \tag{13}$$

This, indeed, can be assumed without loss of generality as shown in the proof of Proposition 3 of [22].

Proposition 3. *Let τ be fulfilling (12) with a function ϱ satisfying (13). Then the function $\gamma : \mathbb{R}^m \to \mathbb{R}$ defined by*

$$\gamma(x) = \inf_{\mathbf{z}\in\mathcal{A}} -\mathbf{q}_0(\mathbf{z}) + \min\{\varrho(|x - \tau(\mathbf{z})|), 2Q\} \tag{14}$$

where $Q = \sup_{\mathbf{z}\in\mathcal{A}} \mathbf{q}_0(\mathbf{z})$ satisfies (8).

2.3 Comments on the Results

As clear by the previous analysis, the desired asymptotic behavior of the system (1) is the one in which the components (w, z) of the overall trajectory evolve on $\mathcal{A}$ and the y component is identically zero. This, in turn, guarantees, by the second part of the Assumption in Section 2.1, that the regulation error (3) is asymptotically vanishing. In order to have this asymptotic desired behavior enforced, a crucial property required to the regulator is to be able to generate any possible asymptotic control input which is needed to keep y identically zero while having (w, z) evolving on $\mathcal{A}$. This, in turn, is what in [4] has been referred to as *internal model property* (with respect to $\mathcal{A}$), namely the property, required to any regulator solving the problem at hand, of reproducing all the "steady state" control inputs needed to keep the regulated error to zero. By bearing in mind (10) and the notation around (8)–(9), it is not hard to see that, in our specific context, the regulator (6) posses the asymptotic internal model property with respect to $\mathcal{A}$ if for any initial condition $\mathbf{z}_0 \in \mathcal{A}$ of the system

$$\begin{aligned} \dot{\mathbf{z}} &= \mathbf{f}_0(\mathbf{z}) \\ y_{\mathbf{z}} &= -\mathbf{q}_0(\mathbf{z}) \end{aligned} \tag{15}$$

yielding a trajectory $\mathbf{z}(t)$, $t \ge 0$, there exists an initial condition $\eta_0 \in \mathbb{R}^m$ of the system

$$\begin{aligned} \dot{\eta} &= F\eta - G\mathbf{q_0}(\mathbf{z}(t)) \\ y_\eta &= \gamma(\eta) \end{aligned} \tag{16}$$

such that the corresponding two output trajectories $y_{\mathbf{z}}(t)$ and $y_\eta(t)$ are such that $y_{\mathbf{z}}(t) = y_\eta(t)$ for all $t \ge 0$. This, indeed, is what is guaranteed by the design formulas (8)–(9). As a matter of fact, by taking $\eta_0 = \tau(\mathbf{z}_0)$, the two formulas (8)–(9) along with the fact that $\mathcal{A}$ is forward invariant for (15), imply that the corresponding state trajectory $\eta(t)$ of (16) is such that $\eta(t) = \tau(\mathbf{z}(t))$ for all $t \geq 0$ and, by virtue of (8), that $y_{\mathbf{z}}(t) = y_\eta(t)$ for all $t \geq 0$. In these terms the triplet $(F, G, \gamma(\,\cdot\,))$ qualifies as an *internal model* able to reproduce all the asymptotic control inputs which are required to enforce a zero regulation error.

Seen from this perspective, Proposition 1 fixes precise conditions under which the asymptotic internal model property can be achieved by a regulator of the form (6). In particular it is interesting to note that, for the function γ to exist, the dimension m of the internal model is required to be sufficiently large with respect to the dimension $s+n$ of the dynamical system (15) whose output behaviors must be replied.

The result previously presented gains further interest in relation to the theory of nonlinear observers recently proposed in [20] and developed in [1], which has represented the main source of inspiration in [23]. In the observation framework of [20], systems (15), (16) are recognized to be the cascade of the "observed" system (15), with state $\mathbf{z}$ and output $y_{\mathbf{z}}$, driving the "observer" (16) whose output $\gamma(\eta)$ is designed to provide an asymptotic estimate of the observed state $\mathbf{z}$. To this purpose, in [1], the map $\gamma(\,\cdot\,)$ is computed as the *left-inverse* of $\tau(\,\cdot\,)$, i.e. such that $\gamma(\tau(\mathbf{z})) = \mathbf{z}$ for all $\mathbf{z} \in \mathcal{A}$, with τ solution of (9). Such a left-inverse, as shown in [1], always exists provided that the dimension of η is sufficiently large (precisely $\dim(\eta) \geq 2\dim(\mathbf{z}) + 2$ as in Proposition 1) and certain *observability conditions* for the system $(\mathbf{f}_0, \mathbf{q}_0)$ hold. To this regard it is interesting to note that, in the context of output regulation, the observability conditions are not needed as the design of $\gamma(\,\cdot\,)$, in order to achieve the internal model property, is done in order to reconstruct the output $\mathbf{q}_0(\mathbf{z})$ of the observed system and not the full state $\mathbf{z}$. This motivates the absence of observability conditions for the system $(\mathbf{f}_0, \mathbf{q}_0)$ in Proposition 1 and, in turn, the absence of *immersion conditions* in the above framework.

3 Essential and Redundant Internal Models

3.1 Essential Regulators

The goal of this part is to enrich the results previously recalled by exploring conditions under which the dimension m of the regulator (6) (fixed, according to Proposition 1, to be $2(s+n)+2$) can be reduced in order to obtain an *essential regulator* preserving the internal model property.

As discussed in Section 2.3, the crucial feature required to the regulator (6) in order to posses the internal model property with respect to $\mathcal{A}$ is that system (16) is able, through its output y_η, to reproduce all the possible output motions of the system (15) with initial conditions taken in the set $\mathcal{A}$, the latter being a compact set satisfying the basic assumption in Section 2.1. From this, it seems natural to approach the problem of identifying an essential regulator by addressing two subsequent issues. First, to address if there exists a *minimal* set $\mathcal{A}_0$ satisfying the basic assumption in Section 2.1. This would lead to identify *steady state trajectories* for (5) which originate essential output behaviors of (15) to be captured by the internal model. Second, to identify conditions under which all the output behaviors of (15) originating from initial conditions in $\mathcal{A}_0$ can be reproduced by the output of a system of the form (16) of minimal dimension (i.e. lower than $2(s+n)+2$). This would lead to identify an *essential internal model* $(F, G, \gamma(\cdot))$ possessing the internal model property with respect to $\mathcal{A}_0$ and thus suitable to obtain an essential regulator of the form (6).

In the next proposition we address the first of the previous issues, by showing the existence of a minimal set satisfying the assumption in Section 2.1 which turns out to be *(forward and backward) invariant* for (5) as precisely formulated in the following. The set in question turns out to be the ω-limit set of the set $W \times Z$ of system (5), denoted by $\omega(W \times Z)$ (see [13]), introduced in [4] in the context of output regulation.

Proposition 4. *Let $\mathcal{A}$ be a set satisfying the assumption in Section 2.1. Then the set $\mathcal{A}_0 := \omega(W \times Z)$ is the* unique invariant *set such that $\mathcal{A}_0 \subseteq \mathcal{A}$ which is asymptotically stable for (5) with a domain of attraction $W \times \mathcal{D}$ with $Z \subset \mathcal{D}$. Furthermore the set in question is minimal, that is there does not exist a compact set $\mathcal{A}_1 \subset \mathcal{A}_0$ which is asymptotically stable for (5) with a domain of attraction of the form $W \times \mathcal{D}$ with $Z \subset \mathcal{D}$.*

Proof. With the notation introduced around (7) in mind and by defining $\mathbf{Z} = W \times Z$, note that, as the positive flow of (5) is bounded, the omega limit set[4] $\omega(\mathbf{Z})$ of the set $\mathbf{Z}$ exists, is bounded and uniformly attracts the trajectories of (5) originating from $\mathbf{Z}$, namely for any $\epsilon > 0$ there exists a $t_\epsilon > 0$ such that $\text{dist}(\mathbf{z}(t, \mathbf{z}), \omega(\mathbf{Z})) \leq \epsilon$ for all $t \geq t_\epsilon$ and $\mathbf{z} \in \mathbf{Z}$ where $\mathbf{z}(t, \mathbf{z})$ denotes the trajectory of $\dot{\mathbf{z}} = \mathbf{f}_0(\mathbf{z})$ at time t passing through $\mathbf{z}$ at time $t = 0$ (see [13]). Furthermore it is possible to prove that $\omega(\mathbf{Z}) \subseteq \mathcal{A}$. As a matter of fact suppose that it is not true, namely that there exists a $\bar{\mathbf{z}} \in \omega(\mathbf{Z})$ and an $\epsilon > 0$ such that $|\bar{\mathbf{z}}|_\mathcal{A} \geq \epsilon$. By definition of $\omega(\mathbf{Z})$, there exist sequences $\{\mathbf{z}_n\}_0^\infty$ and $\{t_n\}_0^\infty$, with $\mathbf{z}_n \in \mathbf{Z}$ and $\lim_{n\to\infty} t_n = \infty$, such that

$$\lim_{n\to\infty} \mathbf{z}(t_n, \mathbf{z}_n) = \bar{\mathbf{z}}\,.$$

This, in particular, implies that for any $\nu > 0$ there exists a $n_\nu > 0$ such that $|\mathbf{z}(t_n, \mathbf{z}_n) - \bar{\mathbf{z}}| \leq \nu$ for all $n \geq n_\nu$. But, by taking $\nu = \min\{\epsilon/2, \nu_1\}$ with ν_1 such that $t_n \geq t_{\epsilon/2}$ for all $n \geq n_{\nu_1}$, this contradicts that $\mathcal{A}$ uniformly attracts the trajectories of (5) from $\mathbf{Z}$ (which, in turn, is implied by asymptotic stability of $\mathcal{A}$ and compactness of $\mathbf{Z}$). This proves that $\omega(\mathbf{Z}) \subset \mathcal{A}$. From this, using the fact that $\mathcal{A}$ is asymptotically stable for $\dot{\mathbf{z}} = \mathbf{f}_0(\mathbf{z})$ and the definition of

[4] We recall that the *ω-limit set* of the set $\mathbf{Z}$, written $\omega(\mathbf{Z})$, is the totality of all points $\mathbf{z} \in \mathbb{R}^{n+s}$ for which there exists a sequence of pairs $(\mathbf{z}_k, t_k)$, with $\mathbf{z}_k \in \mathbf{Z}$ and $t_k \to \infty$ as $k \to \infty$, such that $\lim_{k\to\infty} \mathbf{z}(t_k, \mathbf{z}_k) = \mathbf{z}$.

ω-limit set of the set $\mathbf{Z}$ (see [13]), it is possible to conclude that $\omega(\mathbf{Z})$ is also asymptotically stable and that the first statement of the proposition holds with $\mathcal{A}_0 = \omega(\mathbf{Z})$.

To prove the second statement of the proposition (namely that $\mathcal{A}_0$ is minimal) suppose that it is not true, that is there exists a closed set $\mathcal{A}_1 \subseteq \mathcal{A}_0$ which is asymptotically stable with a domain of attraction containing $\mathbf{Z}$. Let $\bar{\mathbf{z}} \in \mathcal{A}_0$ and $\epsilon > 0$ such that $|\bar{\mathbf{z}}|_{\mathcal{A}_1} = 2\epsilon$. By assumption, $\mathcal{A}_1$ uniformly attracts trajectories of (5) originating from $\mathbf{Z}$ which implies that there exists a $t_\epsilon > 0$ such that $|\mathbf{z}(t,\mathbf{z})|_{\mathcal{A}_1} \leq \epsilon$ for any $\mathbf{z} \in \mathbf{Z}$ and for all $t \geq t_\epsilon$. Now set

$$\mathbf{z}^\star = \mathbf{z}(-(t_\epsilon + 1), \bar{\mathbf{z}})$$

and note that $\mathbf{z}^\star \in \mathcal{A}_0 \subseteq \mathbf{Z}$, as $\mathcal{A}_i$ is invariant, and $\bar{\mathbf{z}} = \mathbf{z}((t_\epsilon + 1), \mathbf{z}^\star)$ by uniqueness of trajectories. But the latter contradicts the fact that $\mathcal{A}_1$ uniformly attracts trajectories of (5) originating from $\mathbf{Z}$ and proves the claim. From this also uniqueness of the invariant set $\mathcal{A}_0$ immediately follows. □

Remark 1. By using the terminology introduced in [4], the set $\mathcal{A}_0 := \omega(W \times Z)$ is precisely the *steady state locus* of (5) with the trajectories of $\mathbf{f}_0|_{\mathcal{A}_0}$ being the *steady state trajectories* of (5). Furthermore, as shown in [4], the triangular structure of (5) leads to a specific structure of $\mathcal{A}_0$. In particular it has been shown in [4] that there exists a (possibly set-valued) upper semi-continuous map $\pi : \mathbb{R}^s \to \mathbb{R}^n$ such that the set $\mathcal{A}_0$ is described as

$$\mathcal{A}_0 = \{(w, z) \in W \times \mathbb{R}^n \; : \; z = \pi(w)\} . \tag{17}$$

◁

Remark 2. Note that, as $\mathcal{A}_0 \subseteq \mathcal{A}$ and $h(w, z, 0) = 0$ for all $(w, z) \in \mathcal{A}$, it turns out that $h(w, z, 0) = 0$ for all $(w, z) \in \mathcal{A}_0$. In particular, this and the claim of the previous proposition yield that the set $\mathcal{A}_0$ fulfills the assumption in Section 2.1. ◁

With this result at hand we pass now to consider the second issue pointed out before, namely the existence of an internal model $(F, G, \gamma(\cdot))$ of dimension lower than $2(s + n) + 2$ having the internal model property with respect to $\mathcal{A}_0$. To this respect it is possible to prove that what determines the dimension of the essential internal model is not the dimension $(s + n)$ of (15) but rather the dimension of the lowest dimensional system able to reproduce, in an appropriate sense, the output behavior of (15). Details are as follows.

Assume the existence of an integer $r < n+s$, of a Riemannian differentiable manifold of dimension r of a compact subset $\mathcal{A}_0'$ of $\mathcal{M}$, of C^1 vector field $\mathbf{f}_0' : \mathcal{M} \to T\mathcal{M}$ which leaves $\mathcal{A}_0'$ backward invariant and of a C^1 function $\mathbf{q}_0' : \mathcal{M} \to \mathbb{R}$, such that for any $\mathbf{z}_0 \in \mathcal{A}_0$ there exists a $\mathbf{z}_0' \in \mathcal{A}_0'$ satisfying

$$\mathbf{q}_0(\mathbf{z}(t, \mathbf{z}_0)) = \mathbf{q}_0'(\mathbf{z}'(t, \mathbf{z}_0')) \qquad \forall t \leq 0.$$

If a triplet $(\mathbf{f}_0', \mathbf{q}_0', \mathcal{A}_0')$ satisfying the previous properties exists, it turns out that the internal model property with respect to $\mathcal{A}_0$ can be achieved by means of a regulator of dimension $m = 2r + 2$. More specifically, it can be proved that there exist an $\ell > 0$ and a set $\mathcal{S} \subset \mathbb{C}$ of zero Lebesgue measure, such that if $(F, G) \in \mathbb{R}^{m\times m} \times \mathbb{R}^{m\times 1}$, with $m = 2r + 2$, is a controllable pair with $\sigma(F) \subset \{\lambda \in \mathbb{C} \,:\, \text{Re}\lambda < -\ell\} \setminus \mathcal{S}$ then there exist a continuous $\tau : \mathcal{A}_0 \to \mathbb{R}^m$ solution of

$$L_{\mathbf{f_0}}\tau(\mathbf{z}) = F\tau(\mathbf{z}) - G\mathbf{q}_0(\mathbf{z}) \qquad \forall \mathbf{z} \in \mathcal{A}_0 \tag{18}$$

and a continuous $\gamma : \mathbb{R}^m \to \mathbb{R}$ solution of

$$\mathbf{q}_0(\mathbf{z}) + \gamma \circ \tau(\mathbf{z}) = 0 \qquad \forall \mathbf{z} \in \mathcal{A}_0 \,. \tag{19}$$

The proof of this claim immediately follows by specializing Proposition 6 in Appendix A by taking $\mathcal{A}_1$, f_1, q_1, n_1 and $\mathcal{A}_2$, f_2, q_2, n_2 in the proposition respectively equal to $\mathcal{A}_0$, $\mathbf{f_0}$, $\mathbf{q}_0$, $n + s$ and $\mathcal{A}_0'$, $\mathbf{f_0}'$, $\mathbf{q}_0'$, r.

Remark 3. Note that the key feature required to the r-dimensional system $\dot{\mathbf{z}}' = \mathbf{f}_0'(\mathbf{z}')$ with output $y_{\mathbf{z}'} = -\mathbf{q}_0'(\mathbf{z}')$ with initial conditions taken in the set $\mathcal{A}_0'$ is to be able to reproduce all the output behaviors (backward in time) of the $(n + s)$-dimensional system (15) with output $y_{\mathbf{z}} = -\mathbf{q}_0(\mathbf{z}(t))$ originating from initial conditions in $\mathcal{A}_0$. ◁

Remark 4. Going throughout the proof of Proposition 6, it turns out that the continuous function γ solution of (18)–(19) coincides with the solution of the equation

$$\mathbf{q}_0'(\mathbf{z}') + \gamma \circ \tau'(\mathbf{z}') = 0 \qquad \forall \mathbf{z} \in \mathcal{A}_0' \tag{20}$$

with the function $\tau' : \mathcal{A}_0' \to \mathbb{R}^m$ satisfying

$$L_{\mathbf{f}_0'}\tau'(\mathbf{z}) = F\tau'(\mathbf{z}') - G\mathbf{q}_0'(\mathbf{z}') \qquad \forall \mathbf{z} \in \mathcal{A}_0' \,. \tag{21}$$

In other words the internal model (F, G, γ) can be tuned by considering, in the design formulas, the reduced-order triplet $(\mathbf{f_0}', \mathbf{q}_0', \mathcal{A}_0')$. According to this, in the following, we will say that the triplet $(\mathbf{f_0}', \mathbf{q}_0', \mathcal{A}_0')$ is *similar* (as far as the design of γ is concerned) to the triplet $(\mathbf{f_0}, \mathbf{q}_0, \mathcal{A}_0)$. ◁

It is interesting to note that a direct application of the previous considerations in conjunction with the results discussed at the end of Remark 1, immediately lead to a reduction of the regulator's dimension with respect to the one conjectured in Proposition 1 (equal to $2(s + n) + 2$). As a matter of fact, assume that the function π in (17) admits a C^2 *selection* $\pi_s(w)$. Set $r = s$ and let $\mathcal{A}_0' \subset \mathbb{R}^s$ be an arbitrary compact set containing W. Furthermore let $\mathbf{f}_0' : \mathbb{R}^s \to \mathbb{R}^s$ be any differentiable function which agrees with $s(\,\cdot\,)$ on W, and define $\mathbf{q}_0'(\,\cdot\,) := q(\,\cdot\,, \pi_s(\,\cdot\,), 0)$. By the structure of $\mathcal{A}_0$ in (17) (with π replaced by π_s) along with the fact that the set W is invariant for (2), it turns out that the triplets $(\mathbf{f}_0', \mathbf{q}_0', \mathcal{A}_0')$ and $(\mathbf{f}_0, \mathbf{q}_0, \mathcal{A}_0)$ are similar (see Remark 4) and thus that the internal model property with respect to $\mathcal{A}_0$ can be achieved by means of a regulator of dimension $m = 2s + 2$.

It must be stressed, though, that the previous considerations highlight only one of the underlying aspects behind the reduction result previously illustrated, namely the fact that only the dimension of the *restricted* dynamics $\mathbf{f}_0|_{\mathcal{A}_0}$ (equal to s in the case the function $\pi(\,\cdot\,)$ in (17) is single valued), and not the full dimension of the dynamics (15) (equal to $n + s$), plays a role in determining the dimension of the regulator. The second fundamental aspect behind the reduction procedure is that possible dynamics of $\mathbf{f}_0|_{\mathcal{A}_0}$ which have no influence on the output behavior of system $(\mathbf{f}_0, \mathbf{q}_0)$ do not affect the dimension of the regulator. This feature can be further explored by making use of standard tools to study local observability decompositions of nonlinear systems as detailed in the following.

In particular assume that $\mathcal{A}_0$ is a smooth manifold (with boundary) of $\mathbb{R}^{n+s}$, denote by ρ its dimension (with $\rho = s$ if the map π in (17) is single valued), and denote by $< \mathbf{f}_0, d\mathbf{q}_0 >$ the minimal co-distribution defined on $\mathcal{A}_0$ which is invariant under $\mathbf{f}_0$ and which contains $d\mathbf{q}_0$, with the latter being the differential of $\mathbf{q}_0$ (see [16]). Furthermore let Q be the distribution defined as the annihilator of $< \mathbf{f}_0, d\mathbf{q}_0 >$, namely

$$Q := < \mathbf{f}_0, d\mathbf{q}_0 >^{\perp} .$$

It is well-known (see [14],[16]) that if, at a point $\bar{\mathbf{z}} \in \mathcal{A}_0$, Q is not singular, it is possible to identify a local change of variables transforming system $(\mathbf{f}_0, \mathbf{q}_0)$ into a special "observability" form. More precisely there exist an open neighborhood $U_{\bar{\mathbf{z}}}$ of $\mathcal{A}_0$ containing $\bar{\mathbf{z}}$ and a (local) diffeomorphism $\Phi : U_{\bar{\mathbf{z}}} \to \mathbb{R}^{\rho}$ which transforms system (15) into the form

$$\begin{aligned} \dot{\chi}_1 &= f_{01}(\chi_1) & \chi_1 &\in \mathbb{R}^{\rho-\nu} \\ \dot{\chi}_2 &= f_{02}(\chi_1, \chi_2) & \chi_2 &\in \mathbb{R}^{\nu} \\ y &= q_{01}(\chi_1)\,, \end{aligned} \tag{22}$$

namely into a form in which only the first $(\rho - \nu)$ state variables influence the output. This representation clearly shows that, locally around $U_{\bar{\mathbf{z}}}$, all the output motions of system (15) can be generated by the system $\dot{\xi} = f_{01}(\xi)$ with output $y_\xi = q_{01}(\xi)$ with dimension $\rho - \nu$. In particular, according to the previous arguments, this suggests that the internal model property, *locally* with respect to $U_{\bar{\mathbf{z}}}$, is potentially achievable by a regulator of dimension $2(\rho - \nu) + 2$. Of course, the local nature of the previous tools prevents one to push further the above reasonings and to be conclusive with respect to the dimension of the regulator possessing the internal model property with respect to the whole $\mathcal{A}_0$. However, it is possible to employ the fact that the co-distribution $Q^{\perp}$ is minimal (which implies that the decomposition (22) is maximal in a proper sense, see [16]), to be conclusive about a *lower bound* on the dimension of any regulator possessing the internal model property with respect to $\mathcal{A}_0$. This is formalized in the next lemma in which we identify a lower bound on the dimension r of any triplet $(\mathbf{f}_0', \mathbf{q}_0', \mathcal{A}_0')$ similar (in the sense of Remark 4) to $(\mathbf{f}_0, \mathbf{q}_0, \mathcal{A}_0)$. The lemma is given under the assumption that there exists a *submersion* $\sigma : \mathcal{A}_0 \to \mathcal{M}$ satisfying

$$\begin{aligned} L_{\mathbf{f}_0}\sigma(\mathbf{z}) &= \mathbf{f}_0'(\sigma(\mathbf{z})) \\ \mathbf{q}_0(\mathbf{z}) &= \mathbf{q}_0'(\sigma(\mathbf{z})) \end{aligned} \tag{23}$$

for all $\mathbf{z} \in \mathcal{A}_0$

Lemma 1. *Let $\mathcal{A}_0$ be a smooth manifold with boundary of dimension ρ and assume the existence of a* regular *point $\bar{\mathbf{z}} \in \mathcal{A}_0$ of the distribution $Q =< \mathbf{f}_0, d\mathbf{q}_0 >^\perp$. Let $\nu < \rho$ be the dimension of Q at $\bar{\mathbf{z}}$. Assume, in addition, the existence of a positive $r \leq \rho$, of a smooth manifold $\mathcal{M}$ of dimension r, of smooth functions $\mathbf{f}_0' : \mathcal{M} \to T\mathcal{M}$ and $\mathbf{q}_0' : \mathcal{M} \to \mathbb{R}$, and of a* submersion *$\sigma : \mathcal{A}_0 \to \mathcal{M}$, which satisfy (23). Then necessarily $r \geq \rho - \nu$.*

Proof. The proof proceeds by contradiction. Suppose that the claim of the lemma is false namely that there exist a positive $r < \rho - \nu$, a triplet $(\mathbf{f_0}', \mathbf{q_0}', \mathcal{M})$ with $\mathcal{M}$ a smooth manifold of $\mathbb{R}^r$ and a submersion $\sigma : \mathcal{A}_0 \to \mathcal{M}$ such that (23) holds for all $\mathbf{z} \in \mathcal{A}_0$. As $\text{rank}(d\sigma(\mathbf{z})/d\mathbf{z}|_{\bar{\mathbf{z}}}) = r$ (since σ is a submersion) it follows that it is always possible to identify a submersion $\lambda : \mathcal{A}_0 \to \mathcal{M}$ such that, by defining

$$\Phi'(\mathbf{z}) = \begin{pmatrix} \Phi_1'(\mathbf{z}) \\ \Phi_2'(\mathbf{z}) \end{pmatrix} := \begin{pmatrix} \sigma(\mathbf{z}) \\ \lambda(\mathbf{z}) \end{pmatrix},$$

$\text{rank}(d\Phi'(\mathbf{z})/d\mathbf{z}|_{\bar{\mathbf{z}}}) = \rho$, namely Φ' qualifies as a local diffeomorphism at $\bar{\mathbf{z}}$. This, in view of (23), guarantees the existence of an open neighborhood $U_{\bar{\mathbf{z}}}'$ of $\mathcal{A}_0$ including $\bar{\mathbf{z}}$ such that system $(\mathbf{f}_0, \mathbf{q}_0)$ in the new coordinates reads locally at $U_{\bar{\mathbf{z}}}'$ as

$$\begin{aligned} \dot{\tilde{\chi}}_1 &= f_1(\tilde{\chi}_1) & \tilde{\chi}_1 &\in \mathbb{R}^r \\ \dot{\tilde{\chi}}_2 &= f_2(\tilde{\chi}_1, \tilde{\chi}_2) & \tilde{\chi}_2 &\in \mathbb{R}^{\rho - r} \\ y &= q_1(\tilde{\chi}_1), \end{aligned} \tag{24}$$

Now partition the change of variables Φ as $\Phi(\mathbf{z}) = \text{col}(\Phi_1(\mathbf{z}), \Phi_2(\mathbf{z}))$ according to (22) and let $\mathbf{z}'$ be a point of $U_{\bar{\mathbf{z}}} \bigcap U_{\bar{\mathbf{z}}}'$ such that $\Phi_1'(\mathbf{z}') = \Phi_1'(\bar{\mathbf{z}})$ and $\Phi_1(\mathbf{z}') \neq \Phi_1(\bar{\mathbf{z}})$ (which is possible as $r < \rho - \nu$). By (24) it turns out that, as long as the trajectories $\mathbf{z}(t, \mathbf{z}')$ and $\bar{\mathbf{z}}(t, \mathbf{z})$ belongs to $U_{\bar{\mathbf{z}}} \bigcap U_{\bar{\mathbf{z}}}'$, the corresponding outputs coincides. This, by minimality of the co-distribution $Q^\perp$ implies that $\Phi_1(\mathbf{z}') = \Phi_1(\bar{\mathbf{z}})$ (see Theorem 1.9.7 in [16]) which is a contradiction. □

3.2 The Potential Advantage of Redundant Regulators

The fact of fulfilling the internal model property with respect to a generic set $\mathcal{A}$ (satisfying the main assumption in Section 2.1), not necessarily coincident with the essential steady state set $\mathcal{A}_0$, inevitably leads to design a regulator (6) which is redundant, namely whose dimension is larger than that is strictly necessary. In more meaningful terms, by bearing in mind the discussion in Section 2.3, the redundancy shows up in the fact that system (16), with $(F, G, \gamma(\cdot))$ having the internal model property with respect to $\mathcal{A} \supset \mathcal{A}_0$, posses the ability of reproducing the output behaviors of (15) generated by trajectories in $\mathcal{A}\backslash\mathcal{A}_0$ which are not, strictly speaking, steady state trajectories.

It's legitimate to wonder what, if there, is the advantage of designing a redundant regulator. The answer to this is given in the next proposition, of interest by its own, in which it is claimed that any "redundant" set is always exponentially stable for (5). The result of this proposition, proved in Appendix B, gains interest in conjunction with Proposition 2 as discussed after the statement.

Proposition 5. *Any compact set $\mathcal{A}$ which is asymptotically stable for (5) and such that $\mathcal{A}_0 \subset \text{int}\mathcal{A}$ is also locally exponentially stable for (5).*

In terms of the framework presented in Section 2, the previous result gains interest in conjunction with Proposition 2 which, besides others, claims that a linear stabilizer $\kappa(\cdot)$ can be obtained if the set $\mathcal{A}$ is locally exponentially stable for[5] (5). In other words the results of Propositions 5 and 2 in relation to the results of Proposition 4 and Remark 2, reveal a trade-off between the simplicity of the stabilizer $\kappa(\cdot)$ and the dimension of the regulator (6). As a matter of fact in Proposition 4 it is claimed that the set $\mathcal{A}$ can be always "shrunk" to obtain a minimal invariant set $\mathcal{A}_0$ instrumental to obtain an essential (low-order) internal model as detailed in the previous section. The possible drawback in this, is that the set $\mathcal{A}_0$ is not guaranteed to be exponentially stable if the set $\mathcal{A}$ is such. This means that a reduced order regulator can be obtained by possibly complicating the function $\kappa(\cdot)$ in (6). On the other hand Proposition 5 asserts that exponential stability can be gained by enlarging a bit the set $\mathcal{A}_0$ but, so doing, necessarily loosing backward invariance as claimed in Proposition 4. This means that a linear function $\kappa(\cdot)$ can be possibly obtained by necessarily accepting a not essential regulator.

The previous considerations highlight a possible benefit in the design of the stabilizer $\kappa(\cdot)$ coming from the redundancy of the regulator, where the redundancy comes from the fact of considering, in the design of the triplet $(F, G, \gamma(\cdot))$, a redundant set $\mathcal{A} \supset \mathcal{A}_0$ instead of the steady state set $\mathcal{A}_0$. At this point one would be tempted to wonder if the redundancy of the regulator can be employed also to obtain a benefit in the design of the function $\gamma(\cdot)$ which, according to the previous arguments, is the true bottleneck in the design procedure of the regulator. A possible answer to this point will be given in the following in which it is assumed fixed a compact set $\mathcal{A} \supseteq \mathcal{A}_0$ which is locally exponentially stable for (5) and it is assumed that the triplet $(\mathbf{f}_0, \mathbf{q}_0, \mathcal{A})$ is similar, in the sense specified below, to a linear system.

[5] Indeed, the extra condition required in Proposition 2 is that the function $\gamma(\cdot)$ is locally Lipschitz. In this paper we do not address this issue and assume it is satisfied.

In the case $\mathcal{A}$ is backward invariant, in the following we simply let $\mathcal{A}_1 = \mathcal{A}$, $f_1 = \mathbf{f}_0$, and $q_1 = \mathbf{q}_0$. In the case $\mathcal{A}$ is not backward invariant, let $\mathcal{A}_1$ be an arbitrary compact set such that $\mathcal{A} \subset \mathtt{Int}(\mathcal{A}_1)$ and $f_1 : \mathbb{R}^{n+s} \to \mathbb{R}^{n+s}$ be an arbitrary differentiable function such that f_1 agrees with $\mathbf{f}_0$ on $\mathcal{A}$ and $f_1(z_1) = 0$ for all $z_1 \in \mathbb{R}^{n+s} \setminus \mathcal{A}_1$. Furthermore, set $q_1 = \mathbf{q}_0$ and note that, by construction, the set $\mathcal{A}_1$ is invariant for $\dot{z}_1 = f_1(z_1)$. Assume now the existence of an integer $r \geq n+s$, of a linear pair $(F_2, Q_2) \in \mathbb{R}^{r \times r} \times \mathbb{R}^{r \times 1}$ and of a compact set $\mathcal{A}_2 \in \mathbb{R}^r$ such that the triplet $(F_2, Q_2, \mathcal{A}_2)$ is similar (in the sense of Remark 4) to the triplet $(f_1, q_1, \mathcal{A}_1)$, that is for all $z_{10} \in \mathcal{A}_1$ there exists a $z_{20} \in \mathcal{A}_2$ such that[6]

$$q_1(z_1(t, z_{10})) = Q_2 e^{F_2 t} z_{20} \qquad \forall t \leq 0 \,.$$

In this setting Proposition 6 in Appendix A, with n_1 and n_2 respectively set to $n+s$ and r, immediately yields that the internal model property with respect to $\mathcal{A}$ can be achieved by means of a *linear* internal model. More specifically, setting $m \geq 2r+2$, Proposition 4 yields that there exists an $\ell > 0$ and a set $\mathcal{S} \subset \mathbb{C}$ of zero Lebesgue measure such that if $(F, G) \in \mathbb{R}^{m \times m} \times \mathbb{R}^m$ is a controllable pair with $\sigma(F) \subset \{\lambda \in \mathbb{C} \,:\, \mathrm{Re}\lambda < -\ell\} \setminus \mathcal{S}$, then there exists a continuous function $\tau_1 : \mathcal{A}_1 \to \mathbb{R}^m$ solution of

$$L_{f_1}\tau_1(z_1) = F\tau_1(z_1) - Gq_1(z_1) \qquad \forall\, z_1 \in \mathcal{A}_1 \tag{25}$$

and a linear function $\Gamma : \mathbb{R}^m \to \mathbb{R}$ satisfying

$$\Gamma\tau_1(z_1) + q_1(z_1) = 0 \qquad \forall\, z_1 \in \mathcal{A}_1 \,. \tag{26}$$

From this, using the fact that f_1 agrees with $\mathbf{f}_0$ on $\mathcal{A}$ and that $q_1 = \mathbf{q}_0$, it turns out that equations (9) and (8) are satisfied with $\tau = \tau_1|_{\mathcal{A}}$ and with $\gamma = \Gamma$. This implies that the regulator (4) having the internal model property with respect to $\mathcal{A}$ can be taken linear.

Remark 5. The previous conditions can be interpreted as an *immersion* of the system $(f_1, q_1, \mathcal{A}_1)$ into a linear system $(F_2, Q_2, \mathcal{A}_2)$ in the sense specified before. In particular the theory in [11] can be used to identify sufficient conditions under which such a immersion exists. It is worth also noting that, in the way in which it is formulated, the existence of such a immersion is affected by the choice of the set $\mathcal{A}_1$ and of the function f_1 which can be arbitrarily chosen as indicated above. This is not the case if the set $\mathcal{A}$ is backward invariant for $\dot{\mathbf{z}} = \mathbf{f}_0(\mathbf{z})$ as the previous considerations are done with $\mathcal{A}_1 = \mathcal{A}$, $f_1 = \mathbf{f}_0$, and $q_1 = \mathbf{q}_0$, and the immersion conditions can be formulated by referring to the "original" triplet $(\mathbf{f}_0, \mathbf{q}_0, \mathcal{A})$. ◁

Remark 6. It is interesting to note that the computation of the (linear) internal model (F, G, Γ) does not require the knowledge of the immersing linear system

[6] Note how the condition in question is satisfied if Assumption 2 of [4] holds (see Lemma 7.1 of [4]).

(F_2, Q_2) but only the knowledge of its dimension r. As a matter of fact, as clear by the previous analysis, the computation of Γ can be carried out in terms of the immersed triplet $(f_1, q_1, \mathcal{A}_1)$ by means of the design formulas (25)–(26) which are known to have a linear solution Γ if $m \geq 2r + 2$. $\triangleleft$

4 Conclusions

In this paper we presented some complementary results of [23] in the context of output regulation for nonlinear systems. Specifically, we presented and discussed results on how to identify internal model-based regulators of minimal dimension preserving the so-called internal model property. The reduction tools consisted in the identification of "essential" steady state dynamics of the regulated plant and on the identification of an "essential" internal model-based regulators. Regarding the first aspect, it has been shown that the crucial tool is the concept of omega-limit set of a set pioneered in [4] in the context of output regulation. As far as the second aspect is concerned, we showed how the crucial step is the identification of observability parts of the steady state-input generator system. The usefulness of "redundant" regulators have been also investigated in terms of design features of the high-gain stabilizer which characterizes the proposed regulator.

A A Reduction Result

Proposition 6. *Let $f_1 : \mathbb{R}^{n_1} \to \mathbb{R}^{n_1}$ and $q_1 : \mathbb{R}^{n_1} \to \mathbb{R}$ be locally Lipschitz functions and let $\mathcal{A}_1$ be a compact backward invariant set for $\dot{z}_1 = f_1(z_1)$. Let $\mathcal{M}$ be a Riemannian differentiable manifold[7] of dimension n_2 and $\mathcal{A}_2$ and $\mathcal{A}_{2e}$ be compact subsets of $\mathcal{M}$ with $\mathcal{A}_2$ subset of the interior $\mathtt{Int}(\mathcal{A}_{2e})$ of $\mathcal{A}_{2e}$. Let $f_2 : \mathcal{M} \to T\mathcal{M}$ be a C^1 vector field which leaves $\mathcal{A}_{2e}$ backward invariant and $q_2 : \mathcal{M} \to \mathbb{R}$ a C^1 function. Assume that, for all $z_1 \in \mathcal{A}_1$, there exists $z_2 \in \mathcal{A}_2$ such that*

$$q_1(\zeta_1(t, z_1)) \equiv q_2(\zeta_2(t, z_2)) \qquad \text{for all } t \leq 0 \,.$$

Set $m = 2n_2 + 2$. There exist an $\ell > 0$ and a set $\mathcal{S} \subset \mathbb{C}$ of zero Lebesgue measure such that, if $(F, G) \in \mathbb{R}^{m\times m} \times \mathbb{R}^m$ is a controllable pair with $\sigma(F) \subset \{\lambda \in \mathbb{C} : Re\lambda < -\ell\}\backslash\mathcal{S}$, then there exists a continuous function $\tau_1 : \mathcal{A}_1 \to \mathbb{R}^m$ solution of

$$L_{f_1}\tau_1(z_1) = F\tau_1(z_1) - Gq_1(z_1) \qquad \forall\, z_1 \in \mathcal{A}_1 \tag{27}$$

and a continuous function $\gamma : \mathbb{R}^m \to \mathbb{R}$ satisfying

$$\gamma \circ \tau_1(z_1) + q_1(z_1) = 0 \qquad \forall\, z_1 \in \mathcal{A}_1 \,.$$

[7] We adopt here the definition [2, Definition III.1.2].

Moreover, if $\mathcal{M} = \mathbb{R}^{n_2}$ and f_2 and q_2 are linear, the above result holds and γ can be chosen linear.

Proof. Because $\mathcal{A}_1$ is compact and backward invariant and q_1 is continuous, we can show, by following the same steps as the ones of the proof of Proposition 1 of [23], that if F is Hurwitz then the function $\tau : \mathcal{A}_1 \mapsto \mathbb{R}^m$ defined as

$$\tau_1(z_1) = \int_{-\infty}^{0} e^{Fs}\, G\, q_1(\zeta_1(s, z_1))ds \tag{28}$$

is well-defined, continuous and solution of (27).

Also our assumptions imply that, for any $z_2 \in \mathcal{A}_{2e}$, the solution $t \in (-\infty, 0] \to \zeta_2(t, z_2) \in \mathcal{A}_{2e}$ is well-defined and $t \in (-\infty, 0] \to q_2(\zeta_2(t, z_2))$ is a bounded function. So, $\tau_2 : \mathcal{A}_{2e} \mapsto \mathbb{R}^m$ defined as

$$\tau_2(\zeta_2) \;=\; \int_{-\infty}^{0} e^{Fs}\, G\, q_2(\zeta_2(s, z_2))ds \tag{29}$$

is well-defined.

If $\mathcal{M} = \mathbb{R}^{n_2}$ and f_2 and q_2 are linear, there exists $\ell > 0$ such that if $\sigma(F) \subset \{\lambda \in \mathbb{C} \;:\; \mathrm{Re}\lambda < -\ell\}$, then this expression makes sense and gives a linear function.

In the case where $\mathcal{M}$ is a more general Riemannian differentiable manifold, we need some more involved steps to show that τ_2 is C^1 on $\mathtt{Int}(\mathcal{A}_{2e})$. To lighten their presentation we replace ζ_2 by ζ, z_2 by z, f_2 by f and q_2 by q. Since q is C^1, it defines a C^0 covector denoted dq satisfying (see [2, Example V.1.4] or [26, p. 150])

$$dq_z(v) \;=\; L_v q(z) \qquad \forall v \in T_z M\;,\; \forall z \in M\;.$$

Here dq_z denote the evaluation of dq at z and $dq_z(v)$ is the real number given by the evaluation of the linear form dq_z at the vector v. Then, let $\underline{\Psi}$ be the contravariant tensor field of order 2 (i.e. the bilinear map) given by the Riemannian metric. Since it is non-degenerate, it defines a covariant tensor field $\overline{\Psi}$ of order 2 (See [2, Exercice V.5.5]) or [10, §3.19] or [26, pp. 414-416]) such that we have the following Cauchy-Schwarz inequality

$$|dq_z(v)| \;\leq\; \underline{\Psi}_z(v, v)\,\overline{\Psi}_z(dq_z, dq_z) \qquad \forall v \in T_z M\;,\; \forall z \in M\;. \tag{30}$$

Also, $\mathcal{A}_{2e}$ being compact, there exists a real number Q such that we have

$$0 \;\leq\; \overline{\Psi}_z(dq_z, dq_z) \;\leq\; Q \qquad \forall z \in \mathcal{A}_{2e}\;. \tag{31}$$

Finally, we note that the one-parameter group action $z \mapsto \zeta(t, z)$ defines the induced one-parameter pushforward map $d\zeta : TM \to TM$, mapping for each t, vectors in $T_z M$ into vectors in $T_{\zeta(t,z)} M$ (see [2, Theorem IV.1.2] or [26, pp. 88-89 and Theorem 3.1]).

With this at hand, by following the arguments in the proof of Proposition 2 of [23], we can prove that τ is C^1 provided there exist real numbers a and ℓ such that we have

$$dq_{\zeta(t,z)}\left(d\zeta_{\zeta(t,z)}v\right) \;\leq\; a\,\exp(\ell|t|)\,\sqrt{\underline{\Psi}_z(v,v)} \qquad \forall v\in T_zM,\ \forall t\leq 0,\ \forall z\in\mathcal{A}_{2e}\,.$$

With (30) and (31), this holds if we have

$$\underline{\Psi}_{\zeta(t,z)}\left(d\zeta_{\zeta(t,z)}v, d\zeta_{\zeta(t,z)}v\right)\leq\exp(\ell|t|)\sqrt{\underline{\Psi}_z(v,v)}\ \ \forall v\in T_zM\,,\forall t\leq 0\,,\forall z\in\mathcal{A}_{2e}. \tag{32}$$

This leads us to evaluate the Lie derivative along f of the contravariant tensor field of order 2 given at the point $\zeta(t,z)$ by $\underline{\Psi}_{\zeta(t,z)}\left(d\zeta_{\zeta(t,z)}\cdot, d\zeta_{\zeta(t,z)}\cdot\right)$ (See [2, Exercise V.2.8] or [10, §3.23.4] or [26, Problem 5.14]). This Lie derivative is a contravariant tensor field of order 2 and therefore, $\underline{\Psi}$ being non-degenerate and $\mathcal{A}_{2e}$ being compact, there exists a positive real number ℓ such that we have

$$-2\ell\,\underline{\Psi}_z\left(d\zeta_z\cdot, d\zeta_z\cdot\right) \;\leq\; L_f\underline{\Psi}_z\left(d\zeta_z\cdot, d\zeta_z\cdot\right) \qquad \forall z\in\mathcal{A}_{2e}\,.$$

From this (32) follows readily and hence τ_2 is C^1 on $\mathtt{Int}(\mathcal{A}_{2e})$ if $\sigma(F)\subset\{\lambda\in\mathbb{C}\,:\,\mathrm{Re}\lambda<-\ell\}$.

Now, coming back to our initial notations, as $\mathcal{M}$ is a differentiable manifold and $\mathcal{A}_2$ is compact, it is possible to cover $\mathcal{A}_2$ with a finite set $\mathcal{I}$ of open sets $\mathcal{O}_i$ each diffeomorphic to $\mathbb{R}^{n_2}$ (see [2, Theorem I.3.6]).

By using off-the-shelf the arguments in the proof of Proposition 2 of [23], it is possible to claim for each of the open set $\mathcal{O}_i$ the existence of a set $\mathcal{S}_i\subset\mathbb{C}$ of zero Lebesgue measure such that, if $\sigma(F)\subset\{\lambda\in\mathbb{C}\,:\,\mathrm{Re}\lambda<-\ell\}\setminus\mathcal{S}_i$, then we have

$$\tau_2(z_{2a})=\tau_2(z_{2b}) \quad\Rightarrow\quad q_2(z_{2a})=q_2(z_{2b}) \qquad \forall\, z_{2a},z_{2b}\in\mathcal{O}_i\,. \tag{33}$$

Since $\mathcal{I}$ is finite, the set $\mathcal{S}=\bigcup_{i\in\mathcal{I}}\mathcal{S}_i$ is of measure zero and, if $\sigma(F)\subset\{\lambda\in\mathbb{C}\,:\,\mathrm{Re}\lambda<-\ell\}\setminus\mathcal{S}$, then we have

$$\tau_2(z_{2a})=\tau_2(z_{2b}) \quad\Rightarrow\quad q_2(z_{2a})=q_2(z_{2b}) \qquad \forall\, z_{2a},z_{2b}\in\mathcal{A}_2\,. \tag{34}$$

With the above and by following the same arguments as the ones used at the end of Proposition 2 of [23] which apply since $\mathcal{A}_2$ is compact, there exists a continuous function $\gamma:\tau_2(\mathcal{A}_2)\subset\mathbb{R}^m\to\mathbb{R}$ satisfying

$$\gamma\circ\tau_2(z_2)+q_2(z_2)=0 \qquad \forall\, z_2\in\mathcal{A}_2\,.$$

As in the proof of Proposition 3 of [23], this function can be extended to all $\mathbb{R}^m$. Clearly if τ_2 and q_2 are linear, then γ is linear.

Finally, pick any $z_1\in\mathcal{A}_1$. By condition (ii) there exist $z_2\in\mathcal{A}_2$ such that $q_1(\zeta_1(t,z_1))=q_2(\zeta_2(t,z_2))$ for all $t\leq 0$ and therefore $\tau_1(z_1)=\tau_2(z_2)$ and $q_1(z_1)=q_2(z_2)$. This implies

$$\gamma\circ\tau_1(z_1)=\gamma\circ\tau_2(z_2)=-q_2(z_2)=-q_1(z_1)$$

which concludes the proof. □

B Proof of Proposition 5

The proof strongly relies on notations and results used in the proof of Theorem 4 in [23] which, for compactness, are not repeated here. We prove the proposition by showing that there exists a compact set $\mathcal{A}_e$ satisfying $\mathcal{A}_0 \subseteq \mathcal{A}_e \subseteq \mathcal{A}$ which is locally exponentially stable for (5). First of all note that, by assumption, there exists a locally Lipschitz Lyapunov function V satisfying the properties of Theorem 4 (with the set $\mathcal{B}$ replaced by $\mathcal{A}_0$) in [23] and in particular

$$\underline{a}(|\mathbf{z}|_{\mathcal{A}_0}) \leq V(\mathbf{z}).$$

Now let

$$r = \min_{\mathbf{z}\in\mathbb{R}^{n+s}\setminus\mathcal{A}} |\mathbf{z}|_{\mathcal{A}_0} > 0 \qquad \text{and} \qquad c_1 = \frac{1}{2}\underline{a}(r) > 0\,,$$

fix

$$\mathcal{A}_e = V^{-1}([0, c_1])\,,$$

and note that $\mathcal{A}_e$ is forward invariant. Moreover $\mathcal{A}_0 \subset \mathcal{A}_e$. We prove now that $\mathcal{A}_e$ is locally exponentially stable by proving the following two facts.

Fact #1 there exists a time T such that $|\mathbf{z}(t,\mathbf{z}_0)|_{\mathcal{A}_e} = 0$ for all $\mathbf{z}_0 \in \mathbf{Z} := W\times Z$ and for all $t \geq T$ (*finite time convergence*).

Fact #2 there exists a constant $L > 0$ such that $|\mathbf{z}(t,\mathbf{z}_0)|_{\mathcal{A}_e} \leq L|\mathbf{z}_0|_{\mathcal{A}_e}$ for all $\mathbf{z}_0 \in \mathbf{Z}$ and for all $t \geq 0$.

To prove fact #1 note that, by property (a) in Theorem 4 of [23] there exists an $a \geq c_1$ such that $\mathbf{Z} \subset V^{-1}([0,a])$ and, by property (c) in the same theorem, there exists $c > 0$ such that $D^+V(\mathbf{z}(t,\mathbf{z}_0)) \leq -cV(\mathbf{z}(t,\mathbf{z}_0))$ for all $\mathbf{z}_0 \in \mathbf{Z}$. By this, using the appropriate comparison lemma, it turns out that

$$V(\mathbf{z}(t,\mathbf{z}_0)) \leq e^{-ct}V(\mathbf{z}_0) \leq e^{-ct}a \qquad \text{for all } t \geq 0 \;\; \mathbf{z}_0 \in \mathbf{Z}$$

by which standard arguments can be used to prove that $\mathcal{A}_e$ is a forward invariant set and that fact #1 holds with $T = 1/c \ln a/c_1$.

To prove fact #2, since $V^{-1}([0,a])$ is a compact set, we can let

$$F = \max_{\mathbf{z}\in V^{-1}([0,a])} |\partial \mathbf{f_0}(\mathbf{z})/\partial \mathbf{z}|\,.$$

Note that, since $V(\mathbf{z}(t,\mathbf{z}_0))$ is non increasing in t, for all $\mathbf{z}_1, \mathbf{z}_2 \in \mathbf{Z}$ and for all $t \geq 0$

$$|\mathbf{z}(t,\mathbf{z}_1) - \mathbf{z}(t,\mathbf{z}_2)| \leq e^{Ft}|\mathbf{z}_1 - \mathbf{z}_2|\,.$$

Now fix $\mathbf{z}_1 \in \mathbf{Z}$ and let $\mathbf{z}_2 \in \mathcal{A}_e$ be such that $|\mathbf{z}_1 - \mathbf{z}_2| = |\mathbf{z}_1|_{\mathcal{A}_e}$. As $\mathcal{A}_e$ is forward invariant, it turns out that $\mathbf{z}(t,\mathbf{z}_2) \in \mathcal{A}_e$ for all $t \geq 0$. Moreover, by fact #1, $\mathbf{z}(t,\mathbf{z}_1) \in \mathcal{A}_e$ for all $t \geq T$. From this $|\mathbf{z}(t,\mathbf{z}_1)|_{\mathcal{A}_e} = 0$ for all $t \geq T$ and

$$|\mathbf{z}(t,\mathbf{z}_1)|_{\mathcal{A}_e} \leq |\mathbf{z}(t,\mathbf{z}_1) - \mathbf{z}(t,\mathbf{z}_2)| \leq e^{FT}|\mathbf{z}_1 - \mathbf{z}_2| \leq e^{FT}|\mathbf{z}_1|_{\mathcal{A}_e}\,.$$

This concludes the proof of fact #2 (taking $L = e^{FT}$) and of the proposition.

References

1. V. Andrieu and L. Praly. On the existence of a Kazantis-Kravaris/Luenberger observer. *SIAM J. Contr. Optimization*, 45:432–456, 2006.
2. W.M. Boothby. *An Introduction to Differentiable Manifolds and Riemannian Geometry*. Academic Press, 1975.
3. C.I. Byrnes, F. Delli Priscoli, A. Isidori, and W. Kang. Structurally stable output regulation of nonlinear systems. *Automatica*, 33:369–385, 1997.
4. C.I. Byrnes and A. Isidori. Limit sets, zero dynamics and internal models in the problem of nonlinear output regulation. *IEEE Trans. on Automat. Contr.*, 48:1712–1723, 2003.
5. C.I. Byrnes and A. Isidori. Nonlinear internal models for output regulation. *IEEE Trans. on Automat. Contr.*, 49:2244–2247, 2004.
6. Z. Chen and J. Huang. Global robust servomechanism problem of lower triangular systems in the general case. *Systems & Control Letters*, 52:209–220, 2004.
7. F. Delli Priscoli. Output regulation with nonlinear internal models. *Systems & Control Letters*, 53:177–185, 2004.
8. F. Delli Priscoli, L. Marconi, and A. Isidori. New approach to adaptive nonlinear regulation. *SIAM J. Contr. Optimization*, 45:829–855, 2006.
9. F. Delli Priscoli, L. Marconi, and A. Isidori. Nonlinear observers as nonlinear internal models. *Systems & Control Letters*, 55:640–649, 2006.
10. B. Doubrovine, S. Novikov, and A. Fomenko. *Modern Geometry-Methods and Applications. Part I: The Geometry of Surfaces, Transformation Groups and Fields*, volume 93 of *Graduate Texts in Mathematics*. Springer Verlag, 2nd edition, 1992.
11. M. Fliess and I. Kupka. A finiteness criterion for nonlinear input-output differential systems. *SIAM J. Contr. Optimization*, 21:721–728, 1983.
12. B.A. Francis and W.M. Wonham. The internal model principle of control theory. *Automatica*, 12:457–465, 1976.
13. J.K. Hale, L.T. Magalhães, and W.M. Oliva. *Dynamics in Infinite Dimensions*. Springer Verlag, New York, 2002.
14. R. Hermann and A.J. Krener. Nonlinear controllability and observability. *IEEE Trans. on Automat. Contr.*, 22:728–740, 1977.
15. J. Huang and C.F. Lin. On a robust nonlinear multivariable servomechanism problem. *IEEE Trans. on Automat. Contr.*, 39:1510–1513, 1994.
16. A. Isidori. *Nonlinear Control Sytems*. Springer Verlag, New York, 3rd edition, 1995.
17. A. Isidori. *Nonlinear Control Systems II*. Springer Verlag, New York, 1st edition, 1999.
18. A. Isidori and C.I. Byrnes. Output regulation of nonlinear systems. *IEEE Trans. on Automat. Contr.*, 25:131–140, 1990.
19. A. Isidori, L. Marconi, and A. Serrani. *Robust Autonomous Guidance: An Internal Model-based Approach*. Limited series Advances in Industrial Control. Springer Verlag, London, 2003.
20. K. Kazantzis and C. Kravaris. Nonlinear observer design using Lyapunov's auxiliary theorem. *Systems & Control Letters*, 34:241–247, 1998.
21. H. Khalil. Robust servomechanism output feedback controllers for feedback linearizable systems. *Automatica*, 30:587–1599, 1994.

22. L. Marconi and L. Praly. Uniform practical nonlinear output regulation. *IEEE Trans. on Automat. Contr.*, 2006. Submitted.
23. L. Marconi, L. Praly, and A. Isidori. Output stabilization via nonlinear Luenberger observers. *SIAM J. Contr. Optimization*, 45:2277–2298, 2007.
24. A. Serrani, A. Isidori, and L. Marconi. Semiglobal output regulation for minimum-phase systems. *Int. J. of Robust and Nonlinear Control*, 10:379–396, 2000.
25. A. Serrani, A. Isidori, and L. Marconi. Semiglobal nonlinear output regulation with adaptive internal model. *IEEE Trans. on Automat. Contr.*, 46:1178–1194, 2001.
26. M. Spivak. *(A Comprehensive Introduction to) Differential Geometry*, volume 1. Publish or Perish, Inc., 2nd edition, 1979.
27. A.R. Teel and L. Praly. Tools for semiglobal stabilization by partial state and output feedback. *SIAM J. Contr. Optimization*, 33:1443–1485, 1995.

Two Global Regulators for Systems with Measurable Nonlinearities and Unknown Sinusoidal Disturbances

Riccardo Marino, Giovanni L. Santosuosso, and Patrizio Tomei

Dipartimento di Ingegneria Elettronica, Università di Roma Tor Vergata, Via del Politecnico 1, 00133 Roma, Italy.

Summary. The class of nonlinear observable systems with known output dependent nonlinearities affected by disturbances generated by an unknown linear exosystem is considered. The output feedback set point regulation problem is addressed by adaptively generating the reference input on line when the exosystem is unknown. Two global solutions to this problem are provided, when only an upper bound on the exosystem order is known: the first one is simpler but is restricted to minimum phase systems while the second one can handle non-minimum phase systems as well and identify the exosystem parameters.

1 Introduction

Output regulation of nonlinear systems has attracted a considerable interest in the last decades after the fundamental contribution [18] given by Alberto Isidori and Christopher Byrnes in 1990. The work [18] (outstanding IEEE TAC paper for the year 1990) poses and solves the output regulation problem for nonlinear systems, which is concerned with having the regulated variables of a given controlled plant to asymptotically track (or reject) all desired trajectories (or disturbances) generated by some fixed autonomous system, called the exosystem. The key contribution in [18] is to show that the problem is solvable if and only if certain nonlinear partial differential equations called 'regulator equations' are solvable; a feedback law achieving local asymptotic output regulation is constructed via the solution of the regulator equations. Alberto Isidori and coworkers have been actively working on the nonlinear regulator problem since 1990 ([3], [2], [33], [17], [34], [4], [5], [21], [8]): his work has been collected in the widely known monographs [15], [16]. In [3] a set of necessary and sufficient conditions is established for the solution of the problem of the asymptotic output regulation under the additional constraint that the regulation strategy is insensitive to small variations of uncertain plant parameters: a solution to this problem is given under the 'immersion assumption', i.e. the control input reference trajectories are assumed to be generated by a neutrally stable linear exosystem. In [33] the special class of nonlinear output feedback systems introduced in [26] with a neutrally stable linear exosystem is considered: global output regulation of the tracking error to zero is achieved for the first time feeding back the regulation error only, assuming that the frequencies of the exosystem are known. In [34] the semiglobal nonlinear output regulation by error feedback is solved for systems with unknown parameters in the linear exosystem when bounds on the frequencies and the disturbances are known by introducing an adaptive internal model: it is remarkable that such a problem was still unsolved at that time for linear systems as well. In [21] a regulation scheme is presented for a large class of nonlinear systems allowed to be non-minimum phase, assuming the exosystem frequencies to be known by taking advantage of the design tools proposed in [17]. In [5] output regulators with nonlinear internal models are designed by dropping the immersion assumption, following the work [4]. The theory of nonlinear adaptive observers is shown in [8] to be effectively used as a powerful tool to improve existing results in the context of adaptive control: local asymptotic output regulation is achieved for a nonlinear system allowed to be non-minimum phase with nonlinear exosystems containing unknown parameters.

Output regulation has also been the object of the research efforts of other groups, both for linear (see [1], [30], [25], [28]) and nonlinear systems (see [14], [31], [32], [9], [10], [11], [12], [35], [13], [7], [6], [22], [29], [20]). In particular, the work in [14] has shown for the first time that the local robust regulation problem for nonlinear systems is solvable if the solution of the regulator equation is polynomial in the exogenous signals and the exosystem that generates them is linear. Several results are available under the plant minimum phase assumption: a global robust state feedback control scheme is presented in [32] for systems affected by unknown sinusoidal disturbances and an unknown noise, following earlier work in [31]. The output regulation problem is addressed in [35] for a class of large-scale nonlinear interconnected systems perturbed by an unknown neutrally stable exosystem via a decentralized error feedback controller. For the same class of systems considered in [33], a global regulation strategy that is adaptive with respect to unknown system parameters is presented in [9] while in [7] the exogenous signals are generated by a nonlinear internal model. When the frequencies of the exosystem are unknown a global output feedback control law which regulates the output to zero is presented in [11] when only an upper bound on the number of exosystem's frequencies is known (see [10] for an extension to output tracking and [12] which allows for unknown parameters in the regulated system). Rejecting and/or tracking exogenous signals for non-minimum phase systems is inherently more difficult. In [22] a "separation principle" approach is followed for the regulation to zero of a class of nonlinear output feedback systems perturbed by unknown frequency sinusoidal disturbances: an exact and independent disturbance estimation algorithm provides exponentially convergent disturbance estimates which are then used to regulate the system output to zero. In [20] the problem of adaptive estimation and rejection of unknown sinusoidal disturbances through measurement feedback for a class of non-minimum phase non-linear MIMO systems is addressed. The algorithms in [22], [8] and [20] require the exact knowledge of the order of the exosystem to guarantee asymptotic regulation to zero of the system output. Following [28], in [23] an exponentially convergent output regulation controller is presented for linear non-minimum phase systems under the assumption that only an upper bound on the number of frequencies disturbing the system is known.

In this paper the class of nonlinear observable systems with known output dependent nonlinearities affected by disturbances generated by an unknown exosystem is considered. The output feedback set point regulation problem is solved by adaptively generating the reference input on line when both the parameters and the order of the exosystem are unknown and the resulting reference control input consists of a biased sum of sinusoids. Two global solutions to this problem are provided when only an upper bound on the exosystem order is known: the first scheme is simpler but is restricted to minimum phase systems while the second one can handle non-minimum phase systems as well and identify the exosystem parameters.

Under the minimum phase assumption, the same problem addressed in this paper can be solved following the approach in [10], [11], [12], while the first proposed algorithm is a special case of a more general one given in [29] that can achieve tracking of arbitrary output signals and allows for unknown parameters in the system to be regulated.

So far, the global exponential output regulation problem for classes of nonlinear systems allowed to be non-minimum phase is unsolved when the exosystem and its order are uncertain.

The second algorithm described in this paper (for systems allowed to be non-minimum phase) extends the results in [22] assuming that only an upper bound on the exosystem order is known. Both the system output and the reference trajectory output are required to be available for measurement: this hypothesis not required in [33], [34], [8], where only the regulation error is available for measurement. On the other hand the second algorithm presented in this paper does not require the knowledge of the exosystem's parameters, as in [33], [9] and [7] nor it requires the knowledge of the exosystem order as in [22], [8], [20]. Set point regulation is considered here while output regulation to zero is addressed in [11], [12], [22], [20]; if the system is minimum phase as in [11], [12] exponential convergence is obtained while asymptotic convergence is obtained in [11], [12]. The key tool is a dynamic algorithm to detect the number of exosystem's frequencies yielding an observation strategy which is adaptive with respect to this number and generates exponential convergent estimates of the exosystem's frequencies.

2 Problem Statement

In this paper we consider the class of nonlinear systems

$$\dot{x} = \Phi(y) + A_n x + bu + Gw; \qquad y = C_n x,; \quad x \in \mathbb{R}^n \tag{1}$$

$$\dot{w} = Rw; \quad w \in \mathbb{R}^r \tag{2}$$

with state $x \in \mathbb{R}^n$, control input $u \in \mathbb{R}$, measurable output $y \in \mathbb{R}; b = [b_1, \ldots, b_n]^T$ is a known constant vector, $\Phi(\cdot)$ is a known smooth vector function of the measurable output y; $A_j \in \mathbb{R}^j \times \mathbb{R}^j$, and $C_j \in \mathbb{R}^j$, are defined as $A_j = \begin{bmatrix} 0 & I_{j-1} \\ 0 & 0 \end{bmatrix}_{j \times j}$, $C_j = \begin{bmatrix} 1 & 0 & \cdots & 0 \end{bmatrix}_{1 \times j}$, j being a positive integer. The exosystem (2) with state $w \in \mathbb{R}^r$ and initial condition $w(0) \in \mathbb{R}^r$ generates the disturbances Gw to be rejected. We address the following control problem.

Problem 1. Consider the system (1) perturbed by the exosystem (2), with known $\Phi(y)$, b and unknown G, R. Assume that R has simple eigenvalues on the imaginary axis including zero, i.e. $\det(R) = 0$. Design a dynamic output feedback control law yielding bounded closed loop trajectories and regulating globally asymptotically the measured output $y \in \mathbb{R}$ of system (1) to the available constant reference y_r for any initial conditions $x(0) \in \mathbb{R}^n$, $w(0) \in \mathbb{R}^r$ of system (1), (2) and for a suitable set of initial conditions of the dynamic controller.

Remark 1. The initial conditions $w(0) \in \mathbb{R}^r$ of the exosystem may not excite all exosystem modes so that the unknown integer r actually represents an upper bound on the exosystem dimension. $\triangleleft$

3 A Regulator for Minimum Phase Systems

In this section we describe a regulation strategy to solve the stated problem assuming that the system (1) is minimum phase and a standard observability condition holds.

(H1) *The vector* $b = [0, \ldots, 0, b_\rho, \ldots, b_n]^T$ *is Hurwitz of degree* ρ *(*ρ *is the known relative degree with* $1 \leq \rho \leq n$*), i.e.* $b_\rho \neq 0$ *and all the zeros of the polynomial* $b_\rho s^{n-\rho} + \cdots + b_{n-1}s + b_n$ *have negative real part.*

(H2) *The pair* $\begin{bmatrix} A_n & G \\ 0 & R \end{bmatrix}$, $\begin{bmatrix} C_n & 0 \end{bmatrix}$ *is observable.*

A constructive procedure for the regulator design goes as follows. By virtue of (H2) there exists a linear unknown change of coordinates $\zeta = \begin{bmatrix} T_1 & T_2 \end{bmatrix} \begin{bmatrix} x \\ \bar{w} \end{bmatrix} \triangleq T \begin{bmatrix} x \\ \bar{w} \end{bmatrix}$, $\zeta \in \mathbb{R}^{n+r}$ such that in the new coordinates system (1), (2) becomes

$$\dot{\zeta} = A_{n+r}\zeta + T_1\Phi(y) + T_1\breve{b}\,(b_\rho u)\,; \qquad y = C_{n+r}\zeta \tag{3}$$

where $\breve{b} = b/\ b_\rho$ so that (to simplify the following design procedure) the ρ-th entry of $\breve{b}$ is 1. We first consider the case $\rho = 1$. Since $\zeta_1 = x_1 = y$, the first row of T_1 is $T_{11} = [1, 0, \ldots, 0]$, thus the first component of the vector $T_1\breve{b}$ is 1. From (3), recalling that we can write $T_1\Phi(y) \triangleq \sum_{i=1}^{q} \beta_i\alpha_i(y)$, where q is a suitable integer, β_i's are unknown parameters and α_i's are known functions, we have

$$\dot{\zeta} = A_{n+r}\zeta + \sum_{i=1}^{q} \beta_i\alpha_i(y) + +T_1\breve{b}\,(b_\rho u)\,; \qquad y = C_{n+r}\zeta. \tag{4}$$

Define the input filtered transformation

$$\begin{cases} \bar{\zeta} \;=\; \zeta - \sum_{i=2}^{n+r} \delta_i \sum_{j=2}^{i} d[j]\mu_{j-1}[i-1] \\ \dot{\mu}[i] = \begin{bmatrix} -\lambda_1 & 1 \ldots & 0 \\ \vdots & \vdots \ldots & \vdots \\ 0 & 0 \ldots & 1 \\ 0 & 0 \ldots & -\lambda_i \end{bmatrix} \mu[i] + \begin{bmatrix} 0 \\ \vdots \\ 0 \\ 1 \end{bmatrix} u, \qquad \mu[i] \in \mathbb{R}^i \\ y[i] = \begin{bmatrix} 1\; 0 \ldots 0 \end{bmatrix} \mu[i], \quad 1 \le i \le n+r-1 \end{cases} \tag{5}$$

in which λ_j, $1 \le j \le n+r-1$ are arbitrary positive reals and $d[n+r] = \begin{bmatrix} 0 \ldots 0\; 1 \end{bmatrix}^T$, $d[j-1] = A_c d[j] + \lambda_{j-1}d[j]$, $2 \le j \le n+r$, and $d = d[1]$. The constants δ_i are given by $\begin{bmatrix} 1\; \delta_2 \cdots \delta_{n+r} \end{bmatrix}^T = \begin{bmatrix} d[1] \cdots d[n+r] \end{bmatrix}^{-1} T_1\breve{b}$. From (4) and (5), we obtain

$$\dot{\bar{\zeta}} = A_{n+r}\bar{\zeta} + d[b_\rho u + \sum_{i=2}^{n+r} \delta_i y[i-1]] + \sum_{i=1}^{q} \beta_i\alpha_i(y); \qquad y = C_{n+r}\bar{\zeta} \tag{6}$$

with $d = [1, d_2, \ldots, d_{n+r}]^T$ a Hurwitz vector such that $s^{n+r-1} + d_2 s^{n+r-2} + \cdots + d_{n+r} = \prod_{i=1}^{n+r-1}(s + \lambda_i)$. The output-filtered transformation

$$\begin{cases} z_1 = \bar{\zeta}_1; \qquad z_j = \bar{\zeta}_j - \sum_{i=1}^{n+r} \xi_{j-1}[i]\beta_i, \quad 2 \le j \le n+r \\ \dot{\xi}[i] = \begin{bmatrix} -d_2 & 1 & \ldots & 0 \\ \vdots & & \ldots & \vdots \\ -d_{n+r-1} & 0 & \ldots & 1 \\ -d_{n+r} & 0 & \ldots & 0 \end{bmatrix} \xi[i] + \left[\begin{array}{c|c} -d_2 & \\ \vdots & I_{n+r-2} \\ -d_{n+r} & \end{array} \right] \alpha_i(y), \; 1 \le i \le q \end{cases} \tag{7}$$

maps (6) into $\dot{z} = A_{n+r}z + [db_\rho u + \sum_{i=2}^{n+r} \delta_i y[i-1] + \sum_{i=1}^{q} \beta_i(\xi_1[i] + \alpha_{i1}(y))]$, where α_{i1} is the first component of α_i. Perform a change of coordinates by setting $\eta_i = z_{i+1} - d_{i+1}z_1$, with $1 \le i \le n+r-1$, which yield

$$\begin{cases} \dot{y} = \eta_1 + d_2 y + \sum_{i=2}^{n+r} \delta_i y[i-1] + \sum_{i=1}^{q} \beta_i(\alpha_{i1}(y) + \xi_1[i]) + b_\rho u \\ \dot{\eta} = \Gamma\eta + \bar{d}y \end{cases} \tag{8}$$

with

$$\Gamma = \begin{bmatrix} -d_2 & 1 \dots 0 \\ \vdots & \vdots \ddots \vdots \\ -d_{n+r-1} & 0 \dots 1 \\ -d_{n+r} & 0 \dots 0 \end{bmatrix}, \quad \bar{d} = \begin{bmatrix} d_3 - d_2^2 \\ \vdots \\ d_{n+r} - d_2 d_{n+r-1} \\ -d_{n+r} d_2 \end{bmatrix}.$$

With reference to (8), we define the following dynamic output feedback adaptive controller ($\tilde{y} = y - y_r$)

$$\begin{cases} u &= \frac{1}{b_\rho}\left[-k\tilde{y} - \hat{\eta}_1 - d_2 y - \sum_{i=2}^{n+r} \hat{\delta}_i y[i-1] - \sum_{i=1}^{q} \hat{\beta}_i(\alpha_{i1}(y) + \xi_1[i])\right] \\ \dot{\hat{\eta}} &= \Gamma\hat{\eta} + \bar{d}y \\ \dot{\hat{\beta}}_i &= c_i(\alpha_{i1}(y) + \xi_1[i])\tilde{y}, \quad 1 \le i \le q \\ \dot{\hat{\delta}}_i &= c_{q+i-1} y[i-1]\tilde{y}, \qquad 2 \le i \le n+r \end{cases} \tag{9}$$

in which c_i, $0 \le i \le n+r+q-1$, are positive adaptation gains. The error dynamics are given by ($\tilde{\eta} = \eta - \hat{\eta}$, $\tilde{\beta}_i = \beta_i - \hat{\beta}_i$, $\tilde{\delta}_i = \delta_i - \hat{\delta}_i$)

$$\begin{cases} \dot{\tilde{y}} &= -k\tilde{y} + \tilde{\eta}_1 + \sum_{i=1}^{q} \tilde{\beta}_i(\alpha_{i1}(y) + \xi_1[i]) + \sum_{i=2}^{n+r} \tilde{\delta}_i y[i-1] \\ \dot{\tilde{\eta}} &= \Gamma\tilde{\eta} \\ \dot{\tilde{\beta}}_i &= -c_i\tilde{y}(\alpha_{i1}(y) + \xi_1[i]), \quad 1 \le i \le q \\ \dot{\tilde{\delta}}_i &= -c_{q+i-1} y[i-1]\tilde{y}, \quad 2 \le i \le n+r. \end{cases} \tag{10}$$

In order to prove the stability of the closed loop system, we consider the function $V = \frac{1}{2}\tilde{y}^2 + \epsilon\tilde{\eta}^T P\tilde{\eta} + \frac{1}{2}\sum_{i=1}^{q} \frac{1}{c_i}\tilde{\beta}_i^2 + \frac{1}{2}\sum_{i=2}^{n+r} \frac{1}{c_{q+i-1}}\tilde{\delta}_i^2$ with $P > 0$ being the matrix solution of $\Gamma^T P + P\Gamma = -Q < 0$ and $\epsilon > 0$. Its time derivative along (10) is such that $\dot{V} = -k\tilde{y}^2 + \tilde{y}\tilde{\eta}_1 - \epsilon\tilde{\eta}^T Q\tilde{\eta}$, which implies by a proper choice of ϵ that for a suitable $c > 0$

$$\dot{V}(t) \le -c\left\|\tilde{y}^T(t),\ \tilde{\eta}^T(t)\right\|^2 \tag{11}$$

so that all the error variables are bounded. Since y_r is constant and bounded we deduce that $y(t)$ and consequently $\xi[i](t)$ are bounded. Now, consider the change of coordinates $y = x_1$, $\bar{\eta}_i = x_{i+1} - \breve{b}_{i+1}x_1$, with $1 \le i \le n-1$. By computing $\dot{\bar{\eta}}$ and recalling that $y(t)$, $w(t)$ are bounded, it can be shown that $\bar{\eta}(t)$ and $x(t)$ are bounded. In particular we have

$$b_\rho u = \frac{dy}{dt} - \breve{b}_2 y - G_1^T w - \Phi_1 - \bar{\eta}_1 \tag{12}$$

and from (5) ($1 \le i \le n+r-1$) we deduce the expression $\sum_{j=0}^{i} d_{n+r-j}[n+r-i]\frac{d^j y[i]}{dt^j} = u$, which substituted in (12) gives $b_\rho \sum_{j=0}^{i} d_{n+r-j}[n+r-i]\frac{d^j y[i]}{dt^j} = \frac{dy}{dt} - \breve{b}_2 y - G_1^T w - \bar{\eta}_1 - \Phi_1$ with ($1 \le i \le n+r-1$). Since $y(t)$, $\bar{\eta}_1(t)$ and $w(t)$ are bounded and the polynomials $s^{n+r-i} + d_{n+r-i+1}[n+r-i]s^{n+r-i+1} + \dots + d_{n+r}[n+r-i]$, $1 \le i \le n+r-1$ are Hurwitz, it follows that $y[i](t)$,

$1 \le i \le n+r-1$, are bounded and from the first equation in (9) $u(t)$ is also bounded. Therefore, from (5), $\mu[i](t)$ are bounded, $1 \le i \le n+r-1$. From (11), we can write

$$\int_0^t \left\| \tilde{y}^T(\tau),\ \tilde{\eta}^T(\tau) \right\|^2 d\tau \le -\frac{1}{c}\int_0^t \dot{V}(\tau) d\tau \le \frac{1}{c} V(0). \tag{13}$$

From (10), $\dot{\tilde{y}}$ and $\dot{\tilde{\eta}}$ are bounded and therefore, applying Barbalat's Lemma (see [27]) from (13) we obtain $\lim_{t\to\infty} \tilde{y}(t) = 0$, while (10) implies that $\tilde{\eta}(t)$ converge exponentially to zero. If the relative degree is $1 < \rho \le n$, we introduce the input-filtered transformation ($\bar{\mu} \in \mathbb{R}^{\rho-1}$)

$$\bar{x} = x - b_\rho \sum_{i=2}^{\rho} \bar{b}[i]\bar{\mu}_{i-1}; \qquad \dot{\bar{\mu}} = \begin{bmatrix} -\bar{\lambda}_1 & 1 & \dots & 0 \\ 0 & -\bar{\lambda}_2 & \dots & 0 \\ \vdots & \vdots & \dots & \vdots \\ 0 & 0 & \dots & 1 \\ 0 & 0 & \dots & -\bar{\lambda}_{\rho-1} \end{bmatrix} \bar{\mu} + \begin{bmatrix} 0 \\ 0 \\ \vdots \\ 0 \\ 1 \end{bmatrix} u \tag{14}$$

with $\bar{b}[\rho] = \breve{b}$, $\bar{b}[i-1] = A_c\bar{b}[i] + \bar{\lambda}_{i-1}\bar{b}[i]$, for $\rho \ge i \ge 2$ and $\bar{\lambda}_i$, $1 \le i \le \rho-1$, arbitrary positive reals. We obtain from (1) and (14)

$$\dot{\bar{x}} = A_n\bar{x} + \bar{b}\left(b_\rho\bar{\mu}_1\right) + \Phi(y) + Gw; \qquad y = C_c\bar{x} \tag{15}$$

where $\bar{b} = \bar{b}[1] = [1, \bar{b}_2, \dots, \bar{b}_n]^T$ is, by construction, an unknown Hurwitz vector such that $s^{n-1} + \bar{b}_2 s^{n-2} + \cdots + \bar{b}_n = (s^{n-\rho} + \breve{b}_{\rho+1}s^{n-\rho-1} + \cdots + \breve{b}_n)\prod_{i=1}^{\rho-1}(s+\bar{\lambda}_i)$. System (15) is in the same form of system (1), with relative degree one with respect to the fictitious input $\bar{\mu}_1(t)$ (and $\bar{b}$ in place of $\breve{b}$). Accordingly, we proceed along the same steps followed in the relative-degree-one case, with the arguments used in [26] (see also [27]) to take into account that now the control input $u(t)$ feeds system (14) yielding the fictitious input $\bar{\mu}_1(t)$.
The previous arguments can be summarized as follows

Proposition 1. *Consider system (1)–(2). If assumptions (H1)–(H2) hold, then the Problem addressed is solvable.*

4 A Regulator for Non-Minimum Phase Systems

In this section we describe a regulation strategy to solve the stated problem without assuming that the system (1) is minimum phase. By relaxing this constraint the following assumptions (H3) and (H4) replace the assumptions (H1) and (H2) in the previous section.

(H3) *The zeroes of the polynomial* $b_1 s^{n-1} + \cdots + b_{n-1}s + b_n$ *and the eigenvalues of the matrix* R *are disjoint sets.*

By virtue of assumption (H3) along with the definition of the matrix A_n there exists a global solution Π, γ to the regulator equations

$$\Pi R = \Phi(y_r) + A_n \Pi + b\gamma + G; \qquad C_n \Pi = q \tag{16}$$

where $q \in \mathbb{R}^r$ is an eigenvector of R associated with the eigenvalue $\lambda = 0$ such that $qw(t) = y_r$ for all $t \geq 0$. If we define $x_r = \Pi w$ as the reference state trajectory and $u_r = \gamma w$ as the reference input, from (16) it follows $\dot{x}_r = \Phi(y_r) + A_n x_r + b u_r + G w$ and $y_r = C_n x_r$. Defining $\tilde{x} = x - x_r$, $\tilde{y} = y - y_r$, $\tilde{\Phi}(\tilde{y}) = \Phi(\tilde{y} + y_r) - \Phi(y_r)$, the regulation error dynamics are given by

$$\dot{\tilde{x}} = A_n \tilde{x} + \tilde{\Phi}(\tilde{y}) + b(u - u_r); \qquad \tilde{y} = C_n \tilde{x}. \tag{17}$$

(H4) *There exists an output feedback dynamic controller*

$$\dot{Y} = \bar{L}(Y, \tilde{y}); \ u_S = \bar{M}(Y, \tilde{y}) \tag{18}$$

with state $Y \in \mathbb{R}^s$, *a global diffeomorphism* $\bar{X} = \Psi(X)$ *with* $X = \left[\tilde{x}^T, \ Y^T\right]^T$, *and a positive definite Lyapunov function* $V\left(\bar{X}\right)$, $V \in C^1$ *such that by setting*

$$F(X) = \begin{bmatrix} A_n \tilde{x} + \tilde{\Phi}(C_n \tilde{x}) + b\bar{M}(Y, C_n \tilde{x}) \\ \bar{L}(Y, C_n \tilde{x}) \end{bmatrix}; \ \bar{F}\left(\bar{X}\right) = \left[\frac{\partial \Psi}{\partial X} F(X, t)\right]_{X = \Psi^{-1}(\bar{X})}$$

and $B = \left[b^T 0\right]^T$, *the following holds: (i)* $C_{n+s}\bar{X} = \tilde{y}$; *(ii)* $\frac{\partial \Psi}{\partial X} B = \bar{B}$ *with* $\bar{B} \in \mathbb{R}^{n+s}$ *constant vector; (iii)* $\alpha_1 \left\|\bar{X}\right\|^2 \leq V\left(\bar{X}\right) \leq \alpha_2 \left\|\bar{X}\right\|^2$; *(iv)* $\frac{\partial V}{\partial \bar{X}} \bar{F}\left(\bar{X}\right) \leq -\alpha_3 \left\|\bar{X}\right\|^2$; *(v)* $\left\|\frac{\partial V}{\partial \bar{X}}\right\| \leq \alpha_4 \left\|\bar{X}\right\|$.

Remark 2. If there exists a controller (18) such that $\dot{X} = F(X)$ is globally exponentially stable and $\frac{\partial F}{\partial X}$ is bounded for all $X \in \mathbb{R}^{n+s}$, then (see [19], Theorem 3.12), then (H4) is satisfied with $\bar{X} = X$ and a suitable $V(X)$ complying with H4-(iii)–(v). ◁

Remark 3. If system (17) complies with hypothesis (H1) i.e. is minimum phase then a diffeomorphism $\bar{X} = \Psi(X)$ and a Lyapunov function $V\left(\bar{X}\right)$ complying with (H4)(i)–(v) can always be constructed iteratively, as shown in [27], Section 6.3, pages 255–269. Hence assumptions (H3) and (H4) are weaker than assumptions (H1) and (H2). ◁

The reference input $u_r(t) = \gamma w(t)$ defined according to (16) is the sum of m biased sinusoids

$$u_r(t) = \kappa_0 + \sum_{h=1}^{m} \kappa_h \sin(\omega_h t + \phi_h) \tag{19}$$

with unknown bias κ_0, unknown magnitudes, κ_h, unknown phases ϕ_h, unknown distinct frequencies ω_h for all $h \in [1, m]$; m is an unknown integer such that $0 \leq m \leq M$, with $2M + 1 = r$ known upper bound. By virtue of (19) defining $\theta = [\theta_1, \ \theta_2, \ \ldots, \theta_m]^T$ so that

$$s^{2m} + \theta_1 s^{2m-2} + \ldots + \theta_{m-1} s^2 + \theta_m = \prod_{i=1}^{m} \left(s^2 + \omega_i^2\right), \tag{20}$$

the reference $u_r(t)$ can be modelled as the output of the following $(2m+1)-$order linear exosystem

$$\dot{\bar{w}} = \begin{bmatrix} S(\theta) & 0 \\ 0 & 0 \end{bmatrix} \bar{w}; \; u_r = \begin{bmatrix} 1 \, 0 \cdots 0 \, 1 \end{bmatrix} \bar{w}, \tag{21}$$

with state $\bar{w} \in \mathbb{R}^{2m+1}$, initial condition $\bar{w}(0) \in \mathbb{R}^{2m+1}$ that excites all oscillatory modes associated with the eigenvalues on the imaginary axis; the latter are parametrized by the vector $\theta \in \mathbb{R}^m$ with $S(\theta) = A_{2m} - [0, \theta_1, 0, \theta_2, 0, \ldots, \theta_m]^T C_{2m}$.

4.1 A Global Filtered Transformation

The aim of this section is to transform (17) and (21) into an "adaptive observer form" (see [24]). The first step is to introduce an auxiliary nonlinear filter ($\hat{x} \in \mathbb{R}^n$)

$$d\hat{x}/dt = [A_n - aC_n]\, \hat{x} + \tilde{\Phi}(\tilde{y}, y_r) + a\tilde{y} + bu, \tag{22}$$

where $a = [a_{n-1}, a_{n-2}, \ldots, a_0]^T$ and $a_i \in \mathbb{R}^+$, $0 \leq i \leq n-1$, are design parameters such that the polynomial $p_a(s) = s^n + a_{n-1}s^{n-1} + \ldots + a_1 s + a_0$ has all its roots with negative real part. By setting $\bar{x} = \tilde{x} - \hat{x}$, from (17), (22), we obtain the linear error dynamics

$$d\bar{x}/dt = [A_n - aC_n]\, \bar{x} - bu_r; \quad \bar{x}_1 = C_n \bar{x} = \tilde{y} - \bar{x}_1. \tag{23}$$

The autonomous system (23), (21) with state $\left[\bar{x}^T, \; \bar{w}^T\right]^T \in \mathbb{R}^{2m+n+1}$ by virtue of (H3) is observable from the available output $\bar{x}_1$ for every θ complying with (20); hence it is transformed into the observer canonical form ($\zeta \in \mathbb{R}^{2m+n+1}$)

$$d\zeta/dt = A_{n+2m+1}\zeta - \bar{a}_0\,[m]\, \bar{x}_1 - \textstyle\sum_{i=1}^{m} \theta_i \bar{a}_i\,[m]\, \bar{x}_1; \qquad \bar{x}_1 = C_{n+2m+1}\zeta \tag{24}$$

with $\bar{a}_i\,[m] \in \mathbb{R}^{n+2m+1}$, $i \in [0, m]$ given by

$$\begin{aligned} \bar{a}_0\,[m] &= [a_{n-1}, ..., a_0, 0, 0, 0, ..., 0, 0, 0]^T, \\ \bar{a}_1\,[m] &= [0, 1, a_{n-1}, ..., a_0, 0, 0, 0, ..., 0]^T, \\ \bar{a}_2\,[m] &= [0, 0, 0, 1, a_{n-1}, ..., a_0, 0, ..., 0]^T, \ldots, \\ \bar{a}_m\,[m] &= [0, 0, 0, ..., 0, 1, a_{n-1}, ..., a_0, 0]^T, \end{aligned} \tag{25}$$

by a linear transformation (which is nonsingular for all θ complying with (20)) expressed in matrix form as $\zeta = T_m(\theta)\left[\bar{x}^T, \; \bar{w}^T\right]$ where the columns $t_i\,[m]$, $i \in [1, n+2m+1]$ of the matrix $T_m(\theta) = [t_1\,[m], \ldots, t_{n+2m+1}\,[m]]$ are

$$\begin{aligned}
t_1\,[m] &= [1, 0, \theta_1, ..., 0, \theta_m, 0, ..., 0]^T\,, \\
t_2\,[m] &= [0, 1, 0, \theta_1, ..., 0, \theta_m, 0, ..., 0]^T\,, \\
&\vdots \\
t_n\,[m] &= [0, ..., 0, 1, 0, \theta_1, ..., 0, \theta_m, 0]^T\,, \\
t_{n+1}\,[m] &= [0, -b_1, ..., -b_n, 0, 0, ..., 0]^T\,, \\
t_{n+2}\,[m] &= [0, 0, -b_1, ..., -b_n, 0, ..., 0]^T\,, \\
&\vdots \\
t_{n+2m}\,[m] &= [0, ..., 0, -b_1, ..., -b_n, 0]^T\,, \\
t_{n+2m+1}\,[m] &= t_{n+1}\,[m] + \sum_{i=1}^{m-1} \theta_i t_{n+2i+1}\,[m] + \theta_m\,[0, \ldots, 0, -b_1, \ldots, -b_n]^T\,.
\end{aligned} \tag{26}$$

Set $\bar{d}\,[m] = [d_1, \ldots, d_{n+2m}]^T \in \mathbb{R}^{n+2m}$ where $d_i \in \mathbb{R}^+$, $1 \leq i \leq n+2m$, are positive real numbers such that all the roots of $p_d(s) = s^{n+2m} + d_1 s^{n+2m-1} + \cdots + d_{n+2m}$ have negative real part. Define as in [24] the filters ($\xi_i \in \mathbb{R}^{n+2m}$, $\mu_i \in \mathbb{R}$, $1 \leq i \leq m$)

$$d\xi_i/dt = \left[A_{n+2m} - \bar{d}\,[m]\,C_{n+2m}\right]\xi_i - [0,\ I_{n+2m}]\,\bar{a}_i\,[m]\,\bar{x}_1, \qquad \mu_i = C_{n+2m}\xi_i. \tag{27}$$

According to [24] the filtered transformation $\bar{\zeta} = \zeta - \begin{bmatrix} 0 \\ \sum_{i=1}^m \xi_i \theta_i \end{bmatrix}$, with $\bar{\zeta} \in \mathbb{R}^{n+2m+1}$, mapping the state vector ζ into a new state vector $\bar{\zeta}$, transforms system (24) into an "adaptive observer" form

$$d\bar{\zeta}/dt = A_{n+2m+1}\bar{\zeta} - \bar{a}_0\,[m]\,\bar{x}_1 + \begin{bmatrix} 1 \\ \bar{d}\,[m] \end{bmatrix} \mu^T\theta; \qquad \bar{x}_1 = C_{n+2m+1}\bar{\zeta}, \tag{28}$$

where $\mu = [\mu_1,\ \mu_2,\ \ldots \mu_m]^T$. The transformation from $\bar{\zeta}$ to $\left[\bar{x}^T\ \bar{w}^T\right]^T$ is given by

$$\begin{pmatrix} \bar{x} \\ \bar{w} \end{pmatrix} = T_m^{-1}(\theta)\left(\bar{\zeta} + \begin{bmatrix} 0 \\ \sum_{i=1}^m \xi_i\theta_i \end{bmatrix}\right). \tag{29}$$

Notice that the map $T_m^{-1}(\theta)$ is well defined since by (H3) the matrix $T_m(\theta)$ is invertible for all θ complying with (20). If $t_m^*(\theta)$ denotes the sum of the $(n+1)$-th and the $(n+2m+1)$-th rows of the adjoint of the matrix $T_m(\theta)$, then the sinusoidal reference $u_r(t) = \left[1\ 0 \cdots 0\ 1\right]\bar{w}$ by (29) can be expressed as

$$u_r(t) = \frac{1}{\det T_m\,(\theta)} t_m^*(\theta)\left(\bar{\zeta}(t) + \begin{bmatrix} 0 \\ \sum_{i=1}^m \xi_i(t)\theta_i \end{bmatrix}\right). \tag{30}$$

4.2 How to Detect the Number of Excited Frequencies

A key step in the regulator design is to build a dynamical system whose residual outputs are related to the number of excited frequencies. To this purpose, three cascaded filters are introduced: the first filter is defined as

$$\begin{cases} d\eta/dt = A_\eta \eta + [0,\ldots,0,1]^T \bar{x}_1 \\ \nu_i = \eta_{2M-2i+4},\ 1 \le i \le M+1 \end{cases} \tag{31}$$

with state $\eta = [\eta_1,\ \eta_2,\ \ldots \eta_{2M+2}]^T \in \mathbb{R}^{2M+2}$, initial condition $\eta(0) \in \mathbb{R}^{2M+2}$, input $\bar{x}_1(t)$ given in (23), output $\nu = [\nu_1,\ \nu_2,\ \ldots \nu_{M+1}]^T$, where $A_\eta = A_{n+2M+2} - [0,\ldots,0,1]^T [\bar{\alpha}_0,\ \bar{\alpha}_1,\ldots,\bar{\alpha}_{2M+1}]$, and the design parameters $\bar{\alpha}_i$, $0 \le i \le 2M+1$ are such that the polynomial $p_{\bar{\alpha}}(s) = s^{2M+2} + \bar{\alpha}_{2M+1}s^{2M+1} + \ldots + \bar{\alpha}_1 s + \bar{\alpha}_0$ has all its roots with negative real part. The dynamical system (31) is introduced to generate the output $\nu \in \mathbb{R}^{M+1}$ whose first i entries $\bar{\nu}_i = [\nu_1(t),\ \nu_2(t),\ \ldots,\nu_i(t)]$ with $i \in [1, M+1]$ can be shown to be persistently exciting if $1 \le i \le m$ and not persistently exciting if $m+1 \le i \le M+1$. The vector $\nu(t) \in \mathbb{R}^{M+1}$ is the input to the second filter

$$\begin{cases} \frac{d\Omega}{dt} = -c_1\Omega + \nu\nu^T,\ \Omega(0) > 0, \\ q_i = |\det(\Omega_i)|^{\frac{1}{i}},\ 1 \le i \le M+1 \end{cases} \tag{32}$$

with state $\Omega \in \mathbb{R}^{M+1} \times \mathbb{R}^{M+1}$ symmetric and positive definite initial condition $\Omega(0) > 0$, outputs $q_i(t)$, $1 \le i \le M+1$, where $\Omega_i \in \mathbb{R}^i \times \mathbb{R}^i$, $1 \le i \le M+1$, denotes the matrix collecting the first $i \times i$ entries of Ω and $c_1 \in \mathbb{R}^+$ is a design parameter. Notice that Ω is symmetric and (32) can be implemented by a filter whose dimension is $(M + 3M + 2)/2$. It can be shown by virtue of the persistency of excitation condition that

$$q_i(t) \ge q_M > 0 \text{ for } 1 \le i \le m, \tag{33}$$

where q_M is a suitable positive real while $q_i(t)$ with $m+1 \le i \le M+1$ are exponentially vanishing. The outputs $q_i(t)$ of filter (32) are the inputs of the third filter

$$d\chi_i/dt = -[\sigma_i(q_i) + \psi(\chi_i)]\chi_i + \tilde{\sigma}_i(q_{M+1}), \tag{34}$$

with $1 \le i \le M$, state $\chi = (\chi_1,\ldots,\chi_M)^T$, in which $\chi_i(0) > 0$; $\sigma_i(\cdot)$ and $\tilde{\sigma}_i(\cdot)$ with $1 \le i \le M$ are suitable class $\mathcal{K}$ functions. The function $\psi(\chi_i)$ depends on a design parameter $\chi_0 \in \mathbb{R}^+$ and is defined as $\psi(\chi_i) = 0$ if $\chi_i \le \chi_0$; $\psi(\chi_i) = 4(\chi_i - \chi_0)^2/\chi_i^2$ if $\chi_0 \le \chi_i \le 2\chi_0$ and $\psi(\chi_i) = 1$ if $2\chi_0 \le \chi_i$. It can be shown that $\chi_i(t)$ for $1 \le i \le m$ are globally vanishing functions and $\chi_i(t) \ge \bar{\chi} > 0$ for all $t \ge 0$ and $m+1 \le i \le M$, where $\bar{\chi}$ is a suitable positive real: hence the cascaded filters allow us to asymptotically determine the number $m \in [0, M]$. The result (see [23]) is summarized below.

Lemma 1. *Consider the cascaded interconnection of the filters (31), (32), (34) with input* $\bar{x}_1 \in \mathbb{R}$ *given by (23), state* $\eta \in \mathbb{R}^{2M+2}$, $\Omega \in \mathbb{R}^{M+1} \times \mathbb{R}^{M+1}$, $\chi \in \mathbb{R}^M$, *and outputs*

$$\begin{cases} \beta_i(t) = 1 & \text{if } q_i(t) > c_2\chi_i(t) \\ \beta_i(t) = q_i / (c_2\chi_i) & \text{if } q_i(t) \le c_2\chi_i(t) \end{cases} \tag{35}$$

where $i \in [1, M]$ *and* $c_2 \in \mathbb{R}^+$ *is a design parameter. The following holds: (i) the state trajectories are bounded for any* $\eta(0) \in \mathbb{R}^{2M+2}$, $\Omega(0) \in \mathbb{R}^{M+1} \times \mathbb{R}^{M+1}$ *such that* $\Omega(0) > 0$, *and any* $\chi_i(0) > 0$, $1 \le i \le M$; *(ii) the functions* $\beta_i(t)$, $i \in [1, M]$, *are such that* $0 \le \beta_i(t) \le 1$, *for all* $t \ge 0$, *and*

$$\begin{cases} \lim_{t\to\infty} \beta_i(t) = 1 \text{ for } 1 \le i \le m, \\ \lim_{t\to\infty} \beta_i(t) = 0 \text{ for } m+1 \le i \le M, \end{cases} \tag{36}$$

where the functions $\beta_i(t)$, $i \in [1, M]$, *tend exponentially to their limits.*

4.3 Adaptive Regulator Design with Frequencies Identification

In this section we design an adaptive observer for system (28) and determine an estimate $\widehat{u}_r(t)$ of the reference $u_r(t)$ via (30); the control input u is the sum of $\widehat{u}_r(t)$ and the function $\bar{M}(Y, y)$ in (18). To this purpose we define the diagonal matrix $\bar{U}(t) \in \mathbb{R}^{n+2M} \times \mathbb{R}^{n+2M}$, with entries $\bar{U}_{i,i}(t) = 1$ for $1 \le i \le n$ and $\bar{U}_{i,i}(t) = \beta_k(t)$, with $k = \left\lceil \frac{i-n}{2} \right\rceil$ for $n+1 \le i \le n+2M$. Consider the vectors $\bar{\beta}(t) = \left[\bar{\beta}_0(t), \ldots, \bar{\beta}_M(t)\right]$ and $\bar{\delta}(t)$ defined as

$$\begin{cases} \bar{\beta}_0(t) = (1 - \beta_1(t)); \ \bar{\beta}_M(t) = \beta_M(t) \\ \bar{\beta}_i(t) = \beta_i(t)(1 - \beta_{i+1}(t)) \text{ for } 1 \le i \le M-1 \end{cases} \tag{37}$$

$$\bar{\delta}(t) = \bar{\beta}_0(t) \begin{pmatrix} \bar{d}\,[0] \\ 0 \end{pmatrix} + \ldots + \bar{\beta}_{M-1}(t) \begin{pmatrix} \bar{d}\,[M-1] \\ 0 \end{pmatrix} + \bar{\beta}_M(t) \left(\bar{d}\,[M]\right) \tag{38}$$

where the entries of the constant vectors $\bar{d}\,[i] \in \mathbb{R}^{n+2i}$, $0 \le i \le M$ are design parameters such that the polynomials $s^{n+2i} + \left[s^{n+2i-1},\ s^{n+2i-2},\ \ldots, 1\right] \bar{d}\,[i]$ are Hurwitz. Notice that by (36)

$$\begin{cases} \lim_{t\to\infty} \bar{\beta}_m(t) = 1 \\ \lim_{t\to\infty} \bar{\beta}_i(t) \ = 0, \forall i \ne m \end{cases}; \quad \lim_{t\to\infty} \bar{U}(t) = \left[\begin{array}{c|c} I_{n+2m} & 0 \\ \hline 0 & 0 \end{array}\right]; \ \lim_{t\to\infty} \bar{\delta}(t) = \begin{bmatrix} \bar{d}\,[m] \\ 0 \end{bmatrix} \tag{39}$$

where the entries of the matrix $\bar{U}(t)$ and of the vector $\bar{\delta}(t)$ tend exponentially to their limits. The matrix $\bar{U}(t)$ and the vector $\bar{\delta}(t)$ are the tools to construct a generalization of the filters (27) that by virtue of (39) are adaptive with respect to the unknown number m: they are defined as

$$\begin{cases} d\widehat{\xi}_i/dt = \bar{U}\left\{\left(A_{n+2M} - \bar{\delta}(t)C_{n+2M}\right)\widehat{\xi}_i - \left([0,\ I_{n+2M}]\,\bar{a}_i\,[M]\right)\bar{x}_1\right\} \\ \qquad -c_3\left(I_{n+2M} - \bar{U}\right)\widehat{\xi}_i, \\ \widehat{\mu}_i = \beta_i C_{n+2M}\widehat{\xi}_i; \quad 1 \le i \le M \end{cases} \tag{40}$$

with state variables $\widehat{\xi}_i \in \mathbb{R}^{n+2M}$, $1 \le i \le M$, arbitrary initial conditions $\widehat{\xi}_i(0) \in \mathbb{R}^{n+2M}$, where $c_3 \in \mathbb{R}^+$ is a positive design parameter and $\bar{a}_i[M]$, $1 \le i \le M$ are defined according to (25) with M in place of m. We consider now an observer for system (28) which is adaptive with respect to the unknown number m of excited frequencies

$$\begin{cases} d\widehat{\zeta}/dt = U(t)\Big\{(A_{n+2M+1} - K(t)C_{n+2M+1})\,\widehat{\zeta} + (K(t) - \bar{a}_0[M])\,\bar{x}_1 \\ \qquad +\delta(t)\sum_{i=1}^{M}\widehat{\mu}_i\widehat{\theta}_i\Big\} - c_4\left[I_{n+2M+1} - U(t)\right]\widehat{\zeta}; \\ d\widehat{\theta}_i/dt = g_i\widehat{\mu}_i(t)\left[\bar{x}_1 - \widehat{\zeta}_1\right] - \bar{g}_i\left[1 - \beta_i(t)\right]\widehat{\theta}_i; \quad 1 \le i \le M \end{cases} \tag{41}$$

with $\widehat{\zeta} \in \mathbb{R}^{n+2M+1}$, $\widehat{\theta} = \left[\widehat{\theta}_1, \widehat{\theta}_2, \ldots, \widehat{\theta}_M\right]^T \in \mathbb{R}^M$, arbitrary initial conditions $\widehat{\zeta}(0) \in \mathbb{R}^{n+2M+1}$, $\widehat{\theta}(0) \in \mathbb{R}^M$, in which $\widehat{\zeta}_1 = C_{n+2M+1}\widehat{\zeta}$, $g_i, \bar{g}_i \in \mathbb{R}^+$, $c_4 \in \mathbb{R}^+$ are design parameters and $\bar{a}_0[M]$ is defined according to (25) with M in place of m; the matrix $U(t) \in \mathbb{R}^{n+2M+1} \times \mathbb{R}^{n+2M+1}$ and the vector $\delta(t) \in \mathbb{R}^{n+2M+1}$ are defined as $U(t) = \begin{bmatrix} 1 & 0 \\ 0 & \bar{U}(t) \end{bmatrix}$, $\delta(t) = \begin{bmatrix} 1 \\ \bar{\delta}(t) \end{bmatrix}$; the vector $K(t) \in \mathbb{R}^{n+2M+1}$ is $K(t) = (A_{n+2M+1} + \lambda I_{n+2M+1})\,\delta(t)$ with $\lambda \in \mathbb{R}^+$ design parameter. Let j be an integer such that $j \in [0, M]$; consider a partition of the vectors $\widehat{\xi}_i \in \mathbb{R}^{n+2M}$ $1 \le i \le M$, $\widehat{\zeta} \in \mathbb{R}^{n+2M+1}$, and $\widehat{\theta} \in \mathbb{R}^M$ into subvectors whose dimension depends on the integer j as follows:

$$\widehat{\xi}_i = \begin{bmatrix} \widehat{\xi}_i^{[j]} \in \mathbb{R}^{n+2j} \\ \breve{\xi}_i^{[j]} \in \mathbb{R}^{2(M-j)} \end{bmatrix}; \quad \widehat{\zeta} = \begin{bmatrix} \widehat{\zeta}^{[j]} \in \mathbb{R}^{n+2j+1} \\ \breve{\zeta}^{[j]} \in \mathbb{R}^{2(M-j)} \end{bmatrix}; \quad \widehat{\theta} = \begin{bmatrix} \widehat{\theta}^{[j]} \in \mathbb{R}^{j} \\ \breve{\theta}^{[j]} \in \mathbb{R}^{M-j} \end{bmatrix}. \tag{42}$$

By virtue of (39) it can be shown (see [23]) that the partition in (42) obtained by setting $j = m$ complies with the following properties.

Lemma 2. *Consider the filters (40) and the observer (41). Set in (42) $j = m$. Then: (i) $(\xi_i - \widehat{\xi}_i^{[m]})$, $\breve{\xi}_i^{[m]}$, with $1 \le i \le m$, $(\bar{\zeta} - \widehat{\zeta}^{[m]})$, $\breve{\zeta}^{[m]}$, $(\theta - \widehat{\theta}^{[m]})$, $\breve{\theta}^{[m]}$, tend exponentially to zero for any initial condition of the systems (28), (40), (41); (ii) $\widehat{\xi}_i \in \mathbb{R}^{n+2M}$, with $1 \le i \le M$, $\widehat{\theta} \in \mathbb{R}^M$ and $\widehat{\zeta} \in \mathbb{R}^{n+2M+1}$ are bounded.*

By virtue of Lemmas 1 and 2 we construct an estimate of the reference $u_r(t)$ in (30). To this purpose, let $j \in [0, M]$ and consider the matrix $T_j(\widehat{\theta}^{[j]}) \in \mathbb{R}^{n+2j+1} \times \mathbb{R}^{n+2j+1}$ whose columns are obtained from (26) with j in place of m and $\widehat{\theta}^{[j]} \in \mathbb{R}^j$ in place of $\theta \in \mathbb{R}^m$. Let $t_j^*(\widehat{\theta}^{[j]})$ be the sum of the $(n+1)$-th and the $(n+2j+1)$-th rows of the adjoint of the matrix $T_j(\widehat{\theta}^{[j]})$ for $j \in [1, M]$ and $t_0^*(\widehat{\theta}^{[0]})$ be the $(n+1)$-th row of the matrix $T_0(\widehat{\theta}^{[0]})$. Set

$$\widehat{\rho}_j(t) = t_j^*(\widehat{\theta}^{[j]})\left(\widehat{\zeta}^{[j]}(t) + \begin{bmatrix} 0 \\ \sum_{i=1}^{M}\widehat{\xi}_i^{[j]}(t)\widehat{\theta}_i(t) \end{bmatrix}\right), \tag{43}$$

with $j \in [0, M]$. Consider the filters

$$dp_j/dt = -c_5 p_j + c_6 \left(1 - \bar{\beta}_j(t)\right) + c_7 \left|\bar{x}_1 - \hat{\zeta}_1\right|, \tag{44}$$

with $j \in [0, M]$, state $p = [p_0, p_1, \ldots p_M] \in \mathbb{R}^{M+1}$ driven by the inputs $(1 - \bar{\beta}_j(t)), j \in [0, M]$, along with the estimation error $\left|\bar{x}_1 - \widehat{\zeta}_1\right|$, where c_5, c_6, c_7 are positive design parameters. Notice that if $p_j(0) > 0$ then all $p_j(t)$ with $j \neq m$ are greater than a positive lower bound, while $p_m(t)$ tends exponentially to zero as t goes to infinity. The estimate $\hat{u}_r$ for the reference $u_r(t)$ in (30) is defined by the adaptive saturation algorithm

$$\begin{aligned} \widehat{u}_r(t) &= \textstyle\sum_{j=0}^{M} \left(\bar{\beta}_j(t)\widehat{u}_j(t)\right) \text{ where} \\ \widehat{u}_j(t) &= \begin{cases} \dfrac{\widehat{\rho}_j(t)}{\det T_j(\widehat{\theta}^{[j]}(t))} & \text{if } \left|\det T_j(\widehat{\theta}^{[j]})\right| > p_j, ; \\ \dfrac{\widehat{\rho}_j(t) \det T_j(\widehat{\theta}^{[j]}(t))}{p_j^2} & \text{if } \left|\det T_j(\widehat{\theta}^{[j]})\right| \leq p_j. \end{cases} \end{aligned} \tag{45}$$

The task of the signals $p_j(t)$ is to avoid the singularities in which $\det T_j(\widehat{\theta}^{[j]}) = 0$, while the functions $\bar{\beta}_j(t)$ by virtue of (36) select the correct reference estimate $\widehat{u}_m(t)$ in the set $[\widehat{u}_0(t), \widehat{u}_1(t), \ldots, \widehat{u}_M(t)]$. The overall compensating control law, which is the sum of a stabilizing part and of a disturbance rejection part, is defined as

$$u = \bar{M}(Y, \tilde{y}) + \hat{u}_r(t). \tag{46}$$

At this point we are ready to state the properties of the regulation strategy introduced in this section.

Proposition 2. *Consider system (1), (2). If assumptions (H3)–(H4) hold, then the Problem addressed is solvable. In particular, the dynamic output feedback regulator (22), (31), (32), (34), (40), (41), (44), (45), (18), (46) guarantees exponentially decreasing output regulation error for any unknown parameters* $\kappa_0 \in \mathbb{R}$, $\kappa_i, \in \mathbb{R}$, $\phi_i \in \mathbb{R}$, $\omega_i \in \mathbb{R}^+$ *with* $i \in [1, m]$ *of the reference input* $u_r(t)$ *given by (19) where* $0 \leq m \leq M$, *with* $2M + 1 = r$ *known upper bound and any regulator initial condition such that* $\det \Omega(0) > 0$, $p_j(0) > 0$, *with* $j \in [0, M]$ *and* $\chi_i(0) > 0$ *with* $i \in [1, M]$.

Proof. (H3) allows to consider the error system (17) affected by the disturbance (21) instead of system (1), while assumption (H4) is used to design the stabilizing part in (46). By Lemmas 1 and 2 the vectors $\hat{x} \in \mathbb{R}^n$, $\eta \in \mathbb{R}^{2M+2}$, $\Omega \in \mathbb{R}^{M+1} \times \mathbb{R}^{M+1}$, $\chi_i \in \mathbb{R}$, $\widehat{\xi}_i \in \mathbb{R}^{2M}$, $1 \leq i \leq M$, $\widehat{\zeta} \in \mathbb{R}^{v+2M+1}$, $\widehat{\theta} \in \mathbb{R}^M$ are bounded, hence also the functions $\widehat{\rho}_j(t)$ with $j \in [0, M]$ defined in (43) are bounded, and we have

$$|\widehat{\rho}_j(t)| \leq \rho_M \text{ for all } t \geq 0, \text{ and } j \in [0, M], \tag{47}$$

where ρ_M is a suitable positive real. Next, we show by virtue of (30) and (44) that the disturbance estimation error $u_r(t) - \widehat{u}_r(t)$ is globally exponentially vanishing. The function $\widehat{u}_r(t) = \sum_{j=0}^{M} \left(\bar{\beta}_j(t)\widehat{u}_j(t)\right)$ is the weighted sum of disturbance estimates for different number of exosystem's frequencies. Consider first the functions $\widehat{u}_j(t)$ with $j \neq m$, $j \in [0, M]$; they are estimates of u_r with the incorrect number of sinusoidal frequencies. From (44), choosing $p_j(0) > 0$ by virtue of (36) we have that $p_j(t) > p^* > 0$, $j \neq m$, $j \in [0, M]$, where p^* is a suitable positive real number. This property guarantees that the functions $\widehat{u}_j(t)$ with $j \neq m$, $j \in [0, M]$ are bounded, so that by (36) each term $\bar{\beta}_j(t)\widehat{u}_j(t)$ with $j \neq m$, $j \in [0, M]$ is exponentially vanishing. Consider now the function $\widehat{u}_m(t)$ representing the estimate of $u_r(t)$ with the correct number of frequencies. From (44) recalling that the functions $(1 - \bar{\beta}_m(t))$ and $\left|\tilde{z}_1(t) - \widehat{\zeta}_1(t)\right|$ are exponentially vanishing by virtue of Lemmas 1 and 2 respectively , we deduce that

$$p_m(0)\exp\left(-a_v t\right) \leq p_m(t) \leq k_p \exp\left(-\lambda_p t\right) \tag{48}$$

for all $t \geq 0$, and suitable positive real numbers k_p and λ_p. Since by Lemma 2 the vector $\left(\theta - \widehat{\theta}^{[m]}(t)\right)$ is exponentially vanishing, then

$$\left|\det T_m(\widehat{\theta}^{[m]})\right| \geq \left|\det T_m(\theta)\right| - k_T \exp\left(-\lambda_s t\right) \tag{49}$$

for suitable positive real numbers k_T and λ_s. Let $k^* = \max\{k_T, k_p\}$ and $\lambda^* = \min\{\lambda_s, \lambda_p\}$; from (49) and (48) we have that $\left|\det T_m(\widehat{\theta}^{[m]})\right| > p_m(t)$ for all $t > t^*$, where $t^* = \dfrac{1}{\lambda^*} \ln\left[\frac{2k^*}{|\det T_m(\theta)|}\right]$, so that by virtue of (44) $\widehat{u}_m(t) = \widehat{\rho}_m(t)/\left|\det T_m(\widehat{\theta}^{[m]}(t))\right|$ for all $t > t^*$ and $|u_r(t) - \widehat{u}_r(t)| \leq k_a \exp\left[-\lambda_u\left(t - t^*\right)\right]$ if $t > t^*$ for suitable positive real numbers k_a and λ_u. Notice that for $0 \leq t \leq t^*$, the function $\tilde{u}_r(t) = u_r(t) - \widehat{u}_r(t)$ is bounded; in fact from (47) and (48) we have $\sup\limits_{0 \leq t \leq t^*} |\tilde{u}_r(t)| \leq \sup\limits_{0 \leq t \leq t^*} |u_r(t)| + \dfrac{\rho_M}{p_m(0)} \exp\left(a_v t^*\right)$. We conclude that $|\tilde{u}_r(t)| \leq k_u \exp\left(-\lambda_u t\right)$ for all $t \geq 0$ and $k_u = \exp\left[\lambda_u t^*\right] \max\{\sup\limits_{0 \leq t \leq t^*} |u_r(t) - \widehat{u}_r(t)|, k_a\}$.

By hypothesis (H4) the closed loop system (17), (18), (46) becomes $\dot{\bar{X}} = \bar{F}\left(\bar{X}\right) + \bar{B}\tilde{u}_r(t)$ with output $\tilde{y} = C_{n+s}\bar{X}$. By computing the time derivative of $V(\bar{X})$ we obtain $\frac{dV}{dt} = \frac{\partial V}{\partial \bar{X}}\bar{F} + \frac{\partial V}{\partial \bar{X}}\bar{B}\tilde{u}_r$. By H4-(iv) and H4-(v) we have $\frac{\partial V}{\partial \bar{X}}\bar{B}\tilde{u}_r \leq \frac{\alpha_3}{2\alpha_4^2}\left\|\frac{\partial V}{\partial \bar{X}}\right\|^2 + \frac{\alpha_4^2}{2\alpha_3}\left\|\bar{B}\right\|^2\tilde{u}_r^2 \leq \frac{\alpha_3}{2}\left\|\bar{X}\right\|^2 + \frac{\alpha_4^2}{2\alpha_3}\left\|\bar{B}\right\|^2\tilde{u}_r^2$ and $\frac{\partial V}{\partial \bar{X}}\bar{F} \leq -\alpha_3\left\|\bar{X}\right\|^2$ respectively. Recalling that by H4-(iii) $-\left\|\bar{X}(t)\right\|^2 \leq -\frac{V(\bar{X}(t),t)}{\alpha_2}$, the substitution of previous inequalities in the general expression of $\frac{dV}{dt}$ yields $\frac{dV}{dt} \leq -\frac{\alpha_3}{2\alpha_2}V(t) + \frac{\alpha_4^2}{2\alpha_3}\left\|\bar{B}\right\|^2\tilde{u}_r^2$. Since $\tilde{u}_r^2(t) \leq k_u^2\exp\left(-2\lambda_u t\right)$, we conclude that the time function $V(\bar{X}(t))$ is globally exponentially vanishing and by H4-(i) the vector $\bar{X}(t)$ and the output $\tilde{y}(t) = y(t) - y_r(t) = C_{n+s}\bar{X}(t)$ are globally exponentially vanishing. □

5 Conclusion

In this paper the set point regulation of the output is considered for a class of systems with output dependent nonlinearities: the systems are allowed to be non-minimum phase and may be affected by disturbances generated by a neutrally stable linear exosystem with unknown parameters and unknown order. Two global regulation schemes have been presented, achieving global convergence to zero of the regulated output, when the order of the exosystem is uncertain and only an upper bound the exosystem order is known. The first algorithm proposed guarantees asymptotic output regulation by taking advantage of the minimum phase assumption, while the second one applies to non-minimum phase systems as well, uses an observer which is adaptive with respect to the number of excited sinusoids and provides exponentially convergent estimates of the exosystem's frequencies.

References

1. M. Bodson and S.C. Douglas. Adaptive algorithms for the rejection of periodic disturbances with unknown frequencies. *Automatica*, 33(12):2213–2221, 1997.
2. C.I. Byrnes, F. Delli Priscoli, and A. Isidori. *Output Regulation of Uncertain Nonlinear Systems.* Birkäuser, Boston, MA, 1997.
3. C.I. Byrnes, F. Delli Priscoli, A. Isidori, and K. Wang. Structurally stable output regulation for nonlinear systems. *Automatica*, 33:369–385, 1997.
4. C.I. Byrnes and A. Isidori. Limit sets, zero dynamics and internal models in the problem of nonlinear output regulation. *IEEE Trans. on Automat. Contr.*, 48:1712–1723, 2003.
5. C.I. Byrnes and A. Isidori. Nonlinear internal models for output regulation. *IEEE Trans. on Automat. Contr.*, 49:2244–2247, 2004.
6. Z. Chen and J. Huang. A general formulation and solvability of the global robust output regulation problem. *IEEE Trans. on Automat. Contr.*, 50(4):448–462, 2005.
7. Z. Chen and J. Huang. Global robust output regulation for output feedback systems. *IEEE Trans. on Automat. Contr.*, 50(1):117–121, 2005.
8. F. Delli Priscoli, L. Marconi, and A. Isidori. Adaptive observers as nonlinear internal models. *Systems & Control Letters*, 55:640–649, 2006.
9. Z. Ding. Global output regulation of uncertain nonlinear systems with exogenous signals. *Automatica*, 37:113–119, 2001.
10. Z. Ding. Adaptive tracking with complete compensation of unknown disturbances for nonlinear output feedback systems. *Proc. Inst. Elec. Eng. Control Theory Appl.*, 149:533–539, 2002.
11. Z. Ding. Global stabilization and disturbance suppression of a class of nonlinear systems with uncertain internal model. *Automatica*, 39(3):471–479, 2003.
12. Z. Ding. Universal disturbance rejection for nonlinear systems in output feedback form. *IEEE Trans. on Automat. Contr.*, 48(7):1222–1227, 2003.
13. J. Huang and C. Chen. A general framework for tackling the output regulation problem. *IEEE Trans. on Automat. Contr.*, 49:2203–2217, 2004.

14. J. Huang and C. Lin. On a robust nonlinear servomechanism problem. *IEEE Trans. on Automat. Contr.*, 39:542–546, 1994.
15. A. Isidori. *Nonlinear Control Sytems.* Springer Verlag, New York, 3rd edition, 1995.
16. A. Isidori. *Nonlinear Control Sytems Vol. II.* Springer Verlag, New York, 1999.
17. A. Isidori. A tool for semiglobal stabilization of ucertain non-minimum-phase nonlinear systems via output feedback. *IEEE Trans. on Automat. Contr.*, 55:1817–1827, 2000.
18. A. Isidori and C.I. Byrnes. Output regulation of nonlinear systems. *IEEE Trans. on Automat. Contr.*, 35:131–140, 1990.
19. H. Khalil. *Nonlinear Systems.* Prentice Hall, Upper Saddle River, NJ, 2nd edition, 1996.
20. W. Lan, B. Chen, and Z. Ding. Adaptive estimation and rejection of unknown sinusoidal disturbances through measurement feedback for a class of non-minimum phase non-linear MIMO systems. *Int. J. Ad. Contr. Sig. Pr.*, 20:77–97, 2006.
21. L. Marconi, A. Isidori, and A. Serrani. Non-resonance conditions for uniform observability in the problem of nonlinear output regulation. *Systems & Control Letters*, 53:281–298, 2004.
22. R. Marino and G. Santosuosso. Global compensation of unknown sinusoidal disturbances for a class of nonlinear non-minimum phase systems. *IEEE Trans. on Automat. Contr.*, 50(11):1816–1822, 2005.
23. R. Marino and G. Santosuosso. Regulation of linear systems with unknown exosystems of uncertain order. *IEEE Trans. on Automat. Contr.*, 2007. To appear.
24. R. Marino, G. Santosuosso, and P. Tomei. Robust adaptive observers for nonlinear systems with bounded disturbances. *IEEE Trans. on Automat. Contr.*, 46(6):967–972, 2001.
25. R. Marino, G. Santosuosso, and P. Tomei. Robust adaptive compensation of biased sinusoidal disturbances with unknown frequency. *Automatica*, 39(10):1755–1761, 2003.
26. R. Marino and P. Tomei. Global adaptive output-feedback control of nonlinear systems. Part I linear parametrization. *IEEE Trans. on Automat. Contr.*, 38:17–32, 1993.
27. R. Marino and P. Tomei. *Nonlinear Control Design – Geometric, Adaptive and Robust.* Prentice Hall, 1995.
28. R. Marino and P. Tomei. Output regulation for linear systems via adaptive internal model. *IEEE Trans. on Automat. Contr.*, 48:2199–2202, 2003.
29. R. Marino and P. Tomei. Adaptive tracking and disturbance rejection for uncertain nonlinear systems. *IEEE Trans. on Automat. Contr.*, 50:90–95, 2005.
30. V. Nikiforov. Adaptive servocompensation of input disturbances. *Proc. of the 13th IFAC World Congress*, pages 175–180, 1996.
31. V. Nikiforov. Adaptive nonlinear tracking with complete compensation of unknown disturbances. *European Journal of Control*, 4:132–139, 1998.
32. V. Nikiforov. Nonlinear servocompensation of unknown external disturbances. *Automatica*, pages 1647–1653, 2001.
33. A. Serrani and A. Isidori. Global robust output regulation for a class of nonlinear systems. *Systems & Control Letters*, 39:133–139, 2000.

34. A. Serrani, A. Isidori, and L. Marconi. Semiglobal nonlinear output regulation with adaptive internal model. *IEEE Trans. on Automat. Contr.*, 46(8):1178–1194, 2001.
35. X. Ye and J. Huang. Decentralized adaptive output regulation for a class of large-scale nonlinear systems. *IEEE Trans. on Automat. Contr.*, 48(8):276–281, 2003.

A Taxonomy for Time-Varying Immersions in Periodic Internal-Model Control

Andrea Serrani

Department of Electrical and Computer Engineering, The Ohio State University, Columbus, OH 43206 - USA

Summary. In extending the solvability of the output regulation problem to encompass more general classes of time-varying exogenous systems, various non-equivalent definitions of observability play a crucial role in the definition and properties of immersion mappings establishing the so-called "internal model property." This paper proposes a classification of the immersion mappings based on the underlying observability property, and describes the connections between different canonical realizations of internal models that fully exploit such properties for robust and adaptive output regulation design in periodic systems.

This paper is humbly dedicated to Alberto Isidori, at once et magister et amicus, *on the joyful occasion of his sixty-fifth birthday.*

1 Notation and Background

Continuous matrix-valued functions $M : \mathbb{R} \to \mathbb{R}^{n\times m}$ are said to be periodic if there exists $T > 0$ such that $M(t+T) = M(t)$ for all $t \in \mathbb{R}$. The smallest such T is said to be the period of $M(\cdot)$. A linear time-varying system

$$\begin{aligned} \dot{x} &= A(t)x + B(t)u \\ y &= C(t)x + D(t)u \end{aligned} \tag{1}$$

with $x \in \mathbb{R}^n$, $u \in \mathbb{R}^m$ and $y \in \mathbb{R}^p$ is said to be periodic if $A(\cdot)$, $B(\cdot)$, $C(\cdot)$, and $D(\cdot)$ are all periodic with the same period. The monodromy matrix of $A(\cdot)$ is defined as $\bar{\Phi}_A := \Phi_A(T,0)$, where $\Phi_A(t,\tau)$ denotes the transition matrix. The eigenvalues of $\bar{\Phi}_A$ are termed the characteristic multipliers of $A(\cdot)$ (see [8].)

Lemma 1 (Sylvester differential equations). *Let $S(t) \in \mathbb{R}^{s\times s}$, $A(t) \in \mathbb{R}^{n\times n}$, and $P(t) \in \mathbb{R}^{n\times s}$ be continuous and periodic matrix-valued functions. Assume that there exist constants $\kappa, \lambda, \mu > 0$, and $\sigma \geq 0$ such that*

$$\|\Phi_A(t,\tau)\| \leq \kappa\, \mathrm{e}^{-\lambda(t-\tau)}\,, \quad \|\Phi_S(t,\tau)\| \geq \mu\, \mathrm{e}^{\sigma(t-\tau)}$$

for all $t, \tau \in \mathbb{R}$. Then, there exists a unique periodic solution of the Sylvester matrix differential equation (SDE)

$$\dot{X}(t) + X(t)S(t) = A(t)X(t) + P(t) \tag{2}$$

Definition 1 (Observability). *A time-varying (not necessarily periodic) representation $(A(\,\cdot\,), C(\,\cdot\,))$ is said to be* completely observable *if for any $t \in \mathbb{R}$ there exists $s > t$ such the observability gramian*

$$W(t,s) = \int_t^s \Phi'_A(\tau,t)C'(\tau)C(\tau)\Phi_A(\tau,t)\mathrm{d}\tau$$

is nonsingular; it is said to be totally observable *if the above holds for every $t \in \mathbb{R}$ and every $s > t$. It is said to be* uniformly completely observable *if there exists positive constants α_1, α_2, δ, and a class-$\mathcal{K}$ function $\gamma(\,\cdot\,)$ such that*

$$\alpha_1 I \le W(t, t+\delta) \le \alpha_2 I \tag{3}$$

$$\|\Phi_A(t,s)\| \le \gamma(|t-s|) \tag{4}$$

hold for all $t, s \in \mathbb{R}$. Finally, the pair $(A(\,\cdot\,), C(\,\cdot\,))$ is said to be uniformly observable *if the observability matrix*

$$N(t) = \big(N'_0(t)\; N'_1(t)\; \cdots N'_{n-1}(t)\big)'$$

is nonsingular for every $t \in \mathbb{R}$, where

$$N_0(t) := C(t)\,, \qquad N_i(t) := \tfrac{\mathrm{d}}{\mathrm{d}t}N_{i-1}(t) + N_{i-1}(t)A(t)\,,\ i \ge 1\,.$$

For *periodic systems*, uniform complete observability is equivalent to complete observability, and is completely characterized by the lower bound on (3). If, in addition, the entries of $(A(\,\cdot\,), C(\,\cdot\,))$ are analytic functions, uniform complete observability is equivalent to total observability. However, uniform observability is a stronger property than uniform complete observability, even for analytic periodic systems [2, 15, 16].

2 Definition of the Problem

We consider in this paper parameterized families of linear time-varying plant models of the form

$$\begin{aligned} \dot{x} &= A(t,\mu)x + B(t,\mu)u + P(t,\mu)w \\ e &= C(t,\mu)x + Q(t,\mu)w\,, \end{aligned} \tag{5}$$

with state $x \in \mathbb{R}^n$, control input $u \in \mathbb{R}$, regulated error $e \in \mathbb{R}$, and parameter vector μ taking values on a given compact set $\mathcal{P} \subset \mathbb{R}^p$. The exogenous input $w \in \mathbb{R}^{n_w}$ is generated by a parameterized family of time-varying exosystem

$$\dot{w} = S(t, \sigma)w \tag{6}$$

with parameter vector σ ranging over a given compact set $\Sigma \subset \mathbb{R}^s$. It is assumed that $A(\cdot, \cdot)$, $B(\cdot, \cdot)$, $C(\cdot, \cdot)$, $P(\cdot, \cdot)$, $Q(\cdot, \cdot)$, and $S(\cdot, \cdot)$ are smooth functions, and that systems (5) and (6) are periodic with period $T > 0$ for all $(\mu, \sigma) \in \mathcal{P} \times \Sigma$. To rule out trivial cases, we assume the following.

Assumption 1. *The characteristic multipliers $\lambda_i(\sigma)$ of $S(\cdot, \sigma)$ are distinct, and satisfy $|\lambda_i(\sigma)| = 1$ for all $i = 1, \ldots, s$ and all $\sigma \in \Sigma$. Furthermore, the pair $(Q(\cdot, \mu), S(\cdot, \sigma))$ is observable for all $\sigma \in \Sigma$ and all $\mu \in \mathcal{P}$.*

While the actual value of the plant parameter vector μ is unknown, we will distinguish explicitly the case in which the exosystem model is uncertain from the one in which the parameter vector σ is known a priori. For the sake of clarity, the dependence on σ will be dropped whenever σ is assumed known.

Problem 1. Assume that σ is known. The *robust periodic output regulation problem* consists in finding a periodic error-feedback compensator of the form

$$\begin{aligned} \dot{\xi} &= F(t)\xi + G(t)e \\ u &= H(t)\xi + K(t)e \end{aligned} \tag{7}$$

with state $\xi \in \mathbb{R}^\nu$, such that, for all $\mu \in \mathcal{P}$:

(i) The origin is a uniformly asymptotically stable equilibrium of the unforced closed-loop system

$$\begin{aligned} \dot{x} &= (A(t, \mu) + B(t, \mu)K(t)C(t, \mu))x + B(t, \mu)H(t)\xi \\ \dot{\xi} &= F(t)\xi + G(t)C(t, \mu)x\,. \end{aligned} \tag{8}$$

(ii) Trajectories of the closed-loop system originating from initial condition $(w_0, x_0, \xi_0) \in \mathbb{R}^{n_w + n + \nu}$ are bounded and satisfy $\lim_{t\to\infty} e(t) = 0$.

The case of uncertain exosystem models (6) will be interpreted as the situation in which a "family" of robust regulation problems, each one corresponding to a given value of $\sigma \in \Sigma$, replaces the above definition.

Definition 2. *A parameterized family of periodic dynamic compensators*

$$\begin{aligned} \dot{\xi} &= F(t, \theta)\xi + G(t, \theta)e \\ u &= H(t, \theta)\xi + K(t, \theta)e\,, \end{aligned} \tag{9}$$

with state $\xi \in \mathbb{R}^\nu$ and parameter vector θ ranging on some set $\Theta \subseteq \mathbb{R}^\varrho$, is said to be a certainty-equivalence robust regulator *if for all $\mu \in \mathcal{P}$:*

(a) the system

$$\dot{x} = (A(t,\mu) + B(t,\mu)K(t,\theta)C(t,\mu))x + B(t,\mu)H(t,\theta)\xi$$
$$\dot{\xi} = F(t,\theta)\xi + G(t,\theta)C(t,\mu)x \tag{10}$$

is uniformly asymptotically stable for all $\theta \in \Theta$,

(b) there exists a continuous assignment $\sigma \mapsto \theta_\sigma$ such that for any given $\sigma \in \Sigma$, the fixed controller

$$\dot{\xi} = F(t,\theta_\sigma)\xi + G(t,\theta_\sigma)e$$
$$u = H(t,\theta_\sigma)\xi + K(t,\theta_\sigma)e \tag{11}$$

solves the robust output regulation problem for (5) *and* (6).

The role of the parameter θ is solely that of selecting, among a continuous family of candidate stabilizing controllers, one that achieves asymptotic regulation in correspondence to a given element in the parameterized set of exosystem models (6).

When the exosystem model is uncertain, the robust regulation problem is approached by seeking a solution based on certainty-equivalence adaptive control, which leads to the following definition.

Problem 2. Assume that σ is unknown. The *adaptive robust periodic output regulation problem* consists in finding a certainty-equivalence controller (9) and a smooth update law $\varphi : \mathbb{R}^\nu \times \mathbb{R}^m \to \mathbb{R}^\varrho$ such that the trajectories of the closed-loop system

$$\begin{aligned}
\dot{w} &= S(t,\sigma)w \\
\dot{x} &= (A(t,\mu) + B(t,\mu)K(t,\hat{\theta})C(t,\mu))x + B(t,\mu)H(t,\hat{\theta})\xi + (P(t,\mu) \\
&\qquad + B(t,\mu)K(t,\hat{\theta})Q(t,\mu))w \\
\dot{\xi} &= F(t,\hat{\theta})\xi + G(t,\hat{\theta})C(t,\mu)x + G(t,\hat{\theta})Q(t,\mu)w \\
\dot{\hat{\theta}} &= \varphi(\xi, e)
\end{aligned} \tag{12}$$

originating from any $(w_0, x_0, \xi_0, \hat{\theta}_0) \in \mathbb{R}^{n_w+n+\nu+\varrho}$ are bounded and satisfy $\lim_{t\to\infty} e(t) = 0$, for all $\mu \in \mathcal{P}$ and all $\sigma \in \Sigma$.

In the formulation of Problem 2, the requirement of uniform asymptotic stability of the unforced closed-loop system has been dropped. This serves the purpose of including the important case in which convergence of $\hat{\theta}(t)$ to θ_σ is not achieved. Convergence of the parameter estimates requires additional hypotheses on the "richness" of the signals in the update law, commonly referred to as *persistence of excitation* (PE), which may lead to unnecessarily restrictive conditions as far as asymptotic regulation of $e(t)$ is concerned.

3 A Taxonomy for Periodic Internal Model Control

An obvious necessary condition for the solvability of Problem 2 is the existence of a certainty-equivalence regulator, which in turn implies solvability of Problem 1 for each fixed $\sigma \in \Sigma$. Consequently, we first address the robust case, and defer the adaptive one to the next section.

Proposition 1. *A robust stabilizing controller* (7) *is a robust regulator if and only if there exist smooth mappings* $\Pi : \mathbb{R} \times \mathbb{R}^p \to \mathbb{R}^{n \times n_w}$, $\Xi : \mathbb{R} \times \mathbb{R}^p \to \mathbb{R}^{\nu \times n_w}$, *and* $R : \mathbb{R} \times \mathbb{R}^p \to \mathbb{R}^{1 \times n_w}$, *periodic in their first argument, satisfying the differential-algebraic equations*

$$\begin{aligned} \dot{\Pi}(t,\mu) + \Pi(t,\mu)S(t) &= A(t,\mu)\Pi(t,\mu) + B(t,\mu)R(t,\mu) + P(t,\mu) \\ 0 &= C(t,\mu)\Pi(t,\mu) + Q(t,\mu) \end{aligned} \tag{13}$$

$$\begin{aligned} \dot{\Xi}(t,\mu) + \Xi(t,\mu)S(t) &= F(t)\Xi(t,\mu) \\ R(t,\mu) &= H(t)\Xi(t,\mu) \end{aligned} \tag{14}$$

for all $t \in [0,T)$ *and all* $\mu \in \mathcal{P}$.

Proof. The result is a simple extension of [19, Prop. 3.2]. □

Equation (14) constitutes the formulation in the periodic setting of the *internal model principle* [9]. In the time-invariant case, an elegant tool for obtaining (14) is provided by the concept of *system immersion.* Loosely speaking, an autonomous system is immersed into another system if any output trajectory of the first system is an output trajectory of the second one. Here, the usefulness of this property lies in the possibility of reproducing the unavailable feedforward control $u = R(t,\mu)w$ as the output of another system, which does not depend on the parameter μ, and satisfies appropriate observability conditions. If this is the case, the system in question can be embedded in the controller to enforce (14), and is therefore regarded as an *internal model* of the exosystem. As opposed to LTI systems, which always admit an immersion into a parameter-independent observable system as a consequence of the Cayley-Hamilton theorem, in the periodic case the situation is much more complicated. A general existence result is not available, and several situations may occur, due to various observability notions. As a result, we propose the following classification.

Definition 3. *The parameterized family of periodic systems*

$$\begin{aligned} \dot{w} &= S(t)w \\ v &= R(t,\mu)w \end{aligned} \tag{15}$$

is immersed into the periodic system $(\Phi(\,\cdot\,), \Gamma(\,\cdot\,))$, *with* $\Phi : \mathbb{R} \to \mathbb{R}^{q \times q}$ *and* $\Gamma : \mathbb{R} \to \mathbb{R}^{1 \times q}$ *smooth functions of their argument, if there exists a smooth mapping* $\Upsilon : \mathbb{R} \times \mathbb{R}^p \to \mathbb{R}^{q \times n_w}$, *periodic in its first argument, satisfying*

$$\begin{aligned} \dot{\Upsilon}(t,\mu) + \Upsilon(t,\mu)S(t) &= \Phi(t)\Upsilon(t,\mu) \\ R(t,\mu) &= \Gamma(t)\Upsilon(t,\mu) \end{aligned} \tag{16}$$

for all $t \in [0,T)$ *and all* $\mu \in \mathcal{P}$. *Furthermore, we say that* (16) *defines (i.) a* regular immersion, *if the pair* $(\Phi(\cdot), \Gamma(\cdot))$ *is uniformly completely observable, (ii.) a* strong immersion, *if it is uniformly observable, and (iii.) a* weak immersion, *if it is detectable and not completely observable.*

Whenever an immersion of the form (16) exists, we will refer to $(\Phi(\cdot), \Gamma(\cdot))$ as either a regular, strong, or weak *internal-model pair*. As mentioned, the reason for the proposed taxonomy lies in the fact that for time-varying systems useful notions of observability are weaker than invertibility of the observability matrix. The advantage in having a strong immersion versus a regular one is that in the former case the pair $(\Phi(\cdot), \Gamma(\cdot))$ is topologically equivalent to canonical forms (see [13, 14]) which are advantageous for controller design.

A necessary and sufficient condition for the existence of a strong immersion is given by following result.

Lemma 2. *The periodic system* (15) *admits a strong immersion if there exist an integer* q *and periodic functions* $a_0(t), a_1(t), \ldots, a_{q-1}(t)$ *such that*

$$L_S^q R(t,\mu) + a_{q-1}(t) L_S^{q-1} R(t,\mu) + \cdots + a_1(t) L_S R(t,\mu) + a_0(t) R(t,\mu) = 0 \tag{17}$$

for all $t \in [0,T)$ *and all* $\mu \in \mathcal{P}$, *where the operator* L_S *is defined recursively as* $L_S^0 R(t) = R(t)$ *and* $L_S^k R(t) = (L_S^{k-1} R(t)) S(t) + \frac{\mathrm{d}}{\mathrm{d}t} L_S^{k-1} R(t)$, $k \geq 1$.

Proof. Sufficiency follows directly from the fact that the mapping

$$\Upsilon(t,\mu) = \begin{pmatrix} R(t,\mu) \\ L_S R(t,\mu) \\ \cdots \\ L_S^{q-1} R(t,\mu) \end{pmatrix}$$

yields an internal-model pair in phase-variable form

$$\Phi_{\mathrm{p}}(t) = \begin{pmatrix} 0 & 1 & \cdots & 0 \\ 0 & 0 & \cdots & 0 \\ \vdots & \vdots & \ddots & \vdots \\ -a_0(t) & -a_1(t) & \cdots & -a_{q-1}(t) \end{pmatrix}, \qquad \Gamma_{\mathrm{p}} = (1\ 0 \cdots 0) \tag{18}$$

which is obviously uniformly observable. Next, assume that a strong internal model pair $(\Phi(\cdot), \Gamma(\cdot))$ exists. Then, $(\Phi(\cdot), \Gamma(\cdot))$ is topologically equivalent to the observer companion form

$$\Phi_{\mathrm{o}}(t) = \begin{pmatrix} -\alpha_{q-1}(t) & 1 & 0 & \cdots & 0 \\ \vdots & \vdots & \vdots & \ddots & \vdots \\ -\alpha_1(t) & 0 & 0 & \cdots & 1 \\ -\alpha_0(t) & 0 & 0 & \cdots & 0 \end{pmatrix}, \qquad \Gamma_{\mathrm{o}} = (1\ 0 \cdots 0) \tag{19}$$

which can be computed from the original internal-model pair by using the dual version of Silverman's algorithm [13]. The observer form can in turn be converted into the phase-variable form by using the observability matrix $W_o(t)$ of $(\Phi_o(t), \Gamma_o)$ as a Lyapunov transformation. As a matter of fact, the equation

$$\dot{W}_o^{-1}(t)W_o(t) + \Phi_p(t)W_o(t) = W_o(t)\Phi_o(t)$$

can be solved recursively for $a_i(t)$, $i = 0, \dots, q-1$, once the functions $\alpha_i(t)$ are known [18]. The existence of a phase-variable form implies (17). □

Although preferable, it may be very difficult (or impossible) to find a strong immersion for a given pair $(S(\cdot), R(\cdot, \mu))$, while a regular immersion may be easier to compute. On the other hand, the reason to introduce weak immersions will be related to the possibility of obtaining non-minimal realizations having special structures to be exploited in adaptive design.

Example 1. Consider a periodic exosystem of the form (15), where

$$S(t) = \begin{pmatrix} 0 & \sin(t) \\ -\sin(t) & 0 \end{pmatrix}, \quad R(\mu) = \begin{pmatrix} \mu_1 & \mu_2 \end{pmatrix}, \quad \mu_1^2 + \mu_2^2 = 1\,.$$

It is easy to check that the pair $\big(S(\cdot), R(\mu)\big)$ is uniformly completely observable for all considered values of μ, but not uniformly observable. However, the pair admits a strong immersion, as Lemma 2 holds with $q = 3$ and

$$a_0(t) = 3\sin(t)\cos(t)\,, \quad a_1(t) = 1 + \sin^2(t)\,, \quad a_2(t) = 0\,.$$

△

For LTI systems, the structure of a realization of least dimension for the immersion (16) is completely determined by the minimal polynomial of the exosystem matrix [9]. As a result, the minimal immersion in phase-variable form (Φ_p, Γ_p) is unique for a given pair $(S, R(\mu))$. This is not the case for periodic systems, as shown in the next example.

Example 2. Consider again the system in Example 1. While the observability matrix of $\big(S(\cdot), R(\mu)\big)$ is singular at $t = 0$ and $t = \pi$, the extended observability matrix

$$N_e(t,\mu) = \begin{pmatrix} R(\mu) \\ R(\mu)S(t) \\ R(\mu)\dot{S}(t) + R(\mu)S^2(t) \end{pmatrix}$$

satisfies $\operatorname{rank} N_e(t,\mu) = 2$, for all $t \in [0, 2\pi)$ and all $\mu : \|\mu\| = 1$. Using Doležal's theorem [17, 1], one can find a periodic nonsingular transformation $P(t) \in \mathbb{R}^{3\times 3}$, which in this case can be chosen independent of μ, such that

$$N_e(t,\mu) = P(t)\begin{pmatrix} \bar{N}_1(\mu) \\ 0 \end{pmatrix}, \quad \bar{N}_1(\mu) = \begin{pmatrix} \mu_1 & \mu_2 \\ -\mu_2 & \mu_1 \end{pmatrix}.$$

Since $\det \bar{N}_1(\mu) = 1$, the inverse of the mapping $\zeta = N_e(t,\mu)\,w$ can be computed as $w = \bar{N}_1^{-1}(\mu)U(t)\zeta$, where $U(t)$ collects the first two rows of $P^{-1}(t)$. Then, one obtains

$$\left[L_S^3 R(\mu)\right] w = \left[L_S^3 R(\mu)\right] \bar{N}_1^{-1}(\mu) U(t)\zeta$$

and, since

$$\left[L_S^3 R(\mu)\right] \bar{N}_1^{-1}(\mu) = \left(-3\cos(t)\sin(t) \;\; -2\sin(t) + \sin(t)\cos^2(t)\right) ,$$

Lemma 2 holds with a different set of functions, namely

$$\begin{aligned} a_0(t) &= \sin(t)\cos(t)\left[\cos^4(t) + 5 - 3\cos^2(t) - 3\sin(t) + 3\sin(t)\cos^2(t)\right] \\ a_1(t) &= \sin(t)\left[3\cos^2(t) + 2\sin(t) - \sin(t)\cos^2(t))\right] \\ a_2(t) &= \sin(t)\cos(t)\left[2 - \cos^2(t) - 3\sin(t)\right] . \end{aligned}$$

△

Finally, the last example concerns an application of regular immersions.

Example 3. Consider an exosystem model with the same dynamics as the one in Example 1, but output map given by

$$R(t,\mu) = \left(\mu_1 + \mu_2\cos(t) \;\; 0\right) , \quad \mu_1^2 + \mu_2^2 = 1 .$$

A simple calculation shows that the pair $(S(\,\cdot\,), R(\,\cdot\,,\mu))$ is completely observable but not uniformly observable. Furthermore, determining whether a strong immersion exists, by using either one of the methods in the previous examples, proves to be computationally intractable, even with the aid of symbolic algebra manipulation tools. However, it is not difficult to see that the two systems with output maps given by $R_1(\mu) = \left(\mu_1 \;\; 0\right)$ and $R_2(t,\mu) = \left(\mu_2\cos(t) \;\; 0\right)$ can be respectively immersed into a 3-dimensional system $(\Phi_1(t), \Gamma_1)$ and a 5-dimensional system $(\Phi_2(t), \Gamma_2)$, both in phase-variable form. Specifically, the pair $(\Phi_1(t), \Gamma_1)$ is the same as the one in Example 1, while for the second one

$$\begin{aligned} &a_0(t) = -21\sin(t)\cos(t) , \quad a_1(t) = 25\cos^2(t) - 8 , \quad a_2(t) = 9\cos(t)\sin(t) \\ &a_3(t) = 7 - \cos^2(t) , \qquad\qquad a_4(t) = 0 . \end{aligned}$$

An immersion for the original system can then be constructed as the parallel interconnection

$$\Phi(t) = \operatorname{diag}\left(\Phi_1(t), \Phi_2(t)\right) , \qquad \Gamma(t) = \left(\Gamma_1 \;\; \Gamma_2\right)$$

which is found to be completely observable, but not uniformly observable.△

3.1 Canonical Parameterizations for Periodic Internal Models

For LTI exosystem models, it has been long recognized that a particular realization of the internal model plays a fundamental role in the solution of a large number of design problems in robust and adaptive regulation [3, 4, 7, 10, 11, 12]. This realization is referred to as a *canonical parametrization* in [12]. A generalization to the periodic case is stated as follows.

Definition 4. *The internal-model pair* $(\Phi(\cdot),\Gamma(\cdot))$ *admits a* canonical parameterization *if there exist a smooth periodic mapping* $M:\mathbb{R}\to\mathbb{R}^{m\times q}$, $m\geq q$, *and a smooth periodic system* $(F_{\rm im}(\cdot),G_{\rm im}(\cdot),H_{\rm im}(\cdot))$ *such that: (i)* $F_{\rm im}(t)\in\mathbb{R}^{m\times m}$ *has all characteristic multipliers in* $|\lambda|<1$, *(ii)* $M(t)$ *has constant rank equal to* q, *and satisfies*

$$\begin{aligned}\dot M(t)+M(t)\Phi(t)&=(F_{\rm im}(t)+G_{\rm im}(t)H_{\rm im}(t))M(t)\\ \Gamma(t)&=H_{\rm im}(t)M(t)\end{aligned}\qquad(20)$$

for all $t\in[0,T)$.

In the LTI case, for a given *observable* (Φ,Γ) the existence of a canonical parametrization is guaranteed for *arbitrary* pairs $(F_{\rm im},G_{\rm im})$ with $F_{\rm im}$ Hurwitz, as long as their controllable subspace has dimension greater or equal than q. This stems from the fact that, under the considered assumptions, the equation

$$M\Phi=F_{\rm im}M+G_{\rm im}\Gamma$$

admits a unique nonsingular solution if $m=q$ and $(F_{\rm im},G_{\rm im})$ is controllable (see [5, Thm. 7-10]). In the time-varying setup, a similar result does not hold for regular immersions, as complete observability of $\big(\Phi(\cdot),\Gamma(\cdot)\big)$ and complete controllability of $(F_{\rm im}(\cdot),G_{\rm im}(\cdot))$ together do not guarantee that the unique periodic solution of the SDE

$$\dot M(t)+M(t)\Phi(t)=F_{\rm im}(t)M(t)+G_{\rm im}(t)\Gamma(t)$$

has constant rank on $[0,T)$. This difficulty can be overcome if one does not insist on looking for arbitrary realizations. For simplicity, we will only consider here the case $m=q$, and defer the case $m>q$ until the next section[1].

Lemma 3. *Let* $\big(\Phi(\cdot),\Gamma(\cdot)\big)$ *be a regular internal-model pair. Set* $F_{\rm im}(t)=-\alpha I-\Phi'(t)$, *where* $\alpha>0$. *Then,* $F_{\rm im}(\cdot)$ *has all characteristic multipliers in* $|\lambda|<1$, *and the unique periodic solution* $\bar M(\cdot)$ *of the SDE*

$$\dot M(t)+M(t)\Phi(t)=F_{\rm im}(t)M(t)+\Gamma'(t)\Gamma(t)$$

is nonsingular for all $t\in[0,T)$.

[1] Note that if $m=q$, the mapping $M(\cdot)$ is necessarily a Lyapunov transformation.

Proof. See [19, Prop. 5.1]. □

Proposition 2. *Let a regular internal-model pair $(\Phi(\cdot), \Gamma(\cdot))$ be given. The choice $F_{\rm im}(\cdot) = -\alpha I - \Phi'(\cdot)$, $G_{\rm im}(\cdot) = \Gamma'(\cdot)$ and $H_{\rm im}(\cdot) = \Gamma(t)\bar{M}^{-1}(\cdot)$, where $\alpha > 0$ is arbitrary, yields a canonical parametrization of $(\Phi(\cdot), \Gamma(\cdot))$. Furthermore, it can be shown that the triplet $(F_{\rm im}(\cdot), G_{\rm im}(\cdot), H_{\rm im}(\cdot))$ is uniformly completely controllable and uniformly completely observable.*

For strong immersions, topological equivalence to the observer canonical form is the key tool in constructing a canonical parametrization.

Proposition 3. *Let a strong internal-model pair $(\Phi(\cdot), \Gamma(\cdot))$ be given. For any constant observable pair $(F_{\rm im}, H_{\rm im})$ with $F_{\rm im} \in \mathbb{R}^{q\times q}$ Hurwitz, there exists a periodic matrix-valued function $G_{\rm im}(\cdot)$ such that $(F_{\rm im}, G_{\rm im}(\cdot), H_{\rm im})$ is a canonical parametrization of $(\Phi(\cdot), \Gamma(\cdot))$.*

Proof. Let $P(\cdot)$ be the unique periodic Lyapunov transformation satisfying

$$\begin{aligned} \dot{P}(t) + P(t)\Phi(t) &= \Phi_{\rm o}(t)P(t) \\ \Gamma(t) &= \Gamma_{\rm o}P(t) \end{aligned}$$

and let the nonsingular matrix $Q \in \mathbb{R}^{q\times q}$ be such that $(QF_{\rm im}Q^{-1}, H_{\rm im}Q^{-1})$ is in observer canonical form. Denote by $\alpha(t) \in \mathbb{R}^q$ the first column of $\Phi_{\rm o}$, and by $B = {\rm col}(b_{q-1}, b_{q-2}, \ldots, b_0)$ the vector of coefficients of the characteristic polynomial $p_F(\lambda) = \lambda^q + b_{q-1}\lambda^{q-1} + \cdots + b_0$ of $F_{\rm im}$. Then, the result holds with $G_{\rm im}(t) := Q^{-1}[\alpha(t) + B]$, and Lyapunov transformation in (20) given by $M(t) = Q^{-1}P(t)$. □

It should be noted that the triplet $(F_{\rm im}, G_{\rm im}(\cdot), H_{\rm im})$ is by construction uniformly observable, but not necessarily uniformly controllable. However, the vector B can be chosen such that $(F_{\rm im}, G_{\rm im}(\cdot))$ is completely controllable.

When a canonical parametrization is available, an internal model of (15) can be conveniently designed as the asymptotically stable system

$$\begin{aligned} \dot{\xi}_1 &= F_{\rm im}(t)\xi_1 + G_{\rm im}(t)u_{\rm im} \\ y_{\rm im} &= H_{\rm im}(t)\xi_1 \,, \end{aligned} \tag{21}$$

under the feedback interconnection $u_{\rm im} = y_{\rm im} + u_{\rm st}$ with a stabilizing controller

$$\begin{aligned} \dot{\xi}_0 &= F_{\rm st}(t)\xi_0 + G_{\rm st}(t)e \\ u_{\rm st} &= H_{\rm st}(t)\xi_0 + K_{\rm st}(t)e \,. \end{aligned}$$

Note that, once we define $\xi = {\rm col}(\xi_0, \xi_1)$, and

$$\begin{aligned} F(t) &= \begin{pmatrix} F_{\rm st}(t) & 0 \\ G_{\rm im}(t)H_{\rm st}(t) & F_{\rm im}(t) + G_{\rm im}(t)H_{\rm im}(t) \end{pmatrix}, & G(t) &= \begin{pmatrix} G_{\rm st}(t) \\ G_{\rm im}(t)K_{\rm st}(t) \end{pmatrix} \\ H(t) &= \begin{pmatrix} H_{\rm st}(t) & H_{\rm im}(t) \end{pmatrix}, & K(t) &= K_{\rm st}(t)\,, \end{aligned}$$

the robust output regulation problem is recast in the framework of Problem 1, whose solution now reposes solely upon the choice of the quadruplet $\big(F_{\rm st}(\,\cdot\,), G_{\rm st}(\,\cdot\,), H_{\rm st}(\,\cdot\,), K_{\rm st}(\,\cdot\,)\big)$ to fulfill condition (i), as equation (14) is solved by $\Xi(t,\mu) = M(t)\Upsilon(t,\mu)$.

3.2 Application to Robust Design for Minimum-Phase Systems

Retaining the structure of the canonical parametrization in the controller greatly simplifies the design of the stabilizer for minimum-phase plant models. Specifically, the class of systems under investigation is characterized as follows.

Assumption 2. *System* (5) *has relative degree* $r(t) = 1$ *(see [6]) and asymptotically stable zero-dynamics, uniformly in* $\mu \in \mathcal{P}$*. Furthermore, the sign of the high-frequency gain* $b(t,\mu) := C(t,\mu)B(t,\mu)$ *is known.*

The assumptions imply that the plant model (5) can be put, by means of a periodic Lyapunov transformation, into the error-system form

$$\begin{aligned} \dot z &= A_{11}(t,\mu)z + A_{12}(t,\mu)e \\ \dot e &= A_{21}(t,\mu)z + a_{22}(t,\mu)e + b(t,\mu)[u - R(t,\mu)w] \end{aligned} \tag{22}$$

where $z \in \mathbb{R}^{n-1}$, $|b(t,\mu)| \geq b_0 > 0$ for all $t \in [0,T)$ and all $\mu \in \mathcal{P}$, and

$$R(t,\mu) = \frac{1}{b(t,\mu)}\,[L_S Q(t,\mu) - A_{21}(t,\mu)\Pi_1(t,\mu) - a_{22}(t,\mu)Q(t,\mu) - P_2(t,\mu)]$$

where $\Pi_1(t,\mu)$ is the unique periodic solution of the SDE

$$\dot\Pi_1(t,\mu) + \Pi_1(t,\mu)S(t) = A_{11}(t,\mu)\Pi_1(t,\mu) + P_1(t,\mu) - A_{12}(t,\mu)Q(t,\mu)\,.$$

Then the following holds.

Proposition 4. *Assume that* $\big(S(\cdot), R(\cdot,\mu)\big)$ *admits a regular immersion* (16)*, with canonical parametrization* $(F_{\rm im}(\,\cdot\,), G_{\rm im}(\,\cdot\,), H_{\rm im}(\,\cdot\,))$ *as in* (20)*. Then, there exists* $k^\star > 0$ *such that for all* $k > k^\star$ *the controller*

$$\begin{aligned} \dot\xi &= \big(F_{\rm im}(t) + G_{\rm im}(t)H_{\rm im}(t)\big)\xi - k\,\mathrm{sign}(b)G_{\rm im}(t)e \\ u &= H_{\rm im}(t)\xi - k\,\mathrm{sign}(b)e \end{aligned}$$

solves the robust periodic output regulation problem (Problem 1).

Proof. The proof follows easily from standard arguments in high-gain feedback design, as it becomes evident once the change of coordinates

$$\chi := \xi - M(t)\Upsilon(t,\mu)w - \frac{1}{b(t,\mu)}G_{\rm im}(t)e$$

is applied [19, Prop. 6.2]. □

An extension to plant models with higher relative degree can be accomplished by resorting to the use of high-gain observers or filtered transformations. Details are omitted, since this extension presents no conceptual difficulty, but an undue complication of the notation.

4 Weak Immersions for Adaptive Robust Regulation

In this section, we turn our attention to the adaptive robust regulation problem (Problem 2), that is, when the parameterized family of exosystem models (6) is considered. The starting point of the analysis is the existence of a family of solutions of the regulator equation (13), parameterized in σ, and the following assumption for the analogous of system (15).

Assumption 3. *The family of exosystem models with output*

$$\begin{aligned} \dot{w} &= S(t,\sigma)w \\ v &= R(t,\sigma,\mu)w \end{aligned} \tag{23}$$

admits a parameterized family of strong internal-model pairs $\big(\Phi(\cdot,\sigma),\Gamma(\cdot,\sigma)\big)$.

Note that a necessary and sufficient condition for the existence of a strong internal-model pair is that Lemma 2 holds with parameter-dependent functions $a_i(\,\cdot\,,\sigma)$, $i=0,\dots,q-1$. For a given family of strong internal-model pairs $\big(\Phi(\,\cdot\,,\sigma),\Gamma(\,\cdot\,,\sigma)\big)$, we look for a particular form of the canonical parametrization that can be exploited for certainty-equivalence design.

Definition 5. *The family of strong internal-model pairs* $\big(\Phi(\,\cdot\,,\sigma),\Gamma(\,\cdot\,,\sigma)\big)$ *admits a canonical parametrization in feedback form if there exist a family of smooth periodic mappings* $M(\,\cdot\,,\theta)\in\mathbb{R}^{m\times q}$, *with* $\theta\in\mathbb{R}^{\varrho}$, $\varrho>s$ *and* $m\geq q$, *and a family of smooth periodic systems* $(F_{\mathrm{im}}(\,\cdot\,),G_{\mathrm{im}}(\,\cdot\,),H_{\mathrm{im}}(\,\cdot\,,\theta))$ *such that: (i)* $F_{\mathrm{im}}(t)\in\mathbb{R}^{m\times m}$ *has all characteristic multipliers in* $|\lambda|<1$, *(ii)* $H_{\mathrm{im}}(t,\theta)=H_{\mathrm{im},1}(t)\theta+H_{\mathrm{im},0}(t)$ *is affine in* θ, *and (iii) there exists a continuous assignment* $\sigma\mapsto\theta_\sigma\in\mathbb{R}^{\varrho}$ *such that* $M(t,\theta_\sigma)$ *has constant rank equal to* q, *and satisfies*

$$\begin{aligned} \dot{M}(t,\theta_\sigma)+M(t,\theta_\sigma)\Phi(t,\sigma) &= (F_{\mathrm{im}}(t)+G_{\mathrm{im}}(t)H_{\mathrm{im}}(t,\theta_\sigma))M(t,\theta_\sigma) \\ \Gamma(t,\sigma) &= H_{\mathrm{im}}(t,\theta_\sigma)M(t,\theta_\sigma) \end{aligned} \tag{24}$$

for all $t\in[0,T)$ *and all* $\sigma\in\Sigma$. *Furthermore, it is said to admit a canonical parametrization in output-injection form if,* mutatis mutandis, *(i)–(iii) above hold for a family of systems of the form* $(F_{\mathrm{im}}(\,\cdot\,),G_{\mathrm{im}}(\,\cdot\,,\theta),H_{\mathrm{im}}(\,\cdot\,))$, *where* $G_{\mathrm{im}}(\,\cdot\,,\theta)$ *is affine in* θ.

It turns out that the existence of a canonical parametrization in feedback form implies the solvability of Problem 2 for the relative degree-one minimum-phase prototype system (22), without any assumption on the persistence of excitation of the exogenous signals. Applying standard arguments, and the periodic version of La Salle's invariance principle, it is indeed possible to prove that the certainty-equivalence adaptive controller

$$\begin{aligned} \dot{\xi} &= \big(F_{\mathrm{im}}(t)+G_{\mathrm{im}}(t)H_{\mathrm{im}}(t,\hat{\theta})\big)\xi-k\,\mathrm{sign}(b)G_{\mathrm{im}}(t)e \\ \dot{\hat{\theta}} &= -\gamma\,H'_{\mathrm{im},1}(t)e \\ u &= H_{\mathrm{im}}(t,\hat{\theta})\xi-k\,\mathrm{sign}(b)e \end{aligned}$$

where $\gamma > 0$ is a gain parameter, solves the adaptive robust output regulation problem for the plant model (22), if the gain k is chosen sufficiently large [20]. Again, the result can be extended to higher relative degree systems, at the expense of a painstaking exercise in backstepping design.

It will be shown that a canonical parametrization in feedback form is guaranteed to exist if the family $\big(\Phi(\,\cdot\,,\sigma),\Gamma(\,\cdot\,,\sigma)\big)$ admits a re-parametrization which is linear in a possibly larger, set of parameters. This allows one to derive a canonical parametrization in output-injection form in a straightforward manner, from which a non-minimal realization yields the requires feedback form. As a result, it is seen the key ingredient in obtaining an internal-model pair that exhibits a structure amenable to certainty-equivalence design is to resorting to a weak immersion rather than a regular one.

To begin with, note that without loss of generality the internal-model pair in Assumption 3 can be taken to be in observer form $\big(\Phi_{\mathrm{o}}(\,\cdot\,,\sigma),\Gamma_{\mathrm{o}}\big)$, with σ-dependent coefficients $\alpha_i(\,\cdot\,,\sigma)$, $i=0,\dots,q-1$ in (19).

Assumption 4. *There exist an integer $\rho \in \mathbb{N}$, a smooth periodic vector-valued function $\beta : \mathbb{R} \to \mathbb{R}^\rho$, and a continuous re-parametrization $\mu \mapsto \theta \in \mathbb{R}^{q\rho}$*

$$\theta' = \begin{pmatrix}\theta'_{q-1} \; \theta'_{q-2} \; \theta'_{q-3} \cdots \theta'_0\end{pmatrix}\,, \qquad \theta_i \in \mathbb{R}^\rho\,, \qquad i=0,\dots,q-1$$

such that $\alpha_i(t,\mu) = \theta'_i\beta(t)\,, \quad i = 0,\dots,q-1\,.$

The assumption allows one to write $\Phi_{\mathrm{o}}(t,\sigma) = \Phi_{\mathrm{b}} - \Theta\beta(t)\Gamma_{\mathrm{o}}$, where the matrix $\Phi_{\mathrm{b}} \in \mathbb{R}^{q\times q}$ is in Brunovsky form, and $\Theta \in \mathbb{R}^{q\times m}$ collects the vectors θ_i. Choose arbitrarily an output-injection gain $L_0 = \begin{pmatrix}l_{q-1} \; l_{q-2} \; \dots \; l_0\end{pmatrix}'$ such that $F := \Phi_{\mathrm{b}} - L_0\Gamma_{\mathrm{o}}$ is Hurwitz, and define $G(t,\theta) := L_0 - \Theta\beta(t)$. Finally, let $H := \Gamma_{\mathrm{o}}$ and note that the triplet $\big(F, G(\,\cdot\,,\theta), H\big)$ is a canonical parametrization in output-injection form of the internal-model pair $\big(\Phi_{\mathrm{o}}(\,\cdot\,,\sigma),\Gamma_{\mathrm{o}}\big)$, as equation (24) holds with $M(\,\cdot\,,\theta_\sigma) = I_q$. Next, we look for a realization of the input/output map of $\big(F, G(\,\cdot\,,\theta), H\big)$ in canonical feedback form. This can not be accomplished by converting the output-feedback form directly into the feedback form by means of a Lyapunov transformation, as this latter is necessarily parameter-dependent. As a matter of fact, it is impossible in general to obtain a *uniform* (that is, minimal) realization of the impulse response of $\big(F, G(\,\cdot\,,\theta), H\big)$ in the desired form $\big(F_{\mathrm{im}}(\,\cdot\,), G_{\mathrm{im}}(\,\cdot\,)), H_{\mathrm{im}}(\,\cdot\,,\theta)\big)$. The fact that F is constant and Hurwitz suggests to look for a *non-minimal* realization instead. Without loss of generality, assume that $\big(F, G(\,\cdot\,,\theta), H\big)$ is a uniform realization of the parameterized family of impulse responses

$$h(t,\tau,\theta) = H\mathrm{e}^{F(t-\tau)}G(\tau,\theta) =: h_0(t-\tau) - h_1(t,\tau,\theta)$$

where $h_0(t-\tau) = H\mathrm{e}^{F(t-\tau)}L_0$ and $h_1(t,\tau,\theta) = H\mathrm{e}^{F(t-\tau)}\Theta\beta(\tau)$. The impulse response $h_0(t-\tau)$ admits the minimal constant realization

$$F_0 = \begin{pmatrix} -l_{q-1} & \cdots & -l_1 & -l_0 \\ 1 & \cdots & 0 & 0 \\ \vdots & \ddots & \vdots & \vdots \\ 0 & \cdots & 1 & 0 \end{pmatrix}, \qquad G_0 = \begin{pmatrix} 1 \\ 0 \\ \vdots \\ 0 \end{pmatrix}, \qquad H_0 = L_0'$$

whereas $h_1(t, \tau, \theta)$ admits a periodic realization of the form

$$F_1 = \begin{pmatrix} -l_{q-1} I_\rho & \cdots & -l_1 I_\rho & -l_0 I_\rho \\ I_\rho & \cdots & 0 & 0 \\ \vdots & \ddots & \vdots & \vdots \\ 0 & \cdots & I_\rho & 0 \end{pmatrix}, \qquad G_1(t) = \begin{pmatrix} \beta(t) \\ 0 \\ \vdots \\ 0 \end{pmatrix}, \qquad H_1(\theta) = \theta' .$$

As a result, the triplet

$$F_{\rm im} = \begin{pmatrix} F_0 & 0 \\ 0 & F_1 \end{pmatrix}, \quad G_{\rm im}(t) = \begin{pmatrix} G_0 \\ G_1(t) \end{pmatrix}, \quad H_{\rm im}(\theta) = \begin{pmatrix} H_0 & -H_1(\theta) \end{pmatrix}$$

is a candidate canonical parametrization in feedback form of the original internal-model pair $(\Phi_{\rm o}(t), \Gamma_{\rm o})$. Note that the dimension of the state-space for $F_{\rm im}$ is $m = q(1+\rho)$. It is left to show that there exists a constant-rank transformation $M(\cdot, \theta)$ yielding (24). Since $\big(F_{\rm im}, G_{\rm im}(\cdot), H_{\rm im}(\theta)\big)$ and $\big(F, G(\cdot, \theta), H\big)$ are realizations of the same impulse response, and the latter one is uniform, Kalman's decomposition theorem [17, 1] applies, and thus there exists a periodic Lyapunov transformation $P(t, \theta) \in \mathbb{R}^{m \times m}$ such that

$$\bar{F}_{\rm im}(t,\theta) := \big[\dot{P}(t,\theta) + P(t,\theta) F_{\rm im}\big] P^{-1}(t,\theta) = \begin{pmatrix} F & 0 & * \\ F_{21}(t,\theta) & F_{22}(t,\theta) & * \\ 0 & 0 & * \end{pmatrix}$$

$$\bar{G}_{\rm im}(t,\theta) := P(t,\theta) G_{\rm im}(t) = \begin{pmatrix} G(t,\theta) \\ G_2(t,\theta) \\ 0 \end{pmatrix}$$

$$\bar{H}_{\rm im}(t,\theta) := H_{\rm im}(\theta) P^{-1}(t,\theta) = \begin{pmatrix} H & 0 & H_3(t,\theta) \end{pmatrix}$$

where the characteristic multipliers of $F_{22}(\cdot, \theta)$ and $F_{33}(\cdot, \theta)$ are in $|\lambda| < 1$. Finally, let $U_2(t, \theta)$ denote the unique periodic solution to the SDE

$$\dot{U}_2(t,\theta) + U_2(t,\theta)\Phi_{\rm o}(t) = F_{22}(t,\theta) U_2(t,\theta) + F_{21}(t,\theta) G_2(t,\theta) H$$

and define $U(t,\theta) = \big(I_q \; U_2'(t,\theta) \; 0\big)'$. Then, it is easy to see that

$$\begin{aligned} \dot{U}(t,\theta) + U(t,\theta)\Phi_{\rm o}(t) &= \big[\bar{F}_{\rm im}(t,\theta) + \bar{G}_{\rm im}(t,\theta)\bar{H}_{\rm im}(t,\theta)\big] U(t,\theta) \\ \Gamma_{\rm o} &= \bar{H}_{\rm im}(t,\theta) U(t,\theta) \end{aligned}$$

and thus the required transformation in (24) is $M(t,\theta) = P^{-1}(t,\theta) U(t,\theta)$.

5 Conclusions

In this paper, we have proposed a classification of the property of system immersion for periodic systems, aimed at underlying the connections between various non-equivalent definitions of systems observability and the existence of robust internal model-based controllers. It has been shown that weak observability properties are related to the possibility of obtaining canonical realizations of internal models that can be used in in certainty-equivalence design to deal with parameter uncertainty on the exosystem model.

References

1. S. Bittanti and P. Bolzern. Stabilizability and detectability of linear periodic systems. *Systems & Control Letters*, 6(2):141–145, 1985.
2. S. Bittanti, P. Colaneri, and G. Guardabassi. H-controllability and observability of linear periodic systems. *SIAM J. Contr. Optimization*, 22(6):889–93, 1984.
3. C.I. Byrnes and A. Isidori. Limit sets, zero dynamics, and internal models in the problem of nonlinear output regulation. *IEEE Trans. on Automat. Contr.*, 48(10):1712–1723, 2003.
4. C.I. Byrnes and A. Isidori. Nonlinear internal models for output regulation. *IEEE Trans. on Automat. Contr.*, 49(12):2244–2247, 2004.
5. C.-T. Chen. *Linear System Theory and Design.* Holt, Rinehart, and Winston, New York, NY, 1984.
6. G. De Nicolao, G. Ferrari-Trecate, and S. Pinzoni. Zeros of continuous-time linear periodic systems. *Automatica*, 34(12):1651–1655, 1998.
7. F. Delli Priscoli. Output regulation with nonlinear internal models. *Systems & Control Letters*, 53(3-4):177–185, 2004.
8. M. Farkas. *Periodic Motions.* Springer Verlag, New York, NY, 1994.
9. B.A. Francis. The linear multivariable regulator problem. *SIAM J. Contr. Optimization*, 15(3):486–505, 1977.
10. J. Huang and Z. Chen. A general framework for tackling the output regulation problem. *IEEE Trans. on Automat. Contr.*, 49(12):2203–2218, 2004.
11. A. Serrani and A. Isidori. Global robust output regulation for a class of nonlinear systems. *Systems & Control Letters*, 39(2):133–139, 2000.
12. A. Serrani, A. Isidori, and L. Marconi. Semi-global nonlinear output regulation with adaptive internal model. *IEEE Trans. on Automat. Contr.*, 46(8):1178–1194, 2001.
13. L. Silverman. Transformation of time-variable systems to canonical (phase-variable) form. *IEEE Trans. on Automat. Contr.*, 11(2):300– 303, 1966.
14. L. Silverman. Synthesis of impulse response matrices by internally stable and passive realizations. *IEEE Trans. Circuit Theory*, 15(3):238– 245, 1968.
15. L.M. Silverman and B.D.O Anderson. Controllability, observability and stability of linear systems. *SIAM J. Contr. Optimization*, 6(1):121–130, 1968.
16. L.M. Silverman and H.E. Meadows. Controllability and observability in time-variable linear systems. *SIAM J. Contr. Optimization*, 5(1):64–73, 1967.
17. L. Weiss and P.L. Falb. Doležal's theorem, linear algebra with continuously parametrized elemens, and time-varying system. *Math. Syst. Theory*, 3:67–75, 1969.

18. Y. Yuksel and J. Bongiorno Jr. Observers for linear multivariable systems with applications. *IEEE Trans. on Automat. Contr.*, 16(6):603–613, 1971.
19. Z. Zhang and A. Serrani. The linear periodic output regulation problem. *Systems & Control Letters*, 55(7):518–529, 2006.
20. Z. Zhang and A. Serrani. Robust regulation with adaptive periodic internal models. *Proc. of the 45rd IEEE Conf. on Decision and Contr.*, 2006.

Paving the Way Towards the Control of Wireless Telecommunication Networks

Francesco Delli Priscoli and Antonio Pietrabissa

Dipartimento di Informatica e Sistemistica "Antonio Ruberti", Università di Roma "La Sapienza", Via Eudossiana 18, 00184 Rome, Italy

Summary. The objective of this chapter is to show how control-based methodologies can be fruitfully used for the development of resource management procedures in communication networks. In particular, we introduce the convergence layer technology independent approach and, within this approach, we describe a model-based design procedure for resource management algorithms. To validate the technology independent approach, the design procedure is then adopted to develop resource management algorithms on different network technologies.

1 Introduction

One of the main challenges of the telecommunication systems which are presently being designed, is an efficient provision of an end-to-end Quality of Service (QoS) tailored to the specific requirements of each connection [15]. In general, a given connection is supported by more than one (wired or wireless) domain[1]. At connection set-up, an *End-to-End QoS Contract* agreed between the user and its operator is "split" (by means of the so-called Bandwidth Broker mechanism which is currently investigated as a solution for QoS support in future IP networking) among the various (wired or wireless) domains supporting the connection. So, for each (wired or wireless) domain supporting a given connection, an intra-domain *QoS Contract* is agreed establishing (i) the characteristics (e.g. in terms of minimum bit rate) of the so-called *Compliant Traffic*, i.e. the traffic, relevant to the connection in question, which has to be *admitted* in the considered domain in whatever traffic condition, (ii) as well as the QoS requirements characterizing such Compliant Traffic (e.g. in terms of maximum delay tolerated by the IP packets and maximum loss probability of the IP packets) within the considered domain, hereinafter referred to as *QoS Constraints*. The above-mentioned splitting is performed in such a way that the respect of the QoS Contracts in the various domains supporting the

[1] By "domain" we mean a subnetwork supported by a specific technology, i.e. by a specific "underlying network" and being handled by a given operator.

connection entails the satisfaction of the End-to-End QoS Contract for the connection in question.

The respect of the QoS Contracts, already a challenging goal in wired domains, is even more challenging in the *wireless* domains in which this goal has to be achieved in conjunction with an efficient exploitation of the available bandwidth which in the wireless domains is a very valuable resource (much more than in the wired ones). As a result, in wireless domains, traffic control strategies can be key factors for respecting the QoS Contracts and, at the same time, efficiently exploiting the available bandwidth.

Two problems have to be coped with: the first is that the current Internet Protocol (IP) only provides best-effort packet delivery service and may consequently be inadequate to allow the respect of the QoS Contracts; in addition, it might make inefficient use of the available bandwidth. The second problem derives from the fact that the various Underlying Networks supporting the domains, in general, avail of different mechanisms, not devoid of remarkable deficiencies, to respect the QoS Contracts. One last aspect to consider is that standardization is well established for the Internet Protocol (IP) layer and for the Underlying Networks in frequent use. Therefore, there is little flexibility in both the IP and the Underlying Network layers for improved IP with wireless access.

In this respect, several recent research projects (e.g. [24]-[12]) are proposing to add a technology-independent layer between the IP layer and the technology dependent Underlying Network layers, hereafter referred to as *Convergence Layer*, which is *transparent* with respect to both the IP layer and the Underlying Network layers, i.e. its insertion between the IP layer and the Underlying Network layers does not modify either the usual IP protocols or the usual Underlying Network protocols (see Fig. 1). The term "underlying" identifies the link-layer transport networks (e.g. Bluetooth, IEEE 802.11, Universal Mobile Telecommunication System (UMTS), etc.) pointing out the fact that such networks have to provide the physical support, over the wired/wireless run, to the "overlying" Convergence Layer and IP layer protocols. The Convergence Layer includes technology independent protocols aiming at recovering the deficiencies of the Underlying Networks. Only those elements of the telecommunication network in which the enhancements in question are actually necessary will be provided with the Convergence Layer. Moreover, in each domain, some (even all) of the functionalities offered by the Convergence Layer can be disabled in case the relevant Underlying Network already provides them in a satisfactory way. Proper technology dependent *drivers* placed between the Convergence Layer and the technology dependent layers will provide for matching the "abstract" technology independent procedures of the former layer with the ones specific of the latter layers. Nevertheless, such drivers will just include interface functionalities, while all intelligent control features will be placed at the Convergence Layer. In light of the above, the overall conceptual layering architecture of the present telecommunication network is shown in Fig. 1 in which, for the sake of simplicity, just four different

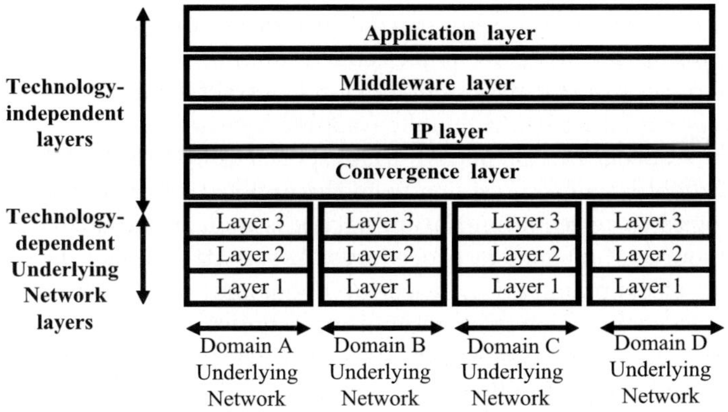

Fig. 1. Overall Layering architecture

domains A,...,D each of them availing of three technology dependent layers are shown.

The advantages of concentrating the intelligent network control functions in the Convergence Layer are quite evident: the same control procedures can be adopted in conjunction with a large variety of Underlying Networks (provided that proper drivers are designed), thus entailing the reuse of the same control algorithms without the need of modifying them for each different Underlying Network (note that a large number of such networks is either already in operation or being designed), or of tailoring them at any upgrade of such Underlying Networks. Moreover, it is evident that the use of the same control algorithms in heterogeneous networks make much easier the interworking among these networks which is one of the most challenging problems of the present global telecommunication network.

It is important remarking that the Convergence Layer is, by definition, a layer including abstract (i.e. technology independent) procedures; in this respect, the designer of this layer has to perform the following steps: (i) to identify the key variables which can describe the network behaviour leaving out of consideration the specific technology, (ii) to model the relationships among these variables in abstract terms, thus identifying the plant, (iii) to seek procedures/algorithms able to lead the plant to achieve the desired performance, thus identifying the plant controller. It is evident that the more research moves towards problem formulation and possible solutions which leave out of consideration the specific technology dependent issues, the more relevance is attributed to approaches and methodologies such as those of system and control theory. In conclusion, Convergence Layer design is very suitable for applying control based methodologies. Highlights of system theory are mathematical modelling of the relevant processes, formulation of management tasks as formal control problems and solving of these problems by appropriate methodologies. Thanks to the Convergence Layer approach, all the above-mentioned highlights can be profitably used in the design of the so-called Resource Management procedures aiming, on the one hand, at assuring the respect of the intra-domain QoS Contracts and, on the other hand, at efficiently exploiting the valuable available bandwidth.

In this respect, in the period 1990–2000, Prof. Alberto Isidori already guessed the potentialities deriving from the application of control theory and methodologies to telecommunication networks and, in particular, to the design of Resource Management procedures. As a matter of fact, he sensed that traditional telecommunication based methodologies (e.g. queue theory), conceived in periods in which telecommunication networks were very far from the current huge complexity, show serious scalability problems; on the contrary, system based modelling and control seem much more suitable "to catch" the complex dynamics of such networks.

The most common Resource Management procedures are Routing, Connection Admission Control (CAC), Dynamic Resource Allocation (DRA), Congestion Control and Scheduling.

- The routing procedure finds the most appropriate (less congested and/or minimum delay,...) path within the considered domain for each connection.
- The CAC procedure decides, at each connection set-up, if the domain can actually afford for the set-up of the new connection with a given QoS Contract. If the connection is accepted by the CAC procedure on a given path decided by the Routing procedure the relevant QoS Contract specifies the Compliant Traffic, as well as the QoS Constraints characterizing such Compliant Traffic.
- The DRA procedure dynamically reallocates, in wireless domains, the available bandwidth trying to match the traffic requests.
- The Congestion Control procedure decides which part of the offered traffic exceeding the Compliant Traffic can be admitted into the domain. In this respect, it should be clear that, in case the domain is idle such procedure tends to admit more traffic than the Compliant one, thus increasing the exploitation of the available bandwidth; conversely, in case the domain is congested such procedure just admits the Compliant Traffic.
- The Scheduling procedure decides the priorities according to which the IP packets have to be transmitted on the available bandwidth, trying to respect the QoS Constraints.

A plenty of the above-mentioned procedures have been designed, implemented and standardized in the various telecommunication domains. In this respect, key problems are the following:

i) most of these procedures are very technology dependent, i.e. they can only work in conjunction with a specific underlying network, being customized on such a specific technology; as a matter of fact, the Convergence Layer approach is very recent and most procedures have been designed without taking into account it;

ii) in general, these procedures are designed so that they work in an uncoordinated fashion one another, in spite of the fact they all aim at pursuing the same goal, namely, respecting the intra-domain QoS Contracts and maximizing bandwidth exploitation; this fact, is due to the huge complexity of keeping these procedures coordinated one another;
iii) in general, these procedures work in an open-loop fashion, being tailored on appropriate network and traffic models, thus entailing an inherent lack of robustness even in consideration of the very rapid evolution of the IP network and traffic characteristics; this is partly due to the fact that only the recent technology enhancements allow a relatively easy monitoring of the feedback variables and hence their utilization in the feedback loop.

The approach introduced by Prof. Isidori in [6], [7] paved the way towards the overcoming of the above-mentioned problems (e.g. see [5]–[10]). As a matter of fact, such a approach is based on the following innovative principles, which cope with the three above-mentioned problems respectively:

i) the modelling of the problem in an abstract, technology independent, control-based fashion and the consequent design of the Resource Management procedures following control theory methodologies. It is evident that this approach is fully compliant with the Convergence Layer approach;
ii) the design of Resource Management procedures trying to jointly play a plurality of roles, thus overcoming the "traditional" role repartitions. In particular, in [6], [7] a Resource Management procedure is designed which includes both the congestion control role, i.e. the role of deciding which IP packets have to be admitted into the considered domain, and the scheduling role, i.e. the role of deciding which IP packets, in the set of the admitted IP ones, have to be served with priority (i.e. have to be actually transmitted). In this respect, the novelty consists in the fact that the controller jointly holds both the congestion control and the scheduling roles. It should be noted that, in most of the previous proposals, these two roles have been dealt with independently each other (e.g. see [13]–[30] for the congestion control role and [19]–[20] for the scheduling role). Note that the fact of simultaneously handling more roles is possible thanks to the relative simplicity with which the domain can be modelled following a control based approach just catching its dominant dynamics;
iii) the handling of the congestion control role in a closed-loop fashion, since the decisions about the admission/rejection of traffic into the wireless domain are based on the present buffer states. In this respect, in the literature [13], the usual way of coping with the congestion control problem was the one of filtering the incoming traffic by means of the so-called Dual Leaky Buckets (DLBs) [13] whose parameters are chosen at connection set-up, i.e. the DLBs operate according to an open-loop approach.

The papers [6], [7] by adopting these principles, present a novel approach to the design of a feedback controller achieving desirable performance with respect to a cost index accounting for the total traffic to be discarded in a considered time period, while respecting the QoS Constraints. Such a novel approach is based on the idea that the controller has to steer the overall system towards "ideal" equilibria at which the most desirable performance is achieved. These ideal equilibria have to be periodically updated since they depend on the present offered traffic which is regarded as the exogenous (not controllable) input.

The following of this chapter will present one of the several results which have been obtained by following the above-mentioned principles formerly introduced by Prof. Isidori. In particular, a model-based design procedure for resource management algorithms in communication networks will be proposed (Section 2) and applied to the problems of Dynamic Resource Allocation in satellite networks (Section 3) and Congestion Control in terrestrial packet networks (Section 4).

2 Model-Based Control of Communication Networks

Model-based control uses the system model to compute the control law, while feedback messages are used to evaluate and correct the model inaccuracies. These methods are widely used in industrial process control, since they address its key problems ([4]): (i) the presence of unknown disturbances; (ii) the unavailability of accurate values of the model parameters; (iii) the presence of constraints on manipulated and controlled variables; (iv) the presence of time-delays. The resource management of communication networks suffers from the same problems listed above: i) hundreds of different network elements (PCs, routers, links, ...), each one running different algorithms and performing different tasks, and their interactions constitute unknown disturbances which cannot be controlled by a single network element; ii) for the same reason, it is impossible to develop an accurate model of the whole telecommunication network and to properly set the parameters, unless we model only the simple constituting elements: buffers and links; iii) manipulated and controlled variables are subject to saturation constraints: e.g., queue lengths and packet transmission rates are non-negative, queue lengths cannot exceed the buffer sizes, and so on; iv) links, algorithm runtimes and queues introduce significant time-delays.

Control-theoretic methods for communication networks are receiving increased attention in the control theory community, as confirmed by several recent special issues of control-theory journals ([3]-[11]). The Congestion Control problem, aimed at controlling the source transmission rate based on the available capacity, has been widely investigated: in [2], the problem is formulated as a stochastic control problem where the controls of different users are subject to different delays; in [28] and [25], an H^∞ controller is designed guaranteeing robust stability with respect to uncertain multiple time-varying delays; in [29], the problem is formulated as a robust tracking control problem. Model-based Congestion Control is examined in [21]-[22]. DRA protocols, aimed at defining the mechanisms required to obtain transmission capacity, are developed in [1], [8], [9], [26]. In [1], the global resource-sharing problem is formulated as an optimization problem involving all terminals and links with a large number of coupled constraints. In [8], a different approach is followed, since the focus is to decouple the Congestion Control algorithm and the DRA. In [9], the scheme of [8] is refined to reduce the queuing delays. With an approach similar to [8], in [26] each terminal uses a local adaptive predictor to forecast the future traffic flow along with a local predictive controller to generate the bandwidth requests.

This section presents the proposed control structure for the development of resource management procedures in communication networks which differ in topology and access technique (*technology independent* approach). The fundamental elements of all telecommunication networks are always the same: buffers and links. The resulting network models are linear, time-invariant (LTI) systems with feedback delays, constituted by integrators, modelling the buffers, feedback delays, modelling the links and the queuing delays, and disturbances, modelling the unknown system behaviour. The idea is that, since these elements are present in any network, a control scheme developed for a certain network can be easily adapted to a different one. Resource management procedures are generally executed by a single element of the network, e.g., by a PC, a router or a satellite terminal. Thus, we are interested in developing the model of the network from the viewpoint of a controller $G(s)$ located in the element itself. We assume that the transfer function of $P(s)$ between the control variable $u(s)$ and the measured variable $q(s)$, representing the system plant, is LTI. For the sake of simplicity, the single-input single-output (SISO) case is considered; the procedure can be extended to the multiple-input multiple-output (MIMO) case by applying the IMC theory for MIMO systems.

2.1 Proposed Control Scheme

The proposed control structure is based on Model Reference Control (MRC), Internal Model Control (IMC) and on the Smith's principle. Given a LTI plant $P(s)$, the idea of MRC is to design a LTI model, named reference model and described by the transfer function $W_{REF}(s)$, representing the desired input-output properties of the closed-loop plant ([16]). For any input $r(s)$, the reference model produces a reference output $q_{REF}(s) = W_{REF}(s)r(s)$, while the actual closed-loop plant produces the output $q(s)$. The objective is to develop a feedback controller $G(s)$ such that $q(s)$ tracks $q_{REF}(s)$. The approach suggested in this section is to design the reference model by utilizing the nominal model of the plant and an appropriate LTI controller, referred to as reference controller and indicated with $C(s)$: let $\tilde{P}(s)$ be the nominal plant model; then, $C(s)$ is a controller producing the control action $u_{REF}(s)$,

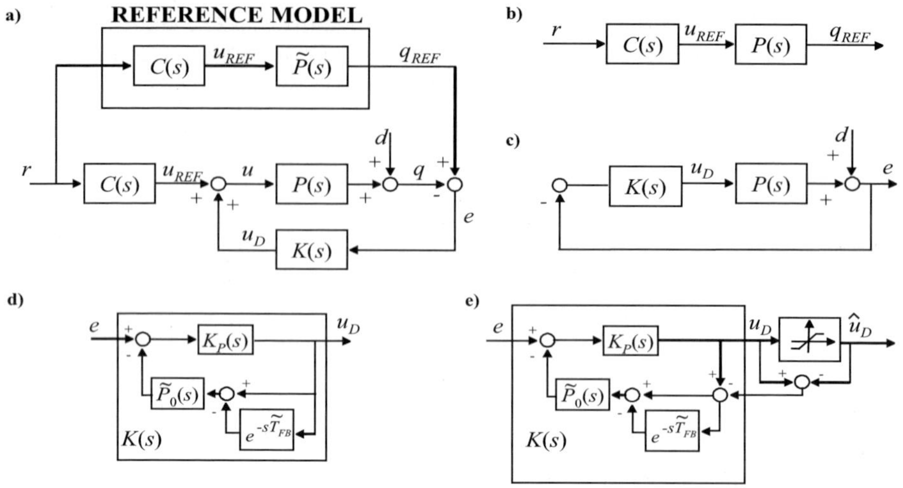

Fig. 2. a) proposed control structure; b) reference system (equivalent system if $d(t) \equiv 0$); c) error system (equivalent system if $d(t) \equiv 0$); d) Smith predictor error controller; e) Smith predictor error controller with anti-windup

which, applied to $\tilde{P}(s)$, achieves the desired output $q_{REF}(s)$ (see the reference model block of Fig. 2a).

Then, IMC is applied; it states that, when no disturbance is present, the closed-loop system behaves as the open-loop one ([23]). Let the reference controller $C(s)$ operate on both the nominal plant model $\tilde{P}(s)$ and on the process plant $P(s)$, as shown in Fig. 2a, and let $d(s)$ model the effect of the disturbances on the output $q(s)$. For a given input $r(s)$, if the nominal plant model is perfect, i.e., $\tilde{P}(s) = P(s)$, and if there are no disturbances, i.e., $d(s) \equiv 0$, it follows that $q(s) \equiv q_{REF}(s)$. The error $e(s) = q(s) - q_{REF}(s)$ expresses the un-modelled effects of $d(s)$ and of the process uncertainties and is used to adjust the control action $u_{REF}(s)$, producing the final control action $u(s)$. As shown in Fig. 2a, the control variable $u(s)$ feeding the actual plant is equal to the sum of $u_{REF}(s)$ and $u_D(s)$, computed by the controller $K(s)$ based on $e(s)$. $K(s)$ is referred to as *error controller.*

The following Theorem 1 straightforwardly derives from the proposed control structure of Fig. 2a.

Theorem 1. *If the model is perfect, i.e.,* $\tilde{P}(s) = P(s)$*, given that the system is LTI, it follows that: (i) the transfer functions* $q_{REF}(s)/r(s)$ *and* $u_{REF}(s)/r(s)$ *of the schemes of Fig. 2a and Fig. 2b are equivalent; (ii) the transfer functions* $e(s)/d(s)$ *and* $u_D(s)/d(s)$ *of the schemes of Fig. 2a and Fig. 2c are equivalent.*

The scheme of Fig. 2b, which is equal to the reference model, is named reference system and depends on $r(s)$ only; the scheme of Fig. 2c is named error system and depends on $d(s)$ only.

Remark 1. If the controller $K(s)$ is such that $e(s)$ in the scheme of Fig. 2c is driven to 0, then the actual output $q(s)$ in the scheme of Fig. 2a tracks $q_{REF}(s)$. ◁

Theorem 2. *Assuming that the process model is perfect, i.e.,* $\tilde{P}(s) \equiv P(s)$*, the control system of Fig. 2a is stable iff the reference and error controllers,* $C(s)$ *and* $K(s)$*, are such that* $P(s)C(s)$ *and* $1/P(s)K(s)$ *are stable.*

Proof. From Theorem 1 it follows that the stability conditions of the overall system (Fig. 2a) can be retrieved by separately examining the transfer functions of the reference system (Fig. 2b), which is $P(s)C(s)$, and of the error system (Fig. 2c), which is $1/P(s)K(s)$. □

Due to the presence of time-delays, the error controller $K(s)$ has to be designed with feedback delay compensation techniques, available within the model-based control framework. The Smith's principle ([26]) has been already adopted (besides process control applications) in control of communication networks ([21], [9]). By applying the standard Smith predictor controller to the error system, the following controller $K(s)$ (shown in Fig. 2d) is obtained:

$$K(s) = \frac{K_P(s)}{1 + K_P(s)\tilde{P}_0(s)(1 - e^{-s\tilde{T}_{FB}})} \tag{1}$$

where $K_P(s)$ is the so-called primary controller, is the estimated delay and $\tilde{T}_{FB}$ is the delay-free part of the process model $\tilde{P}(s) = \tilde{P}_0(s)e^{-s\tilde{T}_{FB}}$.

2.2 Robustness with Respect to Time-Varying Delays and Anti-Windup

Finally, two problems must be addressed: i) the presence of time-varying delays in many communication control problems; ii) the windup problem, which arises from the facts that network models have integrators modelling buffers, and that the proposed model-based control explicitly uses the network models in the controller. For both the identified problems, the proposed framework can exploit the favorable characteristics of IMC.

As analyzed in [23], to achieve robust stability and performance, it is often possible to de-tune the primary controller of the IMC controller. With reference to the error system control scheme of Fig. 2c, a sufficient condition for robust stability is given by the following Theorem 3.

Theorem 3. *([23]) Assuming that (i)* $K(s)$ *stabilizes* $\tilde{P}(s)$*, and that (ii) the number of unstable poles of the process model* $\tilde{P}(s)$ *is equal to the number of unstable poles of actual process* $P(s)$*; then, a sufficient condition for robust stability of the closed-loop control system of Fig. 2c is the following:*

$$\|W_m(j\omega)\tilde{T}(j\omega)\|_\infty = \sup_\omega |W_m(j\omega)\tilde{T}(j\omega)| < 1 \tag{2}$$

where $W_m(j\omega)$ *is the upper-bound of the multiplicative uncertainty function and* $\tilde{T}(j\omega)$ *is the complementary sensitivity function of the system model.*

Note that equation (2) defines a conservative stability bound, since it is based on the uncertainty upper-bound $W_m(j\omega)$. A general procedure to achieve robust stability and robust performance within the IMC framework is given in [23].

IMC properties are useful also when dealing with anti-windup: the well-known IMC anti-windup scheme (see [31], [18]) proves to be adequate in most cases. Basically, the IMC anti-windup scheme states that the actuator and process non-linearity have to be included in the model ([4]), as exemplified in Fig. 2e for the case of actuator saturation. This way the IMC properties still hold.

2.3 Design Procedure

In conclusion, once an appropriate LTI network model is developed, the controller $G(s)$ is determined by performing the following two steps:

1. The first step consists in the design of the reference controller $C(s)$ and of the error controller $K(s)$ by considering the nominal models, without uncertainties and saturation non-linearity. In particular:
   1a Neglecting the un-modelled disturbances, i.e., considering the reference system of Fig. 2b, the reference controller $C(s)$ is designed with the aim of obtaining the desired output $q_{REF}(s)$ based on an optimal criterion.
   1b Considering the disturbance only, i.e., considering the error system of Fig. 2c, the error controller $K(s)$ is designed aimed at driving the output error $e(s)$ to zero (so that $q(s)$ tracks $q_{REF}(s)$, see Remark 1). The Smith's principle (Fig. 2d) is used to compensate the feedback delay.
2. The second step takes into account the uncertainties and saturation non-linearity. In particular:
   2a Robustness with respect to model uncertainties is dealt with by appropriately de-tuning the controller, as described in [23].
   2b IMC anti-windup schemes are added to overcome windup problems caused by saturations.

A key feature of the procedure is that the two controllers, $C(s)$ and $K(s)$, are uncoupled, have different functionality and are developed independently.

3 Dynamic Resource Allocation (DRA) Algorithm for Satellite Networks

In this section, a DRA procedure for satellite networks is developed by using the proposed design procedure. DRA procedures are required when the transmission media is shared, as in satellite networks, and defines the rules by which the terminals request the transmission capacity to the network control centre ([8]). The traffic source, which sends packets to the terminal, is modelled by a non-negative input bit rate $r_{IN}(t)$ with

$$r_{IN}(t) > 0 \qquad \forall t \,. \tag{3}$$

Since it can be measured by the terminal, $r_{IN}(t)$ is a measured disturbance. The buffer in the source terminal, which collects the packets waiting for transmission, is modelled by an integrator. Let $q(t)$ denote the queue length in this buffer; the variation of the queue length is given by the input rate $r_{IN}(t)$ minus the output rate $r_{OUT}(t)$, namely

$$\dot{q}(t) = r_{IN}(t) - r_{OUT}(t) \,. \tag{4}$$

Based on the available measures $r_{IN}(t)$ and $q(t)$, the controller $G(s)$ in the source terminal computes the capacity requests $r_{REQ}(t)$. The request arrives in the network control centre, which computes the capacity allocation based on the available link capacity. The allocation is sent back to the terminal. In geostationary satellite networks, the time interval between the transmission of a capacity request and the receiving of the associated capacity allocation is constant and equal to 500 ms; this interval constitutes the feedback delay T_{FB}. If the network is not congested, the assigned bit rate is equal to the requested bit rate: $r_{OUT}(t) = r_{REQ}(t - T_{FB})$; conversely, if the network is congested, the control centre assigns less capacity: $r_{OUT}(t) < r_{REQ}(t - T_{FB})$. Thus, the network control centre and the transmission delay can be modelled as a delay block cascaded to an additive disturbance $d(t)$, defined as follows

$$d(t) = r_{REQ}(t - T_{FB}) - r_{OUT}(t) \,. \tag{5}$$

Since, as above-mentioned, $r_{REQ}(t - T_{FB}) r_{OUT}(t)$, the following holds true

$$d(t) \geq 0 \qquad \forall t \,. \tag{6}$$

The process model is considered as perfect: $\tilde{P}(s) = P(s) = \frac{1}{s} e^{-sT_{FB}}$; the delay-free process model is $\tilde{P}_0(s) = P(s) = \frac{1}{s}$.

The objective of the DRA procedures is twofold.

i) The terminals must use the whole requested capacity (full link utilization). Thus a proper amount of packets must be accumulated in the buffer: if the buffer is empty when the allocation is received, the unused allocated capacity is wasted. In the model, this corresponds to having $q(t) \geq 0 \, \forall t$.
ii) To reduce the queuing delay, $q(t)$ must be minimized while respecting (i).
iii) In case of congestion, $q(t)$ grows regardless of the controller; the normal behaviour must be recovered at congestion end (*congestion recovery*).

Fig. 3 shows the model and the controller developed with the proposed procedure. The first step is the development of the reference controller $C(t)$ neglecting the disturbance, i.e., setting $d(t) \equiv 0$. The measured disturbance

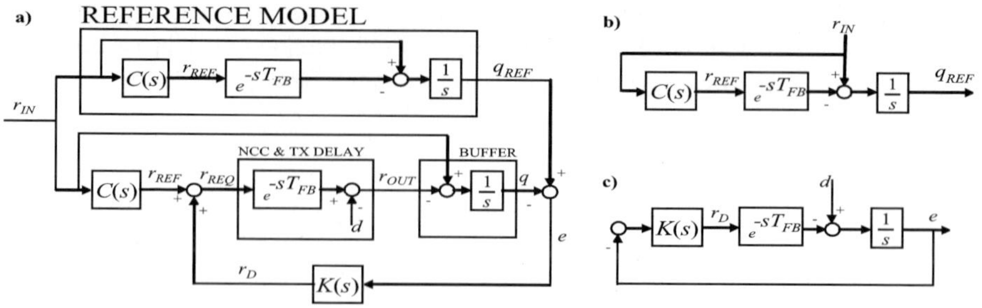

Fig. 3. a) proposed control system; b) reference system; c) error system

$r_{IN}(s)$ is not neglected (Fig. 2b). The following Lemma demonstrates that the controller $C(t) \equiv 1$ meets the control objective (i) (the proof can be found in [9]).

Lemma 1. *([9]) By setting $d(t) \equiv 0$, the proportional controller $C(t) \equiv 1$ is such that $q_{REF}(t) \geq 0 \; \forall t, \forall r_{IN}(t)$ (objective (i)). Moreover, the obtained $q_{REF}(t)$ is the minimum one guaranteeing that $q_{REF}(t) \geq 0 \; \forall t$ (objective (ii)).*

The second step is the development of the Smith predictor controller $K(s)$ for the error scheme obtained by considering $r_{IN}(t) \equiv 0$ (Fig. 3c)

$$K(s) = \frac{K_P(s)}{1 + K_P(s)(P_0(s) - P(s))} = \frac{K}{1 + K(1 - e^{-sT_{FB}})/s} \tag{7}$$

where the primary controller $K_P(s) = K$ is proportional. The controller (7) is such that the error $e(t)$ is driven to 0 when $d(t)$ is 0, as shown by Lemma 2.

Lemma 2. *By setting $K > 0$, the error controller (7) is such that, if at a time t_C a congestion terminates (i.e.,$d(t) = 0 \; t > t_C$), $e(t)$ is exponentially driven to 0, with time constant $\tau = 1/K$. If $d(t) \geq 0 \; \forall t$, it follows that $e(t) \geq 0 \quad \forall t$.*

Proof. The transfer function between $e(s)$ and $d(s)$ is the following:

$$\frac{e(s)}{d(s)} = \frac{1 - e^{-sT_{FB}}}{s} + \frac{e^{-sT_{FB}}}{s + K} \tag{8}$$

which, considering that $K > 0$, proves the first part of the Lemma. The inverse Laplace transform of (8) is:

$$e(t) = \int_{t-T_{FB}}^{t} d(\tau)d\tau + \int_{0}^{t-T_{FB}} e^{-K(t-T_{FB}-\tau)}d(\tau)d\tau \tag{9}$$

which, since $K > 0$ and $d(t) \geq 0 \, \forall t$, proves the second part of the Lemma. □

The overall scheme of Fig. 3a is obtained by applying the reference and the error controllers $C(s)$ and $K(s)$. The following control action is determined:

$$r_{REQ}(t) = R_{IN}(t) + K\left(q(t) - \int_{t-T_{FB}}^{t} r_{REQ}(\tau)d\tau\right) . \tag{10}$$

Theorem 4 follows from Theorem 1, Lemma 1 and Lemma 2.

Theorem 4. *By setting $K > 0$, since $d(t) \geq 0 \, \forall t$ (see equation (6)), the control action (10) meets the control objectives:*

i) $q(t) \geq 0 \, \forall t, \forall r_{IN}(t)$ *(*full link utilization efficiency*).*
ii) If $d(t) \equiv 0$ (congestion-less case), $q(t)$ is the minimum guaranteeing (i).
*iii) If at time t_C a congestion terminates, i.e., $d(t) = 0 \; \forall t > t_C$, $q(t)$ is driven to $q_{REF}(t)$ exponentially, with time-constant $\tau = 1/K$ (*congestion recovery*).*

Remark 2. The control action (10) is the same control action obtained in [9]. The control law (10) has been selected for the implementation in the hardware demonstrator of a satellite network [27], developed by the SATIP6 project [17], financed by the European Union within the 5th Framework Programme. ◁

4 Congestion Control for Terrestrial Networks

In this section, a Congestion Control procedure for terrestrial packet networks is developed by using the proposed design procedure. Congestion control procedures define the rules by which the source terminals adjust their transmission rates to avoid congestions of the network node buffers. In Asynchronous Transfer Mode (ATM) end-to-end Congestion Control protocols, the source terminal exchanges control messages with the destination terminal. The flow of data packets between the source and the destination, referred to as connection, crosses a certain number of network nodes and is characterized by a maximum transmission rate, the Peak Cell Rate (PCR). The PCR is communicated to the path nodes during setup. Control packets are transmitted by the source to the destination, which transmits them back to the source, communicating the maximum queue length q(t) among the buffers of the path nodes (*bottle-neck queue length*). Under the assumption that control packets have strict priority over data packets, the transmission delay of the control messages from the bottle-neck node to the destination and back to the source, denoted with T_{BW}, is constant. The controller $G(t)$ in the source computes the transmission rate $r_S(t)$ based on the PCR of the connection and on the bottle-neck queue length $q(t)$.

The rate $r_S(t)$ is non-negative and limited by the PCR of the connection, namely

$$0 \leq r_S(t) \leq PCR \qquad \forall t\,. \tag{11}$$

Assuming that each connection has a reserved buffer in each crossed node (*per-connection buffering*), if no congestion occurs, all the queue lengths of the buffers in the path are equal to 0; conversely, if a node is congested (*bottle-neck node*), its queue length grows, whereas the queue length in the other nodes is still equal to 0. Two consequences follow:

i) The transmission delay T_{FW} from the source to the bottle-neck node is constant; thus, the feedback delay $T_{FB} = T_{FW} + T_{BW}$ of the system, measured during connection setup, is constant;
ii) Only the buffer of the bottle-neck node has to be modelled ([21]). The bottle-neck buffer is modelled by an integrator; the variation of the queue length is given by the input rate $r_{IN}(t)$ minus the depletion rate of the bottle-neck node buffer $r_{OUT}(t)$, that is

$$\dot{q} = r_{IN}(t) - r_{OUT}(t) \tag{12}$$

$r_{IN}(t)$ is the rate of the packets arriving from the source given by

$$r_{IN}(t) = r_S(t - T_{FW})\,. \tag{13}$$

Assuming that the connection starts at $t = 0$, the rate available in the bottle-neck node $r_{OUT}(t)$ is equal to or less then PCR for $t \geq T_{FW}$ (at time $t = T_{FW}$ the first packet of the connection is received). Then, the reduction of the available rate due to concurrent traffic is modelled by an additive disturbance, $d(t)$, defined as follows

$$d(t) = PCRu_{-1}(t - T_{FW}) - r_{OUT}(t)\,. \tag{14}$$

Since $d(t)$ is the portion of transmission rate which is unavailable, it follows that

$$0 < d(t) < PCR \qquad \forall t\,. \tag{15}$$

$d(t)$ is regarded as an unmeasured disturbance, whereas $PCR\,u_{-1}(t - T_{FW})$ is regarded as a known disturbance. Under the final assumption that the sources are persistent, i.e., that they can always transmit at full rate (PCR), the objective of the Congestion Control procedures is twofold.

i) $q(t)$ must be kept lower than the buffer size S (*overflow avoidance*).
ii) When no congestion is occurring, the source rate $r_S(t)$ must be equal to PCR (*congestion recovery*).

The control scheme of Fig. 4 a) results from the application of the control structure proposed in Section 2.

The first step of the design procedure is the development of the reference controller $C(t)$, neglecting the unmeasured disturbances. The reference system, obtained with $d(t) \equiv 0$, is shown in Fig. 4b. The following

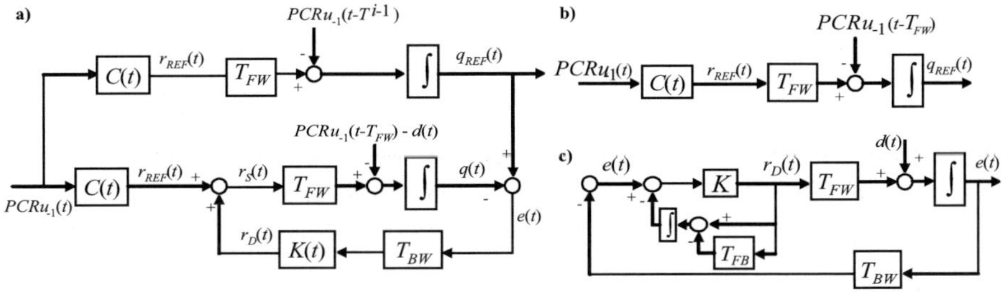

Fig. 4. Congestion control: a) model and control structure; b) reference system; c) error system

Lemma demonstrates that the controller $C(t) \equiv 1$ meets the control objectives.

Lemma 3. *The reference controller* $C(t) \equiv 1$ *achieves the minimum* $q_{REF}(t)$ *guaranteeing the* full link utilization *efficiency at any time t (i.e.,* $q_{REF}(t) \geq 0 \; \forall t$*). Furthermore, the reference rate* $r_{REF}(t)$ *is equal to* $PCRu_{-1}(t)$*.*

Proof. The obtained reference queue length is always equal to 0, which proves the first part of the Lemma. The second part is evident from Fig. 4b. □

The second step of the design procedure is the development of the error controller $K(t)$ neglecting the the measured disturbance. The Smith predictor controller with a proportional primary controller, whose transfer function is given by equation (7), is selected (see Fig. 4c). The following Lemma holds (the proof is similar to the proof of Lemma 2).

Lemma 4. *By setting* $K > 0$ *and assuming* $0 \leq d(t) \leq PCR \; \forall t$*, the rate* $r_D(t)$ *computed by* $K(s)$ *is such that:*

i) $0 < e(t) < PCR(T_{FB} + 1/K)$*;*
ii) if $d(t) = 0 \; \forall t > t_C$*,* $e(t)$ *is exponentially driven to* 0*, with time-constant* $\tau = 1/K$*.*

Theorem 5 follows from Theorem 1, Lemma 3 and Lemma 4.

Theorem 5. *By setting* $K > 0$ *and the buffer size* $S > PCR(T_{FB} + 1/K)$*, the control action (10) is such that:*

i) $0 \leq q(t) \leq S$ *(*full link utilization efficiency *and* overflow avoidance*);*
ii) if at time t_C *a congestion terminates, i.e.,* $d(t) = 0 \; \forall t > t_C$*,* $q(t)$ *is driven to* $q_{REF}(t) \equiv 0$ *exponentially, with time-constant* $\tau = 1/K$ *(*congestion recovery*).*

Remark 3. By applying the *reference* and the *error controllers* to the scheme of Fig. 4a, the following control action is determined

$$\begin{aligned} r_S(t) &= r_{REF}(t) + r_D(t) = r_{REF}(t) + K\left[-q(t) + \int_{t-T_{FB}}^{t} r_D(\tau)d\tau\right] \\ &= K\left\{\frac{r_{REF}}{K} - q(t) + \int_{t-T_{FB}}^{t} [r_S(\tau) - r_{REF}(\tau)]d\tau\right\} = \\ &= K\left\{PCR\left[\frac{u_{-1}}{K} + \int_{t-T_{FB}}^{t} u_{-1}(\tau)d\tau\right] - q(t) + \int_{t-T_{FB}}^{t} r_S(\tau)d\tau\right\}. \end{aligned} \tag{16}$$

If the buffer size S is equal to $PCR(T_{FB}+1/K)$, for $t > T_{FB}$ the control action (16) becomes equivalent to the control action of [21]. ◁

Within the proposed framework, the Congestion Control scheme can be extended to the case of a time-varying delay. We assume that the delay is the only process uncertainty:

$$\tilde{P}(j\omega) = \tilde{P}_0(j\omega)e^{-j\omega\tilde{T}_{FB}} = P_0(j\omega)e^{-j\omega\tilde{T}_{FB}}. \tag{17}$$

Theorem 6 follows from Theorem 3.

Theorem 6. *By using the delay estimate $\tilde{T}_{FB}$ in the Smith predictor controller $K(s)$ (as in Fig. 3d), and by setting the primary controller $K = \lambda/\delta$, where $\lambda \in (0, 1.45]$ is the* detuning parameter *and $\delta = \max_t\{T_{FB}(t) - \tilde{T}_{FB}\}$ is the maximum time-delay uncertainty, the system of Fig 4a is robustly stable.*

Proof. By considering Theorem 3 and the *error system* of Fig. 4c, given that $P_0(s) = \tilde{P}_0(s)$, $\tilde{T}(j\omega)$ and $W(j\omega)$ are computed as follows:

$$\begin{aligned} \tilde{T}(j\omega) = T(j\omega) &= \frac{C_0(j\omega)P_0(j\omega)}{1 + C_0(j\omega)P_0(j\omega)} = \frac{\lambda/\delta}{(j\omega + \lambda/\delta)} \\ W(j\omega) &= \frac{P(j\omega) - \tilde{P}(j\omega)}{\tilde{P}(j\omega)} = e^{-j\omega\delta} - 1. \end{aligned} \tag{18}$$

The following upper bound of $W(j\omega)$ is straightforwardly determined

$$W_m(j\omega) = \begin{cases} |e^{-j\omega\delta} - 1| \text{ if } \omega < \pi/\delta \\ 2 \text{ if } \omega \geq \pi/\delta. \end{cases} \tag{19}$$

Since $W_m(j\omega)$ is a high-pass filter and $\tilde{T}(j\omega)$ is a low-pass filter with pole $p = -\lambda/\delta$, there exists a value of λ such that the sufficient condition (2) is met. In the considered case, the maximum value (numerically computed) is $\lambda = 1.45$. □

Remark 4. Even if the proposed controller is more conservative with respect to controllers developed with other approaches, it has a great advantage in its simplicity, (which is a key characteristic for the sake of implementation) allowing a straightforward extension to an adaptive robust Congestion Control scheme: by estimating on-line the delay and of the delay uncertainty - robust estimators for time-delay systems are available in the literature ([14]) -, it is

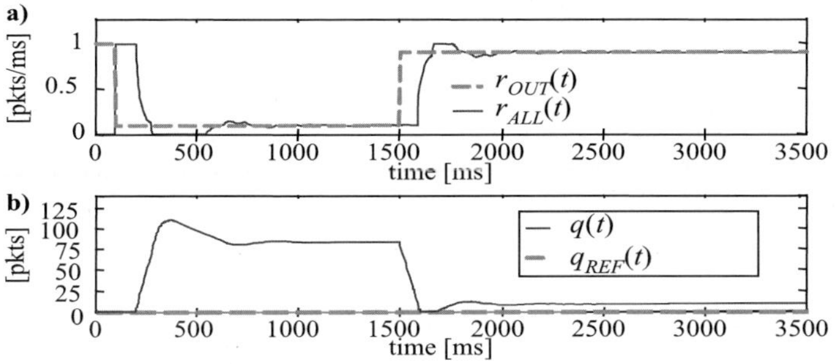

Fig. 5. Robust Congestion Control scheme with anti-windup: a) available rate $r_{OUT}(t) = PCR\, u_{-1}(t) - d(t)$ and rate allocation $r_S(t)$; b) queue length $q(t)$

sufficient to set the controller gain according to Theorem 6. For comparison purposes, let us consider the H^∞ approach of [28]: in this case, given the estimates, the computation of the controller (to be executed every time the estimates are updated) would imply to solve an optimization problem. $\triangleleft$

Finally, we consider the anti-windup scheme[2] of Fig. 2e. The actuator constraint (11) states that $0 \leq r_S(t) \leq PCR$; in the IMC theory, it is sufficient to limit $r_D(t)$ between $-PCR$ and 0, since $r_S(t) = PCR\, u_{-1}(t) + r_D(t)$.

Fig. 5 shows the results of a numerical simulation, with the following parameters: $PCR = 1\, pkt/ms$, $T_{FB} = 100\, ms$, $\tilde{T}_{FB} = 70\, ms$, $\delta = 30\, ms$, $d(t) = cu_{-1}(t - 100\, ms) - du_{-1}(t - 1500\, ms)]$, $c = 0.9\, pkts/ms$, $d = 0.8\, pkts/ms$, $\lambda = 1.35$. The figure shows that the control signal $r_{ALL}(t)$ is always between 0 and PCR and tracks the bottle-neck rate $r_{OUT}(t) = PCR\, u_{-1}(t) - d(t)$, and that the queue length is stabilized.

5 Conclusions

According to the most recent trends in telecommunications, control-based methodologies are particularly suited to deal with resource management procedures. Following this trend, we introduced the convergence layer *technology independent* approach and, within this approach, described a model-based design procedure for resource management algorithms. The key point of the proposed procedure is that it is based on a generic network model, i.e., it is technology independent; in fact, the examples shown in this work are relevant to different network technologies. In particular, the Dynamic Resource Allocation (DRA) protocol for satellite network of [9] and the Congestion Control for packet networks of [21] were obtained by applying the proposed procedure with the objectives of *full link utilization* and *congestion avoidance*,

[2] We will consider the actuator saturation only; process saturation (i.e., buffer over- and under-flows) are easily dealt with as in [9].

respectively. Moreover, the schemes were extended to the case of variable time-delays. In [10], the presented procedure was successfully applied also to the *ad hoc* Wireless LAN scenario to develop a DRA scheme integrated with a Congestion Control scheme which, at the same time, meets the objectives of *full link utilization* and *congestion avoidance*. The on-going work aims at providing QoS guarantees to the controlled traffic (in terms of maximum delay and losses) and at developing a robust adaptive framework for variable-delay scenarios.

References

1. G. Açar and C. Rosenberg. Weighted fair bandwidth-on-demand (WFBoD) for geostationary satellite networks with on-board processing. *Computer Networks*, 39(1):5–20, 2002.
2. E. Altman, T. Başar, and R. Srikant. Congestion control as a stochastic control problem with action delays. *Automatica*, 35(12):1937–1950, 1999.
3. V. Anantharam and J. Walrand. Special issue on control methods for communication. *Automatica*, 35(12), 1999.
4. R.D. Braatz. *Internal model control.* Control Handbook, W.S. Levine Ed. CRC Press, 1995.
5. C. Bruni, F. Delli Priscoli, G. Koch, and S. Vergari. Traffic management in a band limited communication network: an optimal control approach. *Int. J. of Control*, 78(16):1249–1264, 2005.
6. F. Delli Priscoli and A. Isidori. A control-engineering approach to traffic control in wireless networks. In *Proc. of the 41st IEEE Conf. on Decision and Contr.*, 2002.
7. F. Delli Priscoli and A. Isidori. A control-engineering approach to integrated congestion control and scheduling in wireless local area networks. *Control Engineering Practice*, 13(5):541–558, 2005.
8. F. Delli Priscoli and A. Pietrabissa. Resource management for ATM-based geostationary satellite networks with on-board processing. *Computer Networks*, 39(1):43–60, 2002.
9. F. Delli Priscoli and A. Pietrabissa. Design of a Bandwidth-on-Demand (bod) protocol for satellite networks modelled as time-delay systems. *Automatica*, 40(5):729–741, 2004.
10. F. Delli Priscoli and A. Pietrabissa. Hop-by-hop congestion control for ad hoc wireless-LAN: A model-based control approach. *Int. J. of Control*, 78(12):877–905, 2005.
11. F. Delli Priscoli and J.P. Thomesse. Special issue on control methods for telecommunication networks. *Control Engineering Practice*, 11(10), 2003.
12. F. Delli Priscoli and A. Vernucci. ATA: A novel call admission control approach for 3G mobile networks based on W-CDMA. In *Proc. of the 5th IST Mobile Communication Summit*, 2004.
13. A.I. Elwalid and D. Mitra. Analysis and design of rate-based congestion control of high speed networks I: Stochastic fluid model, access regulation. *Queuing Systems, Theory and Applications*, 9:29–64, 1991.

14. A. Fattouh, O. Sename, and J.M. Dion. A LMI approach to robust observer design for linear time-delay systems. In *Proc. of the 33rd IEEE Conf. on Decision and Contr.*, 1994.
15. L. Georgiadis, L. Guerin, V. Peris, and K.N. Sivarajan. Efficient network QoS provisioning based on per node traffic shaping. *IEEE/ACM Transactions on Networking*, 4(4):482–501, 1996.
16. P. Ioannou. *Model Reference Adaptive Control.* Control Handbook, W.S. Levine Ed. CRC Press, 1996.
17. Satellite broadband multimedia system for IPv6 (SATIP6) project. Contract IST-2001-34344, 2001.
18. M. Kothare, P.J. Campo, M. Morari, and C.N. Nett. A unified framework for the study of anti-windup designs. *Automatica*, 30(12):1869–1883, 1994.
19. C. Liu and J. Layland. Scheduling algorithms for multiprogramming in a hard-time environment. *Journal of ACM*, 20(1):46–61, 1973.
20. S. Lu, V. Bharghavan, and R. Srikant. Fair scheduling in wireless packet networks. *IEEE Trans. on Networking*, 7(4):473–489, 1999.
21. S. Mascolo. Congestion control in high-speed communication networks. *Automatica*, 35(12):1921–1935, 1999.
22. P. Mishra, H. Kanakia, and S. Tripathi. On hop-by-hop rate based congestion control. *IEEE ACM Transactions on Networking*, 4(2):224–239, 1996.
23. M. Morari and E. Zafiriou. *Robust Process Control.* Prentice Hall, Englewood Cliffs, New Jersey, 1989.
24. L. Munoz, M. Garcia, P. Mahonen, Z. Selby, D. Melpignano, and G. Orphanos. Wireless IP based on WAL concept: WINE. *Multiradio Multimedia Communications*, 2000.
25. L. Munyas-Elmas and A. Iftar. Stability margins for a rate-based flow control problem in multiple bottleneck networks. In *Proc. of the 16th IFAC World Congress*, 2005.
26. Z.J. Palmor. *Time-delay compensation - Smith predictor and its modifications.* Control Handbook, W.S. Levine Ed. CRC Press, 1996.
27. A. Pietrabissa, T. Inzerilli, O. Alphand, P. Berthou, M. Mazzella, E. Fromentin, T. Gayraud, and F. F. Lucas. Validation of a QoS architecture for DVB/RCS satellite networks via a hardware demonstration platform. *Computer Networks*, 49(6):797–815, 2005.
28. P.F. Quet, B. Ataşlar, A. İftar, H. Özbay, S. Kalyanaraman, and T. Kang. Rate-based flow controllers for communication networks in the presence of uncertain time-varying multiple time-delays. *Automatica*, 38(6):917–928, 2002.
29. S. Tarbouriech, C. T. Abdallah, and M. Ariola. Bounded control of multiple-delay systems with applications to ATM networks. In *Proc. of the 40th IEEE Conf. on Decision and Contr.*, 2001.
30. G. Wu, E. Chong, and R. Givan. Burst-level congestion control using hindsight optimization. *IEEE Trans. on Automat. Contr.*, 47(6):979–991, 2002.
31. A. Zheng, M. Kothare, and M. Morari. Antiwindup design for internal model control. *Int. J. of Control*, 60(5):1015–1024, 1994.

Nonlinear Synchronization of Coupled Oscillators: The Polynomial Case

Jung-Su Kim and Frank Allgöwer

Institute for Systems Theory and Automatic Control, University of Stuttgart, Germany

Summary. This paper presents a feedback method to achieve synchronization of coupled identical oscillators which are characterized by polynomial vector fields. Here, synchronization means asymptotic coincidence of the states of all the systems. Even though their models are identical, the state trajectories of the identical systems are different because of different initial conditions. Unlike other approaches where just a linear damping term is added to each system in order to achieve synchronization, we design nonlinear coupling functions between the subsystems in such a way that stability of the error dynamics between any two models results. To do that, a certain dissipation inequality and sum of squares as a computational tool are used. Finally, two examples are presented to illustrate the proposed method.

This work is dedicated to Professor Alberto Isidori on the occasion of his 65th birthday.

1 Introduction

Collective motion has received a lot of attention in many areas [19, 18] over the last years. In particular, synchronization is a topic of great interest in many engineering, scientific, and biological systems such as fireflies flashing and fish schooling [18], pacemaker cells in the heart and gene clocks [9], synchronous behavior of neurons [14], and coordinated motion in robotics [10]. Mathematically, synchronization can be seen as the asymptotic coincidence of the state vectors of two (or more) systems [1]. There is a large literature on the analysis and synthesis of synchronization of dynamical systems [19, 10, 18] and in particular synchronization of coupled oscillators [13, 18, 9]. This paper is concerned with the synthesis problem, i.e. the design of inputs (called coupling function) for each oscillator such that the difference between their state trajectories converges to zero.

Synchronization of coupled oscillators is of particular interest because it appears popularly in neuroscience and biochemical networks [14, 9, 17]. Neurons or biochemical reactions modeled as coupled oscillators interact with each other through the coupling function in order to show synchronous behavior. Consequently, elucidating the interaction is an important step in order to understand the nature and develop engineering applications. In the literature, there are several mathematical models which are mainly used in research concerned with coupled oscillators: Hodgkin-Huxley, Kuramoto, Lorenz, Fitzhugh-Nagumo, Hindmarsh-Rose, Goodwin, and the repressilator model [14, 18, 17, 9]. Note that the Lorenz, Fitzhugh-Nagumo, and Hindmarsh-Rose models are in polynomial form, i.e. the right hand side of their ordinary differential equation model consists of polynomial functions. We present a method to design a nonlinear coupling function for synchronization of such polynomial oscillators. The polynomial property enables us to employ an efficient numerical method which facilitates the design.

Existing design methods for synchronization have several drawbacks. Firstly, the resulting coupling functions are of the form of linear feedback or linear feedback with variable gain. Since the model is nonlinear, it is not natural that the coupling function is linear. Moreover, it turned out in our simulation studies that the synchronization of the linearly coupled oscillators are not robust against external disturbances. Secondly, many papers handle synchronization between only two models. Thirdly, no efficient computational tools are employed for designing the coupling function. Most of all, no systematic control theoretic method is used to design the coupling function. This motivates us to devise a novel nonlinear coupling function leading to synchronization.

This paper is devoted to design the coupling function of coupled polynomial oscillators in order to achieve synchronization. To do that, we consider the error dynamics of two oscillators. Then, the inputs (i.e. coupling functions) to the oscillators are designed such that the error dynamics are stabilized. In order to design those inputs, a dissipation inequality is considered, which leads to stability of the error dynamics, and the sum of squares technique is employed to solve the inequality. After that the method is extended to the multiple oscillators case with a particular interconnection. Finally, the proposed method is applied to two different oscillators in order to show its effectiveness.

2 Preliminaries

In this section, some terminology and mathematical tools needed in the paper are introduced, i.e. the definition of the synchronization is given and some stability concepts and basics of graph theory are reviewed. The synchronization problem under consideration can be stated as follows:

Definition 1 *Suppose that there are N identical subsystems*

$$\dot{x}_i = f(x_i, u_i), \ i = 1, \cdots, N,$$

where $x_i \in \mathbb{R}^n$ is the state and $u_i \in \mathbb{R}^m$ the input of the ith subsystem. Furthermore, the N subsystems potentially have different initial conditions. The synchronization problem is to design control inputs u_i, also called coupling functions, such that the following two conditions are satisfied.

C1. The difference between the states of any two subsystems converges to zero, i.e. $x_i(t) - x_j(t) \to 0$ $(i \neq j,\ i, j = 1, \cdots, N)$ for $t \to \infty$.
C2. All states are bounded i.e. $\|x_i(t)\| < \infty$, $i = 1, \cdots, N$ for $t \geq 0$.

In order to analyze the synchronization problem in Definition 1, the following stability concept, which can take external inputs into account, is useful.

Definition 2 *[4] Consider the system $\dot{x} = f(x, w)$, where $x \in \mathbb{R}^n$ is the state and $w \in \mathbb{R}^m$ the external input. The system is said to be input-to-state stable (ISS) if there exist a class $\mathcal{KL}$ function $\beta(\,\cdot\,,\,\cdot\,)$ and a class $\mathcal{K}$ function $\gamma(\,\cdot\,)$ such that the solution of the system $\dot{x} = f(x, w)$ satisfies*

$$\|x(t)\| \leq \beta(\|x(0)\|, t - t_0) + \gamma(\sup_{t_0 \leq \tau \leq t} \|w(\tau)\|), \tag{1}$$

where $\|\cdot\|$ denotes the standard Euclidean norm.

Lemma 1 *[5] If a nonlinear system $\dot{x} = f(x)$ is globally exponentially stable, then the disturbed system $\dot{x} = f(x) + w$ is ISS with respect to w.*

Remark 1. From Definition 2 and Lemma 1 it follows that the state of an ISS system is bounded if the external input is bounded and that the state of an ISS system converges to zero if the external input does. ◁

Finally, we introduce some terminology appearing in graph theory. A graph consists of two types of elements, namely vertices sometimes also called nodes (each oscillator in our case) and edges (interconnections in our case). A graph is said to be *acyclic* if there is no path along which one can return to the starting node. A *tree* is a connected acyclic graph. A *spanning tree* of a graph is just a subgraph that contains all the vertices and is a tree. Sometimes it is convenient to consider one vertex of the tree as special; such a vertex is then called the root of this tree. Given a root, a partial ordering can be defined on the tree as follows: given two vertices i and j, $i \leq j$ whenever i is part of the (unique) path from the root to j. For every node i in the spanning tree except for the root, there is one unique node j satisfying $j \leq i$ and j is directly connected to i. This node is called the parent of node i and conversely node i is a child of node j. The root has no parent node. In this paper, $p(i)$ denotes the index of the parent of node i. Any node who has no child nodes is called a leaf of the tree. For details see [2] and the references therein.

3 Feedback Synchronization of Coupled Oscillators

In this paper, the synchronization problem for coupled oscillators is considered. The ith oscillator is described by

$$\dot{x}_i = f(x_i) + Gu_i, \tag{2}$$

where $x_i \in \mathbb{R}^n$ is the state and $u_i \in \mathbb{R}$ is the coupling function of the ith oscillator. Furthermore, it is assumed that $f(\cdot) : \mathbb{R}^n \rightarrow \mathbb{R}^n$ is a polynomial vector field and that the state trajectory of $\dot{x}_i = f(x_i)$ shows globally bounded oscillatory behavior. In the literature, there are many oscillator models that are in the form of (2), e.g. Fitzhugh-Nagumo [16], Hindmarsh Rose [14], Lorenz [20] model.

3.1 Synchronization of Two Oscillators

First we consider the synchronization problem between two oscillators, i.e. $N = 2$. Although the two oscillators are identical, the trajectories are different from each other because of their different initial conditions. The error dynamics between the two oscillators can be written in linear-like form as

$$\dot{e}_{12} = A(x_1, x_2)e_{12} + Gu_{12}, \tag{3}$$

where $e_{12} = x_1 - x_2$, $u_{12} = u_1 - u_2$. $A(x_1, x_2)$ is defined appropriately from $f(x_1) - f(x_2)$[1]. It is worthwhile to note that for polynomial oscillators the elements of $A(x_1, x_2)$ are also polynomials. It is clear that if u_1 and u_2 are designed such that their difference becomes a stabilizing input for the resulting error dynamics then synchronization is achieved. The next lemma is instrumental in designing such inputs.

Lemma 2 *[3] If there exist a positive definite matrix Q and a polynomial matrix $M(x_1, x_2)$ such that the inequality*

$$\theta^T[A(x_1, x_2)Q + GM(x_1, x_2)]\theta < -\varepsilon\theta^T\theta \tag{4}$$

holds for all nonzero θ and a positive constant ε, then $u_{12} = M(x_1, x_2)Q^{-1}e_{12}$ is a globally exponentially stabilizing control for the error dynamics (3). △

Inequality (4) is a dissipation inequality since it can be rewritten as $\dot{V} \leq -\varepsilon V$ with $V = \frac{1}{2}\theta^T Q^{-1}\theta$ and $\theta = Qe$. In general inequality (4) is very difficult to solve. However, if the oscillators are described by polynomial functions, it is possible to solve inequality (4) for a reasonable problem size using efficient numerical methods, e.g. sum of squares techniques [15, 12]. If we obtain the stabilizing input u_{12} for the error dynamics using Lemma 2, then we have an indefinite equation

$$u_1 - u_2 = u_{12} \tag{5}$$

for the two unknown inputs u_1 and u_2, that has indefinitely many solutions. The following theorem proposes a method to determine the two inputs u_1 and u_2 using u_{12} in order to solve the synchronization problem between the two oscillators.

[1] The details of $A(x_1, x_2)$ depend of course on the model under consideration. For an example, see Section 4.

Theorem 1 *Suppose that the oscillator (2) is ISS with respect to u_i and that the stabilizing input u_{12} for the error dynamics is designed via inequality (4). Then the synchronization problem of two oscillators is solved if the two coupling functions are determined as*

$$u_1 = \delta u_{12}, \ u_2 = (1 - \delta)u_{12}, \tag{6}$$

where $\delta \in [0, 1]$.

Proof. Note that the resulting u_1 and u_2 satisfy equation (5) and go to zero as u_{12} does. Distributing the inputs like this implies that $C1$ is fulfilled because of the global exponential stability of the error dynamics. Since the error dynamics is globally exponentially stable, its input u_{12} also converges to zero. In light of the fact that each oscillator is ISS with respect to its input, global boundedness of each oscillator follows from convergence of its input. Therefore, $C2$ is also fulfilled. Since both conditions are satisfied, the synchronization problem is solved by the coupling functions in (6). □

Remark 2. The distribution method in (6) is just an example. Of course, there are other possibilities to distribute the stabilizing input u_{12} to each input u_i [6]. ◁

Remark 3. If two inputs are determined as $u_1 = 0$ and $u_2 = -u_{12}$ then the resulting systems become $\dot{x}_1 = f(x_1)$ and $\dot{x}_2 = f(x_2) - Gu_{12}$ i.e. one system with input and the other without input. Such distribution implies that there is a leader (x_1 model) in the sense that the other model (x_2 model) follows the leader. This setup is quite similar to the observer problem by viewing the x_1 model as the observed system and the x_2 model as the observer. That is why the synchronization problem can be viewed as a generalized observer problem [8]. ◁

3.2 Synchronization of Multiple Oscillators

In the previous section, the synchronization problem for two oscillators was considered. In the case of multiple oscillators ($N > 2$), the coupling functions for synchronization heavily rely on the interconnection topology. In this section, it is shown that the proposed method is applicable to the multiple oscillators case where the oscillator interconnection is in spanning tree form. To this end, we propose a solution for an example at first and then present a solution for the general case.

As an example, we consider the particular interconnection in Fig. 1. In view of the two oscillators case, the interconnection in Fig. 1 results in the following seven error variables e_{12}, e_{23}, e_{34}, e_{45}, e_{36}, e_{67}, e_{68}, where $e_{ij} = x_i - x_j$. Note that there exist associated error dynamics as in (3) for each of these error variables (e.g. $\dot{e}_{23}=A(x_2, x_3)e_{23}+Gu_{23}$) and indefinite equations

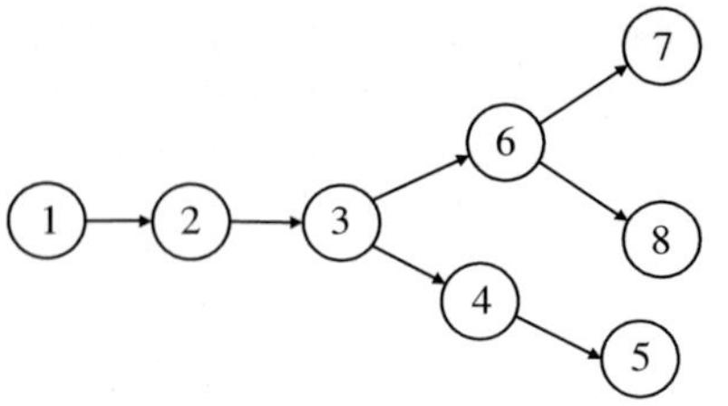

Fig. 1. An interconnection topology among multiple models. The *arrows* denote how the state information is exchanged between models

$$u_{12} = u_1 - u_2,\ u_{23} = u_2 - u_3,\ u_{34} = u_3 - u_4, \tag{7a}$$

$$u_{45} = u_4 - u_5,\ u_{36} = u_3 - u_6,\ u_{67} = u_6 - u_7,\ u_{68} = u_6 - u_8. \tag{7b}$$

All inputs u_{ij} can be determined in analogy to the previous section through the solution of dissipation inequalities in the form of (4). Suppose that these u_{ij} are distributed as

$$\begin{aligned} &u_1 = 0,\ u_2 = -u_{12},\ u_3 = -u_{23},\ u_4 = -u_{34} \\ &u_5 = -u_{45},\ u_6 = -u_{36},\ u_7 = -u_{67},\ u_8 = -u_{68}. \end{aligned} \tag{8}$$

This distribution does not solve the equations (7a)–(7b). The resulting coupling functions u_i lead to the following error dynamics

$$\dot{e}_{12} = A(x_1, x_2)e_{12} + Gu_{12} \tag{9a}$$

$$\dot{e}_{23} = A(x_2, x_3)e_{23} + Gu_{23} - Gu_{12}, \tag{9b}$$

$$\dot{e}_{34} = A(x_3, x_4)e_{34} + Gu_{34} - Gu_{23}, \tag{9c}$$

$$\dot{e}_{45} = A(x_4, x_5)e_{45} + Gu_{45} - Gu_{34}, \tag{9d}$$

$$\dot{e}_{36} = A(x_3, x_6)e_{36} + Gu_{36} - Gu_{23}, \tag{9e}$$

$$\dot{e}_{67} = A(x_6, x_7)e_{67} + Gu_{67} - Gu_{36}, \tag{9f}$$

$$\dot{e}_{68} = A(x_6, x_8)e_{68} + Gu_{68} - Gu_{36}. \tag{9g}$$

Note that error dynamics (9b)–(9g) are not globally exponentially stable because of the additional inputs $-Gu_{ij}$. However, the error dynamics (9b)–(9g) without the additional input $-Gu_{ij}$ are exponentially stable, e.g. $\dot{e}_{23} = A(x_2, x_3)e_{23} + Gu_{23}$ in (9b) is globally exponentially stable. Hence, by considering the additional inputs $-Gu_{ij}$ in (9b)–(9g) as disturbances to the corresponding exponentially stable dynamics, we can prove the stability of the error dynamics (9) as follows: Since e_{12} and therefore also u_{12} converge to zero for $t \to \infty$, it follows that both e_{23} and u_{23} go to zero for $t \to \infty$ because of Lemma 1. Since this argument can be applied repeatedly to the error dynamics (9c)–(9g), convergence of all error variables to zero can be established, i.e. $C1$ is fulfilled. Moreover, since all coupling functions u_i go to zero, $C2$ is also satisfied due to the ISS property of each oscillator. This means that the coupling functions u_i determined from (8) solve the synchronization problem for the interconnection depicted in Fig. 1. On the basis of the result for this example, a general solution is presented in the next theorem for the synchronization problem of multiple oscillators with a spanning tree form interconnection.

Theorem 2 *The following procedure solves the synchronization problem for multiple oscillators in the spanning tree form interconnection.*

1. *Label each system from 1 to N with 1 being the root of the tree*
2. *Determine the error dynamics and compute the corresponding stabilizing inputs u_{ij} using Lemma 2*
3. *Distribute the inputs u_{ij} to each coupling function u_i as follows*

$$\begin{aligned} u_1 &= 0, \\ u_j &= -u_{p(j)j}, \ j = 2, \cdots, N. \end{aligned}$$

Proof. Note that the distribution in the theorem always results in

$$\dot{e}_{12} = A(x_1, x_2)e_{12} + Gu_{12},$$

which is exponentially stable. So e_{12} and u_{12} converge to zero exponentially. For the jth model ($j = 2, \cdots, N$), the error dynamics can be written as

$$\dot{e}_{p(j)j} = A(x_{p(j)}, x_j)e_{p(j)j} + Gu_{p(j)j} - Gu_{p(p(j))p(j)}. \tag{10}$$

From an inductive argument, convergence of the last term $Gu_{p(p(j))p(j)}$ follows from that of u_{12}. Therefore $e_{p(j)j}$ also goes to zero in an asymptotic fashion which means the fulfillment of $C1$. Since all coupling functions u_i, ($i = 2, \cdots, N$) go to zero, all states are bounded because of the ISS property of each oscillator. This completes the proof. □

This analysis seems physically plausible for the considered interconnection in the sense that the convergence of the state of one oscillator to the previous oscillator depends on the convergence of the previous oscillator to its parent[2]. In other words, the synchronization is propagated from the root to all leaves of the tree. Hence, the theorem generalizes the approach for the given multiple oscillators case in spanning tree form.

Remark 4. In view of the whole procedure to design the coupling function for synchronization, the most important step is to determine the stabilizing input u_{12} for the error dynamics in (3). Lemma 2 provides an efficient way to find the stabilizer of the error dynamics provided all systems are given by polynomial right hand sides. If one is able to find a stabilizing feedback for the error dynamics then the proposed method to the synchronization problem is of course also applicable for other models which are possibly not in polynomial form. ◁

[2] The term previous is understood in the sense of the partial ordering on the tree.

4 Examples

In this section, we apply the proposed synchronization method to two particular coupled oscillators, namely the Fitzhugh-Nagumo oscillator and the Goodwin oscillator. The first example shows the solution for the polynomial case and discusses how some of the assumptions in Section 3.1 can be released. The second example addresses the approach for non-polynomial oscillators.

4.1 Synchronization of Fitzhugh-Nagumo Oscillators

Consider the synchronization of two Fitzhugh-Nagumo (FN) oscillators

$$\begin{aligned} \dot{x}_{i1} &= -x_{i1}^3 + (a+1)x_{i1}^2 - ax_{i1} - x_{i2} + I_a + u_i \\ \dot{x}_{i2} &= bx_{i1} - \gamma x_{i2}, \ i = 1, 2, \end{aligned} \tag{11}$$

where x_{ij} denotes jth element of the state of oscillator i, all other variables are parameters. This type of oscillator is commonly used for neuron research [16, 7].

In order to apply the proposed method, we first need to check ISS of (11). It is not easy to show ISS of the FN oscillator in the sense of its original definition. However, ISS of the model is mainly used in order to show global boundedness of the state trajectory in the previous section. Actually, what is necessary is the converging-input bounded-state (CIBS) property to show fulfillment of $C2$. For the FN oscillator, we can prove CIBS by showing that the model is ISS outside some ball in $\mathbb{R}^2$ using a quadratic Lyapunov function. A similar proof can be found in [11]. So, the FN oscillator has the CIBS property and we can apply the design method presented in Theorem 1.

For two FN oscillators, the error dynamics are written as

$$\begin{aligned} \dot{e}_1 &= -e_1 t_2 + (a+1)e_1 t_1 - e_1 - e_2 + u_{12} \\ \dot{e}_2 &= be_1 - \gamma e_2, \end{aligned} \tag{12}$$

where $e_1 = x_{11} - x_{21}$, $e_2 = x_{12} - x_{22}$, $t_1 = x_{11} + x_{21}$, and $t_2 = x_{11}^2 + x_{11}x_{21} + x_{21}^2$. This equation can be represented in the simple linear-like form as in (3) with

$$A(x_1, x_2) = \begin{bmatrix} -t_2 + (a+1)t_1 - 1 & -1 \\ b & -\gamma \end{bmatrix}, \ G = \begin{bmatrix} 1 \\ 0 \end{bmatrix}.$$

Following Lemma 2, we can solve the corresponding dissipation inequality using SOSTOOLS [15] which gives the stabilizing input for the error dynamics (12) as follows

$$u_{12} = (0.309 + t_2)e_1 + (0.32676 + 0.209t_1)e_2. \tag{13}$$

Finally, applying the input distribution presented in Theorem 1, we can obtain the coupling functions for two FN oscillators. Fig. 2 results from applying

$u_1 = u_2 = 0$ and Fig. 3 from applying the designed coupling function using (13) and (6). Fig. 2(a) shows asynchronous behavior because the initial conditions of the two models are different, i.e. $(x_{11}(0), x_{12}(0)) = (3.5, 2.5)$ and $(x_{21}(0), x_{22}(0)) = (1, 0.7)$. On the other hand, Fig. 3 demonstrates that the proposed method successfully results in synchronization. Fig. 4(a) shows that five FN oscillators with the interconnection depicted in Fig. 4(b) are synchronized by the method in Theorem 2. As in the previous case, the following different initial conditions are used in the simulation:

$$(x_{11}(0), x_{12}(0), x_{21}(0), x_{22}(0), x_{31}(0), x_{32}(0), x_{41}(0), x_{42}(0), x_{51}(0), x_{52}(0)) =$$

$$(3.5, 2.5, 1, 0.7, 1.5, 1.7, 4.0, 2.2, 3, 3.5).$$

4.2 Synchronization of Two Goodwin Oscillators

The proposed method for the design of the coupling function can be applied in a straightforward manner when the oscillator models have polynomial right hand sides. With the example of the synchronization of two Goodwin oscillators ([17]) we show that the method can also be applied to the non-polynomial case in some instances. This is particularly important because many oscillators appearing in biochemical networks are commonly modeled as rational systems, i.e. the vector fields in the right hand side of the ordinary differential equation model are rational functions. As mentioned in Remark 4, the important aspect in achieving synchronization is to stabilize the resulting error dynamics. In this example, it is shown that the proposed method (error dynamics stabilization and distribution) can also be applied to oscillators with rational vector fields if the resulting error dynamics can be exponentially stabilized.

Consider the following simplified Goodwin oscillator [17]

$$\dot{x}_{i1} = -b_1 x_{i1} + \frac{1}{1 + x_{i3}^{17}} + u_i \tag{14a}$$

$$\dot{x}_{i2} = -b_2 x_{i2} + b_2 x_{i1} \tag{14b}$$

$$\dot{x}_{i3} = -b_3 x_{i3} + b_3 x_{i2}, \; i = 1, 2. \tag{14c}$$

Also here x_{ij} denotes the jth component of the state vector of oscillator i. Note that this oscillator is not a polynomial model but a rational model because of the term $\frac{1}{1+z_{i3}^{17}}$ in the first equation. This oscillator model is commonly used to describe enzyme kinetics in biological systems, where x_{ij} describe three biochemical products. Note that this model is a nonnegative system; the model is invariant in the positive orthant of the state space with nonnegative initial conditions. It is easy to see ISS of this model considering its structure. Equation (14a) is ISS with respect to the last two terms $\frac{1}{1+x_{i3}^{17}} + u_1$ because of the linear stable term $-b_1 x_{i1}$ and the second term $\frac{1}{1+x_{i3}^{17}}$ bounded by 1. Therefore, the first equation is ISS with respect to the input u_i. Moreover,

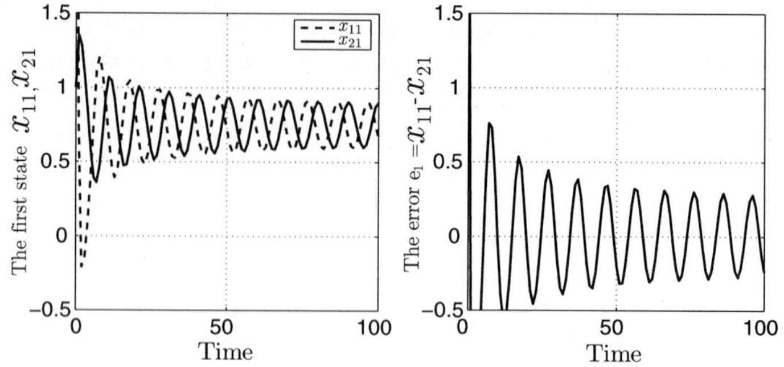

(a) Asynchronous behavior (b) Error

Fig. 2. Two uncoupled Fitzhugh-Nagumo oscillators with different initial conditions and $u_1 = u_2 = 0$

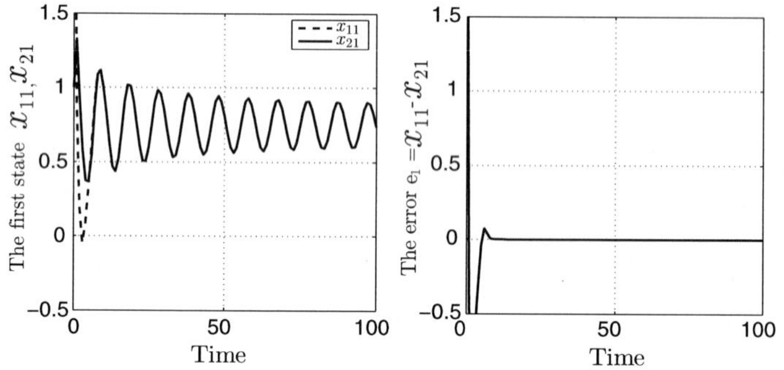

(a) Synchronous behavior (b) Error

Fig. 3. Two Fitzhugh-Nagumo oscillators: the synchronized case

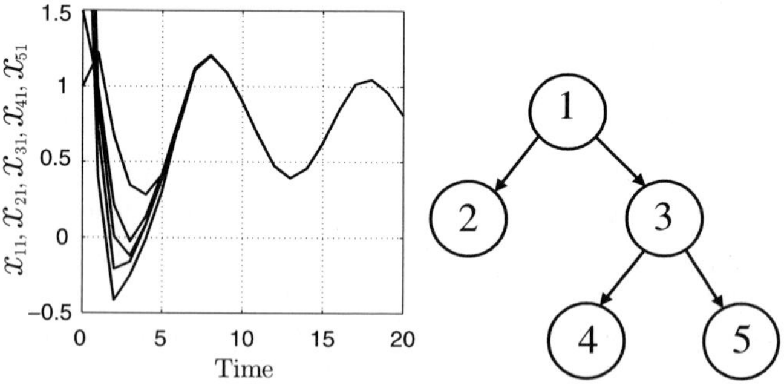

(a) Synchronous behavior (b) Interconnection

Fig. 4. Five Fitzhugh-Nagumo oscillators: the synchronized case the second and third equations are also ISS with respect to x_{i1} and x_{i2}, respectively. Therefore, the model is ISS with respect to the input.

The error dynamics can be written as in (3) with

$$A(x_1, x_2) = \begin{bmatrix} -b_1 & 0 & -t_1 \\ b_2 & -b_2 & 0 \\ 0 & -b_3 & b_3 \end{bmatrix}, \; G = \begin{bmatrix} 1 \\ 0 \\ 0 \end{bmatrix}$$

where $t_1 = \frac{1}{1+x_{13}^{17}} - \frac{1}{1+x_{23}^{17}}$. The following is a stabilizing input for the resulting error dynamics

$$u_{12} = t_1$$

because the closed-loop becomes $\dot{e}_{12} = A_c e_{12}$ where

$$A_c = \begin{bmatrix} -b_1 & 0 & \\ b_2 & -b_2 & 0 \\ 0 & -b_3 & b_3 \end{bmatrix}, e_{12} = x_1 - x_2, b_1 = b_2 = b_3 = 0.5$$

with A_c being Hurwitz. By distributing u_{12} to each coupling function u_1 and u_2 as in (6), synchronization is obtained. Similarly to the previous example,

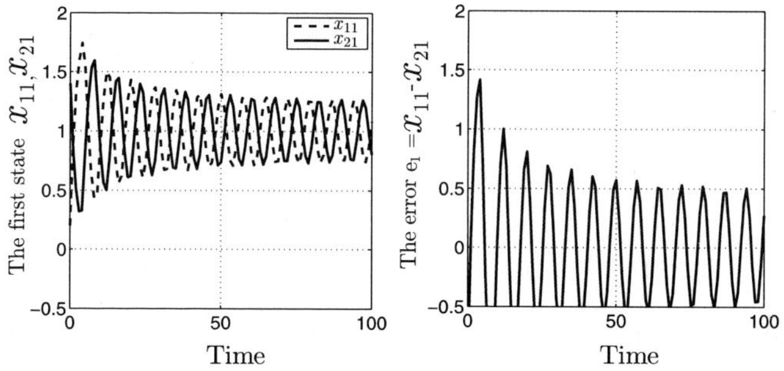

(a) Asynchronous behavior (b) Error

Fig. 5. Two uncoupled Goodwin oscillators with different initial conditions

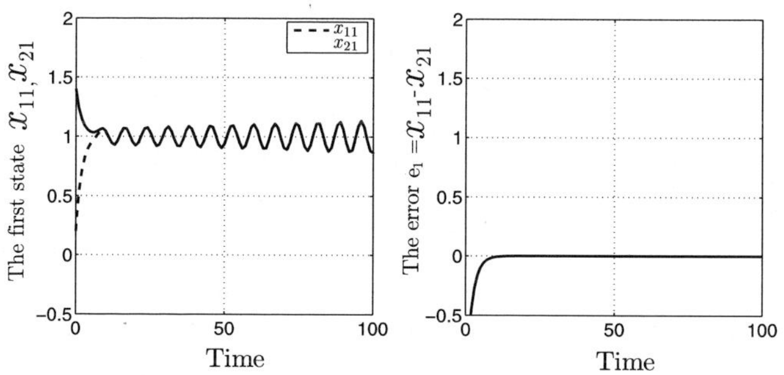

(a) Synchronous behavior (b) Error

Fig. 6. Two Goodwin oscillators: the synchronized case

Fig. 5 and Fig. 6 show the effectiveness of the method where the initial conditions $(x_{11}(0), x_{12}(0), x_{13}(0)) = (0.2, 0.3, 0.5)$ and $(x_{21}(0), x_{22}(0), x_{23}(0)) = (1.4, 1.5, 1.7)$ are used.

5 Summary and Outlook

In this paper, a nonlinear feedback design is proposed for the synchronization of coupled oscillators described by polynomial models. Unlike most previous results in which a linear damping term is used, we design the nonlinear coupling functions in such a way that they stabilize the error dynamics. First, the synchronization problem for the two oscillators case is presented. It was shown that the coupling function can be derived by solving a particular dissipation inequality which can be solved in the polynomial case using sum of squares (SOS) techniques. For a multiple oscillators case, a solution was proposed if the interconnection is in the spanning tree form. Finally, the proposed method is applied to two oscillator models, namely the Fitzhugh-Nagumo and the Goodwin oscillators.

In this paper, a particular multiple oscillators case is handled. So, the solution to the general multiple oscillators case with arbitrary interconnection is still waiting for answers. In this paper, availability of the full state is assumed. So a feedback scheme for synchronization which uses only the output of the oscillator is an interesting problem.

References

1. I.I. Blekhman, A.L. Fradkov, H. Nijmeijer, and A.Y. Pogromsky. On self-synchronization and controlled synchronization. *Systems & Control Letters*, 31:299–306, 1997.
2. R. Diestel. *Graph Theory*. Springer Verlag, 2005.
3. C. Ebenbauer. *Polynomial Control Systems: Analysis and Design via Dissipation Inequalities and Sum of Squares*. Ph.D dissertation, University of Stuttgart, Germany, 2005.
4. A. Isidori. *Nonlinear Control Systems, Vol. II*. Springer Verlag, 1999.
5. H.K. Khalil. *Nonlinear Systems*. Prentice Hall, 2002.
6. J. S. Kim and F. Allgöwer. A nonlinear synchronization scheme for multiple Hindmarsh-Rose models. Submitted for publication, 2007.
7. D. Mishra, A. Yadav, S. Ray, and P.K. Kalra. Controlling synchronization of modified Fitzhugh-Nagumo neurons under external electrical stimulation. *NeuroQuantology*, 4(1):50–67, 2006.
8. H. Nijmeijer and I.M.Y. Mareels. An observer looks at synchronization. *IEEE Trans. on Circuits and Systems-I: Fundamental Theory and Applications*, 44:882–890, 1997.
9. J.G. Ojalvo, M.B. Elowitz, and S. Strogatz. Modeling a synthetic multicellular clock: Repressilators coupled by quorum sensing. In *Proc. Natl. Acad. Sci. U.S.A.*, volume 101, pages 10955–10960, 2004.

10. R. Olfati-Saber and R.M. Murray. Consensus problems in networks of agents with switching topology and time-delays. *IEEE Trans. on Automat. Contr.*, 49(9):1520–1533, 2004.
11. W. T. Oud and I. Tyukin. Sufficient conditions for synchronization of Hindmarsh and Rose neurons: passivity-based-approach. In *Proc. of the 6th IFAC Symposium in Nonlinear Control Systems*, Stuttgart, 2004.
12. P.A. Parrilo. *Structured Semidefinite Programs and Semialgebraic Geometry Methods in Robustness and Optimization.* Ph.D dissertation, California Institute of Technology, 2000.
13. L. Pecora and M. Barahona. Synchronization of oscillators in complex networks. *Chaos and Complexity Letters*, 1:61–91, 2005.
14. R.D. Pinto, P. Varona, A.R. Volkovskii, A. Szucs, H.D.I. Abarbanel, and M.I. Rabinovich. Synchronous behavior of two coupled electronic neurons. *Physical Review E*, 62(2):2644–2656, 2000.
15. S. Prajna, A. Papachristodoulou, P. Seiler, and P.A. Parrilo. *SOSTOOLS and its control applications, In Positive Polynomials in Control.* Lecture Notes in Control and Information Sciences. Springer Verlag, 2005.
16. G.-B. Stan. *Global Analysis and Synthesis of Oscillations: a Dissipativity Approach.* Ph.D dissertation, University of Liege, Belgium, 2005.
17. G.-B. Stan, A.O. Hamadeh, J. Goncalves, and R. Sepulchre. Output synchronization in networks of cyclic biochemical oscillators. In *Proc. of the 2007 Amer. Contr. Conf.*, New York, 2007.
18. S.H. Strogatz. From Kuramoto to crawford: Exploring the onset of synchronization in populations of coupled oscillators. *Physica D*, 143:1–20, 2000.
19. S.H. Strogatz. Exploring complex networks. *Nature*, 410:268–276, 2001.
20. Y. Wang, A.H. Guan, and H.O. Wang. Feedback and adaptive control for the synchronization of Chen system via a single variable. *Physics Letters A*, 312:34–40, 2003.

Part V

Geometric Methods

Disturbance Decoupling for Open Quantum Systems: Quantum Internal Model Principle

Narayan Ganesan and Tzyh-Jong Tarn

Electrical and Systems Engineering, Washington University in St. Louis, MO-63130, USA

Summary. Decoherence, which is caused due to the interaction of a quantum system with its environment plagues all quantum systems and leads to the loss of quantum properties that are vital for quantum computation and quantum information processing. In this chapter we propose a novel strategy using techniques from systems theory specifically classical disturbance decoupling to completely eliminate decoherence and also provide conditions under which it can be done so. A novel construction employing an auxiliary system, the bait, which is instrumental to decoupling the system from the environment will be found useful. Almost all the earlier work on decoherence control employ density matrix and stochastic master equations to analyze the problem. Our approach to decoherence control involves the bilinear input affine model of quantum control system which lends itself to various techniques from classical control theory, but with non-trivial modifications to the quantum regime. The elegance of this approach yields interesting results on open loop decouplability and Decoherence Free Subspaces (DFS). Additionally, the feedback control of decoherence may be related to disturbance decoupling for classical input affine systems, which entails careful application of the methods by avoiding all the quantum mechanical pitfalls. The two concepts are contrasted and an improved theory of disturbance decoupling for general input affine systems is developed. In the process of calculating a suitable feedback the system has to be restructured due to its tensorial nature of interaction with the environment, which is unique to quantum systems. Finally the results are also shown to be superior to the ones obtained via master equations.

1 Introduction

In this chapter we address the problem of control of decoherence in open quantum systems via a classical state feedback. While most of earlier work in literature deals with analyzing the behavior of the reduced density matrix of an open quantum system and designing controls so as to counteract the effects of environment on the density matrix of the system, we present a different approach to this problem. This approach that not only helps us get to the root of the problem but also helps design the solution. A bilinear input affine system that describes the state dynamics of a quantum system not only helps us get further insight than the density matrix master equation but also offers conditions for controlling decoherence when modeled as a disturbance decoupling problem. The structure of the system and the few similarities it bore to classical non-linear disturbance decoupling, originally put forward by Isidori et al. [11], [10], aroused our interest in this line of approach. However many of the properties and methods applicable to classical systems like direct sum of vector spaces, classical additive noises, real vector spaces were now undermined by tensor product interaction, quantum noise and complex projective spaces. Nevertheless the approach seemed promising and we have shown that this not only yields results that are characteristically different but also qualitatively superior to the ones already present on decoherence control. We have now learnt a great deal about behavior of open quantum systems and have also stumbled on a few interesting results regarding the nature of control hamiltonians and the Internal Model Principle analog for quantum systems that is first of its kind in the literature. The above results might prove important in their own right and in due course of time could influence the design of future quantum control systems.

2 Previous Work in the Literature

Decoherence is the process by which quantum systems lose their coherence information by coupling to the environment. The quantum system entangles to the states of the environment and the system density matrix can be diagonalized in a preferred basis states for the environment, dictated by the model of the interaction hamiltonian [27]. Decoherence is now the biggest stumbling block towards exploitation of quantum speedup [19] using finite quantum systems in information processing. Many authors have addressed the control and suppression of decoherence in *open quantum systems* by employing a variety of open loop and feedback strategies. The effect of decoherence suppression under arbitrarily fast open-loop control was studied by Viola et al. [24], [25]. Another method along similar lines for control of decoherence by open-loop multipulses was studied by Uchiyama et al. [23]. A very illustrating example of decoherence of single qubit system used in quantum information processing and its effective control using pulse method was worked out by Protopopescu et al [20]. Shor[21] and Calderbank [2] also came up with interesting error-correction schemes for detecting and reducing effects of decoherence on finite quantum registers. Recently many authors have also studied the application of feedback methods in control of decoherence [3], [6]. Technological advances enabling manipulation, control of quantum systems and recent advances in quantum measurements using weak coupling, non-demolition principles [1] etc, has opened up avenues for employing feedback based control strategies for quantum systems [26], [12], [6].

In this chapter we analyze the efficacy of feedback methods in eliminating decoherence. A wave function approach as opposed to density matrices for the control equation is adopted which represents the system in an input-affine form and greatly enables one to exploit methodologies from systems theory. We first analyze what it means for a complex scalar function to be invariant of certain parameters. The generality of the treatment adopted here makes all types of quantum systems amenable to the results.

3 Mathematical Preliminaries

A pioneering effort to study quantum control systems using bilinear input affine model was carried out by Huang et al. [7]. The model has since found various applications and is found extremely useful in analyzing the controllability properties of a quantum system on the state space of analytic manifolds [18] which can be seen to exploit previous results on controllability of finite dimensional classical systems by Sussmann and Jurdjevic [22] and further exploiting the results by Kunita [13], [14]. In this chapter we will explore the conditions for a scalar function represented by a quadratic form to be invariant under the dynamics of the above model(with the additional assumption of time-varying vector fields) in the presence of a perturbation or interaction hamiltonian. Such a formalism can be seen to readily related to decoherence in open quantum systems where in a perturbative hamiltonian that couples the system and environment can be seen to play the role of a *disturbance.* However it will also be seen that the aforementioned is not quite similar to classical disturbance decoupling problem and one should be extremely careful in adapting the classical results to decoherence control in open quantum systems.

Let

$$\frac{\partial \xi(t,x)}{\partial t} = [H_0 \otimes \mathcal{I}_e(t,x) + \mathcal{I}_e \otimes H_e(t,x) + H_{SB}(t,x) + \sum_{i=1}^{r} u_i(t) H_i \otimes \mathcal{I}_e(t,x)]\xi(t,x) \tag{1}$$

be the quantum control system corresponding to an open quantum system interacting with the environment;
$\mathcal{H}_s$ be the system's Hilbert space;
$\mathcal{H}_e$ be the environment's Hilbert space;
$\mathcal{H}_s$ could be finite or infinite dimensional and $\mathcal{H}_e$ is generally infinite dimensional;
$\xi(t,x)$ be the wave function of the system and environment.
H_0 and H_e are skew Hermitian operators corresponding to the drift Hamiltonian of the system and environment while H_i's correspond to the control Hamiltonian of the system. H_{SB} governs the interaction between the system

and the environment. The above operators are assumed to be time varying and dependent on the spatial variable. Consider a scalar function (typically the expected value of an observable) of the form,

$$y(t, \xi) = \langle \xi(t, x) | C(t, x) | \xi(t, x) \rangle \tag{2}$$

where again $C(t, x)$ is assumed to be time-varying operator acting on system Hilbert space. The above is the general form of a time dependent quantum system and we wish to study the invariance properties of the function $y(t, \xi)$ with respect to the system dynamics.

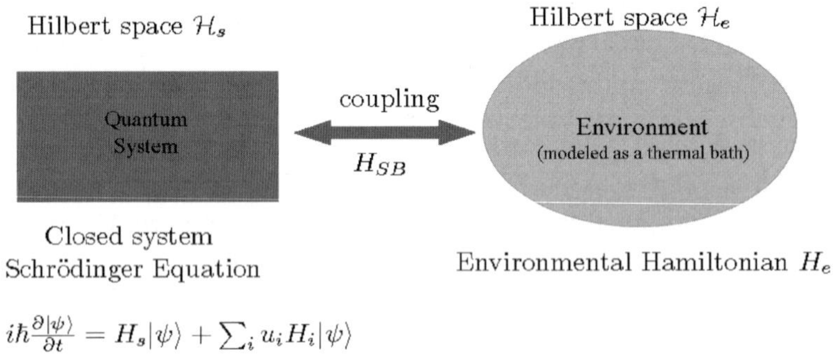

Fig. 1. An Open quantum system interacting with the environment via H_{SB}

Let $y(t, \xi) = f(t, x, u_1, \cdots, u_r, H_{SB})$ be a complex scalar map of the system as a function of the control functions and interaction Hamiltonian over a time interval $t_0 \leq t \leq t_1$. The function is said to be invariant of the interaction Hamiltonian if

$$f(t, x, u_1, \cdots, u_r, H_{SB}) = f(t, x, u_1, \cdots, u_r, 0) \tag{3}$$

for all admissible control functions $u_1, \cdots, u_r$ and a given interaction Hamiltonian H_{SB}.

The output equation. It can be seen that a suitable value of the operator C could yield the off-diagonal terms of the density matrix of the system as the output y. The above output equation takes a quadratic form in the state ξ of the combined system and the environment. Some of the possible physical of implications of the output equation are as follows.

(i) An expected value of a physical observable or an observation. The operator C could also be a non-demolition observable in which case $y(t)$ is the output of the measurement performed on the system.
(ii) By a suitable choice of the operator C the value $y(t)$ can now be thought of as a complex functional representing the coherence between the states

of interest. For example $C = |s_i\rangle\langle s_j| \otimes \mathbb{I}_e$ can be seen to yield the coherence between the orthogonal states of the system $|s_i\rangle$ and $|s_j\rangle$. For the pure state $\xi = \sum c_i|s_i\rangle$, $y(t) = c_i^* c_j$ and for the completely mixed state $\xi = \sum c_i|s_i\rangle|e_i\rangle$ where $|e_i\rangle$ are the orthogonal states of the environment, a similar calculation yields $y = 0$.

(iii)The operator C could also be a general linear operator, an example of which is discussed in the section on DFS later.

The analysis of time-varying systems carried out here assumes in general that the component Hamiltonian operators carry explicit time dependence which is not under the control of an external agent. And we do so by introducing a time invariant system in the augmented state space domain $\mathcal{M}' = \mathcal{M} \oplus \mathbb{R}$. A similar scheme was also used by Lan et al. [15] to study controllability properties of such time-varying quantum systems.

Let $x_1 = t$, the new equation governing the evolution of the system can be written as

$$\frac{\partial}{\partial t}\begin{pmatrix} x_1 \\ \xi(t,x) \end{pmatrix} = \begin{pmatrix} 1 \\ (H_0(x_1,x) + H_e(x_1,x))\xi(t,x) \end{pmatrix} + \begin{pmatrix} 0 \\ u_i H_i(x_1,x)\xi(t,x) \end{pmatrix} + \begin{pmatrix} 0 \\ H_{SB}(x_1,x)\xi(t,x) \end{pmatrix} \tag{4}$$

with

$$y(t,\xi) = \langle \xi(t,x)|C(t,x)|\xi(t,x)\rangle. \tag{5}$$

The vector fields

$$K_0 = \begin{pmatrix} 1 \\ (H_0 + H_e)\xi(x,t) \end{pmatrix}, K_i = \begin{pmatrix} 0 \\ H_i\xi(x,t) \end{pmatrix}$$

and

$$K_I = \begin{pmatrix} 0 \\ H_{SB}\xi(x,t) \end{pmatrix}$$

corresponding to drift, control and interaction can be identified to contribute to the dynamical evolution. The above problem statement can be visualized as the state trajectories corresponding to different interaction vector fields K_I intersecting isobars of $y(t)$ at the same time instants.

The following lemma [4] provides the basic conditions necessary for invariance of the output equation with respect to the interaction vector field.

Lemma 1. *Consider the quantum control system (4) and suppose that the corresponding output given by equation (5) is invariant under given H_{SB}. Then for all integers $p \geq 0$ and any choice of vector fields $X_1, \cdots, X_p$ in the set $\{K_0, K_1, \cdots, K_r\}$ wc havc*

$$L_{K_I} L_{X_1} \cdots L_{X_p} y(t,\xi) = 0 \quad \textit{for all } t, \xi. \tag{6}$$

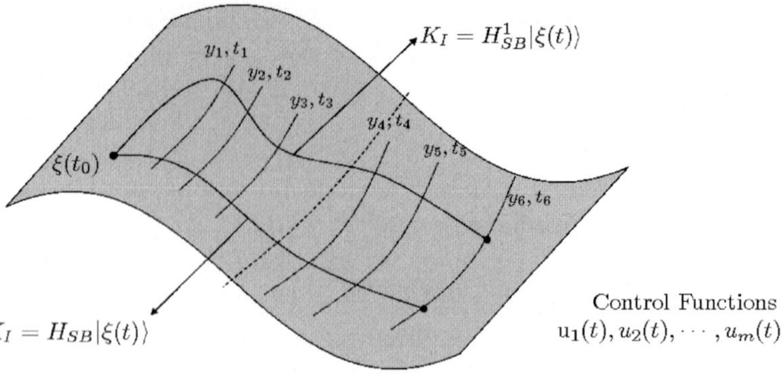

Fig. 2. The above is a geometric representation of the problems statement. The trajectories corresponding to two interaction Hamiltonians H_{SB} and H'_{SB} and a given set of control functions $u_1, \cdots, u_r$ are shown

The sufficient condition for output invariance however requires a stronger condition of analyticity of the system.

Lemma 1 implies that the necessary conditions for output invariance are,

$$\begin{aligned} L_{K_I} y(t,\xi) &= 0 \\ L_{K_I} L_{K_{i_0}} \cdots L_{K_{i_n}} y(t,\xi) &= 0 \end{aligned} \tag{7}$$

for $0 \leq i_0, \cdots, i_n \leq r$ and $n \geq 0$, where $K_0, \cdots, K_r$ are the vector fields of the augmented system and K_I, the interaction vector field. In addition, the following lemma[4] which ties the sufficiency of the above conditions to analytic property of the system can be stated thus.

Lemma 2. *Suppose the system (4) is analytic, then y is invariant under given H_{SB} if and only if (6) is satisfied.*

4 Invariance for the Quantum System

With the preceding mathematics preliminaries in place we can now apply the above conditions to the quantum system with careful consideration of the nature of the complex functional and the analytic manifold. We can now state the condition for output invariance with respect to a perturbation or interaction Hamiltonian, the proof and motivation for which is presented in detail in [4].

Theorem 1. *Let $\mathcal{C}_0 = C(t)$ and for $n = 1, 2, \cdots$*

$$\tilde{\mathcal{C}}_n = span\{ad^j_{H_i} \mathcal{C}_{n-1}(t) | j = 0, 1, \ldots; i = 1, \ldots, r\}$$

$$\mathcal{C}_n = \left\{ \left(ad_H + \frac{\partial}{\partial t} \right)^j \tilde{\mathcal{C}}_n ; j = 0, 1, \cdots \right\}.$$

Define a distribution of quantum operators, $\tilde{\mathcal{C}}(t) = span\{\mathcal{C}_1(t), \cdots, \mathcal{C}_n(t), \cdots\}$. *The output equation (5) of the quantum system is decoupled from the environmental interactions if and only if*

$$[\tilde{\mathcal{C}}(t), H_{SB}(t)] = 0. \tag{8}$$

A few applications of the theorem including the result on DFS, which was originally proposed by Lidar et al. [16] by analysis of Markovian master equation for open quantum systems that naturally gives rise to subspaces that are immune to the effects of decoherence namely dissipation and loss of coherence, can be derived as a special case of the above condition [4]. In this chapter we will however be more interested on the feedback aspects of decoherence control.

Decoherence of a collection of 2-level systems in the presence of control. For a collection of 2-level systems interacting with a bath of oscillators the corresponding Hamiltonian is

$$H = \frac{\omega_0}{2}\sum_{j=1}^{N}\sigma_3^{(j)} + \sum_k \omega_k b_k^\dagger b_k + \sum_k \sum_{j=1}^{N} \sigma_3^{(j)}(g_k b_k^\dagger + g_k^* b_k) .$$

It can be seen that for $C = |000\rangle\langle 000| + |001\rangle\langle 001| + |010\rangle\langle 100| + |011\rangle\langle 101|$, for a 3 qubit system the invariance condition (8) is satisfied, meaning coherence between basis states $|001\rangle, |010\rangle$ etc of DFS is preserved under interaction Hamiltonian $H_{SB} = \sum_k \sum_{j=1}^N \sigma_3^{(j)}(g_k b_k^\dagger + g_k^* b_k)$ where the system is assumed to interact through the collective operator $\sum_j \sigma_3^{(j)}$ and g_k's describe coupling to the mode k of the environment. However in the presence of the external *symmetry breaking* control Hamiltonians $H_i = u_i \sigma_1^{(i)}$, the invariance condition is no longer satisfied for the operator C as $[[C, \sigma_1^{(i)}], \sigma_3^{(j)}] \neq 0$ and hence the coherence between the states is not preserved anymore. This is because of the transitions outside DFS caused by the control Hamiltonians. We will devote the rest of the chapter to studying feedback control of decoherence and its physical implications.

5 Feedback Control

In this section we analyze feedback based control methods for decoupling or rendering the system invariant of the interaction in case it is not already so (as in the previous example). We assume a classical state feedback of the form $u = \alpha(\xi) + \beta(\xi).v$ and derive conditions for decouplability, where α, β are $r \times 1$ vector and $r \times r$ matrix respectively of scalar functions depending on state $|\xi\rangle$ of the system

$$\frac{\partial}{\partial t}\begin{pmatrix} x_1 \\ \xi(t,x) \end{pmatrix} = \begin{pmatrix} 1 \\ (H_0 + H_e + \sum \alpha_i H_i)(x_1,x)\xi(t,x) \end{pmatrix} + \begin{pmatrix} 0 \\ \sum v_i \sum \beta_{ij} H_j(x_1,x)\xi(t,x) \end{pmatrix} + \begin{pmatrix} 0 \\ H_{SB}(x_1,x)\xi(t,x) \end{pmatrix} \quad (9)$$

where again the following vector fields $\tilde{K}_0, \tilde{K}_i$ can be identified as the modified drift and control vector fields and K_I, the original disturbance vector field.

As stated before the necessary and sufficient conditions for a scalar function $y(t)$ of the system to be invariant of the interaction vector field can now be written as

$$L_{K_I} y(t) = 0$$
$$L_{K_I} L_{\tilde{K}_{i_0}} \cdots L_{\tilde{K}_{i_n}} y(t) = 0 \quad (10)$$

for $0 \leq i_0, \cdots, i_n \leq r$ and $n \geq 0$, which after calculation of the Lie derivatives explicitly can now be hypothesized as

$$[\tilde{\mathcal{C}}(t), H_{SB}] \subset \tilde{\mathcal{C}}(t) \quad (11)$$

where the operator distribution $\tilde{\mathcal{C}}$ is as defined before. The problem of feedback based decoupling is now stated entirely in terms of the open loop Hamiltonians and parameters. Hence it is to be noted that operators that form $\tilde{\mathcal{C}}$, are open loop system Hamiltonians whose domain of operation is the system Hilbert space $\mathcal{H}_s$. However that of the interaction Hamiltonian H_{SB} and its commutator with $\tilde{\mathcal{C}}(t)$ is both system and environment. In other words, in order for the feedback to be an effective tool in solving the decoherence problem, the control Hamiltonians H_i have to act non-trivially on both the Hilbert spaces which would enable all the operators in (11) to act on system-environment Hilbert space. In the light of the above conclusion the statement in theorem (1) can be augmented with the following condition. For the distribution of quantum operators, $\tilde{\mathcal{C}}(t) = \Delta\{C_1(t), C_2(t), \cdots, C_n(t), \cdots\}$ as defined in theorem (1), the output equation (2) of the quantum system is decoupled from the environmental interactions if and only if the following hold.
Case (I). Open Loop:

$$[\tilde{\mathcal{C}}(t), H_{SB}(t)] = 0. \quad (12)$$

Case (II). Closed Loop: the *necessary* conditions for closed loop control are

$$[C, H_{SB}] = 0 \qquad \text{and} \qquad [\tilde{\mathcal{C}}(t), H_{SB}(t)] \subset \tilde{\mathcal{C}}(t).$$

In this chapter we will be primarily concerned with designing feedback for quantum systems of the form $u = \alpha(\xi) + \beta(\xi)v$ where α and β are real vector and a full rank real matrix of the state (or its estimate thereof) of dimension $1 \times r$ and $r \times r$ respectively. We examine a few systems of interest with control Hamiltonians, that might be decoupled via feedback of the above form.

A Single Qubit System. Consider a single qubit spin-1/2 system coupled to a bath of infinite harmonic oscillators through an interaction Hamiltonian H_{SB}. The Hamiltonian of the system+bath can be written as

$$H = \frac{\omega_0}{2}\sigma_z + \sum_k \omega_k b_k^\dagger b_k + \sum_k \sigma_z (g_k b_k^\dagger + g_k^* b_k)$$

where $k = 0, 1, 2 \cdots$ and the system is acted upon by the free Hamiltonian H_0 and the decoherence Hamiltonian H_{SB}. As is well known there is a rapid destruction of coherence between $|0\rangle$ and $|1\rangle$ according to the decoherence function given by [20]. In order to cast the above problem in the present framework we consider a bilinear form of an operator C that monitors coherence between the basis states. Considering C to be the non-hermitian operator $|0\rangle\langle 1|$ we have a function $y(t)$ given by $y(t) = \langle \xi(t)|C|\xi(t)\rangle$ that monitors coherence between the states $|0\rangle$ and $|1\rangle$. The problem now reduces to analyzing the applicability of the theorem 1 to the given system. It can be seen right away that the condition $[\tilde{\mathcal{C}}, H_{SB}] \neq 0$ for the distribution $\tilde{\mathcal{C}}$ defined previously, as calculated in the previous chapter. This implies that the coherence is not preserved under free dynamics or in presence of open loop control. In order to eliminate this decoherence by feedback we now assume the system to be acted upon by suitable control Hamiltonians $\{H_1, \cdots, H_r\}$ and corresponding control functions $\{u_1, \cdots, u_r\}$. As we pointed out earlier the necessary condition is relaxed to $[\tilde{\mathcal{C}}, H_{SB}] \subset \tilde{\mathcal{C}}$, with the operators C and H_{SB} still required to commute with each other $[C, H_{SB}] = 0$. For the single qubit example the second condition fails to hold, thus leaving the system unable to be *completely* decoupled and hence vulnerable to decoherence even in the presence of closed loop and feedback control.

Two Qubit Case The corresponding 2-qubit control system can be written as

$$\frac{\partial|\xi(t)\rangle}{\partial t} = \left(\sum_{j=1}^{2} \frac{\omega_0}{2}\sigma_z^{(j)} + \sum_k \omega_k b_k^\dagger b_k\right)|\xi(t)\rangle + \sum_{k,j} \sigma_z^{(j)}(g_k b_k^\dagger + g_k^* b_k)|\xi(t)\rangle$$
$$+ u_1(t)\sigma_x^{(1)} + u_2(t)\sigma_y^{(1)} + u_3(t)\sigma_x^{(2)} + u_4(t)\sigma_y^{(2)} \qquad (13)$$

where $j = 1, 2; k = 0, 1, 2 \cdots$ and the above system satisfies the basic necessary condition $[C, H_{SB}] = 0$ but not the stronger condition provided in Case(II) of the theorem. Hence the system would eventually leave the DFS and is susceptible to decoherence in the presence of arbitrary control, in other words, not open-loop decoupled from H_{SB}. In order to analyze the conditions in the presence of a classical state feedback $u = \alpha(\xi(t)) + \beta(\xi(t)).v$, the corresponding conditions (II) of the theorem are to be examined. As described before, due to the locality of the control Hamiltonians H_i acting on the system, the necessary condition specified in theorem (1) would not be satisfied non-trivially unless the distribution $\tilde{\mathcal{C}}$(the operators H_i) acted non-trivially on both $\mathcal{H}_s$ and $\mathcal{H}_e$. The above form cannot be achieved by control Hamiltonians acting only on the system. However the situation can be salvaged if one considered a "bait" qubit whose rate of decoherence or the environmental interaction can be modulated externally at will and the *bait* qubit is now allowed to interact with our qubits of interest through an Ising type coupling. Physically this amounts to a coherent qubit with controllable environmental interaction. The

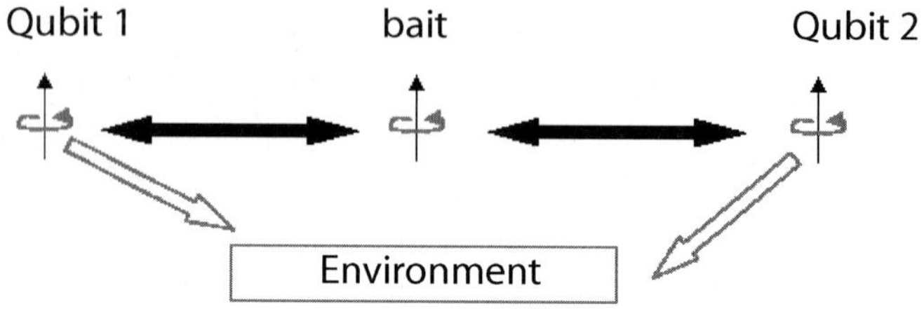

Fig. 3. The 2 Qubit system is allowed to interact with another qubit, the *bait* whose interaction with the thermal bath is controlled

Schrödinger equation for the above system can now be written as

$$
\begin{aligned}
i\hbar\frac{\partial|\xi(t)\rangle}{\partial t} &= \left(\sum_{j=1}^{2}\frac{\omega_0}{2}\sigma_z^{(j)} + \sum_k \omega_k b_k^\dagger b_k\right)\xi(t) + \sum_{k,j}\sigma_z^{(j)}(g_k b_k^\dagger + g_k^* b_k)\xi(t) \\
&+ (u_1\sigma_x^{(1)} + u_2\sigma_y^{(1)} + u_3\sigma_x^{(2)} + u_4\sigma_y^{(2)} + \frac{\omega_0}{2}\sigma_z^{(b)} + u_5\sigma_x^{(b)} + u_6\sigma_y^{(b)} \\
&+ u_7 J_1\sigma_z^{(1)}\sigma_z^{(b)} + u_8 J_2\sigma_z^{(2)}\sigma_z^{(b)} + u_9\sum_k \sigma_z^{(b)}(w_k b_k^\dagger + w_k^* b_k))\xi(t)
\end{aligned}
$$

where $j = 1, 2; k = 0, 1, 2\cdots$ and $u_1, \cdots, u_9$ are the control functions assumed to be dependent on time t and $\sigma_x, \sigma_y, \sigma_z$ are regular hermitian operators with the superscripts denoting the sub-system of interest. The coefficients of controls u_7 and u_8 are generated by the Ising type coupling between qubits 1, 2 and the bait with the corresponding coupling constants J_1 and J_2 respectively. The last term in the above control system is due to the interaction of the bait qubit with the environment whose interaction enters the system in a controllable way, hence can be treated as a separate control Hamiltonian. With the above construction it can be verified that for the distribution $\tilde{\mathcal{C}}$, as per calculations outlined in [5] the Lie bracket of $[\tilde{\mathcal{C}}, H_{SB}]$ is contained $\tilde{\mathcal{C}}$ which brings us one step closer to decoupling the coherence between the qubits from H_{SB}.

6 Invariant Subspace Formalism

In this section we explore an alternative formalism that complements the above approach and extremely helpful in feedback based decoupling. Consider the following necessary conditions for closed loop decouplability, viz.

$$L_{K_I} y(t) = 0 \qquad \text{and} \qquad L_{K_I} L_{\tilde{K}_0} y(t) = 0. \tag{14}$$

Hence $L_{\tilde{K}_0} L_{K_I} y(t) = 0$, where $\tilde{K}_0$ and $\tilde{K}_i$ are closed loop vector fields. The above equations after subtraction imply $L_{[\tilde{K}_0, K_I]} y(t) = 0$. The above equation in conjunction with other necessary conditions imply $L_{[\tilde{K}_0, K_I]} L_{\tilde{K}_j} y(t) = 0$ and $L_{\tilde{K}_j} L_{[\tilde{K}_0, K_I]} y(t) = 0$, from which it can again be concluded that $L_{[[\tilde{K}_0, K_I], \tilde{K}_j]} y(t) = 0$. In fact the above pattern of equations could be extended to any number of finite Lie brackets to conclude that

$$L_{[[\cdots[K_I, \tilde{K}_{i_1}], \tilde{K}_{i_2}]\cdots \tilde{K}_{i_k}]} y(t) = 0 \tag{15}$$

which leads us to define a set of vector fields or distribution Δ that share the same property, $K_\nu \in \Delta$ s.t $L_{K_\nu} y(t) = 0$. It is observed immediately that $K_I \in \Delta$. Such a distribution Δ is said to belong to $\ker(dy(\xi, t))$. And from the necessary conditions listed above the distribution is observed to be invariant under the closed loop control and drift vector fields $\tilde{K}_0, \cdots, \tilde{K}_m$, (i.e) $\forall K_\nu \in \Delta$,

$$[K_\nu, \tilde{K}_i] \in \Delta, \forall i \in 0, \cdots, m.$$

We will henceforth refer to the distribution as the *invariant distribution.* It is also to be noted that the above calculations are reversible and the original necessary and sufficient conditions can be derived starting from the invariant distribution. Hence the necessary and sufficient conditions for open loop decouplability can now be restated in terms of the invariant distribution following the definition.

Definition 1. *A distribution is said to controlled invariant on the analytic manifold D_ω if there exists a feedback pair (α, β), α, vector valued and β, matrix valued functions such that*

$$[\tilde{K}_0, \Delta](\xi) \subset \Delta(\xi) \qquad \textit{and} \qquad [\tilde{K}_i, \Delta](\xi) \subset \Delta(\xi) \tag{16}$$

where, $\tilde{K}_0 = K_0 + \sum_{j=1}^r \alpha_j K_j$ and $\tilde{K}_i = \sum_{j=1}^r \beta_{ij} K_j$.

It is now possible to simplify the conditions for a controlled invariant distribution with the following lemma which relates to the open loop vector fields.

Lemma 3. *An involutive distribution Δ defined on the analytic manifold $\mathcal{D}_\omega$ is invariant with respect to the closed loop vector fields $(\tilde{K}_0, \tilde{K}_1, \cdots, \tilde{K}_r)$ for some suitable feedback parameters $\alpha(\xi)$ and $\beta(\xi)$ if and only if*

$$[K_0, \Delta] \subset \Delta + G \qquad \textit{and} \qquad [K_i, \Delta] \subset \Delta + G \tag{17}$$

where G is the distribution created by the control vector fields

$$G = \textit{ span } \{K_1, \cdots, K_r\}. \tag{18}$$

At this point it is possible to express the necessary and sufficient conditions for the feedback control system $(\tilde{K}_0, \tilde{K}_1, \cdots, \tilde{K}_r)$ to be decoupled from the interaction vector field K_I by combining the above results. Moreover the conditions can be expressed entirely in terms of the open loop vector fields and the controlled invariant distribution without ever having to involve the feedback parameters $\alpha(\xi)$ and $\beta(\xi)$. We state the central theorem for decoupling of quantum systems via classical feedback without proof.

Theorem 2. *The output $y(t,\xi) = \langle\xi|C(t)|\xi\rangle$ can be decoupled from interaction vector field K_I via suitable feedback (α,β) if and only if there exists an involutive distribution Δ such that*

$$[K_0, \Delta] \subset \Delta + G \qquad [K_i, \Delta] \subset \Delta + G$$

and $\Delta \subset ker(dy)$.

The proof, along with other background, could be found in the recent exposition by the authors [5]. It is clear by now that the existence of the invariant distribution Δ is essential to performing the decoupling and hence we state an algorithm in order to arrive at the much sought after invariant distribution, the general idea being: Start out by assigning Δ to the whole of null space of $y(t)$ and iteratively remove the part of the distribution that does not satisfy the invariance conditions (16).
Step 1: Let $\Delta_0 = \ker(dy(t,\xi))$.
Step 2: $\Delta_{i+1} = \Delta_i - \{\delta \in \Delta_i : [\delta, K_j] \notin \Delta_i + G, \forall 0 \leq j \leq r\}$
Step 3: Maximal invariant distribution is such that $\Delta^* = \Delta_{i+1} = \Delta_i$.

We can perform the computation in the dual space $T^*_\xi(M)$ instead which has computational advantages and arrive at the following algorithm.
Step 1: Let $\Omega_0 = \text{span}(dy(t,\xi))$.
Step 2: $\Omega_{i+1} = \Omega_i + L_{K_0}(\Omega_i \cap G^\perp) + \sum_{j=1}^r L_{K_i}(\Omega_i \cap G^\perp)$.
Step 3: The Algorithm converges to $\Omega^* = \Omega_{i+1} = \Omega_i$.

7 Extension to Control Algebra

In the previous sections we provided a state feedback given by the vector $\alpha(\xi)$ and matrix $\beta(\xi)$ which were assumed to be analytical functions of the state ξ. In particular, the analyticity is required for the proof of necessity as well as sufficient conditions. However, the class of analytic functions is too restrictive in terms of feedback that can actually be implemented on the system. For example, by rapid pulses which are arbitrarily strong and fast one can generate lie bracket of the vector control vector fields which can act as a new control to the system available for feedback. In the light of non-analytic feedback it might be necessary to modify the conditions that guarantee decouplability of the system. Another approach which is sufficiently general would be to use the theory already developed for analytic feedback to systems whose control vector fields belong to the control algebra of the original system,(i.e) we propose to use the system, where $\hat{K}_i \in \{K_1, \cdots, K_r\}_{LA} = \mathcal{G}$. The theory of analytic feedback can now be extended to controls from the control algebra instead of just the original set of controls. Hence we can restate the conditions for decouplability in terms of the control algebra, which follows directly from the previous theorem.

Lemma 4. *The output $y(t)$ is decouplable via analytic feedback functions $\alpha(\xi)$ and $\beta(\xi)$ from the interaction vector field K_I if and only if there exists a controllability invariant distribution Δ, i.e.*

$$[\Delta, \mathcal{G}] \subset \Delta \oplus \mathcal{G} \qquad \textit{and} \qquad [\Delta, \mathcal{C}] \subset \Delta \oplus \mathcal{G} \tag{19}$$

where $\mathcal{C} = \{ad^j_{K_i} K_0, i = 1, \cdots, r; j = 0, 1 \cdots\}$ and $\mathcal{G} = \{K_1, \cdots, K_r\}_{LA}$.

The above lemma just states a condition and does not provide an explicit formulation of the application of feedback. In order to provide the analytic feedback we consider a modified system with additional control vector fields generated from the original system. Consider the following modified system with finite dimensional control algebra $\mathcal{G}$

$$\frac{\partial \xi(t)}{\partial t} = K_0|\xi(t)\rangle + \sum_{i=1}^{m} u_i \hat{K}_i|\xi(t)\rangle + K_I|\xi(t)\rangle \tag{20}$$

where the vector fields $\hat{K}_i \in \mathcal{G}$ which are generated by the vector fields of the original system are such that $\mathcal{G} = \text{span}\{\hat{K}_1, \cdots, \hat{K}_m\}$, (i.e) the set of vector fields $\hat{K}_i$, not necessary a linearly independent set form a vector space basis for $\mathcal{G}$. This is a required condition as the analytic feedback functions which can only generate utmost linear combinations of the existing control vector fields, (i.e) $\text{span}\{K_1, \cdots, K_r\}$ is inadequate to leverage the set of all possible controls. Hence it is necessary to modify the original system in order to utilize the repertoire of all possible controls for efficient feedback control. It is also to be noted that in so doing we do not alter the set of reachable or controllable set of the original system, but altering the output decouplability instead which is an observability property of the system.

7.1 Examples

As an example of the above formalism consider a single qubit coupled to the environment

$$\begin{aligned}\frac{\partial \xi(t)}{\partial t} &= \frac{\omega_0}{2}\sigma_z \xi(t) + \sum_k \omega_k b_k^\dagger b_k \xi(t) + u_1 \sigma_x \xi(t) + u_2 \sigma_y \xi(t) \\ &+ \sum_{k=0}^{\infty} \sigma_z (g_k b_k^\dagger + g_k^* b_k) \xi(t)\end{aligned}$$

with the output, $y(t) = \langle\xi(t)|C|\xi(t)\rangle$. When we check against the necessary condition that $\sum_k \sigma_z(g_k b_k^\dagger + g_k^* b_k)\xi(t) \in \ker(dy(t))$, which the single qubit system fails to satisfy, the conclusion that a single qubit system is not decouplable coincides with the results obtained earlier by operator algebra.

Now, consider the two-qubit system

$$\frac{\partial\xi(t)}{\partial t} = \left(\sum_{j=1}^{2} \frac{\omega_0}{2}\sigma_z^{(j)} + \sum_k \omega_k b_k^\dagger b_k\right)\xi(t) + \sum_{k,j} \sigma_z^{(j)}(g_k b_k^\dagger + g_k^* b_k)\xi(t) \quad (21)$$
$$+ u_1(t)\sigma_x^{(1)}\xi(t) + u_2(t)\sigma_y^{(1)}\xi(t) + u_3(t)\sigma_x^{(2)}\xi(t) + u_4(t)\sigma_y^{(2)}\xi(t)$$

where $j = 1,2; k = 0,1,2\cdots$ and which has a DFS of dimension two, span$\{|01\rangle, |10\rangle\}$ (incidentally, which implies $K_I \in \ker(dy)$), the states within which remain coherent in the absence of controls. The real problem arises in the presence of symmetry breaking perturbations or control Hamiltonians. Hence the problem at hand is to render the states coherent(or the output $y(t) = \langle\xi(t)|C|\xi(t)\rangle$ invariant) even in the presence of arbitrary control. It can be seen [5] that the interaction vector field in deed belongs to $K_I = \sum_{j,k} \sigma_z^{(j)}(g_k b_k^\dagger + g_k^* b_k)\xi(t) \in \ker(dy(t)$, where $j = 0,1$ and $k = 0,1,\cdots$, but fails to satisfy $[K_i, K_I] \in \Delta + G$ or $\ker(dy) + G$ or $\ker(dy) + \mathcal{G}$,(i.e) neither belongs to the span of the control vector fields, control algebra generated by the above vector fields or the controllability invariant distribution Δ.

Now consider the two qubit system with bait, which was discussed in the last section with the schematic and dynamic equation provided earlier.

The control system governing the mechanics of the system is given by,

$$\frac{\partial|\xi(t)\rangle}{\partial t} = \left(\sum_{j=1}^{2} \frac{\omega_0}{2}\sigma_z^{(j)} + \sum_k \omega_k b_k^\dagger b_k\right)\xi(t) + \sum_{k,j} \sigma_z^{(j)}(g_k b_k^\dagger + g_k^* b_k)\xi(t)$$
$$+ (u_1(t)\sigma_x^{(1)} + u_2(t)\sigma_y^{(1)} + u_3(t)\sigma_x^{(2)} + u_4(t)\sigma_y^{(2)} + \frac{\omega_0}{2}\sigma_z^{(b)} \quad (22)$$
$$+ u_5\sigma_x^{(b)} + u_6\sigma_y^{(b)} + u_7 J_1\sigma_z^{(1)}\sigma_z^{(b)} + u_8 J_2\sigma_z^{(2)}\sigma_z^{(b)})\xi(t)$$
$$+ u_9 \sum_k \sigma_z^{(b)}(w_k b_k^\dagger + w_k^* b_k)\xi(t)$$

where $j = 1,2; k = 0,1,2\cdots$ with $\sigma_{x|y|z}$ now skew hermitian and the same output equation as before. It is seen that $K_I \in \ker(dy(t))$ and

$$[K_i, K_I] = [\sigma_{x|y}^{(1)}\xi, \sum_j \sigma_z^{(j)}(g_k b_k^\dagger + g_k^* b_k)\xi] = c.\sum_k \sigma_{y|x}^{(1)}(g_k b_k^\dagger + g_k^* b_k)|\xi\rangle$$

now belongs to the control algebra generated by the additional vector fields introduced by the *bait* system. Hence the system which was designed in order to meet the necessary condition, $[\tilde{\mathcal{C}}, H_{SB}] \subset \tilde{\mathcal{C}}$, given by the observation space formalism is also seen to meet the conditions given by tangent space or controllability invariant distribution formalism. A rather interesting scenario arises as the drift vector field K_0 is a part of $\mathcal{G}$ and the interaction vector field K_I which is a part of the invariant subspace $\Delta \subset \ker(dy(t))$, is already contained within the control algebra, (i.e) $K_I \in \mathcal{G}$. The necessary and sufficient conditions for decouplability using feedback are easily satisfied as $[K_I, \hat{K}_i] \in \mathcal{G} \forall \hat{K}_i \in \mathcal{G}$ and $[K_I, K_0] \in \mathcal{G}$. Hence,

$$[\Delta, \hat{K}_i] \subset \Delta \oplus \mathcal{G} \qquad \text{and} \qquad [\Delta, K_0] \subset \Delta \oplus \mathcal{G} \tag{23}$$

and the invariant subspace Δ can now be guaranteed to exist and at least one dimensional equal to span$\{K_0\}$. Hence existence of feedback and decouplability is guaranteed for the system whose linear combination of control vector fields is also its control algebra $\mathcal{G}$. Such a system can be obtained from the above system by *restructuring* its control vector fields.

7.2 The Control System

By restructuring the above control system it is possible to generate control vector fields from the algebra $\mathcal{G}$. The actual control system whose linear combination of the control vector field equals its control algebra must be obtained by trial and error. It can be seen that for raw system with bait as described above the control algebra is infinite dimensional by the virtue of the infinite dimensional environment. Hence the only way the span of control vector fields could be made equal to the control algebra is by considering a system with countable control vector fields, which is practically not feasible. Another approach to ensure the system has a finite dimensional algebra is to truncate the states of the environment to certain finite dimension. The above approximation is valid in light of coherent optical states [17] whose higher energy states occur with vanishingly small coefficients. Hence with the above approximation carried out on a single mode environment we arrive at the following control system with *linearly independent* control vectors obtained by restructuring the original two qubit with bait whose control algebra and the span of control vector fields coincide [5]

$$\begin{aligned}\frac{\partial|\xi(t)\rangle}{\partial t} &= \left(\sum_{j=1}^{2} \frac{\omega_0}{2}\sigma_z^{(j)} + \sum_k \omega_k b_k^\dagger b_k\right)|\xi(t)\rangle + \sum_k \sigma_z^{(j)}(gb^\dagger + g^* b)|\xi(t)\rangle \qquad (24)\\ &+ \sum_{i=0}^{5} u_{1i}\sigma_x^{(1)}(wb^\dagger + w^* b)^i|\xi(t)\rangle + \sum_{i=0}^{5} u_{2i}\sigma_y^{(1)}(wb^\dagger + w^* b)^i|\xi(t)\rangle \\ &+ \sum_{i=0}^{5} u_{3i}\sigma_x^{(2)}(wb^\dagger + w^* b)^i|\xi(t)\rangle + \sum_{i=0}^{5} u_{4i}\sigma_y^{(2)}(wb^\dagger + w^* b)^i|\xi(t)\rangle\,.\end{aligned}$$

For the control system described above where the control vector fields $\{K_{ji}\}$, $0 \leq i \leq 5$ and $1 \leq j \leq 4$, span the entire control algebra and hence

$$[\Delta, K_{ji}] \subset \Delta + \mathcal{G}, 0 \leq i \leq 5 \text{ and } 1 \leq j \leq 4 \tag{25}$$

where $\mathcal{G} = \{K_1, \cdots, K_{24}\}_{LA} = \text{span}\{K_1, \cdots, K_{24}\}$. The above discussion can be summarized in the form of a theorem,

Theorem 3. *For the system given by (24) which is a control system on the analytic manifold $\mathcal{S}_H \subset \mathcal{H}_s \otimes \mathcal{H}_e$, with the output equation (2) there exist feedback parameters $\alpha(\xi)$ and $\beta(\xi)$ such that under the feedback control of the form $u = \alpha(\xi) + \beta(\xi).v$, we have the following invariance condition satisfied*

$$f(t, x, v_1, \cdots, v_r, K_I) = f(t, x, v_1, \cdots, v_r, K_I')$$

for any two interaction vector fields K_I and K_I' generated by interaction Hamiltonians H_{SB} and H_{SB}' and where $f \equiv y(t) = \langle \xi(t)|C(t)|\xi(t)\rangle$, is the map of the coherence functional for $0 \leq t \leq T$.

The purpose of the bait qubit and an induced interaction with the environment was to *enlarge* the control algebra as the control provided by the original two qubit system(13) was not sufficient to perform feedback decoupling. The construction now allows us to get a handle on the environment and enables us carry out feedback decoupling. A detailed procedure to determine the actual feedback parameters $\alpha(\xi)$ and $\beta(\xi)$ to completely decouple is discussed in [5]. Hence the system is completely decoupled even in the presence of symmetry breaking control Hamiltonians via classical state feedback.

The following table is helpful in noting the above decouplability results.

	Open Loop	Closed Loop	Closed Loop Restructured
Single Qubit	NO	NO	NO
Two Qubit	NO	NO	NO
Two Qubit or higher with bait qubit	NO	NO	*YES**

*-The system can be completely decoupled under the additional assumption of a finite dimensional environment

We note that the conditions for decouplability from Open loop to Closed loop to Closed Loop Restructured are progressively relaxed. Hence a system that is not Closed Loop Restructured decouplable cannot be Closed Loop or Open Loop decoupled. It is known from classical control theory that feedback can modify the observability properties of any system but not the controllability properties. It is the *observability of the decoherence* that we intend to modify in the above work by modeling it as a disturbance decoupling problem thus rendering the decoherence acting on the system unobservable on the states of interest. However in order to accomplish the goals we had to introduce additional couplings and a bait subsystem that were not a part of the system initially.

7.3 Operator Conditions for Decouplability

We had analyzed the decouplability of a given system via operators C and $H_1, \cdots, H_r$ and arrived at the necessary conditions

$$[C, H_{SB}] = 0 \qquad \text{and} \qquad [\mathcal{C}, H_{SB}] \subset \mathcal{C}. \tag{26}$$

However we weren't able to proceed as far as providing a sufficient condition for feedback decouplability via the operator or observation space formalism. Such a formalism was instrumental in designing a bait system that satisfied the above condition. Even though original bait system which satisfied the above conditions was not feedback decouplable, the restructured system (24) was. Hence it is not an unreasonable conjecture that the above conditions imply that at least one member of the class of all restructured control systems of the original system can be completely decoupled via feedback. Even though we weren't able to prove the above conjecture it is a well posed problem and could be studied further to yield insights into feedback decouplability.

8 Quantum Internal Model Principle

The fig.(4) outlines the schematic of control system for the decoupling problem, where the coherence measure of the open quantum system and the corresponding closed system must identically be equal to zero. In order to decouple the output from the environment one needs to determine the feedback coefficients $\alpha(\xi)$ and $\beta(\xi)$ where both depend on the combined state of the system and environment. Hence one needs to have a good estimate of the system as well as the environment itself for successful implementation of feedback decoupling. In other words the state observer must include a model of the environment which would enable us estimate its state. At this point, the important differences between classical and quantum decoupling problems can be understood at the outset. The necessary condition in terms of the operator algebra $[\tilde{\mathcal{C}}, H_{SB}] \subset \tilde{\mathcal{C}}$ was instrumental in design of the bait subsystem. However the structure of the system needed to be altered in order to,

(*i*) Artificially induce coupling between qubits 1, 2 and the environment with the help of the bait.

(*ii*) Generate vector fields in higher power of the environment operator to as to generate linearly independent vectors.

Hence it was necessary to modify the core system in more ways than one in order to perform decoupling. Hence, even though environment is an undesirable interaction the higher powers of the same helped us generate linearly

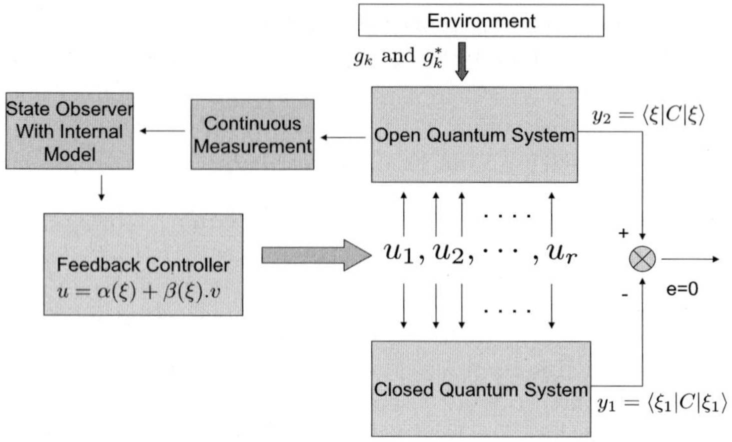

Fig. 4. The difference between coherence measures from the open quantum system and the closed quantum systems must vanish

independent vectors in the tangent space, which was absolutely necessary for decoupling. Hence the environmental coupling here befits the description of *necessary evil.* In classical dynamic feedback [8] the design of controller depends on the exosystem. In contrast the state observer/estimator needs to know the model of environment in order to estimate the combined state ξ and calculate the feedback. Hence the model discussed above could be thought of as the Internal Model Principle analog of quantum control systems. In addition classical output regulation problem concerns with following a reference signal in the presence of environmental disturbance that depends on a prescribed exosystem. On the other hand the disturbance decoupling problem focuses on eliminating the effects of the environment.

9 Classical and Quantum Disturbance Decoupling

In this section we will highlight a few more important differences between the decoherence control in quantum systems and disturbance decoupling of classical input affine systems in $\mathbb{R}^n$.

(i) Classical noise is additive, $\dot{x} = f(x)+u_ig_i(x)+wp(x)$ and operate on the same vector space. Whereas quantum noise is tensorial. The *noise* parameter g_k and g_k^* dictate the coupling between the environment and the system, (i.e), $K_I = (\sigma_z^{(1)} + \sigma_z^{(2)}) \otimes (g_k^* b_k + g_k b_k^\dagger)|\xi\rangle$ corresponds to the classical noise vector $p(x)$, and it can be easily seen that there is no noise operating on the system in the classical sense. Hence decoherence is not classical noise.

(ii) Vector spaces in quantum control systems are over complex fields. This increases the dimensionality by 2 fold in many instances where linearly combination has to be taken. Hence in order to generate every vector in a vector space of n independent states, we require $2n$ linearly independent vectors.

(iii) The necessary and sufficient conditions impose restrictions on the form of control Hamiltonian that could help decouple the system. From the conditions derived in the previous and current chapter, it is impossible to decouple one part of the system from the other unless our control Hamiltonians operate on the both the Hilbert spaces non-trivially (i.e) $H_i \in \mathcal{B}(\mathcal{H}_A \otimes \mathcal{H}_B)$, the set of linear operators in the joint Hilbert space of both the systems. It was in light of this condition that the bait system was originally introduced.

(iv) Distributions need not necessarily be non-singular. For instance the tangent space of an $\mathfrak{su}(2)$ system is spanned by $\sigma_z|\xi\rangle, \sigma_x|\xi\rangle, \sigma_y|\xi\rangle, \mathbb{I}|\xi\rangle$, where $\xi = c_0|0\rangle + c_1|1\rangle$ and the operators are again assumed to be skew hermitian counterparts of hermitian $\sigma_z, \sigma_x, \sigma_y$. Even though the four vectors are linearly independent for almost all non-zero values of c_0 and c_1 the distribution is not non-singular. Consider $|\xi\rangle = |0\rangle$ and the corresponding tangent vectors are $-i|0\rangle, i|1\rangle, -|1\rangle, i|0\rangle$, whose real linear combination is rank deficient. Hence it can be seen that the vector $|0\rangle$ *does not* belong to tangent space $T_{|0\rangle}$ at the point $|\xi\rangle = |0\rangle$. In general the tangent vectors at point ξ is different from that of another point ξ_1. One of the most serious implications is that we cannot find a linear map that transforms the distribution Δ to a constant d dimensional distribution

$$T.\Delta = \begin{bmatrix} I^{d\times d} \\ 0 \end{bmatrix}$$

at every point ξ, an approach that was used in Isidori [9] to greatly simplify finding commuting vectors $|v_1\rangle, \cdots |v_n\rangle$ in an n dimensional tangent space. The commuting vectors were just taken to be the co-ordinate basis at every point x.

10 Conclusion

We analyzed the conditions for eliminating the effects of decoherence on quantum systems whose coherence can be monitored in the form of a complex scalar output equation. The results hold globally on the analytic manifold. It was seen that the analysis performed on the analytic manifold yielded conditions in terms of the operators acting on the system in order for it to be decoupled. The conditions and a step by step procedure to calculate a classical deterministic feedback under which the 2-qubit system could be successfully decoupled from decoherence was presented. But in order for a Quantum System to be decoupled from decohering interaction the necessary condition $[C, H_{SB}] = 0$ must be satisfied which translates to existence of DFS or "decoupled in the absence of control". Hence, the results of this chapter imply that the coherence can be preserved even in the presence of symmetry breaking control Hamiltonians which help in the controllability of the system. As mentioned before the analysis carried out in the bilinear form not only helped us learn about the control Hamiltonians helpful in decoupling the system but also provided a solution under which the system would be completely decoupled as opposed to partial decoupling. Such a control strategy would be immensely helpful in performing decoherence free quantum computation thus enabling us to exploit the computational speed up provided by quantum parallelism.

References

1. V. B. Barginsky, Y.I. Vorontsov, and K.S. Thorne. Quantum nondemolition measurements. *Science*, 209(4456):547, 1980.
2. A.R. Calderbank and P.W. Shor. Good quantum error-correcting codes exist. *Phys. Rev. A*, 54:1098, 1996.
3. A.C. Doherty, K. Jacobs, and G. Jungman. Information, disturbance and Hamiltonian feedback control. *Phys. Rev. A*, 63-062306, 2001.
4. N. Ganesan and T.J. Tarn. Control of decoherence in open quantum systems using feedback. *Proc. of the 44rd IEEE Conf. on Decision and Contr.*, pages 427–433, 2005.
5. N. Ganesan and T.J. Tarn. Decoherence control in open quantum system via classical feedback. *Phys. Rev. A*, (75(032323)), 2007.
6. D.B. Horoshko and S.Y. Kilin. Decoherence slowing via feedback. *Journal of Modern Optics*, 44(11/12):2043, 1997.
7. G.M. Huang, T.J. Tarn, and J.W. Clark. On the controllability of quantum mechanical systems. *J. Math. Phys*, 24(11):2608, 1983.
8. J. Huang. *Nonlinear Output Regulation, Theory and Application.* SIAM, Philadelphia, 2004.
9. A. Isidori. *Nonlinear Control Sytems.* Springer Verlag, New York, 3rd edition, 1995.
10. A. Isidori, A.J. Krener, C. Gori Giorgi, and S. Monaco. Locally (f, g)-invariant distributions. *Systems & Control Letters*, 1:12–15, 1981.
11. A. Isidori, A.J. Krener, C. Gori Giorgi, and S. Monaco. Nonlinear decoupling via feedback: A differential geometric approach. *IEEE Trans. on Automat. Contr.*, 26:331–345, 1981.
12. K. Jacobs. How to project qubits faster using quantum feedback. *Phys. Rev. A*, 67-030301(R), 2003.
13. H. Kunita. Supports of diffusion processes and controllability problems. *Proc. Int. Sym. on SDE*, pages 163–185, 1976.
14. H. Kunita. On the controllability of nonlinear systems with applications to polynomial systems. *Appl. Math. Optim.*, 5(89), 1979.
15. C. Lan, T.J. Tarn, Q.S. Chi, and J.W. Clark. Analytic controllability of time-dependent quantum control systems. *J. Math. Phys*, 2005.
16. D. A. Lidar, I.L. Chuang, and K.B. Whaley. Decoherence-free subspaces for quantum computation. *Phys. Rev. Lett.*, 15(12):2594, 1998.
17. W.H. Louisell. *Quantum Statistical Properties of Radiation.* John Wiley & Sons, New York, 1973.
18. E. Nelson. Analytic vectors. *Math. Ann.*, 70:572, 1959.
19. M.A. Nielsen and I.L. Chuang. *Quantum Computation and Quantum Information.* Cambridge University Press, 2000.

20. V. Protopopescu, R. Perez, C. D'Helon, and J. Schmulen. Robust control of decoherence in realistic one-qubit quantum gates. *J. Phys A: Math. Gen.*, 36:2175, 2003.
21. P. Shor. Scheme for reducing decoherence in quantum computer memory. *Phys. Rev. A*, 52:2493, 1995.
22. H.J. Sussman and V. Jurdjevic. Controllability of nonlinear systems. *J. Diff Eqns*, 12(95), 1972.
23. C. Uchiyama and M. Aihara. Multipulse control of decoherence. *Phys. Rev. A*, 66, 2002.
24. L. Viola, E. Knill, and S. Lloyd. Dynamical decoupling of open quantum systems. *Phys. Rev. Lett.*, 82(12):2417, 1999.
25. L. Viola, S. Lloyd, and E. Knill. Universal control of decoupled quantum systems. *Phys. Rev. Lett.*, 83(23):4888, 1999.
26. S. Wallentowitz. Quantum theory of feedback of bosonic gases. *Phys. Rev. A*, 66-032114, 2002.
27. W.H. Zurek. Pointer basis of quantum apparatus: Into what mixture does the wave packet collapse? *Phys. Rev. D*, 24:1516–1525, 1981.

Controller and Observer Normal Forms in Discrete-Time

Salvatore Monaco[1] and Dorothée Normand-Cyrot[2]

[1] Dip. di Inf. e Sistemistica, Università di Roma "La Sapienza", V. Eudossiana 18, 00184 Rome, Italy
[2] Laboratoire des Signaux et Systèmes, CNRS, Supélec, F-91192 Gif-sur-Yvette, France

Summary. Up to what extent is it possible to simplify the nonlinearities of a given discrete-time system through transformations involving feedbacks or output-injections? Solutions to these problems, at the basis of several control and state estimation design procedures, take advantage of a combined use of geometric and algebraic methods, two major setups in nonlinear control theory strongly influenced by the fundamental work of Alberto Isidori. Links between these frameworks are illustrated in the present paper which is written to celebrate his 65th birthday.

With our sincere gratitude, Happy Birthday Alberto.

1 Introduction

Starting from the eighties, differential geometry and algebraic methods revealed to be powerful tools in the study of nonlinear control systems [8]. Amongst others, the method of normal forms, introduced in the control literature in [12],[10] stands in setting and solving a given problem following a stepwise procedure which works out on successive polynomial approximations. The idea, which finds its origins in pure mathematics making reference to the Cartan's method of equivalence or the Poincaré forms, is widely renewed when applied to control systems. This justified an increasing effort of research (see [9, 1, 5, 14, 18, 7, 17, 6]).

Motivated by its interest for observer design [13, 19, 11, 3], the approach has been more recently used to solve the dual problem of reducing by output-injections a given system to its linear observer form [2].

While the normal form approach can be similarly developed for both cases of vector fields (differential dynamical systems) and maps (discrete-time systems), such a parallelism becomes difficult when dealing with forced dynamical systems. Manipulations over difference equations induce compositions of functions and the design of feedback or output-injection involve intricate problems of inversions of maps. Even if many analogies can be set, specific studies are necessary in discrete time.

Given a nonlinear discrete-time system, we ask the two questions: up to what extent is it possible to linearize the system through coordinates change and feedback? up to what extent is it possible to linearize the system through coordinates change and output-injection? In the present paper, normal forms are studied in the formal context of asymptotic series expansions. Approximations have to be understood as referred to homogeneous polynomial approximations of increasing degree. In fact, the problem is set and solved step-by-step. At each step, a certain degree of approximation can be gained and the remaining terms – those which cannot be cancelled – specify the obstruction to the reduction into linear forms so defining what will be called *controller and observer normal forms.* These terms are referred to as the resonance terms and the equations to be solved are referred to as the homological equations.

The study of the resonant structure is performed starting from a differential/difference representation of a discrete-time system introduced in [16] and assumed linearly controllable and observable. The investigation is developed following [17] where "extended controller" normal forms are introduced and computed making reference to approximate feedback linearization. A parallel study is performed for observer normal forms and approximate linearization by output-injection. Two types of normal forms are described depending if one privileges cancellation in the drift term or in the controlled vector fields. The normal forms are computed step-by-step for increasing degrees of approximation. The method, based on the solvability of the homological equations in terms of coordinates change and feedback and/or output-injection, is directly constructive either to design a linearizing controller or a state-estimator with linear error dynamics. Moreover, working in such a differential context, makes it possible to stress a link between the presence of resonance terms which cannot be cancelled and the obstructions to the geometric solutions.

The paper is organized as follows. Section 2 introduces the context and sets the faced problems. Sections 3 and 4 deal with homogeneous linear equivalences by feedback and output-injection; necessary and sufficient conditions are given and the the homological equations are derived. On these bases, two types of controller and observer normal forms are described.

Notation - The state variables ζ and/or x belong to $\mathcal{X}$, an open set of R^n and the control variables v and/or u belong to $\mathcal{U}$, a neighborhood of zero in R. All the involved objects, maps, vector fields, control systems are analytic on their domains of definition, infinitely differentiable admitting convergent Taylor series expansions. A vector field on $\mathcal{X}$, analytically parameterized by u, $G(x,u) \in T_x\mathcal{X}$ defines a u-dependent differential equation of the form $\frac{dx^+(u)}{du} = G(x^+(u),u)$ where the notation $x^+(u)$ indicates that the state evolution is a curve in R^n, parameterized by u; $G(x,u)$ is complete, an absolutely continuous solution exists for all u. A R^n-valued mapping $F(.,u) : x \to F(x,u)$, denotes a forced discrete-time dynamics while $F : x \to F(x)$ and/or $F(.,0)$ denotes unforced evolutions. Given a generic map on $\mathcal{X}$, its evaluation at a point x is denoted indifferently by "(x)" or "$\big|_x$". $J_x F\big|_{x=0} = \frac{dF(x)}{dx}\big|_{x=0}$ indicates the Jacobian of the function evaluated at $x = 0$. Given a vector field G on $\mathcal{X}$ and assuming that F is a diffeomorphism on $\mathcal{X}$, F_*G denotes the transport of G along F, defined as the vector field on $\mathcal{X}$ verifying $F_*G\big|_F = (J_xF)G$; analogously F_*^pG denotes the transport of G along F^p verifying $F_*^pG|_{F^p} = (J_xF^p)G$. The upperscript $(.)^{[m]}$ stands for the homogeneous term of degree m of the Taylor series expansion of the function or vector field into parentheses; $(.)^{(\geq m)}$ stands for the terms of order greater or equal to m. Analogously, $R^{[m]}(.)$ (resp. $R^{\geq m}(.)$) stands for the space of vector fields or functions whose components are polynomials of degree m (resp. formal power series of degree $\geq m$) in the indeterminates. The results are local in nature and convergence problems are not addressed so that the involved transformations (coordinates change, feedback, output-injection) as well as the proposed solutions described by their asymptotic expansions will be referred to as *formal* ones.

2 Context and Problem Settlement

We consider a discrete-time system, controllable and observable in the first approximation around the equilibrium pair $(x = 0, u = 0)$, described by the differential/difference representation – DDR – introduced in [16].

$$\zeta^+ = F(\zeta) \tag{1}$$

$$\frac{d\zeta^+(v)}{dv} = G(\zeta^+(v), v); \quad \zeta^+(0) = \zeta^+ \tag{2}$$

$$y = h(\zeta)\,. \tag{3}$$

In (1)–(3) $F(0) = 0$ and G admit the Taylor-type expansion around $v = 0$, $G(.,v) := G_1(.) + \sum_{i\geq 1} \frac{v^i}{i!} G_{i+1}(.)$, with $G_1(.) := G(.,0)$, $G_1(0) \neq 0$ and, for $i \geq 1$, $G_{i+1}(.) := \frac{\partial^i G(.,v)}{\partial v^i}\big|_{v=0}$.

With respect to the usual representation of a discrete-time dynamics in the form of a map the following comments are in order.

- A nonlinear difference equation in the form of a map, $\zeta \to F(\zeta, v)$, can be recovered integrating (2) between 0 and v_k with initialization at (1), $\zeta^+(0) = \zeta^+ = F(\zeta_k)$; one has

$$\zeta_{k+1} = \zeta^+(v_k) = F(\zeta_k, v_k) = F(\zeta_k) + \int_0^{v_k} G(\zeta^+(w), w))dw.$$

At the same time, one has

$$y_{k+1} = h(\zeta^+(v_k)) = h(F(\zeta_k)) + \int_0^{v_k} L_{G(.,w)}h(\zeta^+(w))\Big|_{F(\zeta_k)} dw$$

where $L_{G(.,w)}h\Big|_{F(\zeta_k)}$ denotes the Lie derivative of h along the vector field $G(.,w)$, evaluated at $F(\zeta_k)$.

- Starting from a difference equation $\zeta \to F(\zeta, v)$, the existence of (1)–(2) follows from the existence of $G(.,v)$ verifying $G(F(.,v),v)) = \frac{\partial F(.,v)}{\partial v}$. The invertibility of $F(.,0)$ is sufficient to prove that $G(.,v)$ can be locally uniquely defined as $G(.,v) := \frac{\partial F(.,v)}{\partial v}|_{F^{-1}(.,v)}$. Sampled dynamics which are characterized by invertible drift term for sufficiently small sampling period do admit such a representation.

There is thus no loss of generality in starting from (1)–(2)–(3) since, under the assumptions of controllability and observability in the first approximation, a possible preliminary transformation can be performed to ensure drift invertibility and the existence of a DDR for both the investigated problems.

Throughout the paper, $\Sigma^{[\infty]}$ will denote the asymptotic expansion of the DDR

$$\zeta^+ = A\zeta + \sum_{m\geq 2} F^{[m]}(\zeta); \quad \zeta^+(0) = \zeta^+ \tag{4}$$

$$\frac{d\zeta^+(v)}{dv} = B + \sum_{m\geq 2}\sum_{i=1}^{m} \frac{v^{i-1}}{(i-1)!} G_i^{[m-i]}(\zeta^+(v)) \tag{5}$$

$$y = C\zeta + \sum_{m\geq 2} h^{[m]}(\zeta) \tag{6}$$

with controllable and observable matrices (A, B, C). For any order of approximation $m \geq 2$, $\Sigma^{[m]}$ will denote its homogeneous approximation of degree m around (A, B, C); i.e.

$$\zeta^+ = A\zeta + F^{[m]}(\zeta); \quad \zeta^+(0) = \zeta^+ \tag{7}$$

$$\frac{d\zeta^+(v)}{dv} = B + \sum_{i=1}^{m} \frac{v^{i-1}}{(i-1)!} G_i^{[m-i]}(\zeta^+(v)). \tag{8}$$

$$y = C\zeta + h^{[m]}(\zeta). \tag{9}$$

Remark 1. **(i)** The $G_i^{[m-i]}$'s, $i \geq 2$ in (8) model nonlinearities with respect to the control variable. Setting $G_i^{[m-i]} = 0$ for $i \geq 2$, (8) reduces to $B + G_1^{[m-1]}(\zeta^+(v))$ and the results obtained are nicely comparable with those obtained in the continuous-time case for input-affine dynamics [18]. **(ii)** We note that (7)–(9) correspond to homogeneous parts of order m around A and C while equation (8) corresponds to the homogeneous part of order $m-1$ around B because, after integration with respect to v, it results to be of order m too. With reference to the representation (4)–(5-(6), the paper studies, through successive polynomial approximations, linearization of $\Sigma^{[\infty]}$ under "suitable" transformations. ◁

2.1 Linear Equivalence by Feedback

Under the assumption of controllability in the first approximation around $(0,0)$ of (1)–(2), a preliminary coordinates change and static state feedback can be applied to transform the matrices A and B into the controllable canonical form

$$A_C = J_\zeta F|_{\zeta=0} = \begin{pmatrix} 0 & 1 & 0 & \dots & 0 \\ 0 & \ddots & \ddots & & \vdots \\ \vdots & & \ddots & \ddots & 0 \\ & & & \ddots & 1 \\ a_0 & \dots & \dots & \dots & a_{n-1} \end{pmatrix}; \quad B_C = G_1(0) = \begin{pmatrix} 0 \\ \vdots \\ 0 \\ 1 \end{pmatrix} \tag{10}$$

so obtaining a representation of the form $\Sigma^{[\infty]}$ (resp. $\Sigma^{[m]}$) denoted $\Sigma_C^{[\infty]}$ (resp. $\Sigma_C^{[m]}$) in the present set up. Moreover, without loss of generality, we assume $a_0 \neq 0$.

A *feedback transformation* $\Gamma^{[\infty]}$ is defined as the successive application of *homogeneous feedback transformations of degree* $m \geq 2$, $\Gamma^{[m]}$; each $\Gamma^{[m]}$ is composed with a coordinates change and a static-state feedback of the form

$$x = \zeta + \phi^{[m]}(\zeta) \tag{11}$$

$$v = \gamma^{[m]}(\zeta, u) = u + \gamma_0^{[m]}(\zeta) + \sum_{i=1}^{m} \frac{u^i}{i!} \gamma_i^{[m-i]}(\zeta) \tag{12}$$

where $\phi^{[m]}$ and the $\gamma_i^{[m-i]}$'s for $i = 0, \dots, m$, are respectively R^n and R-valued mappings. $\Gamma^{[m]}$ does not modify the linear part (A_C, B_C) of $\Sigma_C^{[m]}$.

What about linearization of Σ_C under the feedback transformations Γ?

2.2 Linear Equivalence by Output-Injection

Under the assumption of observability in the first approximation around $(0,0)$ of (1)–(3), a preliminary coordinates change and output-injection can be applied to transform the matrices (A, C) into the observable canonical form (A_O, C_O)

$$A_O = J_\zeta F|_{\zeta=0} = \begin{pmatrix} 0 & 0 & 0 & \dots & a_0 \\ 1 & \ddots & \ddots & & \vdots \\ \vdots & & \ddots & \ddots & \vdots \\ & & & \ddots & \vdots \\ 0 & \dots & \dots & 1 & a_{n-1} \end{pmatrix}; \qquad C_O = (0, \dots, 0, 1) \tag{13}$$

so obtaining a representation of the form $\Sigma^{[\infty]}$ (resp. $\Sigma^{[m]}$) denoted by $\Sigma_O^{[\infty]}$ (resp. $\Sigma_O^{[m]}$) in the present set up. Moreover, without loss of generality, we assume $a_0 \neq 0$.

An *output transformation* $H^{[\infty]}$ is defined as the successive application of *homogeneous output transformations of degree* $m \geq 2$, $H^{[m]}$. Each $H^{[m]}$, is described by the coupled action of an homogeneous coordinates change (11) and an homogeneous output-injection of degree m, composed with two parts

$$\alpha^{[m]}(y) \quad \text{and} \quad \beta^{[m-1]}(y, v) = \beta_1^{[m-1]}(y) + \sum_{i=2}^{m} \frac{v^{i-1}}{(i-1)!} \beta_i^{[m-i]}(y)$$

where the $\alpha^{[m]}$ and the $\beta_i^{[m-i]}$'s for $i = 1, ..., m$, are R^n-valued mappings which act over the DDR as follows

$$\zeta^+ \rightarrow \zeta^+ + \alpha^{[m]}(y) \tag{14}$$

$$\frac{d\zeta^+(v)}{dv} \rightarrow \frac{d\zeta^+(v)}{dv} + \beta^{[m-1]}(y, v). \tag{15}$$

$H^{[m]}$ does not modify the linear part (A_O, B, C_O) of $\Sigma_O^{[m]}$.

What about linearization of Σ_O under the output transformations H ?

2.3 Some Definitions

Definition 1. *$\Sigma^{[m]}$ is locally linear equivalent by feedback (resp. by output-injection) if there exists an homogeneous transformation $\Gamma^{[m]}$ (resp. $H^{[m]}$) which brings $\Sigma^{[m]}$ into the linear system (A_C, B_C, C) (resp. (A_O, B, C_O)) modulo terms in $R^{\geq m+1}(\zeta, v)$.*

Definition 1 refers to the problem of *homogeneous linear equivalence by feedback at degree m* – **HLEF(m)** – (resp. *homogeneous linear equivalence by output-injection at degree m* – **HLEI(m)** –).

Definition 2. *$\Sigma^{[\infty]}$ is locally linear equivalent by feedback (resp. by output-injection) if there exists a transformation $\Gamma^{[\infty]}$ (resp. $H^{[\infty]}$) which brings $\Sigma^{[\infty]}$ into the linear system (A_C, B_C, C) (resp. (A_O, B, C_O)). When the equivalence holds modulo terms in $R^{\geq M+1}(\zeta, v)$, an approximated solution is obtained.*

Definition 2 refers to the problem of *linear equivalence by feedback* – **LEF** – and *approximate linear equivalence by feedback up to degree M* – **ALEF(M)** – (resp. *linear equivalence by output-injection* – **LEI** – and *approximate linear equivalence by output-injection up to degree M* – **ALEI(M)** –).

Up to what extent is it possible to simplify the nonlinearities of $\Sigma^{[\infty]}$ and thus to achieve linearization through these transformations ? The problem is solved step-by-step. For each degree of approximation $m \geq 2$, we look for an homogeneous transformation under which $\Sigma^{[m]}$ is simplified at most as possible while leaving unchanged the linear part and parts of degree $< m$.

The resulting representations define the normal forms, as given in the following definition.

Definition 3. *The iterated application of the $\Gamma^{[m]}$'s (resp. the $H^{[m]}$'s), each one for eliminating at most as possible parts of degree equal to m, give the general structures which cannot be more simplified. They will be called controller (observer) normal forms.*

Roughly speaking normal forms are linked to the problem of maximally simplifying the given system making use of feedback or output-injection.

3 Controller Normal Forms

As noted, controller normal forms are in relation with the problem of feedback linearization. From [15], (see also different approaches proposed in [4]), the following results are recalled.

3.1 Geometric Conditions

Theorem 1. *Linear equivalence by feedback. The* **LEF** *problem is solvable for (1)–(2) if and only if*

(i) *span*$(G_2, G_3, ...) \subset$ *span*(G_1);
(ii) *the distribution* $(G_1, \ldots, F_*^{n-2}G_1)$ *is involutive around* 0.

Proposition 1. *Homogeneous linear equivalence by feedback at degree m. The* **HLEF(m)** *problem is solvable for (1)–(2) if and only if*

(i) *span*$(G_2^{[m-2]}, ..., G_m^{[0]}) \subset$ *span*(B);
(ii) *the distribution* $(B + G_1^{[m-1]}, \ldots, A^{n-2}B + (F_*^{n-2}G_1)^{[m-1]}$ *is involutive around* 0 *modulo terms in* $R^{\geq m-1}(\zeta)$.

3.2 Homological Equations

Given $\Sigma_C^{[m]}$, the **HLEF(m)** problem is solved by computing $\Gamma^{[m]}$ which cancels the terms of degree m in (7) and of degree $m-1$ in (8). The computation goes through the solution of homogeneous algebraic equations, the so-called *m-th degree homological equations.* The terms which cannot be eliminated are named the *m-th degree resonance terms* which define the *m-th degree normal form.* Thanks to the introduced formalism, approximate feedback linear equivalence can be reported to the solvability of homological equations of increasing degree.

Let us work out the action of $\Gamma^{[m]}$ over $\Sigma_C^{[m]}$. First, the coordinates change $\phi^{[m]}$ transforms $(F^{[m]}, G_i^{[m-i]})$ into $(\bar{F}^{[m]}, \bar{G}_i^{[m-i]})$ below

$$\bar{F}^{[m]}(.) = F^{[m]}(.) + \phi^{[m]}(A_C.) - A_C\phi^{[m]}(.) \tag{16}$$

$$\bar{G}_1^{[m-1]}(.) = G_1^{[m-1]}(.) + \frac{d\phi^{[m]}(.)}{d\zeta}B_C \tag{17}$$

$$\bar{G}_i^{[m-i]}(.) = G_i^{[m-i-1]}(.); \quad i = 2, ..., m. \tag{18}$$

The feedback action further transforms (16) to (18) into

$$\tilde{F}^{[m]}(.) = F^{[m]}(.) + \phi^{[m]}(A_C.) - A_C\phi^{[m]}(.) + \gamma_0^{[m]}(.)B_C$$

$$\sum_{i=1}^{m} \frac{u^{i-1}}{(i-1)!}\tilde{G}_i^{[m-i]}(.) = \frac{d\phi^{[m]}(.)}{d\zeta_n} + \sum_{i=1}^{m} \frac{u^{i-1}}{(i-1)!}G_i^{[m-i]}(.)$$
$$+ \sum_{i=1}^{m} \frac{u^{i-1}}{(i-1)!}\gamma_i^{[m-i]}(A_C^{-1}. - vA^{-1}B_C)B_C$$

because, up to an error in $R^{\geq m}(\zeta)$

$$\zeta = A_C^{-1}\zeta^+(v) - vA_C^{-1}B_C$$
$$dv = du + \sum_{i=1}^{m} \frac{u^{i-1}}{(i-1)!}\gamma_i^{[m-i]}(\zeta)du$$
$$= du + \sum_{i=1}^{m} \frac{u^{i-1}}{(i-1)!}\gamma_i^{[m-i]}(A_C^{-1}\zeta^+(v) - vA_C^{-1}B_C)du$$

and up to an error in $R^{\geq m+1}(\zeta, v)$

$$\zeta^+(v)\big|_{v=\gamma_0^{[m]}} = \zeta^+(0) + \gamma_0^{[m]}B$$
$$x^+(v) = \zeta^+(v) + \phi^{[m]}(\zeta^+(v)).$$

In conclusion, $\Gamma^{[m]}$ brings the system $\Sigma_C^{[m]}$ into $\tilde{\Sigma}_C^{[m]}$ described by

$$x^+ = A_C x + F^{[m]}(x) + \phi^{[m]}(A_C x) - A_C \phi^{[m]}(x) + \gamma_0^{[m]}(x) B_C \tag{19}$$

$$\frac{dx^+(u)}{du} = \frac{d\phi^{[m]}(x^+(u))}{dx_n} + \sum_{i=1}^{m} \frac{u^{i-1}}{(i-1)!} G_i^{[m-i]}(x^+(u))$$
$$+ \sum_{i=1}^{m} \frac{u^{i-1}}{(i-1)!} \gamma_i^{[m-i]}(A_C^{-1} x^+(u) - u A_C^{-1} B_C) B_C. \tag{20}$$

Remark 2. We deduce from (20), setting $u = 0$, that

$$\tilde{G}_1^{[m-1]}(.) = G_1^{[m-1]}(.) + \frac{d\phi^{[m]}(.)}{d\zeta_n} + \gamma_1^{[m-1]}(A_C^{-1}.) B_C \tag{21}$$

while for $i \geq 2$, $\tilde{G}_i^{[m-i]}$ and $G_i^{[m-i]}$ differ from their last component only. To write down the expression of $\tilde{G}_i^{[m-i]}$ in terms of $G_i^{[m-i]}$ involve the expansion with respect to u of $\gamma_i^{[m-i]}(A_C^{-1}\zeta - uA_C^{-1}B_C)$. ◁

The following result is an immediate consequence of the equalities (19)–(20).

Proposition 2. *The* **HLEF(m)** *problem is solvable if and only if there exist* $(\phi^{[m]}, \gamma_i^{[m-i]}; i = 0, ..., m)$*, satisfying*

$$-F^{[m]}(\zeta) = \phi^{[m]}(A_C\zeta) - A_C\phi^{[m]}(\zeta) + \gamma_0^{[m]}(\zeta) B_C \tag{22}$$

$$-\sum_{i=1}^{m} \frac{u^{i-1}}{(i-1)!} G_i^{[m-i]}(\zeta) = \frac{d\phi^{[m]}(\zeta)}{d\zeta_n} + \sum_{i=1}^{m} \frac{u^{i-1}}{(i-1)!} \gamma_i^{[m-i]} A_C^{-1}(\zeta - uB_C) B_C. \tag{23}$$

Equations (22)–(23) are referred to as the *controller homological equations at degree m.*

Proposition 1 and Proposition 2 provide two equivalent criteria for the solvability of the **HLEF(m)** problem. While the explicit computation of the homogeneous transformation $\Gamma^{[m]}$ requires through Proposition 1 to solve partial a partial derivative equation, it requires through Proposition 2 to solve algebraic equations. This is a main constructive aspect of the normal forms approach.

3.3 The Normal Forms

Two types of normal forms, denoted in the sequel $\Sigma_{NFA}^{[m]}$ and $\Sigma_{NFB}^{[m]}$ respectively, can be obtained depending on how the computations are performed.

$\boldsymbol{\Sigma_{NFA}^{[m]}}$ – Looking at the *first-type of controller homological equations* (22) rewritten componentwise, we have

$$-F_1^{[m]}(\zeta) = \phi_1^{[m]}(A_C\zeta) - \phi_2^{[m]}(\zeta) \tag{24}$$
$$\vdots$$
$$-F_{n-1}^{[m]}(\zeta) = \phi_{n-1}^{[m]}(A_C\zeta) - \phi_n^{[m]}(\zeta) \tag{25}$$
$$-F_n^{[m]}(\zeta) = \phi_n^{[m]}(A_C\zeta) - \sum_{i=0}^{n-1} a_i \phi_{i+1}^{[m]}(\zeta) + \gamma_0^{[m]}(\zeta) \tag{26}$$

where $F_i^{[m]} : R^n \to R$ (resp. $\phi_i^{[m]} : R^n \to R$) indicates the i-th component of $F^{[m]}$ (resp. $\phi^{[m]}$), that is an homogeneous polynomial of order m in the variables $(\zeta_1, ..., \zeta_n)$. From (24) to (25), we immediately deduce that for $j = 2, ..., n$, $\phi_j^{[m]}$ can be used to cancel all the terms of order m into the corresponding $(j-1)$-th component of the drift so getting the solution

$$\phi_j^{[m]}(\zeta) = \phi_{j-1}^{[m]}(A_C\zeta) + F_{j-1}^{[m]}(\zeta).$$

From (26), we immediately deduce that $\gamma_0^{[m]}$ can be used to cancel all the terms of order m into the last component of the drift so getting the solution

$$\gamma_0^{[m]}(\zeta) = -F_n^{[m]}(\zeta) - \phi_n^{[m]}(A_C\zeta) + \sum_{i=0}^{n-1} a_i \phi_{i+1}^{[m]}(\zeta).$$

All the terms in the drift have been cancelled and all the coefficients in $\phi_1^{[m]}(\zeta)$ are kept free.

$\mathbf{\Sigma_{NFB}^{[m]}}$ – Looking now at the *second-type of controller homological equations* (23), indicating by $G_{i;j}^{[m-i]}$, the j-th component of $G_i^{[m-i]}$, we easily verify that for $i = 1, ..., m$, $\gamma_i^{[m-i]}(A_C^{-1}\zeta)$ can be used to cancel the last component $G_{in}^{[m-i]}(\zeta)$ of $G_i^{[m-i]}(\zeta)$ while its remaining $n-1$ components cannot be modified except that of $G_1^{[m-1]}$. More precisely, from (21), $G_1^{[m-1]}$ has to satisfy for $i = 1, ..., n-1$, the equality

$$-G_{1;i}^{[m-1]}(\zeta) = \frac{d\phi_i^{[m]}(\zeta)}{d\zeta_n}. \tag{27}$$

After some easy, even quite tedious, manipulations the following theorem can be proved.

Theorem 2. *For any degree $m \geq 2$ and neglecting higher degree terms, any homogeneous discrete-time dynamics $\Sigma_C^{[m]}$ can be transformed under homogeneous feedback transformation $\Gamma^{[m]}$ into one of the two controller normal forms below.*
The first type of normal form (linearity of the drift) – $\mathbf{\Sigma_{NFA}^{[m]}}$:

$$\begin{aligned}
x^+ &= A_C x \\
\frac{dx_1^+(u)}{du} &= \sum_{i=2}^{m} \frac{u^{i-1}}{(i-1)!} G_{i;1}^{[m-i]}(x^+(u)) \\
\frac{dx_2^+(u)}{du} &= x_1^+(u) Q_{2;1}^{[m-2]}(x_1^+(u), ..., x_n^+(u)) + \sum_{i=2}^{m} \frac{u^{i-1}}{(i-1)!} G_{i;2}^{[m-i]}(x^+(u)) \\
&\cdots \\
\frac{dx_{n-1}^+(u)}{du} &= \sum_{i=1}^{n-2} x_i^+(u) Q_{n-1,i}^{[m-2]}(x_i^+(u), ..., x_n^+(u)) + \sum_{i=2}^{m} \frac{u^{i-1}}{(i-1)!} G_{i;n-1}^{[m-i]}(x^+(u)) \\
\frac{dx_n^+(u)}{du} &= 1.
\end{aligned}$$

The second type of normal form $(G_1 = B)$ *–* $\mathbf{\Sigma^{[m]}_{NFB}}$*:*

$$\begin{aligned} x_1^+ &= x_2 + x_n x_1 F_{1;1}^{[m-2]}(x_1, \ldots, x_n) \\ &\ldots \\ x_{n-2}^+ &= x_{n-1} + \sum_{i=1}^{n-2} x_n x_i F_{n-2;i}^{[m-2]}(x_i, \ldots, x_n) \\ x_{n-1}^+ &= x_n \\ x_n^+ &= \sum_{i=0}^{n-1} a_i x_{i+1} \\ \frac{dx_p^+(u)}{du} &= \sum_{i=2}^{m} \frac{u^{i-1}}{(i-1)!} G_{i;p}^{[m-i]}(x^+(u)); \quad p = 1, ..., n-1 \\ \frac{dx_n^+(u)}{du} &= 1. \end{aligned}$$

Remark 3. **(i)** By construction, the homogeneous m-th degree normal forms are unique modulo transformations of degree m. **(ii)** By construction, the homogeneous m-th degree normal forms defined above are equivalent through transformations of degree m and modulo approximations in $R^{\geq m+1}(x, u)$ to those defined in [1], [5], [14], [7], in the usual formalism. $\triangleleft$

As an homogeneous transformation of a given degree does not modify the lower degree terms, applying the results of Theorem 2 to each successive homogeneous part of degree m and increasing the degree, Theorem 3 below describes the normal forms of a nonlinear discrete-time dynamics through feedback transformations.

Theorem 3. *The nonlinear discrete-time dynamics* $\Sigma^{[\infty]}$ *can be transformed under feedback transformation* $\Gamma^{[\infty]}$ *into a dynamics exhibiting one of the two controller normal forms below.*
The first type of normal form (linearity of the drift) – $\mathbf{\Sigma^{[\infty]}_{NFA}}$*:*

$$\begin{aligned} x^+ &= A_C x \\ \frac{dx_1^+(u)}{du} &= \sum_{i=2}^{\infty} \frac{u^{i-1}}{(i-1)!} G_{i;1}(x^+(u)) \\ \frac{dx_2^+(u)}{du} &= x_1^+(u) Q_{2;1}(x_1^+(u), ..., x_n^+(u)) + \sum_{i=2}^{\infty} \frac{u^{i-1}}{(i-1)!} G_{i;2}(x^+(u)) \\ &\ldots \\ \frac{dx_{n-1}^+(u)}{du} &= \sum_{i=1}^{n-2} x_i^+(u) Q_{n-1,i}(x_i^+(u), ..., x_n^+(u)) + \sum_{i=2}^{\infty} \frac{u^{i-1}}{(i-1)!} G_{i;n-1}(x^+(u)) \\ \frac{dx_n^+(u)}{du} &= 1 \end{aligned}$$

where $Q_{j;i}(x_i,...,x_n)$ *is a formal series defined by the formal summation*

$$Q_{j;i}(x_i,...,x_n) = \sum_{m=0}^{\infty} Q_{j;i}^{[m]}(x_i,...,x_n).$$

The second type of normal form ($G_1 = B_C$) – $\mathbf{\Sigma_{NFB}^{[\infty]}}$*:*

$$\begin{aligned}
x_1^+ &= x_2 + x_n x_1 F_{1;1}(x_1,\ldots,x_n) \\
&\ldots \\
x_{n-2}^+ &= x_{n-1} + \sum_{i=1}^{n-2} x_n x_i F_{n-2;i}(x_i,\ldots,x_n) \\
x_{n-1}^+ &= x_n \\
x_n^+ &= \sum_{i=0}^{n-1} a_i x_{i+1} \\
\frac{dx_p^+(u)}{du} &= \sum_{i=2}^{\infty} \frac{u^{i-1}}{(i-1)!} G_{i;p}(x^+(u)); \quad p = 1,...,n-1 \\
\frac{dx_n^+(u)}{du} &= 1
\end{aligned}$$

where $F_{j;i}(x_i,...,x_n)$ *is a formal series defined by the formal summation*

$$F_{j;i}(x_i,...,x_n) = \sum_{m=0}^{\infty} F_{j;i}^{[m]}(x_i,...,x_n).$$

Remark 4. Assuming the $G_i's$ equal to zero for $i \geq 2$ in Σ^{∞}, discrete-time and continuous-time controller normal forms exhibit strongly comparable structures. More precisely, $\Sigma_{NFB}^{[\infty]}$ is the discrete-time equivalent of the continuous-time extended controller normal form introduced in [10], [9] while $\Sigma_{NFA}^{[\infty]}$ is the discrete-time equivalent of the continuous-time dual normal form introduced in [9], [18]. ◁

If Theorem 3 holds true with a summation over m starting at M, approximate feedback linearization at degree M hods and the linearizing feedback is obtained. Does M coincide with the maximum degree at which Proposition 1 holds true ? Unfortunately, this is not necessarily the case since, while homogeneous normal forms at degree m are uniquely defined, the normal forms are not, because homogeneous transformation of degree m, which cannot change the terms of degree m, can change terms of degree higher than m and thus the homogeneous normal forms of degree higher than m. It can only be said that, if Proposition 1 holds till M^*, then there exists a normal form with a summation starting at $m = M^*$. The evaluation of such an M^* can be linked to the solvability of the homological equations in terms of a set of suitable polynomial, the so called invariants as shown in the continuous-time [9], [18] and the discrete-time contexts [17].

4 Observer Normal Forms

Observer normal forms are linked to linear equivalence by output-injection; this problem can be rephrased in terms of equivalence through coordinates change to the canonical observer form, a linear dynamics with a nonlinear output-injection, [13]. In the usual formalism of maps it takes the form

$$\begin{aligned} \zeta_{k+1} &= A_O^{'}\zeta_k + \psi(y_k, v_k) \\ y &= C_O\zeta \end{aligned} \tag{28}$$

with the pair of matrices $(A_O^{'}, C_O)$ in the *canonical observer form* (13) with $a_i = 0$ for $i = 0, ..., n-1$, and

$$\psi(y,v) = \begin{pmatrix} a_0 \\ \vdots \\ a_{n-2} \\ a_{n-1} \end{pmatrix} y + \begin{pmatrix} b_0 \\ \vdots \\ b_{n-2} \\ b_{n-1} \end{pmatrix} v + \psi^{[\geq 2]}(y,v). \tag{29}$$

In the present context, the *DDR canonical observer form* is

$$\zeta^+ = A_O^{'}\zeta + \alpha(y) \tag{30}$$

$$\frac{d\zeta^+(v)}{dv} = G_O(\zeta_1^+(v), v) = B + G_O^{\geq 1}(\zeta_1^+(v), v); \quad \zeta^+(0) = \alpha(y) \tag{31}$$

$$y = C_O\zeta.$$

It is a matter of computation to verify the following statement.

Proposition 3. *The discrete-time canonical observer form (28) admits the DDR canonical observer form (30), (31) and viceversa.*

We just note that $\psi(y,v)$ in (29) can be recovered by integrating (31); i.e.

$$\psi(y,v) = \alpha(y) + \int_0^v G_O(\zeta_1^+(w), w)dw \quad \text{with} \quad \zeta_1^+(0) = \alpha_1(y)$$

and the output injection defined in (14)–(15) can be rewritten as

$$\alpha^{\geq 2}(y) = \psi^{\geq 2}(y,0) \quad \text{and } \beta^{\geq 1}(y,v) = G_O^{\geq 1}(\psi_1(y,v),v) = \frac{\partial \psi^{\geq 2}(y,v)}{\partial v}.$$

The lemma below immediately follows from Definitions 1 and 2.

Lemma 1. *The –* **HLEI(m)** *– problem is solvable iff there exists an homogeneous coordinates change of degree m which transforms $\Sigma^{[m]}$ into*

$$\zeta^+(0) = A_O\zeta + \alpha^{[m]}(y) \tag{32}$$

$$\frac{d\zeta^+(v)}{dv} = B + G_O^{[m-1]}(\zeta_1^+(v), v) \tag{33}$$

$$y = C_O\zeta$$

modulo terms in $R^{\geq m+1}(\zeta, v)$.

The homogeneous output-injection at degree m being described by the pair $(\alpha^{[m]}(y), \beta^{[m-1]}(y,v)$. with $G_O^{[m-1]}(\zeta_1^+(v), v) = \beta^{[m-1]}(\frac{\zeta_1^+(v)-b_0 v}{a_0}, v)$ or equivalently $\beta^{[m-1]}(y,v) = G_O^{[m-1]}(a_0 y + b_0 v)$.

Lemma 2. *The –* **LEI** *– problem is solvable iff there exists a coordinates change which transforms $\Sigma^{[\infty]}$ into the DDR canonical observer form (30)–(31). If the equivalence holds modulo terms in $R^{\geq M+1}(\zeta, v)$, an approximated solution is obtained.*

Let us now recall the geometric conditions ensuring equivalence through output-injection to the DDR canonical observer form.

4.1 Geometric Conditions

As in the continuous-time case, given $\Sigma^{[\infty]}$, an instrumental tool is the vector field $r_{d1}(\zeta)$ (see [3]), solution of

$$\begin{pmatrix} dh \\ \vdots \\ d(h \circ F^{n-2}) \\ d(h \circ F^{n-1}) \end{pmatrix} \Bigg|_{\zeta} r_{d1}(\zeta) = \begin{pmatrix} 0 \\ \vdots \\ 0 \\ 1 \end{pmatrix} \tag{34}$$

and the $n-1$ vector fields $r_{di} := F_* r_{di-1} = F_*^{i-1} r_{d1}$ defined for $i = 2, \cdots, n$, as the iterated transport of r_{d1} along F.

Theorem 4. *Linear equivalence by output-injection. The* **LEI** *problem is solvable for (1)–(2) if and only if, given r_{di} for $i = 1, ..., n$, the conditions below hold true locally around* 0

- A_{d2}: $[r_{d1}, r_{di}](\zeta) = 0$ *for* $i = 2, ..., n$;
- A_{d3}: $[G_p, r_{di}](\zeta) = 0$ *for* $i = 2, ..., n$ *and* $p \geq 0$.

Observability in the first approximation ensures the existence and uniqueness of r_{d1} satisfying (34). A_{d2} requires the nilpotency at the first order of the distribution generated by the vector fields $(r_{d1}, \cdots r_{dn})$ and guarantees the existence of a coordinates change as well as the specific structure of (30) while A_{d3} guarantees the specific structure of (31).

The vector fields $r_{di}(\zeta)$ can be described by their asymptotic expansions $r_{di}(\zeta) = r_{di} + \sum_{m \geq 1} r_{di}^{[m]}(\zeta)$ with constant parts r_{di} satisfying for $i = 1, ..., n$

$$\begin{pmatrix} C \\ \vdots \\ CA^{n-2} \\ CA^{n-1} \end{pmatrix} r_{d1} = \begin{pmatrix} 0 \\ \vdots \\ 0 \\ 1 \end{pmatrix}; \qquad r_{di} := A^{i-1} r_{d1}.$$

Referring now to the same property at degree m yields the following statement.

Proposition 4. *Homogeneous linear equivalence by output-injection at degree m. The* **HLEI(m)** *problem is solvable for (1)-(2) if and only if given $r_{di}^{[m]}$'s, the conditions below hold true locally around* 0

- $A_{d2}^{[m]}$: $[r_{d1} + r_{d1}^{[m-1]}(\zeta), r_{di} + r_{di}^{[m-1]}(\zeta)] = 0$ *for* $i = 2, ..., n$ *modulo terms in* $R^{\geq m-1}$
- $A_{d3}^{[m]}$: $[B + G_p^{[m-1]}(\zeta), r_{di} + r_{di}^{[m-1]}(\zeta)] = 0$ *for* $i = 2, ..., n$ *and* $p \geq 0$ *modulo terms in* $R^{\geq m-1}$.

4.2 Homological Equations

Hereafter, how the differential representation $\Sigma_O^{[m]}$ is transformed under homogeneous output transformation is discussed; the computations are performed modulo an error of order $m+1$ in the state and control variables (error in $R^{\geq m+1}$). The transformation $H^{[m]}$ being composed with two parts, $Id+\phi^{[m]}(.)$ acts as a usual coordinates change so that $(F^{[m]}(.), G_i^{[m-i]}(.), i = (1, ..., m))$ are transformed into $(\bar{F}^{[m]}(.), \bar{G}_i^{[m-i]}(.), i = (1, ..., m))$ below

$$\begin{aligned}
\bar{F}^{[m]}(.) &= F^{[m]}(.) + \phi^{[m]}A_O. - A_O\phi^{[m]}(.) \\
\bar{G}_1^{[m-1]}(.) &= G_1^{[m-1]}(.) + \frac{d\phi^{[m]}}{dx}B \\
\bar{G}_i^{[m-i]}(.) &= G_i^{[m-i]}(.); \quad i = 2, ..., m \\
\bar{h}^{[m]}(.) &= h^{[m]}(.) - C_O\phi^{[m]}(.)
\end{aligned} \tag{35}$$

because, up to an error in $R^{\geq m+1}$, $\zeta = x - \phi^{[m]}(x)$. The output-injection further transforms (35) into

$$\tilde{F}^{[m]}(.) = \bar{F}^{[m]}(.) + \alpha^{[m]}(C_O.); \quad \tilde{h}^{[m]}(.) = \bar{h}^{[m]}(.)$$

$$\sum_{i=1}^{m} \frac{v^{i-1}}{(i-1)!}\tilde{G}_i^{[m-i]}(.) = \sum_{i=1}^{m} \frac{v^{i-1}}{(i-1)!}\left(\bar{G}_i^{[m-i]}(.) + \beta_i^{[m-i]}(y)\right).$$

In conclusion, $H^{[m]}$ brings $\Sigma_O^{[m]}$ into $\tilde{\Sigma}_O^{[m]}$ below

$$\begin{aligned}
x^+ &= A_Ox + F^{[m]}(x) + \phi^{[m]}(A_Ox) - A_O\phi^{[m]}(x) + \alpha^{[m]}(C_Ox) \\
\frac{dx^+(v)}{dv} &= \frac{d\phi^{[m]}}{dx}(x^+(v))B + \sum_{i=1}^{m} \frac{v^{i-1}}{(i-1)!}G_i^{[m-i]}(x^+(v)) \\
&\quad + \sum_{i=1}^{m} \frac{v^{i-1}}{(i-1)!}\beta_i^{[m-i]}(\frac{x_1^+(v) - vb_0}{a_0}) \\
y &= C_Ox + h^{[m]}(x) - C_O\phi^{[m]}(x)
\end{aligned}$$

with

$$\tilde{G}_1^{[m-1]}(x) := \frac{d\phi^{[m]}(x)}{dx}B + G_1^{[m-1]}(x) + \beta_1^{[m-1]}(\frac{x_1}{a_0}) \tag{36}$$

because, up to an error in $R^{\geq 2}$, $x_1^+(v) = a_0 y + v b_0$. The computation of the $G_i^{[m-i]}(x^+(v))$'s for $i \geq 2$ involves the preliminary expansion with respect to v of $\beta_i^{[m-i]}(\frac{x_1^+(v)-vb_0}{a_0})$. With reference to the problem of of linear equivalence by output-injection we easily deduce the result below.

Proposition 5. *The* **HLEI(m)** *problem is solvable if and only if there exist* $(\phi^{[m]}(.), \alpha^{[m]}(.), \beta_i^{[m-i]}(.))$ *satisfying*

$$-F^{[m]}(\zeta) = \phi^{[m]}(A_O\zeta) - A_O\phi^{[m]}(\zeta) + \alpha^{[m]}(C_O\zeta) \quad (37)$$

$$-\sum_{i=1}^{m} \frac{u^{i-1}}{(i-1)!} G_i^{[m-i]}(\zeta) = \frac{d\phi^{[m]}(\zeta)}{d\zeta} B + \sum_{i=1}^{m} \frac{v^{i-1}}{(i-1)!} \beta^{[m-i]} C_O A_O^{-1}(\zeta - vB) \quad (38)$$

$$h^{[m]}(\zeta) = C_O\phi^{[m]}(\zeta). \quad (39)$$

Equations (37)–(38)–(39) are referred to as *observer homological equations at degree m.*

Proposition 4 and Proposition 5 provide two equivalent criteria for the solvability of the **HLEI(m)** problem. While the explicit computation of the homogeneous transformation $H^{[m]}$ requires through Proposition 4 to find r_{d1} solving a partial derivative equation, it requires through Proposition 5 to solve algebraic equations. Once again, this represents a main constructive aspect of the normal forms approach.

4.3 The Normal Forms

Two types of normal forms, denoted in the sequel by $\Sigma_{ONFA}^{[m]}$ and $\Sigma_{ONFB}^{[m]}$ can be obtained depending on how the computations are performed. We preliminarily note that (39) can be exactly solved setting $\phi_n^{[m]}(\zeta) = h^{[m]}(\zeta)$. Rewriting now (37) componentwise, we get

$$\begin{aligned}
-F_1^{[m]}(.) &= \phi_1^{[m]}(A_O.) - a_0\phi_n^{[m]}(.) + \alpha_1^{[m]}(\zeta_n) \\
-F_2^{[m]}(.) &= \phi_2^{[m]}(A_O.) - \phi_1^{[m]}(.) - a_1\phi_n^{[m]}(.) + \alpha_2^{[m]}(\zeta_n) \\
&\cdots \\
-F_n^{[m]}(.) &= \phi_n^{[m]}(A_O.) - \phi_{n-1}^{[m]}(.) - a_{n-1}\phi_n^{[m]}(.) + \alpha_n^{[m]}(\zeta_n)
\end{aligned}$$

from which we deduce that all the nonlinearities of the type ζ_n^m in any component $F_i^{[m]}(.)$ can be cancelled through an adequate choice of $\alpha_i^{[m]}(\zeta_n)$. Further on, each component $\phi_i^{[m]}(.)$ for $i = 1, ..., n-1$ can be used to solve the second to the last equation; the terms $\phi_{in...n}\zeta_n^m$ remaining free for $i = 1, ..., n-1$.

It results that two type of normal forms can be described depending if one chooses to use these remaining freedom degrees to cancel terms in the first component of $F^{[m]}(.)$ or if one chooses to cancel terms in (38). This will be differentiated later on. Let us now put in light in the equations (38) its first part for $i = 1$

$$-G_1^{[m-1]}(\zeta) = \frac{d\phi^{[m]}(\zeta)}{d\zeta}B + \beta_1^{[m-1]}(\frac{\zeta_1}{a_0})$$

from which we deduce that all the terms in ζ_1^{m-1} in $G_1^{[m-1]}(\zeta)$ can be cancelled through an adequate choice of $\beta_1^{[m-1]}(\frac{\zeta_1}{a_0})$ from row 1 to n. To exactly describe the homological equation satisfied by the $G_i^{[m-i]}(\zeta)$'s, $i = 2, ..., m$ is more involved. It is however sufficient to note that

$$-G_i^{[m-i]}(\zeta) = \chi_i(\beta_1^{[m-1]}(\frac{-vb_0}{a_0}), ..., \beta_{i-1}^{[m-i+1]}(\frac{-vb_0}{a_0})) + \beta_i^{[m-i]}(\frac{\zeta_1}{a_0})$$

where each $\chi_i(\beta_1^{[m-1]}, ..., \beta_{i-1}^{[m-i+1]})$ is a linear combination of its arguments and depends on the previous $\beta_j^{[m-j]}$. From this, we deduce that iteratively for each $i = 2, ..., m$, all the terms in $\zeta_1^{[m-i]}$ in $G_i^{[m-i]}$ can be cancelled through an adequate choice of $\beta_j^{[m-i]}(\frac{\zeta_1}{a_0})$ from row $j = 1$ to n. When $m = i$, all the terms in the constant vector field $G_m^{[0]}$ can be cancelled by an adequate choice of $\beta_m^{[0]}$; i.e. setting

$$\beta_m^{[0]} = -G_m^{[0]} - \sum_{i=1}^{m-1} \frac{(m-1)!}{(i-1)!}\beta_i^{[m-i]}(\frac{-b_0}{a_0}).$$

Let us now differentiate the study. Taking in mind that the coefficients $\phi_{in...n}$ are kept free for $i = 1, ..., n-1$, let us use these coefficients to cancel either terms in $G_1^{[m-1]}(\zeta)$ or in $F_1^{[m]}(\zeta)$.

The *first type of normal forms* privileges cancellation of nonlinearities in the drift. To do so, considering again the first type of homological equations, we note that, due to the particular structure of the matrix A_O, it is possible to use the remaining coefficients $\phi_{in....n}$ for $i = 1, ..., n-1$ to cancel $n-1$ terms in $\zeta_i\zeta_n^{m-1}$ in $F_1^{[m]}(\zeta)$ for $i = 1, ..., n-1$. More precisely, each $\phi_{in.....n}$ is used to cancel the corresponding coefficient of $\zeta_i\zeta_n^{m-1}$ in $F_1^{[m]}(\zeta)$. Arguing so, the first class of normal form $\mathbf{\Sigma}_{\mathbf{ONFA}}^{[\mathbf{m}]}$ is obtained.

The *second type of normal forms* privileges cancellation of nonlinearities in $G_1^{[m-1]}(\zeta)$. Considering again the homological equation associated with $G_1^{[m-1]}(\zeta)$, we note that full cancellation of $G_1^{[m-1]}(\zeta)$ is possible from row 1 to $n-1$ through and adequate choice of $\frac{d\phi_j^{[m]}(\zeta)}{d\zeta_p}$ for $j = 1, ..., n-1$ because $\phi_n(\zeta)$ has been used to cancel nonlinear parts in the output mapping and for a fixed $p \in (1, ..., n)$ associated to the corresponding $b_{p-1} \neq 0$ (at least one b_i

is non zero due to the controllability assumption). Arguing so, terms of the form $\zeta_p F_{j+1}^{m-1}(\zeta)$ except ζ_1^m if $p = 1$ are resonance terms in components 2 to n of the drift. We conclude that all the nonlinearities are resonance terms in the last component $G_{1n}^{[m-1]}(\zeta)$ except terms in ζ_1^{m-1} so getting the normal form below $\mathbf{\Sigma_{ONFB}^{[m]}}$.

Theorem 5. *For any degree $m \geq 2$ and neglecting higher degree terms, any homogeneous discrete-time dynamics $\Sigma_O^{[m]}$ can be transformed under homogeneous output transformation $H^{[m]}$ into one of the two homogeneous normal forms at degree m below.*
The first type of observer normal form of degree m; $\mathbf{\Sigma_{ONFA}^{[m]}}$:

$$\begin{aligned} x^+ &= A_O x + \left(F_1^{[m]}(x), 0, ..., 0\right)^T \\ \frac{dx^+(v)}{dv} &= B + G_1^{[m-1]}(x^+(v)) + \sum_{i=2}^{m-1} \frac{v^{i-1}}{(i-1)!} G_i^{[m-i]}(x^+(v)) \\ y &= x_n \end{aligned}$$

where $F_1^{[m]}(x)$ is a polynomial of degree m without terms in $x_i x_n^{m-1}$ for $i = 1, ..., n$; for $i = 1, ..., m-1$, $G_i^{[m-i]}(x)$ is a vector of polynomials of degree $m - i$ without terms in x_1^{m-i}; moreover $G_m^{[0]} = 0$.
The second type of observer normal form of degree m; $\mathbf{\Sigma_{ONFBa}^{[m]}}$:

$$\begin{aligned} x^+ &= A_O x + \left(F_1^{[m]}(x), F_2^{[m]}(x), ..., F_n^{[m]}(x)\right)^T \\ \frac{dx^+(v)}{dv} &= B + \left(0, ..., 0, G_{1n}^{[m-1]}(x^+(v))\right)^T + \sum_{i=2}^{m-1} \frac{v^{i-1}}{(i-1)!} G_i^{[m-i]}(x^+(v)) \\ y &= x_n \end{aligned}$$

$\mathbf{\Sigma_{ONFBa}^{[m]}}$ *($b_p \neq 0$ for a fixed $p \in (0, ..., n-2)$): $F_1^{[m]}(x)$ is a polynomial of degree m without terms in $x_i x_n^{m-1}$ for $i = 1, ..., n$; for $j = 2, ..., n$, $F_j^{[m]}(x) = x_p F_j^{[m-1]}(x_1, ..., x_n)$.* $\mathbf{\Sigma_{ONFBb}^{[m]}}$*($b_{n-1} \neq 0$): $F_1^{[m]}(x)$ is a polynomial of degree m without terms in x_n^m; for $j = 2, ..., n$, $F_j^{[m]}(x) = x_n F_j^{[m-1]}(x_1, ..., x_n)$ without terms in x_n^m; Moreover, for both forms $G_i^{[m-i]}(x)$ is a vector of polynomials of degree $m - i$ without terms in x_1^{m-i} for $i = 2, ..., m-1$; the last row $G_{1n}^{[m-1]}(x)$ does not contain terms in x_1^{m-1}; $G_m^{[0]} = 0$.*

Applying iteratively the result of Theorem 5 to each homogeneous part of degree m, starting at $m = 2$ and increasing the degree, Theorem 6 below describes the observer normal forms of a discrete-time system in its DDR.

Theorem 6. *The nonlinear discrete-time system $\Sigma^{[\infty]}$ can be transformed under $H^{[\infty]}$ into a system exhibiting one of the two observer normal forms below.*
The first type of observer normal form; $\mathbf{\Sigma_{ONFA}^{[\infty]}}$:

$$x^+ = A_O x + \Big(F_1(x), 0, ..., 0\Big)^T$$
$$\frac{dx^+(v)}{dv} = B + G_1(x^+(v)) \sum_{i\geq 2} \frac{v^{i-1}}{(i-1)!} G_i(x^+(v))$$
$$y = x_n$$

where $F_1(x) = \sum_{m=2}^{\infty} F_1^{[m]}(x)$ *is a formal series which does not contain terms in* $x_i F_1^{\geq 1}(x_n)$ *for* $i = 1, ..., n$*; for* $i \geq 1$, $G_i(x) = \sum_{m=1}^{\infty} G_i^{[m]}(x)$ *is a vector of formal series which does not contain terms in* $G_i^{\geq 1}(x_1)$.
The second type of observer normal form; $\mathbf{\Sigma_{ONFB}^{[\infty]}}$:

$$x^+ = A_O x + \Big(F_1(x), F_2(x), ..., F_n(x)\Big)^T$$
$$\frac{dx^+(v)}{dv} = B + \Big(0, ..., 0, G_{1n}(x^+(v))\Big)^T + \sum_{i\geq 2} \frac{v^{i-1}}{(i-1)!} G_i(x^+(v))$$
$$y = x_n$$

$\mathbf{\Sigma_{ONFBa}^{[\infty]}}$ *(*$b_p \neq 0$ *for a fixed* $p \in (0, ..., n-2)$*):* $F_1(x) = \sum_{m=2}^{\infty} F_1^{[m]}(x)$ *is a formal series which does not contain terms in* $x_i F_1^{\geq 1}(x_n)$ *for* $i = 1, ..., n$*; for* $j = 2, ..., n$, $F_j(x) = x_p F_j^{\geq 1}(x_1, ..., x_n)$. $\mathbf{\Sigma_{ONFBb}^{[\infty]}}$ *(*$b_{n-1} \neq 0$*):* $F_1(x) = \sum_{m=2}^{\infty} F_1^{[m]}(x)$ *is a formal series which does not contain terms in* $F_1^{\geq 2}(x_n)$*; for* $j = 2, ..., n$, $F_j(x) = x_n F_j^{\geq 1}(x_1, ..., x_n)$ *without terms in* $F_j^{\geq 2}(x_n)$*. For both forms,* $G_{1n}(x) = \sum_{m=1}^{\infty} G_{1n}^{[m]}(x)$ *is a formal series which does not contain terms in* $G_{1n}^{\geq 1}(x_1)$*; for* $i \geq 2$, $G_i(x) = \sum_{m=1}^{\infty} G_i^{[m]}(x)$ *is a vector of formal series which does not contain terms in* $G_i^{\geq 1}(x_1)$.

5 Conclusions

The paper describes controller and observer normal forms at any order m for nonlinear discrete-time dynamics controllable and observable in the first approximation. Nonlinear discrete-time dynamics are described as coupled differential/difference equations rather than in the usual form of a map. Such a representation makes it possible a link between the resonance terms contained in these forms and the obstruction to the geometric properties ensuring feedback linearization or observer design with linear error dynamics.

References

1. J.P. Barbot, S. Monaco, and D. Normand-Cyrot. Quadratic forms and approximated feedback linearization in discrete time. *Int. J. of Control*, 67:567–586, 1997.

2. I. Belmouhoud, M. Djemai, and J.P. Barbot. Observability quadratic normal form for discrete-time systems. *IEEE Trans. on Automat. Contr.*, 50(7):1031–1037, 2005.
3. C. Califano, S. Monaco, and D. Normand-Cyrot. On the observer design in discrete time nonlinear systems. *Systems & Control Letters*, 49:255–265, 2003.
4. J.W. Grizzle. *Feedback Linearization of Discrete-Time Systems*, volume 83 of *Lecture Notes in Control and Info. Sciences.* Springer Verlag, Berlin, 1986.
5. B. Hamzi, J.P. Barbot, S. Monaco, and D. Normand-Cyrot. Nonlinear discrete-time control of systems with a Naimark-Sacker bifurcation. *Systems & Control Letters*, 44:245–258, 2001.
6. B. Hamzi, A.J. Krener, and W. Kang. The controlled center dynamics of discrete-time control bifurcations. *Systems & Control Letters*, 55:585–596, 2006.
7. B. Hamzi and I.A. Tall. Normal forms for nonlinear discrete-time control systems. *Proc. of the 42nd IEEE Conf. on Decision and Contr.*, pages 1357–1361, 2003.
8. A. Isidori. *Nonlinear Control Sytems.* Springer Verlag, New York, 3rd edition, 1995.
9. W. Kang. Extended controller form and invariants of nonlinear control systems with a single input. *Journal of Math. Syst. Estimation and Control*, 6:27–51, 1996.
10. W. Kang and A.J. Krener. Extended quadratic controller normal form and dynamic state feedback linearization of nonlinear systems. *SIAM J. Contr. Optimization*, 30:1319–1337, 1992.
11. N. Kazantzis and C. Kravaris. Discrete-time nonlinear observer design using functional equations. *Systems & Control Letters*, 42:81–94, 2001.
12. A.J. Krener. Approximate linearization by state feedback and coordinate change. *Systems & Control Letters*, 5:181–185, 1984.
13. A.J. Krener and A. Isidori. Linearization by output-injection and nonlinear observers. *Systems & Control Letters*, 3:47–52, 1983.
14. A.J. Krener and L. Li. Normal forms and bifurcations of discrete-time nonlinear control systems. *SIAM J. Contr. Optimization*, 40:1697–1723, 2002.
15. S. Monaco and D. Normand-Cyrot. A unifying representation for nonlinear discrete-time and sampled dynamics. *Journal of Math. Syst. Estimation and Control*, 5(1):103–105, 1995.
16. S. Monaco and D. Normand-Cyrot. Nonlinear discrete-time representations, a new paradigm. Perspectives in Control, a tribute to Ioan Doré Landau, pages 191–205. Springer Verlag, London, 3rd edition, 1998.
17. S. Monaco and D. Normand-Cyrot. Normal forms and approximated feedback linearization in discrete time. *Systems & Control Letters*, 55:71–80, 2006.
18. I.A. Tall and W. Respondek. Feedback classification of nonlinear single-input control systems with controllable linearization: normal forms, canonical forms and invariants. *SIAM J. Contr. Optimization*, 41:1498–1531, 2003.
19. X.H. Xia and W.B. Gao. Nonlinear observer design by observer error linearization. *SIAM J. Contr. Optimization*, 27:199–216, 1989.

A Geometric Approach to Dynamic Feedback Linearization

Stefano Battilotti and Claudia Califano

Dipartimento di Informatica e Sistemistica "Antonio Ruberti", Università di Roma "La Sapienza", Via Eudossiana 18, 00184 Rome, Italy

Summary. The paper deals with dynamic feedback linearization of multi input continuous time affine systems. The geometric properties of a dynamic feedback linearizable system as well as those of the compensator which achieves linearization are here enlightened. On the basis of these geometric properties an algorithm for the computation of a dynamic feedback obtained from the composition of regular static state feedback laws and integrators is proposed. Our result is based on the geometric approach introduced by Isidori and coworkers in 1981 for dealing with nonlinear control problems.

1 Introduction

The use of a geometric framework for addressing control problems was first introduced, in the sixties, in the pioneering works of Morse and Wonham. In these papers control problems such as disturbance decoupling and noninteracting control were formulated and solved in the framework of linear geometry, using mathematical tools such as linear vector spaces and matrix theory. In [13], inspired by the approach of Morse and Wonham, the authors introduced the use of differential geometry and the concept of distributions for formulating and studying the problem of nonlinear noninteracting control, opening the way to a solution to this problem. These geometric concepts were also used to successfully address the static state feedback linearization problem and related problems, first investigated in ([5]), and solved in ([16], [11], [18], [19], [15], [17], [12], [21], [2], [6]).

The use of differential geometry played a fundamental role also for seeking dynamic solutions which were first considered in [14], [20] and [7]. In [8], sufficient conditions were given for the solvability of the problem via prolongations and diffeomorphism. A different approach based on algebraic techniques was instead proposed in [9], [10], where differentially flat systems were introduced. Necessary and sufficient conditions for the solvability of the problem were given in [1]. However these conditions are not constructive thus not allowing a direct computation of the dynamic compensator.

Recently in [3], the use of differential geometry allowed to propose an algorithm for the computation of a dynamic compensator consisting of prolongations, in the case of two input continuous affine systems. The general multi input case was instead considered in [4].

In the present paper we extend the results proposed in [4] by considering regular dynamic compensators for multi input continuous time affine systems.

2 Preliminaries

Consider the continuous time analytic system

$$\dot{x} = f(x) + \sum_{i=1}^{m} g_i(x)u_i \tag{1}$$

where $x \in \mathbb{R}^n$, $f(x)$ and $g_i(x)$, $i = 1, \cdots, m$ are smooth maps defined on a open set of $\mathbb{R}^n$. The following notation will be used: given two smooth vector fields f and g_i, $ad_f g_i := [f, g_i] = \frac{\partial g_i}{\partial x} f - \frac{\partial f}{\partial x} g_i$, and $ad_f^k g_i = ad_f(ad_f^{k-1} g_i)$. We denote by $g = (g_1, \cdots, g_m)$, by $\mathcal{G}_i := \text{span}\{g, \cdots, ad_f^i g\}$, by $\bar{\mathcal{G}}_i$, the involutive closure of $\mathcal{G}_i$. Let us recall that the involutivity and constant dimensionality of the distributions $\mathcal{G}_i$ together with the controllability of the given dynamics are necessary and sufficient conditions for linear static feedback equivalence.

Assume now that (1) is linearizable over an open and dense set $\mathcal{U}_0 \ni (x_0, 0)$ with a regular dynamic feedback of the form

$$\begin{aligned} \dot{\zeta} &= \eta(x, \zeta) + \delta(x, \zeta)v \\ u &= \alpha(x, \zeta) + \beta(x, \zeta)v\,. \end{aligned} \tag{2}$$

In the present section we first enlighten some properties of the linearizable dynamics (1) and then those of the regular dynamic feedback (2). These properties will allow us to define an algorithm for the computation of a solution.

2.1 The Original Dynamics Properties

Assume that the dynamics (1) is not static feedback linearizable. Then there exists an index k such that the distributions $\mathcal{G}_{k+j}$ are involutive and of constant dimension on an open and dense subset $\mathcal{U}_0$, for any $j \geq 0$ whereas $\mathcal{G}_{k-1}$ is not involutive. Let ρ_{k-1} be the dimension of $\mathcal{G}_{k-1}$ and $\rho_{k-1} + s$ the dimension of its involutive closure $\bar{\mathcal{G}}_{k-1}$, with $s \leq m$. Then there exist s independent vector fields τ_i, $i = 1, \cdots, s$ which belong to $\mathcal{G}_k$ such that

$$\bar{\mathcal{G}}_{k-1} = \mathcal{G}_{k-1} + \text{span}\{\tau_i,\ i = 1, \cdots, s\}\,. \tag{3}$$

Let us introduce the following definitions.

Definition 1. *Let $k > 0$ be the greatest index such that $\mathcal{G}_{k+l}$ is involutive for any $l \geq 0$ whereas $\mathcal{G}_{k-1}$ is not involutive. Let $\bar{\mathcal{G}}_{k-1} = \mathcal{G}_{k-1} + \mathrm{span}\{\tau_1, \cdots, \tau_s,\ s \leq m\}$ be its involutive closure. The Non-Characteristic set NC^k is given by*

$$NC^k = \{(ad_f^l g_{s_j}, ad_f^r g_{s_t}), s_j, s_t \in [1,m],\ l, r, \leq k-1 : [ad_f^l g_{s_j}, ad_f^r g_{s_t}] \notin \mathcal{G}_{k-1}\}$$

The j-th channel is said to be k-eligible if there exists at least one pair $(ad_f^l g_j, ad_f^r g_{s_t}) \in NC^k$.

Definition 2. *Let $k > 0$ be the greatest index such that $\mathcal{G}_{k+l}$ is involutive for any $l \geq 0$ whereas $\mathcal{G}_{k-1}$ is not involutive. Let $\bar{\mathcal{G}}_{k-1} = \mathcal{G}_{k-1} + \mathrm{span}\{\tau_1, \cdots, \tau_s, s \leq m\}$ be its involutive closure. Then the j-th k-eligible channel is said to be k-unlocked if $ad_f^k g_j \notin \bar{\mathcal{G}}_{k-1}$ or $ad_f^k g_j \in \bar{\mathcal{G}}_{k-1}$ with $ad_f^k g_j \in \mathcal{G}_{k-1} + \mathrm{span}\{ad_f^k g_l,\ l = 1, \cdots, m,\ l \neq j\}$. The j-th k-eligible channel is k-locked if it is not k-unlocked.*

We recall the following result proven in [4], which was at the basis of the algorithm proposed in the same work for computing a set of prolongation indices in order to linearize a given dynamics by adding integrators.

Proposition 1. *Let $k > 0$ be the greatest index such that $\mathcal{G}_{k+i}$ is involutive for any $i \geq 0$ whereas $\mathcal{G}_{k-1}$ is not involutive. Assume that its involutive closure $\bar{\mathcal{G}}_{k-1}$ is given by $\bar{\mathcal{G}}_{k-1} = \mathcal{G}_{k-1} + \mathrm{span}\{\tau_1, \cdots, \tau_s\} = \mathcal{G}_{k-1} + \mathrm{span}\{ad_f^k g_1, \cdots, ad_f^k g_s\}$. Then for any $i \geq 0$ the distribution*

$$\mathcal{G}_{k-1+i} + \mathrm{span}\{ad_f^{k+i} g_1, \cdots, ad_f^{k+i} g_s\}$$

is involutive and of constant dimension.

The next proposition generalizes the previous result, by allowing the use of regular static state feedback.

Proposition 2. *Let $k > 0$ be the greatest index such that $\mathcal{G}_{k+i}$ is involutive for any $i \geq 0$ whereas $\mathcal{G}_{k-1}$ is not involutive. Assume that its involutive closure $\bar{\mathcal{G}}_{k-1}$ is given by $\bar{\mathcal{G}}_{k-1} = \mathcal{G}_{k-1} + \mathrm{span}\{\tau_1, \cdots, \tau_s\}$. Then for any index j such that $ad_f^k g_j \notin \bar{\mathcal{G}}_{k-1}$ or $ad_f^k g_j \in \bar{\mathcal{G}}_{k-1}$ with $ad_f^k g_j \in \mathcal{G}_{k-1} + \mathrm{span}\{ad_f^k g_l,\ l = 1, \cdots, m,\ l \neq j\}$, there exist a regular static state feedback $u = \beta(x)v$ such that denoting by $\tilde{g} = g\beta$, $\forall t \geq 0$ the distribution $\mathcal{G}_{k-1+t} + \mathrm{span}\{ad_f^{k+t} \tilde{g}_l,\ l = 1 \cdots, m,\ l \neq j\}$ is involutive and of constant dimension, and $\bar{\mathcal{G}}_{k-1} \subseteq \mathcal{G}_{k-1} + \mathrm{span}\{ad_f^k \tilde{g}_l,\ l = 1 \cdots, m,\ l \neq j\}$.*

Proof. Assume first that $ad_f^k g_j \notin \bar{\mathcal{G}}_{k-1}$. Denote by ρ_j the rank of $\mathcal{G}_j$. Let $i \geq 0$ be the greatest index such that $n = \rho_{k+i} > \rho_{k+i-1}$. By assumption there exist $m_1 = \rho_{k+i} - \rho_{k+i-1}$ independent functions λ_l such that denoting by $\lambda^1 = (\lambda_1^1, \cdots, \lambda_{m_1}^1)^T$, $d\lambda^1\, \mathcal{G}_{k+i-1} = 0$ while $d\lambda^1 (ad_f^{k+i} g_1 \cdots ad_f^{k+i} g_m) \neq 0$, with full row rank. Let $m_2 = \rho_{k+i-1} - \rho_{k+i-2} - m_1$ and compute the m_2 functions

$\lambda^2 = (\lambda_1^2, \cdots, \lambda_{m_2}^2)^T$, such that $((\lambda^1)^T, (L_f\lambda^1)^T, (\lambda^2)^T)^T$ are independent, $d\lambda^2\, \mathcal{G}_{k+i-2} = 0$ while

$$\begin{pmatrix} dL_f\lambda^1 \\ d\lambda^2 \end{pmatrix} (ad_f^{k+i-1} g_1 \cdots ad_f^{k+i-1} g_m) \neq 0$$

with full row rank. Let $m_i = \rho_{k+1} - \rho_k - (i-1)m_1 - (i-2)m_2 - \cdots m_{i-1}$ and compute the m_i functions $\lambda^i = (\lambda_1^i, \cdots, \lambda_{m_i}^i)^T$, such that

$$((\lambda^1)^T, \cdots, (L_f^{i-1}\lambda^1)^T, \cdots, (\lambda^{i-1})^T, (L_f\lambda^{i-1})^T, (\lambda^i)^T)^T$$

are independent, $d\lambda^i\, \mathcal{G}_k = 0$ while

$$\begin{pmatrix} dL_f^{i-1}\lambda^1 \\ \vdots \\ d\lambda^i \end{pmatrix} (ad_f^{k+1} g_1 \cdots ad_f^{k+1} g_m) \neq 0$$

with full row rank. Denote by $\bar{\rho}_{k-1}$ the rank of $\bar{\mathcal{G}}_{k-1}$, the involutive closure of $\mathcal{G}_{k-1}$, and let $m_{i+1} = \rho_k - \bar{\rho}_{k-1} - im_1 - \cdots - m_i$. Compute the m_{i+1} functions $\lambda^{i+1} = (\lambda_1^{i+1}, \cdots, \lambda_{m_{i+1}}^{i+1})^T$, such that

$$((\lambda^1)^T, \cdots, (L_f^i\lambda^1)^T, \cdots, (\lambda^i)^T, (L_f\lambda^i)^T, (\lambda^{i+1})^T)$$

are independent, $d\lambda^{i+1}\, \bar{\mathcal{G}}_{k-1} = 0$ while

$$\begin{pmatrix} dL_f^i\lambda^1 \\ \vdots \\ d\lambda^{i+1} \end{pmatrix} (ad_f^k g_1 \cdots ad_f^k g_m) = \begin{pmatrix} a_{11} & \cdots & a_{1m} \\ \vdots & \ddots & \vdots \\ a_{s1} & \cdots & a_{sm} \end{pmatrix} \neq 0 \tag{4}$$

with full row rank $s = m_1 + \cdots + m_{i+1}$. Let r be the greatest index such that $d(L_f^r\lambda_t^h)\, ad_f^k g_j = a_{lj} \neq 0$. Set

$$\begin{aligned} u_i &= v_i, \qquad i \neq j \quad i = 1, \cdots, m \\ u_j &= v_j - \sum_{\substack{i=1 \\ i \neq j}}^{m} \frac{a_{li}}{a_{lj}} v_i \end{aligned}$$

which corresponds to set on the closed loop system

$$\begin{aligned} \tilde{g}_i &= g_i - \frac{a_{li}}{a_{lj}} g_j, \qquad i \neq j \quad i = 1, \cdots, m \\ \tilde{g}_j &= g_j. \end{aligned}$$

Correspondingly $d(L_f^r\lambda_t^h)\, ad_f^k \tilde{g}_l = 0$, $l \neq j$, while $d(L_f^r\lambda_t^h)\, ad_f^k \tilde{g}_j \neq 0$. By assumption $ad_f^k \tilde{g}_j \notin \bar{\mathcal{G}}_{k-1}$. Assume now that for some index $i_1 \geq 0$ the distribution

$$\mathcal{G}_{k-1+i_1} + \text{span}\{ad_f^{k+i_1}\tilde{g}_1, \cdots, ad_f^{k+i_1}\tilde{g}_{l\neq j}, \cdots, ad_f^{k+i_1}\tilde{g}_m\} \tag{5}$$

is not involutive. For $0 \leq i_1 \leq r$ consider the set of independent functions $((L_f^{i-i_1}\lambda^1)^T, \cdots, (\lambda^{i-i_1})^T)^T$. By construction for $l \neq j$, $d(L_f^{r-i_1}\lambda_t^h)ad_f^{k+i_1}\tilde{g}_l = 0$, while $d(L_f^{r-i_1}\lambda_t^h)ad_f^{k+i_1}\tilde{g}_j \neq 0$. Consequently there should exist a Lie bracket

$$\begin{aligned}&[ad_f^{k+i_1}\tilde{g}_{l\neq j}, ad_f^{s_2}\tilde{g}_{i_2}] = \\ &\quad \alpha ad_f^{k+i_1}\tilde{g}_j|_{mod\ \mathcal{G}_{k-1+i_1}+\text{span}\{ad_f^{k+i_1}\tilde{g}_1,\cdots,ad_f^{k+i_1}\tilde{g}_{l\neq j},\cdots,ad_f^{k+i_1}\tilde{g}_m\}}\end{aligned}$$

where if $s_2 = k + i_1$, $i_2 \neq j$. We should have that

$$dL_f^{r-i_1}\lambda_t^h[ad_f^{k+i_1}\tilde{g}_{l\neq j}, ad_f^{s_2}\tilde{g}_{i_2}] = dL_f^{r-i_1}\lambda_t^h\ \alpha ad_f^{k+i_1}\tilde{g}_j \neq 0$$

which is absurd since

$$\begin{aligned}d(L_f^{r-i_1}\lambda_t^h)[ad_f^k\tilde{g}_{l\neq j}, ad_f^{s_2}\tilde{g}_{i_2}] &= d(d(L_f^{r-i_1}\lambda_t^h)\ ad_f^{s_2}\tilde{g}_{i_2})\ ad_f^{k+i_1}\tilde{g}_{l\neq j} \\ &\quad - d(d(L_f^{r-i_1}\lambda_t^h)\ ad_f^{k+i_1}\tilde{g}_{l\neq j})\ ad_f^{s_2}\tilde{g}_{i_2} = 0.\end{aligned}$$

Finally for $r < i_1 \leq i$, by construction $\begin{pmatrix} dL_f^{i-i_1}\lambda^1 \\ \vdots \\ d\lambda^{i-i_1}ad_f^{k+i_1} \end{pmatrix} g_j = 0$, which immediately proves the involutivity of the distribution (5). The case $i_1 > i$ is trivial.

Consider now the case in which $ad_f^k g_j \in \bar{\mathcal{G}}_{k-1}$ with $ad_f^k g_j \in \mathcal{G}_{k-1} + \text{span}\{ad_f^k g_l,\ l = 1, \cdots, m,\ l \neq j\}$, so that $ad_f^k g_j = \sum_{\substack{i=1\\i\neq j}}^m \alpha_i ad_f^k g_i|_{mod\ \mathcal{G}_{k-1}}$. Set

$$\begin{aligned}u_l &= v_l - \sum_{\substack{i=1\\i\neq j}}^m \alpha_i v_j, \qquad l \neq j,\ l = 1, \cdots, m, \\ u_j &= v_j\end{aligned}$$

so that on the closed loop system

$$\begin{aligned}\tilde{g}_l &= g_l, \qquad l \neq j,\ l = 1, \cdots, m, \\ \tilde{g}_j &= g_j - \sum_{\substack{i=1\\i\neq j}}^m \alpha_i g_i\end{aligned}$$

and correspondingly $ad_f^k\tilde{g}_j \in \mathcal{G}_{k-1}$, which ensures the involutivity and constant dimensionality of the distributions

$$\mathcal{G}_{k-1+i} + \text{span}\{ad_f^k\tilde{g}_{l+i},\ l = 1, \cdots, m,\ l \neq j\} \qquad i \geq 0.$$

□

2.2 The Dynamic Feedback Properties

Lemma 1. *If in (2) $\beta(x,\zeta)$ is invertible over an open and dense set $\mathcal{U}_0 \ni (x_0,0)$ then (1) is static feedback equivalent to a linear system over $\mathcal{U}_0$.*

Proof. Since $\beta(x,\zeta)$ is invertible over an open and dense set $\mathcal{U}_0 \ni (x_0,0)$, then we can consider the static state feedback $v = \beta(x,\zeta)^{-1}(w - \alpha(x,\zeta))$. The obtained closed loop dynamics

$$\begin{aligned} \dot{x} &= f(x) + g(x)w \\ \dot{\zeta} &= \bar{\eta}(x,\zeta) + \bar{\delta}(x,\zeta)w \end{aligned} \tag{6}$$

will be static feedback equivalent to a linear system. Consequently the distributions $\mathcal{G}_i^e$ defined on the extended system must be involutive and of constant dimension locally around $(x_0,0)$. Let us now note that since

$$F = \begin{pmatrix} f(x) \\ \bar{\eta}(x,\zeta) \end{pmatrix}, \quad G_i^e = \begin{pmatrix} g_i(x) \\ \bar{\delta}_i(x,\zeta) \end{pmatrix}, \; i = 1,\cdots,m,$$

consequently $ad_F^j G_i(.) = \begin{pmatrix} ad_f^j g_i(x) \\ * \end{pmatrix}$, $i = 1,\cdots,m$, $j \geq 0$. Consider now two elements $ad_F^{j_1} G_{i_1}(.)$, $ad_F^{j_2} G_{i_2}(.)$, $j_1, j_2 \leq j$, which belong to the distribution $\mathcal{G}_j^e$. The Lie bracket

$$\begin{aligned} [ad_F^{j_1} G_{i_1}, ad_F^{j_2} G_{i_2}] &= \begin{pmatrix} [ad_f^{j_1} g_{i_1}, ad_f^{j_2} g_{i_2}](x) \\ * \end{pmatrix} \\ &\in \text{span} \left\{ \begin{pmatrix} g_i(x) \\ * \end{pmatrix}, \cdots, \begin{pmatrix} ad_f^j g_i(x) \\ * \end{pmatrix}, \; i = 1,\cdots,m \right\} \end{aligned}$$

which implies that $[ad_f^{j_1} g_{i_1}, ad_f^{j_2} g_{i_2}] \in \mathcal{G}_j$, $\forall j_1, j_2 \leq j$, i.e. the involutivity of the $\mathcal{G}_j$'s. Moreover the constant dimensionality of $\mathcal{G}_j^e$ implies the constant dimensionality of $\mathcal{G}_j$ over an open and dense set $\mathcal{U}_0' \subset \mathcal{U}_0$ so that (1) is static feedback linearizable on $\mathcal{U}_0'$ which ends the proof. □

The previous result can be used to point out some properties of the class of dynamic feedback laws which can be considered in order to achieve linearization. As we will show hereafter if we consider a dynamic compensator of minimal dimension in appropriate coordinates it can be written as a combination of a feedback which depends only on the state variables of the given system plus an integrator.

Lemma 2. *Assume that (1) is dynamic feedback linearizable with the regular dynamic feedback (2) of dimension ν. Let $\rho =$ rank $\beta(x,\zeta) \leq m$. Then there exists a diffeomorphism such that in the new coordinates, and after a possible reordering of the inputs, (2) can be written as*

$$
\begin{aligned}
\dot{\chi}_i &= \bar{\eta}_i(x,\chi) + \bar{\delta}_i(x,\chi)v, \qquad i = a, b \\
u_a &= \bar{\alpha}_a(x,\chi) + \bar{\beta}_a(x,\chi)v \\
u_b &= \chi_a + \bar{M}(x,\chi)\left(\bar{\alpha}_a(x,\chi) + \bar{\beta}_a(x,\chi)v\right)
\end{aligned} \tag{7}
$$

with χ_a *of dimension* $m - \rho$ *and correspondingly* χ_b *of dimension* $\nu - m + \rho$.

Proof. By assumption in (2), $\rho = \text{rank } \beta(x,\zeta) \le m$. Moreover the dynamic feedback (2) is regular so that

$$
\text{rank}\left(\frac{\partial \alpha}{\partial \zeta} \;\middle|\; \beta\right) = m.
$$

Consequently, after a possible reordering of the inputs, there exists a partition of the input vector $(u_a^T, u_b^T)^T$ with u_a of dimension ρ and u_b of dimension $m-\rho$ such that the feedback $u = \alpha(x,\zeta) + \beta(x,\zeta)v$ can be rewritten as

$$
\begin{aligned}
u_a &= \alpha_a(x,\zeta) + \beta_a(x,\zeta)v \\
u_b &= \alpha_b(x,\zeta) + M(x,\zeta)u_a
\end{aligned}
$$

with β_a of full row rank ρ and rank $\dfrac{\partial \alpha_b}{\partial \zeta} = m - \rho$. We can then consider the coordinates change $\chi_a = \alpha_b(x,\zeta)$, and χ_b such that $(x^T, \chi_a^T, \chi_b^T)^T$ is an independent coordinates set. In these coordinates (2) reads (7). □

Proposition 3. *If the regular dynamic feedback (7) achieves linearization for (1), then also the regular dynamic feedback*

$$
\begin{aligned}
\dot{\chi}_i &= \bar{\eta}_i(x,\chi) + \bar{\delta}_i(x,\chi)v \qquad i = a, b \\
u_a &= \bar{\alpha}_a(x,\chi) + \bar{\beta}_a(x,\chi)v \\
u_b &= \chi_a + \bar{M}(x,0)\left(\bar{\alpha}_a(x,\chi) + \bar{\beta}_a(x,\chi)v\right)
\end{aligned} \tag{8}
$$

achieves linearization.

Proof. The proof, which is omitted for space reasons, is based on the analysis of the linear approximations of the closed–loop system obtained by first considering the dynamic feedback (7) and then the dynamic feedback (8). For the two closed loop systems the same output functions achieve defined relative degree $(r_1, \cdots, r_m)$ with $\sum_{i=1}^{m} r_i = n + \nu$. □

Let us finally note that the dynamic feedback (8) can be rewritten as a regular static state feedback plus an integrator, i.e.

$$
\begin{aligned}
&u_a = w_a, \quad u_b = \bar{M}(x,0)w_a + \chi_a \\
&\dot{\chi}_a = w_b
\end{aligned} \tag{9}
$$

and a residual dynamics

$$\begin{aligned}\dot{\chi}_b &= \bar{\eta}_b(x,\chi) + \bar{\delta}_b(x,\chi)v \\ w_a &= \bar{\alpha}_a(x,\chi) + \bar{\beta}_a(x,0)v \\ w_b &= \bar{\eta}_a(x,\chi) + \bar{\delta}_a(x,\chi)v\,.\end{aligned} \tag{10}$$

Iterating the procedure on the residual dynamics (10) we can rewrite the dynamic feedback (8) as the composition of a dynamic feedback given by a chain of regular static state feedback laws and integrators, which characterize a compensator of minimal order plus a residual dynamics.

Hereafter we discuss the static state feedback and the dynamic extension actions, in order to understand the different steps of the proposed algorithm. To this end, let us first recall that dynamic feedback laws are required when for some index k, the distribution $\mathcal{G}_{k-1}$ is not involutive, while for any $i \geq 0$, $\mathcal{G}_{k+i}$ is involutive and of constant dimension on an open and dense set $\mathcal{U}_0$. As a consequence the involutive closure $\bar{\mathcal{G}}_{k-1} := \mathcal{G}_{k-1} + \mathrm{span}\{\tau_1, \cdots, \tau_s\} \subseteq \mathcal{G}_k$.

The Static State Feedback Action

Assume that the j-th channel is k-unlocked, so that it may be extended in order to achieve involutivity. The static state feedback is then used in order to guarantee that for any $i_1 \geq 0$ the distributions

$$\mathcal{G}_{k-1+i_1} + \mathrm{span}\{ad_f^{k+i_1} g_{l \neq j},\ l = 1, \cdots, m\}$$

are involutive and of constant dimension on an open and dense set $\mathcal{U}_0$. According to the proof of Proposition 2 we recognize two different kind of static state feedback whether $ad_f^k g_j \notin \bar{\mathcal{G}}_{k-1}$ or $ad_f^k g_j \in \bar{\mathcal{G}}_{k-1}$ with $ad_f^k g_j \in \mathcal{G}_{k-1} + \mathrm{span}\{ad_f^k g_{l \neq j},\ l = 1, \cdots, m\}$.

Direction feedback

This feedback is used when $ad_f^k g_j \notin \bar{\mathcal{G}}_{k-1}$. According to the proof of Proposition 2 compute the decoupling matrix (4). Let the l-th row correspond to the output with maximal relative degree with respect to the j-th input. Set

$$\begin{aligned}u_i &= v_i, & i \neq j, \quad i = 1, \cdots, m \\ u_j &= v_j - \sum_{\substack{i=1\\ i\neq j}}^{m} \frac{a_{li}}{a_{lj}} v_i\end{aligned}$$

which corresponds to set on the closed-loop system

$$\begin{aligned}\tilde{g}_i &= g_i - \frac{a_{li}}{a_{lj}} g_j, & i \neq j, \quad i = 1, \cdots, m \\ \tilde{g}_j &= g_j.\end{aligned} \tag{11}$$

Reduction feedback

This feedback is used when $ad_f^k g_j \in \bar{\mathcal{G}}_{k-1}$ with $ad_f^k g_j = \sum_{\substack{i=1\\ i\neq j}}^{m} \alpha_i ad_f^k g_i|_{mod\ \mathcal{G}_{k-1}}$.

Set

$$u_l = v_l - \sum_{\substack{i=1\\ i\neq j}}^{m} \alpha_i v_j, \qquad l \neq j,\ l = 1,\cdots,m,$$
$$u_j = v_j \tag{12}$$

which corresponds to set on the closed–loop system

$$\tilde{g}_l = g_l, \qquad l \neq j,\ l = 1,\cdots,m,$$
$$\tilde{g}_j = g_j - \sum_{\substack{i=1\\ i\neq j}}^{m} \alpha_i g_i\,. \tag{13}$$

The Dynamic Extension Action

The dynamic extension may be used in two different situations, which correspond respectively to the case in which there is at least one k-unlocked channel, and to the case in which there are no k-unlocked channels. These situations are discussed hereafter.

One unlocked channel – In this case there exists at least one channel which is k-unlocked. Let j be such a channel and assume that the direction feedback or the reduction feedback has been used depending whether $ad_f^k g_j \notin \bar{\mathcal{G}}_{k-1}$ or $ad_f^k g_j \in \bar{\mathcal{G}}_{k-1}$ with $ad_f^k g_j \in \bar{\mathcal{G}}_{k-1} + \text{span}\{ad_f^k g_{l\neq j},\ l = 1,\cdots,m\}$. Set

$$u_j = \chi_{j1}, \qquad \dot{\chi}_{j1} = \bar{v}_j \tag{14}$$

which corresponds to set an integrator on the input j. The extended system

$$\dot{x} = f(x) + \tilde{g}_j \chi_{j1} + \sum_{\substack{i=1\\ i\neq j}}^{m} \tilde{g}_i v_i$$
$$\dot{\chi}_{j1} = \bar{v}_j$$

is characterized by the following set of distributions

$$\mathcal{G}_l^e = \text{span}\{\frac{\partial}{\partial \chi_{j1}}, \tilde{g}\frac{\partial}{\partial x}, \cdots, ad_f^{l-1}\tilde{g}\frac{\partial}{\partial x}\} + \text{span}\{ad_f^l \tilde{g}_i \frac{\partial}{\partial x},\ i = 1,\cdots,m,\ i \neq j\}$$

As a consequence due to Proposition 2, we have that the distribution $\mathcal{G}_{k+i_1}^e$ is involutive for any $i_1 \geq 0$.

No unlocked channels – In this case there are no k-unlocked channels. We can then seek (if there exist) for the smallest index $i \leq n$ such that

$$\mathcal{G}_{k-1} + \text{span}\{ad_f^k g_{j_1}, \cdots, ad_f^{k+i} g_{j_1},\ j_1 = 1,\cdots,m,\ j_1 \neq j\} \equiv \mathcal{G}_{k+i},$$

and the j-th channel is eligible. Set

$$u_j = \chi_{j1}, \qquad \begin{cases} \dot{\chi}_{j1} = \chi_{j2} \\ \quad \vdots \\ \dot{\chi}_{ji} = \bar{v}_j \end{cases} \tag{15}$$

which corresponds to set i integrators on the j-th input and after the reduction feedback is used, leads to the one k-unlocked channel situation.

3 The Algorithm

We now propose an algorithm for the computation of a dynamic feedback which renders the extended system equivalent to a linear system. The algorithm is based on the results of the previous section.

The Dynamic Feedback Linearization Algorithm

Step 0. Let k be the first index such that $\mathcal{G}_{k+i}$ is involutive for any $i \geq 0$ and $\mathcal{G}_{k-1}$ is not involutive and compute its involutive closure $\bar{\mathcal{G}}_{k-1}$.

Step 1. Compute the Noncharacteristic set NC^k. If there is at least a k-unlocked channel go to Step 2, else compute the smallest index $i \leq n$ such that

$$\mathcal{G}_{k-1} + \text{span}\{ad_f^k g_{j_1}, \cdots, ad_f^{k+i} g_{j_1}, \; j_1 = 1, \cdots, m, \; j_1 \neq j\} \equiv \mathcal{G}_{k+i},$$

and the j-th channel is eligible. If such an index does not exist the algorithm ends, else apply the dynamic extension (15) on the j-th channel and go back to Step 0.

Step 2. Consider the set $\mathcal{I} = \{i \in [1, m] : \text{the } i\text{-th channel is } k\text{-unlocked}\}$ of k-unlocked channels and define recursively

$$\mathcal{A}^{k-1} := \{(ad_f^{k-1} g_i, ad_f^r g_{s_t}) : \; [ad_f^{k-1} g_i, ad_f^r g_{s_t}] \notin \mathcal{G}_{k-1}, i \in \mathcal{I}\}$$

$$\vdots$$

$$\mathcal{A}^l := \mathcal{A}^{l+1} \cup \{(ad_f^l g_i, ad_f^r g_{s_t}) : \; [ad_f^l g_i, ad_f^r g_{s_t}] \notin \mathcal{G}_{k-1}, i \in \mathcal{I}\}, \; l < k-1 \,.$$

Let $\hat{l}$ be the first index such that $\mathcal{A}^{\hat{l}+1} \not\equiv \mathcal{A}^0$, while $\mathcal{A}^{\hat{l}} \equiv \mathcal{A}^0$ and consider the index set

$$\mathcal{I}^{\hat{l}} := \{i \in \mathcal{I} : \; (ad_f^{\hat{l}} g_i, ad_f^r g_{s_t}) \in \mathcal{A}^{\hat{l}}\} \,.$$

Let i_1 be the smallest index in $\mathcal{I}^{\hat{l}}$.

Step 3. If $ad_f^k g_{i_1} \notin \bar{\mathcal{G}}_{k-1}$ apply the direction feedback (11) and go to Step 5.

Step 4. If $ad_f^k g_{i_1} \in \mathcal{G}_{k-1} + \text{span}\{ad_f^k g_{l \neq i_1}, \; l = 1, \cdots, m\}$ and $ad_f^k g_{i_1} \notin \mathcal{G}_{k-1}$ apply the reduction feedback (12), compute the modified vector fields $ad_f^t \tilde{g}_{i_1}$, $t = 0, \cdots, k$, and go back to Step 1.

Step 5. Set

$$u_{i_1} = \zeta_{i1}, \quad \dot{\zeta}_{i1} = v_{i_1}.$$

Go back to Step 0.

4 Some Examples

We end by proposing two examples which enlighten the different situations that can be encountered. In the first example the involutivity of the distributions is lost due to a Lie bracket which involves both channels. The direction feedback is used to solve the problem. In the second example the reduction feedback is used, and then since on the modified system there are no unlocked channels a dynamic extension is considered in order to solve the problem.

Example 1. Consider the system

$$\begin{aligned} \dot{x}_1 &= x_2 + x_3x_5, & \dot{x}_2 &= x_3 + x_1x_5, & \dot{x}_3 &= u_1 + x_2x_5, \\ \dot{x}_4 &= x_5, & \dot{x}_5 &= x_6, & \dot{x}_6 &= u_1 + u_2. \end{aligned}$$

Step 0. The distributions $\mathcal{G}_i$, are given by

$$\mathcal{G}_0 = \text{span}\left\{\frac{\partial}{\partial x_3} + \frac{\partial}{\partial x_6}, \frac{\partial}{\partial x_6}\right\},$$

$$\mathcal{G}_1 = \mathcal{G}_0 + \text{span}\left\{-x_5\frac{\partial}{\partial x_1} - \frac{\partial}{\partial x_2} - \frac{\partial}{\partial x_5}, \frac{\partial}{\partial x_5}\right\},$$

and $\mathcal{G}_2 \equiv \mathbb{R}^6$ with

$$\begin{aligned} ad_f^2 g_1 &= (1 + x_3 - x_6)\frac{\partial}{\partial x_1} + (x_1 + x_5^2)\frac{\partial}{\partial x_2} + (x_2 + x_5)\frac{\partial}{\partial x_3} + \frac{\partial}{\partial x_4} \\ ad_f^2 g_2 &= x_3\frac{\partial}{\partial x_1} + x_1\frac{\partial}{\partial x_2} + x_2\frac{\partial}{\partial x_3} + \frac{\partial}{\partial x_4}. \end{aligned}$$

The distribution $\mathcal{G}_1$ is not involutive, since

$$\tau_1 = [ad_f g_1, ad_f g_2] = -\frac{\partial}{\partial x_1} = \gamma(ad_f^2 g_1 - ad_f^2 g_2)|_{mod\ \mathcal{G}_1}.$$

Its involutive closure $\bar{\mathcal{G}}_1 = \mathcal{G}_1 + \text{span}\{\frac{\partial}{\partial x_1}\}$.
Step 1. We have $NC^2 = (ad_f g_1, ad_f g_2)$. Both channels are 2-unlocked.
Step 2. $\mathcal{I} = \{1, 2\}$, $A^1 \equiv NC^2$ and $\mathcal{I}^1 = \{1, 2\}$. We thus choose $i_1 = 1$.
Step 3. Since $ad_f^2 g_1 \notin \bar{\mathcal{G}}_1$, according to Proposition 2 we get $\lambda = x_4$ and correspondingly $L_g L_f^2 \lambda\ u = u_1 + u_2$. Consequently we apply the direction feedback

$$u_1 = v_1 - v_2, \quad u_2 = v_2$$

which corresponds to set on the closed–loop system $\tilde{g}_1 = g_1$ and $\tilde{g}_2 = g_2 \quad g_1$.
Step 5. We apply the dynamic extension $v_1 = \zeta_1$, $\dot{\zeta}_1 = w_1$, $v_2 = w_2$.

The extended dynamics is static feedback equivalent to a linear system as it can be easily verified. △

Example 2. Consider the system

$$\begin{array}{lll} \dot{x}_1 = u_1, & \dot{x}_2 = x_1, & \dot{x}_3 = x_2 + x_6 + x_2 u_1 \\ \dot{x}_4 = u_2 + x_1 u_3, & \dot{x}_5 = x_4, & \dot{x}_6 = x_5 + x_4 x_2 \\ \dot{x}_7 = u_3 & & \end{array}$$

Step 0. The distributions $\mathcal{G}_i$ are given by

$$\mathcal{G}_0 = \text{span}\left\{\frac{\partial}{\partial x_1} + x_2\frac{\partial}{\partial x_3}, \frac{\partial}{\partial x_4}, x_1\frac{\partial}{\partial x_4} + \frac{\partial}{\partial x_7}\right\}$$

$$\mathcal{G}_1 = \mathcal{G}_0 + \text{span}\left\{-\frac{\partial}{\partial x_2} + x_1\frac{\partial}{\partial x_3}, -\frac{\partial}{\partial x_5} - x_2\frac{\partial}{\partial x_6}, -x_1\frac{\partial}{\partial x_5} - x_1 x_2\frac{\partial}{\partial x_6}\right\}$$

$$\mathcal{G}_2 = \mathcal{G}_1 +$$

$$\text{span}\left\{\frac{\partial}{\partial x_3} + x_4\frac{\partial}{\partial x_6}, x_2\frac{\partial}{\partial x_3} + (1 - x_1)\frac{\partial}{\partial x_6}, x_2 x_1\frac{\partial}{\partial x_3} + (x_1 - x_1^2)\frac{\partial}{\partial x_6}\right\} \equiv \mathbb{R}^7 .$$

The distribution $\mathcal{G}_1$ is not involutive since

$$[g_1, ad_f g_1] = 2\frac{\partial}{\partial x_3} = \tau_1 \in \text{span}\{ad_f^2 g_1 + \frac{x_4}{1 - x_1} ad_f^2 g_2\}$$

and

$$[ad_f g_1, ad_f g_2] = \frac{\partial}{\partial x_6} = \tau_2 \in \text{span}\{x_2 ad_f^2 g_1 + ad_f^2 g_2\}.$$

$$[ad_f g_1, ad_f g_3] = x_1\frac{\partial}{\partial x_6} = x_1 \tau_2.$$

We thus have that $\bar{\mathcal{G}}_1 = \mathcal{G}_1 + \text{span}\{ad_f^2 g_1, ad_f^2 g_2\} = \mathcal{G}_2$.
Step 1. We have $NC^2 = \{(g_1, ad_f g_1), (ad_f g_1, ad_f g_2), (ad_f g_1, ad_f g_3)\}$ and $\mathcal{I} = 3$.
Step 2. $\mathcal{A}^1 = (ad_f g_1, ad_f g_3) \equiv \mathcal{A}^0$.
Step 4. Since $ad_f^2 g_3 = x_1 ad_f^2 g_2$, we apply the Reduction Feedback (12):

$$u_1 = v_1, \quad u_2 = v_2 - x_1 v_3, \quad u_3 = v_3$$

which corresponds to set on the closed–loop system

$$\tilde{g}_1 = g_1, \ \tilde{g}_2 = g_2, \ \tilde{g}_3 = g_3 - x_1 g_2 = \frac{\partial}{\partial x_7} .$$

We compute $ad_f \tilde{g}_3 = 0$ and go back to Step 1.
Step 1. We have now $NC^2 = \{(g_1, ad_f g_1), (ad_f g_1, ad_f g_2)\}$. Channels one and two are locked, while channel three is not eligible. We must then look for the first index i such that $\mathcal{G}_1 + \text{span}\{ad_f^2 g_j, \cdots, ad_f^{2+i} g_j\} \equiv \mathbb{R}^7$, $j = 1, 2$. Let us

compute $ad_f^3 g_1$ and $ad_f^3 g_2$. We have

$$ad_f^3 g_1 = -x_4 \frac{\partial}{\partial x_3}, \quad ad_f^3 g_2 = (x_1 - 1)\frac{\partial}{\partial x_3}.$$

Consequently

$$\bar{\mathcal{G}}_1 = \mathcal{G}_1 + \text{span}\{ad_f^2 g_2, ad_f^3 g_2\}.$$

It is then necessary to put an integrator on the first input channel we have

$$\begin{array}{llll} \dot{x}_1 = \zeta_1, & \dot{x}_2 = x_1, & \dot{x}_3 = x_2 + x_6 + x_2\zeta_1, & \dot{x}_4 = v_2 \\ \dot{x}_5 = x_4, & \dot{x}_6 = x_5 + x_4 x_2, & \dot{x}_7 = v_3 & \dot{\zeta}_1 = v_1, \end{array}$$

and we go back to Step 0.

Step 0. For the extended system we get

$$\begin{aligned}
\mathcal{G}_0 &= \text{span}\left\{\frac{\partial}{\partial \zeta_1}, \frac{\partial}{\partial x_4}, \frac{\partial}{\partial x_7}\right\} \\
\mathcal{G}_1 &= \mathcal{G}_0 + \text{span}\left\{-\frac{\partial}{\partial x_1} - x_2\frac{\partial}{\partial x_3}, -\frac{\partial}{\partial x_5} - x_2\frac{\partial}{\partial x_6}\right\} \\
\mathcal{G}_2 &= \mathcal{G}_1 + \text{span}\left\{\frac{\partial}{\partial x_2} - x_1\frac{\partial}{\partial x_3}, x_2\frac{\partial}{\partial x_3} + (1 - x_1)\frac{\partial}{\partial x_6}\right\} \\
\mathcal{G}_3 &= \mathcal{G}_2 + \text{span}\left\{-(1 + 2\zeta_1)\frac{\partial}{\partial x_3} - x_4\frac{\partial}{\partial x_6}, (1 + 2x_1)\frac{\partial}{\partial x_3} - \zeta_1\frac{\partial}{\partial x_1}\right\} \equiv \mathbb{R}^8.
\end{aligned}$$

The distribution $\mathcal{G}_2$ is not involutive due to the Lie bracket

$$\tau_1 = [ad_f g_1, ad_f^2 g_1] = 2\frac{\partial}{\partial x_3} \in \text{span}\{ad_f^3 g_2\}_{\text{mod}\mathcal{G}_2}.$$

Step 1. $NC^3 = (ad_f g_1, ad_f^2 g_1)$ and $\mathcal{I} = \{1\}$.
Step 2. $\mathcal{A}^2 \equiv NC^3$ and $\mathcal{I}^2 = \{1\}$.
Step 4. Since $ad_f^3 g_1 = -\frac{1+2\zeta_1}{1+2x_1} ad_f^3 g_2|_{mod\ \mathcal{G}_2}$ we set

$$\hat{g}_1 = \tilde{g}_1 + \tilde{g}_2 \frac{1 + 2\zeta_1}{1 + 2x_1}, \quad \hat{g}_2 = \tilde{g}_2, \quad \hat{g}_3 = \tilde{g}_3$$

which corresponds to the reduction feedback

$$v_1 = w_1, \quad v_2 = w_2 + \frac{1 + 2\zeta_1}{1 + 2x_1} w_1, \quad v_3 = w_3$$

and we go back to Step 1.

Step 1div2. Since the second channel is not involved in any Lie bracket we still have that $NC^3 = (ad_f \hat{g}_1, ad_f^2 \hat{g}_1)$, $\mathcal{A}^2 \equiv NC^3$ and $\mathcal{I}^2 = \{1\}$. Moreover $ad_f^3 \hat{g}_1 \in \mathcal{G}_2$ so that we go to Step 5.

Step 5. We set

$$w_1 = \zeta_2, \quad \dot{\zeta}_2 = \tilde{w}_1 .$$

The closed–loop system is then given by

$$\begin{aligned}
&\dot{x}_1 = \zeta_1,\ \dot{x}_2 = x_1,\ \dot{x}_3 = x_2 + x_6 + x_2\zeta_1 \\
&\dot{x}_4 = w_2 + \frac{1+2\zeta_1}{1+2x_1}\zeta_2,\ \dot{x}_5 = x_4,\ \dot{x}_6 = x_5, \\
&\dot{x}_7 = w_3,\ \dot{\zeta}_1 = \zeta_2,\ \dot{\zeta}_2 = \tilde{w}_1\,.
\end{aligned}$$

Accordingly

$$\begin{aligned}
\mathcal{G}_0 &= \operatorname{span}\left\{\frac{\partial}{\partial\zeta_2}, \frac{\partial}{\partial x_4}, \frac{\partial}{\partial x_7}\right\} \\
\mathcal{G}_1 &= \mathcal{G}_0 + \operatorname{span}\left\{-\frac{\partial}{\partial\zeta_1} - \frac{1+2\zeta_1}{1+2x_1}\frac{\partial}{\partial x_4}, -\frac{\partial}{\partial x_5}\right\} \\
\mathcal{G}_2 &= \mathcal{G}_1 + \operatorname{span}\left\{\frac{\partial}{\partial x_1} + x_2\frac{\partial}{\partial x_3} + \frac{2\zeta_1(1+2\zeta_1)}{(1+2x_1)^2}\frac{\partial}{\partial x_4} + \frac{1+2\zeta_1}{(1+2x_1)}\frac{\partial}{\partial x_5}, \frac{\partial}{\partial x_6}\right\} \\
\mathcal{G}_3 &\equiv \mathbb{R}^9
\end{aligned}$$

which are involutive and of constant dimension thus ensuring that the extended system is feedback equivalent to a linear system. △

5 Conclusions

In the present paper we have analyzed the geometric properties of a dynamic feedback linearizable system as well as those of the compensator which achieves linearization. On the basis of these geometric properties an algorithm for the computation of a dynamic feedback obtained from the composition of regular static state feedback laws and integrators has been proposed. The optimal choice of the input channel on which it is necessary to set integrators will be the objective of future work.

Acknowledgments

The authors wish to thank Alberto Isidori for his illuminating ideas which have accompanied us in these past years.

References

1. E. Aranda-Bricaire, C.H. Moog, and J.B. Pomet. A linear algebraic framework for dynamic feedback linearization. *IEEE Trans. on Automat. Contr.*, 40:127–132, 1995.
2. S. Battilotti. *Noninteracting Control with Stability for Nonlinear Systems.* Springer Verlag, 1994.

3. S. Battilotti and C. Califano. Further results on dynamic feedback linearization. *Proc. of the 2003 European Control Conference*, 2003.
4. S. Battilotti and C. Califano. A constructive condition for dynamic feedback linearization. *Systems & Control Letters*, 52:329–338, 2004.
5. R.W. Brockett. Feedback invariants for nonlinear systems. *IFAC*, pages 1115–1120, 1978.
6. C. Califano, S. Monaco, and D. Normand-Cyrot. On the feedback linearization problem. *Systems & Control Letters*, 36:61–67, 1999.
7. B. Charlet, J. Lévine, and R. Marino. On dynamic feedback linearization. *Systems & Control Letters*, pages 143–151, 1989.
8. B. Charlet, J. Lévine, and R. Marino. Sufficient conditions for dynamic feedback linearization. *SIAM J. Contr. Optimization*, 29:38–57, 1991.
9. M. Fliess, J. Lévine, P. Martin, and P. Rouchon. On differentially flat nonlinear systems. *Proc. of the 2nd IFAC Symposium in Nonlinear Control Systems*, pages 408–412, 1992.
10. M. Fliess, J. Lévine, P. Martin, and P. Rouchon. Flatness and defect of nonlinear systems: introductory theory and examples. *Int. J. of Control*, 61:1327–1361, 1995.
11. L.R. Hunt, R. Su, and G. Meyer. Design for multi-input nonlinear systems. In R.S. Millman R.W. Brockett and H. Sussmann, editors, *Differential Geometric Control Theory*, pages 268–298. Birkhauser, 1983.
12. A. Isidori. *Nonlinear Control Systems.* Springer Verlag, London, UK, 3rd edition, 1995.
13. A. Isidori, A.J. Krener, C. Gori Giorgi, and S. Monaco. Decoupling via feedback: a differential geometric approach. *IEEE Trans. on Automat. Contr.*, pages 331–345, 1981.
14. A. Isidori, C.H. Moog, and A. De Luca. A sufficient condition for full linearization via dynamic state feedback. *Proc. of the 25th IEEE Conf. on Decision and Contr.*, pages 203–208, 1986.
15. B. Jakubczyk. Feedback linearization of discrete-time systems. *Systems & Control Letters*, 9:441–446, 1987.
16. B. Jakubczyk and W. Respondek. On linearization of control systems. *Bull. Acad. Polonaise Sci.*, 28:517–522, 1980.
17. H.G. Lee, A. Arapostathis, and S.I. Marcus. On the linearization of discrete-time systems. *Int. J. of Control*, 45:1783–1785, 1987.
18. R. Marino. On the largest feedback linearizable subsystem. *Systems & Control Letters*, pages 345–351, 1986.
19. S. Monaco and D. Normand-Cyrot. On the immersion under feedback of a multidimensional discrete time nonlinear system into a linear system. *Int. J. of Control*, 38:245–261, 1983.
20. S. Monaco and D. Normand-Cyrot. Minimum phase nonlinear discrete-time systems and feedback stabilization. *Proc. of the 26th IEEE Conf. on Decision and Contr.*, pages 979–986, 1987.
21. H. Nijmeijer and A.J. van der Schaft. *Nonlinear Dynamical Control Systems.* Springer Verlag, 1990.

Part VI

Asymptotic Analysis

The Steady-State Response of a Nonlinear Control System, Lyapunov Stable Attractors, and Forced Oscillations

Chris I. Byrnes[1] and David S. Gilliam[2]

[1] Washington University in St. Louis, MO-63130, USA
[2] Texas Tech University, Lubbock, TX 79409, USA

Summary. In this paper, we study the existence of periodic solutions to periodically forced systems in both the equilibrium and the nonequilibrium cases. In the equilibrium case, we prove an averaging theorem. Then, we develop the rigorous theory of the steady-state response of a nonlinear control system in order to derive some positive results on periodic forcing of autonomous systems with global Lyapunov attractors. Finally, on this occasion it is so much of a pleasure to thank Alberto Isidori for all he taught us and for all the joy we shared, and continue to share, in our research. Tanti auguri!

1 Averaging

In his plenary lecture at the 2001 IEEE CDC, Dennis Bernstein expressed the desirability of having Lyapunov-theoretic proof of the Averaging Theorem. While it turns that a Lyapunov-assisted proof makes things easier, this is not the principle purpose of this paper but rather serves as a neat pedagogical introduction to some of the more subtle results on the existence of "steady state" responses of nonlinear dynamical systems, as pioneered in [6], and discussed later in the paper.

Consider the system

$$\dot{x} = f(x) + \varepsilon p(x, t, \varepsilon),$$

where $x \in \mathbb{R}^n$ and $p(x, t+T, \varepsilon) = p(x, t, \varepsilon)$ for some (period) $T > 0$. The nicest way to analyze a periodically forced system is to use what the Russians [8] have classically called the "toroidal cylinder", which is simply $\mathbb{R}^n \times S^1$, with coordinates (x, τ). On the toroidal cylinder, we consider the autonomous vector field $\widetilde{f}$ defined by

$$\begin{aligned} \dot{x} &= f(x) + \varepsilon p(x, \tau, \varepsilon), \\ \dot{\tau} &= 1\,. \end{aligned} \tag{1}$$

We now assume that the origin is locally exponentially stable for the unforced system (i.e., for $\varepsilon = 0$). By the converse theorem of Lyapunov for locally exponentially stable equilibria, there exists a Lyapunov function V for $\dot{x} = f(x)$ which is quadratic-like in the sense that there exist [4] positive constants $(\alpha_1, \alpha_2, \alpha_3, \alpha_4)$ such that

$$\alpha_1 \|x\|^2 \le V(x) \le \alpha_2 \|x\|^2, \tag{2}$$

$$\dot{V}(x) \le -\alpha_3 \|x\|^2, \tag{3}$$

$$\|\text{grad}V(x)\| \le \alpha_4 \|x\|. \tag{4}$$

These equalities are important in understanding the topology of the sublevel and level sets of V. As far as we are aware, the first serious study of the topological questions was presented in [9], who proved that if a Lyapunov function was proper (the inverse image of compact sets are compact) then the sublevel sets are contractible and the level sets are homotopic to spheres. Since we shall not use these results (although one could) in our proof, we shall not go into more detail. Indeed, we mention them only for the sake of scholarship. The topological result we need is as stated below and is one that seems to be of independent interest. We believe it is well-known but, since we haven't seen a proof of it, we shall provide one. Recall that $V(0) = 0$ and $\text{grad}V(0) = 0$ and that the Hessian of V at 0, $D^2V(0)$ is positive semidefinite, since 0 is a minimum of V. In this case, 0 is said to be a nondegenerate critical point provided the Hessian is nonsingular at 0.

Lemma 1. *Suppose* 0 *is a locally exponentially stable equilibrium for* $\dot{x} = f(x)$ *and that* V *is a Lyapunov function satisfying* (2)–(4). *Then, for* c *sufficiently small,* $V^{-1}(c)$ *is diffeomorphic to a sphere in* $\mathbb{R}^n$ *and* $V^{-1}(-\infty, c]$ *is diffeomorphic to a closed disc.*

Proof. The left-hand side of (2) trivially implies that $D^2V(0)$ is positive definite and therefore, 0 is a nondegenerate critical point. Being a minimum, the Morse Lemma [7] implies that, near the origin, there is a smooth change of coordinates in which $V(x)$ is expressible as

$$x_1^2 + x_2^2 + \cdots + x_n^2$$

from which our claims follow immediately. □

We now consider the perturbed system.

Theorem 1. *(Averaging) If* 0 *is a locally exponentially stable equilibrium for the system* (1) *when* $\varepsilon = 0$*, then for* $\varepsilon \ll \infty$ *there exists a locally exponentially stable periodic orbit* $\gamma_\varepsilon(t)$ *of period* T *whose amplitude is* $O(\varepsilon)$.

Proof. As before, we will consider the augmented vector field $\widetilde{f}$ defined by (1) but, in order to take advantage of the compactness of the sublevel sets of V, we will view $\widetilde{f}$ as evolving on a truncation of the toroidal cylinder, viz., the compact "Lyapunov can" defined as $M_c = V^{-1}(-\infty, c] \times S^1$.

For $\varepsilon = 0$, the Lyapunov can is clearly positively invariant and contains the periodic orbit $\gamma_0 = \{(0, \tau)\}$. We note that, for $\varepsilon = 0$, the Poincarè map, $\mathcal{P}$, on the submanifold (with boundary) $\{(x, 0)\}$ is in fact the time-T map Φ_T for the autonomous system. In particular, at $x = 0$ and for $\varepsilon = 0$, we have

$$D\mathcal{P}(0,0) = D\Phi_T(0) = \exp TDf(0).$$

Therefore, γ_0 is exponentially stable.

We now consider $\varepsilon \geq 0$ and claim M_c can be rendered positively invariant. Indeed, on M_c we have

$$\dot{V} = L_f V + \varepsilon L_p V \leq -\alpha_3 \|x\|^2 + \varepsilon \alpha_4 \|x\| \|p\|.$$

Setting $\|p\| = A$ and $\alpha_4 A = \beta$, we obtain

$$\dot{V} \leq -\alpha_3 \|x\|^2 + \varepsilon \beta \|x\|.$$

Taking $\varepsilon << \infty$, we can conclude that $\widetilde{f}$ points inwards on ∂M_c. Denote the time t map of $\widetilde{f}$ with initial condition (x_0, τ_0) by $\widetilde{\Phi}_t(x_0, \tau_0)$.

Since M_c is positively invariant, and since $(\widetilde{f}, d\tau) = 1 > 0$, for any initial condition $(x_0, 0)$, there is a minimum time $\widetilde{T}(x_0, 0) > 0$, such that $\widetilde{\Phi}_{\widetilde{T}}((x_0, 0)$ lies in in the interior of the closed disk $V^{-1}(-\infty, c] \times \{0\}$ in M_c. In fact $\widetilde{T} = T$. In particular, the Poincaré map

$$\mathcal{P} : V^{-1}(-\infty, c] \times \{0\} \to V^{-1}(-\infty, c] \times \{0\}$$

is defined and has a fixed point, by the Brouwer Fixed Point Theorem. Therefore, the perturbed system, for $\varepsilon \ll \infty$, has a periodic orbit of period T in M_c which, after setting $t = \tau$, is a periodic orbit of the perturbed system.

Taking a sequence $(c_n) \to 0$ as $n \to \infty$, the same argument, mutatis mutandis, we find that there exists a sequence $\varepsilon_n \to 0$ as $n \to \infty$, such that for any $\varepsilon \leq \varepsilon_n$, M_{c_n} is positively invariant. shows This shows that there exist periodic orbits of arbitrarily small amplitude evolving in a nested sequence of compact (tubular) neighborhoods of γ_0.

However, to conclude the existence of a stable periodic orbit of small amplitude, we first further augment the perturbed system by adding the state ε which evolves according to $\dot{\varepsilon} = 0$. We will denote the time-T map for this augmented system by $\widetilde{\Phi}^{\varepsilon}_T$ and note that for any periodic orbit γ in M_c the Poincaré map along γ coincides with the time T map of the augmented system; i.e.,

$$\mathcal{P}^{\varepsilon}(x, 0, \varepsilon) = \Phi^{\varepsilon_T}(x, \varepsilon).$$

In particular, $\mathcal{P}^{\varepsilon}(x, 0, \varepsilon)$ is smooth in the initial data $(x, 0, \varepsilon)$. We now apply the Implicit Function Theorem to the fixed point equation

$$x - \mathcal{P}^{\varepsilon}(x, 0, \varepsilon) = 0$$

to find a smooth branch $x(\varepsilon)$ of fixed points passing through the point $(0, 0, 0)$; i.e. such that $x(0) = 0$. Explicitly, differentiating with respect to x we obtain

$$I - D\mathcal{P}^{\varepsilon}(x, 0, \varepsilon) = I - \Phi_T^{\varepsilon}(x, \varepsilon).$$

Computing determinants and evaluating at $(x, \varepsilon) = (0, 0)$ we obtain

$$\det(I - D\mathcal{P}^0(0, 0, 0)) = \det(I - \exp\{TDf(0)\} \neq 0.$$

From this we conclude that there exists a smooth branch of fixed points $x(\varepsilon)$ passing through $x(0) = 0$ and therefore there exist a smooth variation of periodic orbits γ_ε of γ_0. Moreover, $D\mathcal{P}^{\varepsilon}(x, 0, \varepsilon)$ varies smoothly with the initial data $(x(\varepsilon), \varepsilon)$; in fact

$$D\Phi_T^{\varepsilon}(x) = \exp\{TDf(x)\} \exp\{\varepsilon Dp(T, x, \varepsilon)\}.$$

For $\varepsilon << \infty$, the second factor on the right side can be made arbitrarily close to 1 and, therefore, we conclude that for $\varepsilon << \infty$ the periodic orbit γ_ε is hyperbolically stable.

Concerning the amplitude estimates, we begin by expressing the smooth branch of fixed points $x(\varepsilon)$ as a constant, $x(0) = 0$, plus a Taylor remainder in ε, from which we see that $\|x(\varepsilon)\| = \mathcal{O}(\varepsilon)$. Next we integrate the perturbed system along the periodic orbit γ_ε, which we parameterize as $x_\varepsilon(t)$. Integrating the perturbed equation along $x_\varepsilon(t)$, we obtain

$$x_\varepsilon(t) = x_\varepsilon(0) + \int_0^t (f(x_\varepsilon(\tau)) + \varepsilon p(\tau, x_\varepsilon(\tau), \varepsilon))d\tau.$$

Since $f(0) = 0$, expressing $f(x_e(\tau))$ in a constant term $(\varepsilon = 0)$ and a Taylor remainder yields

$$\int_0^t f(x_\varepsilon(\tau))d|\tau = \varepsilon \int_0^t R(x_\varepsilon(\tau))d\tau$$

which in norm is $\mathcal{O}(\varepsilon)$ for $0 \leq t \leq T$. The norm of the remaining integral is clearly $\mathcal{O}(\varepsilon)$ for $0 \leq t \leq T$ so that, by the triangle inequality, $\|x_\varepsilon(t)\| = \mathcal{O}(\varepsilon)$ for $0 \leq t \leq T$. □

2 Limit Sets

In general, the problem of determining periodic solutions of periodically varying nonlinear systems can be seen as a special case of the more general problem of characterizing the "steady-state" response of a nonlinear system to specific classes of (periodic or even non-periodic) forcing inputs.

Traditionally, the idea of a separation between steady-state and transient response stems from the observation that, in any finite-dimensional time-invariant linear system, (i) the forced response to an input which is a polynomial or exponential function of time normally includes a term which is a polynomial (of degree not exceeding that of the forcing input) or an exponential function (with an exponent whose rate of change is the same as that of the forcing input) of time, and (ii) if the unforced system itself is asymptotically stable, this term is the unique function of time to which the actual response converges as the initial time tends to $-\infty$ (regardless of what the state of the system at the initial time is).

For nonlinear control systems, the concept of steady-state response has only recently been formalized ([2], [3]). In the nonlinear case, we would still want to have an analogue of item (i) stated above, but any statement such as "the response is equal to the sum of the steady-state response and the transient response" can never serve as a definition since two of the three terms are undefined. We now understand that that some form of uniformity in the decay of the transient response, similar to Lyapunov stability, is a key – at least for bounded sets of initial data. And the formalization we have made also makes item (ii) just as precise, by focusing on solutions that are bounded not only forward in time but also backward in time. We now illustrate these issues in the one-dimensional case.

Example 1. Consider the scalar dynamical system

$$\dot{x} = x - x^3.$$

Every solution is bounded forward in time so for each x_0 the ω-limit set $\omega(x_0)$, as defined by G. D. Birkhoff in [1], is non-empty. For any bounded interval B one might think that the steady-state response to initial conditions in B would be

$$\psi(B) = \bigcup_{x_0 \in B} \omega(x_0),$$

i.e. the union of the ω-limits set of all points of B. In the one-dimensional case, all ω limits are equilibria. Here there are three equilibria, $\{-1, 0, 1\}$, of which 0 is asymptotically unstable and the other two are asymptotically stable. Therefore, $\Psi(B) = \{-1, 0, 1\}$, for any interval B containing $[-1, 1]$, but $\Psi(B)$ is not uniformly attractive even for B bounded. Moreover, there are uncountably many points that are the limits of initial data with the initial time tending to $-\infty$, as the following Fig. 1 illustrates.

That is, taking an x_0 with $0 < x_0 < 1$ and a decreasing sequence (x_n) starting with x_0 and converging to 0 gives a sequence of times t_n such that

$$\Phi_{t_n}(x_n) = x_0.$$

Letting $-t_n \to -\infty$ gives a sequence of times and a sequence of initial data (x_n) satisfying item (ii) discussed above.

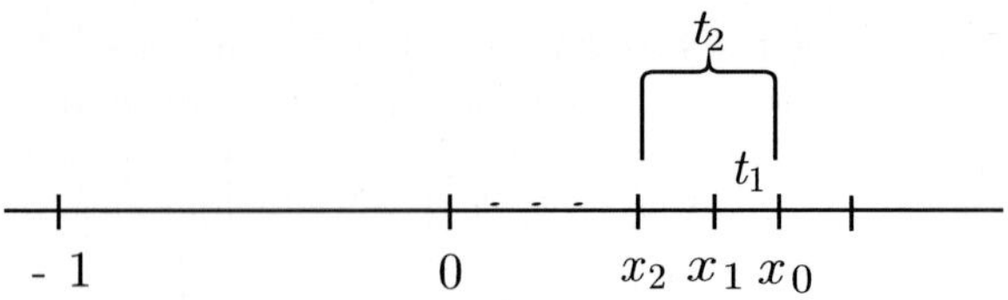

Fig. 1. A picture of Example 1

This suggests taking $[-1,1]$ as the steady-state response of the dynamical system. One should also note that the above argument proves that $\{-1,0,1\}$ is not uniform attractive, while $[-1,1]$ is a globally attractive, Lyapunov stable attractor. We also note that it consists of all trajectories which are bounded both forward and backward in time. △

In general , but not always, we will have to be satisfied with semiglobally attractivity, but we can retain Lyapunov stability. We now formalize what we saw in the above example.

Consider an *autonomous* ordinary differential equation

$$\dot{x} = f(x) \tag{5}$$

with $x \in \mathbb{R}^n$, $t \in \mathbb{R}$, and let

$$\phi : (t,x) \mapsto \phi(t,x)$$

define its flow [5]. Suppose the flow is forward complete. The *ω-limit set* of a subset $B \subset \mathbb{R}^n$, written $\omega(B)$, is the totality of all points $x \in \mathbb{R}^n$ for which there exists a sequence of pairs (x_k, t_k), with $x_k \in B$ and $t_k \to \infty$ as $k \to \infty$, such that

$$\lim_{k\to\infty} \phi(t_k, x_k) = x\,.$$

In case $B = \{x_0\}$ the set thus defined, $\omega(x_0)$, is precisely the ω-limit set. As in Example 1, with a given set B, is it is also convenient to associate the set

$$\psi(B) = \bigcup_{x_0 \in B} \omega(x_0)$$

i.e. the union of the ω-limits set of all points of B. Clearly, by definition

$$\psi(B) \subset \omega(B)\,,$$

but, as Example 1 shows, the equality may not hold.

G.D.Birkhoff has shown that, if $\phi(t,x_0)$ is bounded in positive time, the set $\omega(x_0)$ is non-empty, compact, invariant, and

$$\lim_{t\to\infty} \operatorname{dist}(\phi(t,x_0), \omega(x_0)) = 0\,.$$

More generally, recall that a set A is said to *uniformly attract* [3] a set B under the flow of (5) if for every $\varepsilon > 0$ there exists a time $\bar{t}$ such that

$$\mathrm{dist}(\phi(t,x),A) \leq \varepsilon, \qquad \text{for all } t \geq \bar{t} \text{ and for all } x \in B.$$

With the above definitions we immediately obtain the following lemma.

Lemma 2. *If B is a nonempty bounded set for which there is a compact set J which uniformly attracts B (thus, in particular, if B is any nonempty bounded set whose positive orbit has a bounded closure), then $\omega(B)$ is nonempty, compact, invariant and uniformly attracts B.*

3 The Steady State Behavior of a Nonlinear System

Consider now again system (5), with initial conditions in a closed subset $X \subset \mathbb{R}^n$. Suppose the set X is *positively invariant*, which means that for any initial condition $x_0 \in X$, the solution $x(t,x_0)$ exists for all $t \geq 0$ and $x(t,x_0) \in X$ for all $t \geq 0$. The motions of this system are said to be *ultimately bounded* if there is a bounded subset B with the property that, for every compact subset X_0 of X, there is a time $T > 0$ such that $\|x(t,x_0)\| \in B$ for all $t \geq T$ and all $x_0 \in X_0$. In other words, if the motions of the system are ultimately bounded, every motion eventually enters and remains in the bounded set B.

Remark 1. Note that, since by hypothesis X is positively invariant, there is no loss of generality in assuming $B \subset X$ in the definition above. Note also that there exists a number M such that $\|x(t,x_0)\| \leq M$ for all $t \geq 0$ and all $x_0 \in B$. In fact, let $\mathrm{Cl}(B)$ denote the closure of B, which is a compact subset of X, and let M_1 the denote the maximum of $\|x\|$ as $x \in \mathrm{Cl}(B)$. By definition of ultimate boundedness, there is a time T such that $\|x(t,x_0)\| \leq M_1$, for all $t \geq T$ and all $x_0 \in \mathrm{Cl}(B)$. Moreover, since $x(t,x_0)$ depends continuously on (t,x_0), there exists a number M_2 such that $\|x(t,x_0)\| \leq M_2$ for all $0 \leq t \leq T$ and all x_0 in $\mathrm{Cl}(B)$. Thus, the property in question is fulfilled with $M = \max\{M_1, M_2\}$. By virtue of this property, one can conclude from Lemma 2 that the set $\omega(B)$ is nonempty and has all the properties indicated in the Lemma itself. Finally, note that, for a system whose motions are ultimately bounded, the set $\omega(B)$ is a unique well-defined set, regardless of how B is taken. In fact, let B' be

[3] Note that, in [5], the property which follows is simply written as

$$\lim_{t\to\infty} \mathrm{dist}(\phi(t,B),A) = 0,$$

with the understanding that

$$\mathrm{dist}(B,A) := \sup_{x\in B} \mathrm{dist}(x,A) = \sup_{x\in B} \inf_{y\in A} \mathrm{dist}(x,y)\,.$$

any other bounded subset of X with the property indicated in the definition of ultimate boundedness. Then, it is not difficult to prove, using the various definitions, that $\omega(B') \subset \omega(B)$. Reversing the role of the two sets shows that $\omega(B) \subset \omega(B')$, i.e. that the two sets in question are identical. ◁

For systems whose motions are ultimately bounded, the notion of steady state can be defined as follows.

Definition 1. *Suppose the motions of system (5), with initial conditions in a closed and positively invariant set X, are ultimately bounded. A* steady state *motion is any motion with initial condition in $x(0) \in \omega(B)$. The set $\omega(B)$ is the* steady state locus *of (5) and the* restriction *of (5) to $\omega(B)$ is the* steady state behavior *of (5).*

4 Examples

The notion thus introduced recaptures the classical notion of steady state for linear systems and provides a new powerful tool to deal with similar issues in the case of nonlinear systems.

Example 2. Consider a nonlinear system

$$\dot{x} = f(x, u) \tag{6}$$

in the neighborhood of a locally exponentially stable equilibrium point. To this end, suppose that $f(0, 0) = 0$ and that the matrix

$$F = \left[\frac{\partial f}{\partial x}\right](0, 0)$$

has all eigenvalues with negative real part. Then, it is well known (see e.g. [4, page 275]) that it is always possible to find a compact subset $X \subset \mathbb{R}^n$, which contains $x = 0$ in its interior and a number $\sigma > 0$ such that, if $\|x_0\| \in X$ and $\|u(t)\| \le \sigma$ for all $t \ge 0$, the solution of (6) with initial condition $x(0) = x_0$ satisfies $\|x(t)\| \in X$ for all $t \ge 0$. Suppose that the input u to (6) is produced by a signal generator of the form

$$\begin{aligned} \dot{w} &= s(w) \\ u &= q(w) \end{aligned} \tag{7}$$

with initial conditions chosen in a compact invariant set W and, moreover, suppose that, $\|q(w)\| \le \sigma$ for all $w \in W$. If this is the case, the set $X \times W$ is positively invariant for

$$\begin{aligned} \dot{w} &= s(w) \\ \dot{x} &= f(x, q(w)), \end{aligned} \tag{8}$$

and the motions of the latter are ultimately bounded, with $B = X \times W$. The set $\omega(B)$ may have a complicated structure but it is possible to show, by means arguments similar to those which are used in the proof of the Center Manifold theorem, that if X and B are small enough the set in question can still be expressed as the graph of a map $x = \pi(w)$. In particular, the graph in question is precisely the center manifold of (8) at $(0,0)$ if $s(0) = 0$ and the matrix

$$S = \left[\frac{\partial s}{\partial w}\right](0)$$

has all eigenvalues on the imaginary axis.

In particular, if for some $w_0 \in W$ the integral curve of $\dot{w} = s(w)$ passing through w_0 at time $t = 0$ is a periodic function of time, the associated steady state response of (6) is the periodic function $x(t) = \pi(w(t))$. △

In the example above the set $\omega(B)$ can be expressed as the graph of a map $x = \pi(w)$. This means that, so long as this is the case, a system of the form (6) has a *unique* well defined *steady state response* to the input $u(t) = q(w(t))$. Of course, in general, this may not be the case and *multiple* steady state responses to a given input may occur. In general, the following property holds.

Lemma 3. *Let W be a compact set, invariant under the flow of (7). Let X be a closed set and suppose that the motions of (8) with initial conditions in $W \times X$ are ultimately bounded. Then, the steady state locus of (8) is the graph of a set-valued map defined on the whole of W.*

Example 3. Consider now the system

$$\begin{aligned} \dot{x} &= y \\ \dot{y} &= x - x^3 - y\Big(-\frac{x^2}{2} + \frac{x^4}{4} + \frac{y^2}{2} + \frac{1}{4} - w\Big) \end{aligned} \tag{9}$$

in which w is a *constant* input generated by the exosystem $\dot{w} = 0$. For any fixed w, this system has three equilibria, at $(x,y) = (0,0)$ and $(x,y) = (\pm 1, 0)$. We show now that, for any fixed w, all trajectories of system (9) are ultimately bounded. In fact, consider the positive semi-definite function

$$V(x,y) = -\frac{x^2}{2} + \frac{x^4}{4} + \frac{y^2}{2} + \frac{1}{4}$$

which is zero only at the two equilibria $(x,y) = (\pm 1, 0)$ and such that, for any $c > 0$, the sets $\Omega_c = \{(x,y) : V(x,y) \le c\}$ are bounded. Note that

$$\dot{V}(x,y) = -y^2(V(x,y) - w)\,.$$

If $w \le 0$, $\dot{V}(x,y) \le 0$ for all (x,y) and therefore, by LaSalle's invariance principle, all trajectories which start in $\mathbb{R}^2$ converge to the largest invariant set contained in the locus where $y = 0$, which only consists of the union of the three equilibria.

If $w > 0$, $\dot{V}(x,y) \le 0$ for all (x,y) in the set $\{(x, y : V(x,y) \ge w\}$. Thus, again by LaSalle's invariance principle, all trajectories which start in the set

$\{(x, y : V(x, y) \geq w\}$ converge to the largest invariant set contained in the locus where either $y = 0$ or $V(x, y) = w$. Since the locus $V(x, y) = w$, the boundary of Ω_w, is itself invariant and the two equilibria $(x, y) = (\pm 1, 0)$ are in Ω_w, it is concluded that all trajectories which start in $\mathbb{R}^2 \setminus \Omega_w$ converge either to the boundary of Ω_w or to the equilibrium $(x, y) = (0, 0)$. On the other hand, the boundary of Ω_w, for $0 < w < 1/4$ consists of two disjoint close curves while for $1/4 \geq w$ it consists of a single closed curve (a "figure eight" for $w = 1/4$).

From this analysis it is easy to conclude what follows. For any pair of compact sets

$$X = \{(x, y) : \max\{|x|, |y| \leq r\} \qquad W = \{w : |w| \leq r\},$$

the positive orbit of $X \times W$ is bounded. Moreover, for large r, if $w \leq 0$, the set

$$\mathcal{SSL}_w = \omega(X \times W) \cup (\mathbb{R}^2 \times \{w\}),$$

i.e. the intersection of $\omega(X \times W)$ with the plane $\mathbb{R}^2 \times \{w\}$ is a 1-dimensional manifold with boundary, diffeomorphic to a closed interval of $\mathbb{R}$. If $0 < w < 1/4$, the set $\mathcal{SSL}_w$ is the union of a 1-dimensional manifolds diffeomorphic to $\mathbb{R}$ and of two disjoint 2-dimensional manifold with boundary, each one diffeomorphic to a closed disc. If $1/4 \leq w$, the set $\mathcal{SSL}_w$ is a 2-dimensional manifold with boundary, diffeomorphic to a closed disc for $1/4 < w$, or to a "filled figure eight" for $w = 1/4$. Different shapes of these sets, for various values of w, are shown in Figure 2. Again, we note that the steady-state locus consists of all trajectories which are bounded both forward and backward in time. △

Example 4. We next consider the unforced van der Pol oscillator. In this case all trajectories are bounded forward in time, and the ω-limits consist of the origin and the limit cycle. As in Example 1, their union is not Lyapunov stable nor is it uniformly attractive for initial data in any bounded set B containing the limit cycle and the equilibrium.

Rather, for such a B, the limit cycle and its interior is $\omega(B)$. We note that $\omega(B)$ is globally attractive and, from the theory of dissipative systems as described in Lemma 1 [6], is Lyapunov stable and consists of all trajectories bounded both forward and backward in time. △

Example 5. Consider the three-dimensional nonlinear system

$$\begin{aligned} \dot{z} &= -(z + w_1)^3 + z + w_1^3 \\ \dot{w}_1 &= w_2 \\ \dot{w}_2 &= -w_1 . \end{aligned}$$

We first note that $z = 0$ is an invariant plane with dynamics of the harmonic oscillator. It is of great interest to understand the transverse (in the

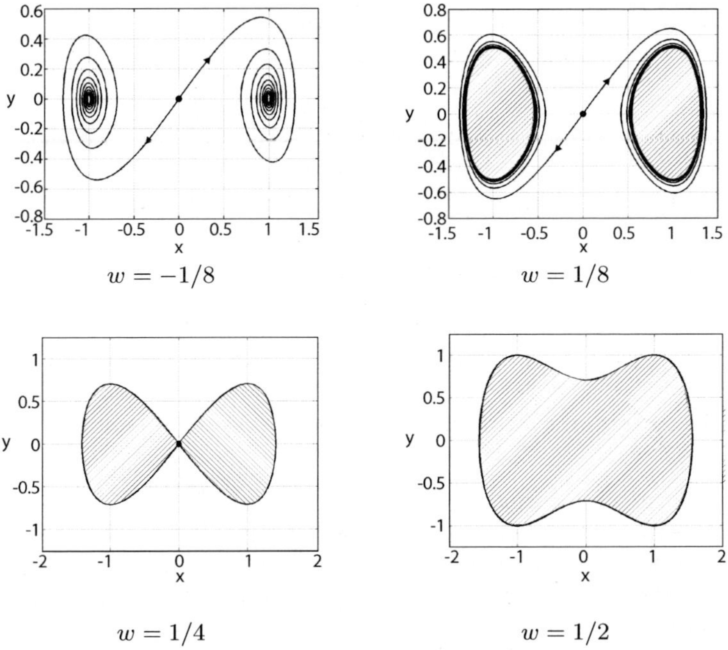

Fig. 2. Steady state locus of Example 3

z-direction) stability of these periodic orbits. For this we use Poincarè's formula for the derivative of the Poincarè map $\mathcal{P}(z_0)$:

$$\begin{aligned} D\mathcal{P}(z_0)\big|_{z=0} &= \exp\left(\int_0^{2\pi} \operatorname{div}(f)\bigg|_{z=0} dt\right) \\ &= \exp\left(\int_0^{2\pi} (-3w_1^2+1)\,dt\right) \\ &= \exp\left(-3A^2\pi + 2\pi\right) \end{aligned}$$

where $A^2 = w_1^2 + w_2^2$.

Therefore, periodic motions in the (w_1, w_2) plane with amplitude A are:

(a) transversely stable if $A > \sqrt{2/3}$;
(b) unstable if $A < \sqrt{2/3}$;
(c) critically stable if $A = \sqrt{2/3}$.

Indeed, $\mathcal{P}$ undergoes a pitchfork bifurcation at the critical amplitude $A_c = \sqrt{2/3}$:

$$D\mathcal{P}_c(0) = 1, \quad D\mathcal{P}_c^2(0) = 0, \quad D\mathcal{P}_c^3(0) = -12\pi.$$

This suggests the existence of multiple periodic orbits.

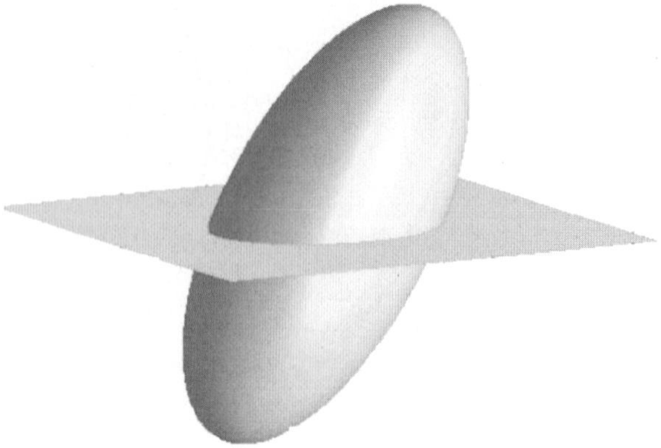

Fig. 3. Structure of the attractor in Example 5: the surface and its interior points

In order to understand the behavior of trajectories far from the plane $z = 0$, we consider an old friend, $V(z, w) = (z^2 + \|w\|^2)$, for which we find $\dot{V} < 0$ when $|z| > 2$. As before, if $A < \sqrt{2/3}$ the Lyapunov can $ø_1^2 + ø_2^2 = A^2$, $|z| \leq 2$ contains a periodic orbit. this we knew, taking $z = 0$, but we also have a positive Lyapunov can $ø_1^2 + ø_2^2 = A^2$, $0 < \varepsilon \leq z \leq 2$, as well as a corresponding negative Lyapunov can.

Therefore, for $A < \sqrt{2/3}$, there exist periodic orbits for both $z > 0$ and $z < 0$. In order to check stability, we compute

$$z^2(t) = \exp\left[-2\int_0^t \left(z^2(\tau) \underbrace{+3w_1(\tau)z(\tau) + 3w_1^2(\tau)}_{=0 \text{ by harmonic balance}} -1\right) d\tau\right] z^2(0).$$

Therefore, if $z(\,\cdot\,, z_0)$ is periodic in t of period $2k\pi$ and $z(t) \neq 0$ then

$$\int_0^{2k\pi} z^2(t)\,dt = k\pi(2 - 3A^2)\,.$$

In particular, periodic orbits with $z(0) \neq 0$ can only exist if

$$(2 - 3A^2) > 0 \quad \text{or} \quad A < \sqrt{2/3}.$$

In this case

$$D\mathcal{P}(z_0) = \exp\big[-2(2 - 3A^2)\pi\big] < 1$$

and therefore each periodic $z(t)$ is hyperbolic and asymptotically stable for $z(t) > 0$ or $z(t) < 0$.

Using index theory, one can see that this implies there exists a unique periodic orbit for $z > 0$ (for $z < 0$, resp.) and for $0 < A < A_c$. and this orbit is exponentially orbitally stable. For the whole cylinder, then, with $|z| \leq 2$ there are three periodic orbits in this range of amplitudes, while for $A = 0$ there are three equilibria at $(0, 0, 0)$ and $(\pm 1, 0, 0)$. △

References

1. G.D. Birkhoff. *Dynamical systems*, volume IX. American Mathematical Society Colloquium Publications, Providence, R.I, 2nd edition, 1966.
2. C. I. Byrnes, D. S. Gilliam, A. Isidori, and J. Ramsey. *On the steady-state behavior of forced nonlinear systems.* Lecture Notes in Control and Inform. Sci. Springer-Verlag, Berlin, 2nd edition, 2003. New trends in nonlinear dynamics and control, and their applications.
3. C.I. Byrnes and I. Isidori. The steady-state response of a nonlinear system: Ideas, tools and applications. *Automatica*, 2006. Submitted.
4. W. Hahn. *Stability of Motion.* Springer -Verlag, New York, 1967.
5. J. K. Hale, L. T. Magalhães, and W. M. Oliva. *Dynamics in Infinite Dimensions.* Springer Verlag, New York, 2002.
6. J.K. Hale. *Asymptotic Behavior of Dissipative Systems*, volume #25. AMS Series: Surv Series, 1988.
7. J. Milnor. *Morse Theory.* Princeton University Press, Princeton, 1963.
8. V.A. Pliss. *Nonlocal problems of the theory of oscillations.* Academic Press, New York, 1966.
9. F.W. Wilson. The structure of the level surfaces of a Lyapunov function. *J. Diff Eqns*, 3:323–329, 1967.

Model Reduction by Moment Matching for Linear and Nonlinear Systems

Alessandro Astolfi[1,2]

[1] Electrical and Electronic Engineering Department, Imperial College London, Exhibition Road, London, SW7 2AZ, UK
[2] Dipartimento di Informatica Sistemi e Produzione, Università di Roma "Tor Vergata", Via del Politecnico 1, 00133, Roma, Italy

This work is dedicated to Alberto Isidori on the occasion of his 65th birthday.

Summary. The model reduction problem by moment's matching for linear and nonlinear systems is discussed. The linear theory is revisited to provide the basis for the development of the nonlinear theory.

1 Introduction

The model reduction problem for linear and nonlinear systems has been widely studied over the past decades. This problem has great importance in applications, because *reduced order models* are often used in analysis and design. This is the case, for example, in the study of mechanical systems, which is often based on models derived from a rigid body perspective that neglects the presence of flexible modes and elasticity; in the study of large scale systems, such as integrated circuits or weather forecast models, which relies upon the construction of simplified models that capture the main features of the system. From a theoretical point of view, the model reduction problem generates important theoretical questions and requires advanced tools from linear algebra, functional analysis and numerical analysis.

The model reduction problem can be simply, and informally, posed as follows. Given a system, described by means of linear or nonlinear differential equations together with an output map, compute a *simpler* system which *approximates* (in a sense to be specified) its behaviour. To render precise this problem formulation it is necessary to define two concepts.

Firstly, the meaning of the approximation. For linear systems one could introduce an approximation error given in terms of the frequency response of a suitably defined *error system*, or in terms of the response of the system for classes of input signals. For example, the methods, known as moment matching methods, which zero the transfer function of the error system for specific frequencies, belong to this class [1]. This approach does not have a direct nonlinear counterpart, despite the recent developments in this direction [13] (see also the early contributions [18, 19, 17]). Alternatively, approximation errors expressed in terms of H_2 or H_∞ norm of the error system have been considered both in the linear case [21, 14, 3] and in the nonlinear case [26]. Finally, approximation errors based on the Hankel operator of the system have been widely considered [11]. This approach leads to the so-called balancing realization problem, which has been also studied in the nonlinear framework [25, 27, 10, 22].

Secondly, the concept of simplicity. For linear systems this is often understood in terms of the dimension of the system, *i.e.* an approximating system is simpler than the model to approximate if its state-space realization has fewer states. For nonlinear systems this dimensional argument may be inappropriate, as one has to take into consideration also the complexity of the functions involved in the state-space representation.

Of course, there are other important issues that have to be clarified, and investigated, in establishing a model reduction theory. In particular, one may require that properties of the model (such as stability or passivity) are retained by the approximation [2], and one has to consider the computational cost associated with the construction of the approximating system. These issues have been widely investigated in the linear framework, see for example the excellent monograph [1], but are largely open for nonlinear systems.

Goal of this work is to develop a theory of model reduction, based on the notion of moment, for nonlinear systems. In this process, we revisit the linear theory, providing new perspectives and results.

This work relies upon the theory of the steady-state response of nonlinear systems, center manifold theory and the tools arising in the theory of output regulation for nonlinear systems. These theories have been partly developed, and their use in control theory has been pioneered, by Alberto Isidori in a series of seminal and groundbreaking papers, see, for example, [18, 19, 6, 7, 16]. They have been communicated to the author during private conversations, providing pre-prints of research in progress and in a series of lectures that, as an undergraduate, he had the fortune to attend in the Spring of 1990. It is remarkable, but not surprising, that these tools have far-reaching applicability.

2 Model Reduction by Moment Matching for Linear Systems – Revisited

2.1 The Notion of Moment

Consider a linear, single-input, single-output[3], continuous-time system described by equations of the form

[3] Similar considerations can be performed for multi-input, multi-output systems.

$$\begin{aligned} \dot{x} &= A\,x + Bu, \\ y &= Cx, \end{aligned} \tag{1}$$

with $x(t) \in \mathbb{R}^n$, $u(t) \in \mathbb{R}$, $y(t) \in \mathbb{R}$, and the associated transfer function

$$W(s) = C(sI - A)^{-1}B. \tag{2}$$

Definition 1. *[1] The 0-moment of system (1) at $s^\star \in \mathbb{C}$ is the complex number*

$$\eta_0(s^\star) = C(s^\star I - A)^{-1}B.$$

The k-moment of system (1) at $s^\star \in \mathbb{C}$ is the complex number

$$\eta_k(s^\star) = \frac{(-1)^k}{k!}\left[\frac{d^k}{ds^k}\left(C(sI - A)^{-1}B\right)\right]_{s=s^\star} = C(s^\star I - A)^{-(k+1)}B$$

Moments can be also characterized, for almost all $s^\star$, in a time-domain setting, as shown in the following statements.

Lemma 1. *Consider system (1) and $s^\star \in \mathbb{C}$. Suppose[4] $s^\star \notin \sigma(A)$. Then*

$$\eta_0(s^\star) = C\Pi,$$

where Π is the (unique) solution of the Sylvester equation

$$A\Pi + B = s^\star \Pi. \tag{3}$$

Proof. By direct computation, equation (3) yields

$$\Pi = (s^\star I - A)^{-1}B,$$

hence $C\Pi = \eta_0(s^\star)$. □

Lemma 2. *Consider system (1) and $s^\star \in \mathbb{C}$. Suppose $s^\star \notin \sigma(A)$. Then*

$$\begin{bmatrix} \eta_0(s^\star) \\ \eta_1(s^\star) \\ \vdots \\ \eta_k(s^\star) \end{bmatrix} = (C\Pi\Psi_k)',$$

where

$$\Psi_k = \mathrm{diag}(1, -1, 1, \cdots, (-1)^k) \in \mathbb{R}^{(k+1)\times(k+1)},$$

and Π is the (unique) solution of the Sylvester equation

$$A\Pi + BL_k = \Pi\Sigma_k, \tag{4}$$

[4] $\sigma(A)$ denotes the spectrum of the matrix A.

with

$$L_k = \begin{bmatrix} 1\, 0 \cdots 0 \end{bmatrix} \in \mathbb{R}^{k+1},$$

and

$$\Sigma_k = \begin{bmatrix} s^\star & 1 & 0 & \cdots & 0 \\ 0 & s^\star & 1 & \cdots & 0 \\ \vdots & \vdots & \ddots & \ddots & \vdots \\ 0 & \cdots & 0 & s^\star & 1 \\ 0 & \cdots & \cdots & 0 & s^\star \end{bmatrix} \in \mathbb{R}^{(k+1)\times(k+1)}.$$

Proof. Let

$$\Pi = \begin{bmatrix} \Pi_0 \, \Pi_1 \cdots \Pi_k \end{bmatrix}$$

and note that equation (4) can be rewritten as

$$\begin{aligned} A\Pi_0 + B &= s^\star \Pi_0, \\ A\Pi_1 &= s^\star \Pi_1 + \Pi_0, \\ &\vdots \\ A\Pi_k &= s^\star \Pi_k + \Pi_{k-1}. \end{aligned}$$

As a result,

$$\begin{aligned} \Pi_0 &= (s^\star I - A)^{-1} B, \\ \Pi_1 &= -(s^\star I - A)^{-2} B, \\ \Pi_2 &= (s^\star I - A)^{-3} B, \\ &\vdots \\ \Pi_k &= (-1)^k (s^\star I - A)^{-(k+1)} B, \end{aligned}$$

hence the claim. □

Remark 1. The pair (L_k, Σ_k) is observable for any $s^\star$. ◁

The main disadvantage of the above results is in the fact that one has to deal with complex matrices and Σ_k and L_k have a special structure. To remove this shortcoming note that moments are coordinates invariant and, by a property of real rational functions,

$$\eta_k(\bar{s}^\star) = \overline{\eta_k(s^\star)}.$$

As a result, the following statements hold.

Lemma 3. *Consider system (1) and $s^\star \in \mathbb{R}$. Suppose $s^\star \notin \sigma(A)$. Then the moments $\eta_0(s^\star)$, ..., $\eta_k(s^\star)$ are in one-to-one relation with the matrix $C\Pi$, where Π is the (unique) solution of the Sylvester equation*

$$A\Pi + BL = \Pi S, \tag{5}$$

with S any non-derogatory[5] *real matrix such that*

$$\det(sI - S) = (s - s^\star)^{k+1}, \tag{6}$$

and L such that the pair (L, S) is observable.

[5] A matrix is non-derogatory if its characteristic and minimal polynomials coincide.

Proof. By observability of the pair (L, S) there is a unique invertible matrix T such that $S = T^{-1}\Sigma_k T$ and $L = L_k T$. As a result, equation (5) becomes

$$A\Pi + BL_k T = \Pi T^{-1}\Sigma_k T,$$

and this can be rewritten as

$$T^{-1}\left(\tilde{A}\tilde{\Pi} + \tilde{B}L_k\right)T = T^{-1}\left(\tilde{\Pi}\Sigma_k\right)T,$$

with

$$\tilde{\Pi} = T\Pi T^{-1}, \qquad \tilde{A} = TA\,T^{-1}, \qquad \tilde{B} = TB.$$

By Lemma 2, and invariance of the moments with respect to the coordinates in the state space, the moments $\eta_0(s^\star), \ldots, \eta_k(s^\star)$ can be univocally expressed in terms of $\tilde{\Pi}$, hence the claim. □

Lemma 4. *Consider system (1) and $s^\star \in \mathbb{C} \setminus \mathbb{R}$. Let $s^\star = \alpha^\star + i\omega^\star$. Suppose $s^\star \notin \sigma(A)$. Then the moments $\eta_0(s^\star)$, $\eta_0(\bar{s}^\star)$, ..., $\eta_k(s^\star)$ and $\eta_k(\bar{s}^\star)$ are in one-to-one relation with the matrix $C\Pi$, where Π is the (unique) solution of the Sylvester equation*

$$A\Pi + BL = \Pi S, \tag{7}$$

with S any non-derogatory real matrix such that

$$\det(sI - S) = \left((s - s^\star)(s - \bar{s}^\star)\right)^{k+1} = \left(s^2 - 2\alpha^\star s + (\alpha^\star)^2 + (\omega^\star)^2\right)^{k+1}, \tag{8}$$

and L such that the pair (L, S) is observable.

Proof. The proof is similar to the one of Lemma 3 hence omitted. □

We complete this section with a property which is instrumental to derive a nonlinear enhancement of the notion of moment.

Theorem 1. *Consider system (1), $s^\star \in \mathbb{C}$ and $k \geq 0$. Assume*[6] *$\sigma(A) \subset \mathbb{C}^-$ and $s^\star \in \mathbb{C}^0$. Let*

$$\dot{\omega} = S\omega, \tag{9}$$

with $\omega(t) \in \mathbb{R}^\kappa$, where

$$\kappa = \begin{cases} k+1 & \text{if } s^\star \in \mathbb{R}, \\ 2(k+1) & \text{if } s^\star \in \mathbb{C} \setminus \mathbb{R}, \end{cases}$$

and S any non-derogatory real matrix with characteristic polynomial as in (6), if $s^\star \in \mathbb{R}$, or as in (8), if $s^\star \in \mathbb{C} \setminus \mathbb{R}$.

Consider the interconnection of systems (1) and (9) with $u = L\omega$, and L such that the pair (L, S) is observable.

Then the moments $\eta_0(s^\star)$, ..., $\eta_k(s^\star)$ are in one-to-one relation with the (well-defined) steady-state response of the output of such interconnected system.

[6] $\mathbb{C}^-$ and $\mathbb{C}^0$ denote the left half of the complex plane and the imaginary axis, respectively.

Proof. We provide a proof which exploits arguments with a nonlinear counterpart (an elementary, alternative, proof can be obtained using Laplace transform arguments). The considered interconnected system is described by

$$\begin{aligned} \dot{\omega} &= S\omega, \\ \dot{x} &= A\ x + BL\omega, \\ y &= Cx. \end{aligned}$$

By the center manifold theorem [8, 15], which is applicable because of the assumptions on $\sigma(A)$ and $\sigma(S)$, this system has a globally well-defined invariant manifold (which is a hyperplane) given by

$$\mathcal{M} = \{(x, \omega) \in I\!R^{n+\kappa} \mid x = \Pi\omega\},$$

with Π the (unique) solution of the Sylvester equation (5). Note that

$$\overbrace{x - \Pi\omega}^{\cdot} = A(x - \Pi\omega),$$

hence $\mathcal{M}$ is attractive. As a result

$$y(t) = C\Pi\omega(t) + Ce^{At}(x(0) - \Pi\omega(0)),$$

where the first term on the right-hand side describes the steady-state response of the system, and the second term on the right-hand side the transient response, which proves the claim. □

2.2 Moment Matching

We are now in a position to define, precisely, the notions of reduced order model and of model reduction by moment matching.

Definition 2. *The system*

$$\begin{aligned} \dot{\xi} &= F\xi + Gu, \\ \psi &= H\xi, \end{aligned} \tag{10}$$

*with $\xi(t) \in I\!R^\nu$ and $\psi(t) \in I\!R$, is a k-*order model at $s^\star$ of system (1) *if system (10) has the same i-moment, with $i = 0, \cdots, k$, at $s^\star$ as (1). In this case, system (10) is said to match the first $k+1$ moments of system (1) at $s^\star$. Furthermore, system (10) is a reduced order model of system (1) if $\nu < n$.*

Theorem 2. *Consider the system (1), the system (10) and $s^\star \in \mathbb{C}$. Suppose $s^\star \notin \sigma(A)$ and $s^\star \notin \sigma(F)$. System (10)* matches *the first $k+1$ moments of (1) at $s^\star$ if and only if*

$$C\Pi = HP, \tag{11}$$

where Π is the (unique) solution of equation (5), P is the (unique) solution of the equation

$$FP + GL = PS, \tag{12}$$

S is as in Lemma 1, and the pair (L, S) is observable.

Proof. The claim is a straightforward consequence of the definition of matching and of the results in Lemmas 3 and 4. □

Remark 2. The results derived so-far are direct consequences of the definition of moment. However, to the best of the author's knowledge, they have not been presented in this form. ◁

2.3 Model Reduction by Moment Matching with Prescribed Eigenvalues

The result established in Lemma 2 can, in principle, be used to solve the model reduction problem by moment matching for system (1) in two steps.

In the former one has to solve the Sylvester equation (5) in the unknown Π. In the latter one has to construct matrices F, G, H and P (possibly with specific properties) such that equations (11) and (12) hold.

This approach is unsatisfactory because it requires the computation of the moments, namely of the matrix $C\Pi$ and hence of Π, whereas most of the existing algorithms [1] are able to achieve moment matching without the need to compute moments. We discuss, and solve, this issue at the end of this section.

In the meanwhile we focus on the second step of the construction, assuming that the matrices F and S have the same dimensions, *i.e.* the order of the reduced model is equal to the number of the moments to match, and we impose the additional constraint that the eigenvalues of the matrix F are given, *i.e.*

$$\sigma(F) = \{\lambda_1, \cdots, \lambda_\nu\},$$

for some given λ_i's such that $\sigma(F) \cap \sigma(S) = \emptyset$.

Let P be any invertible matrix such that condition (11) holds, for some selection of the matrix H[7] and set

$$G = P\Delta,$$

with Δ such that

$$\sigma(S - \Delta L) = \sigma(F), \tag{13}$$

Note that, by observability of the pair (L, S), there is a matrix Δ such that condition (13) holds. Finally, let

$$F = P(S - \Delta L)P^{-1}.$$

It is straightforward to conclude that this procedure yields the required matrices. Note finally that the matrix P could be selected to assign a special structure to the reduced model and, in particular, to the matrix F.

[7] There are several matrices P achieving this goal, in particular one could pick $P = I$ and $H = C\Pi$.

To avoid the computation of the moments, *i.e.* of the matrix Π, one could proceed as follows. Consider system (1), $s^\star \in \mathbb{C}$ and construct a reduced order model achieving moment matching at $s^\star$ with any efficient algorithm that does not require the computation of the moments, see [1, 4, 9, 12, 20]. This yields a reducer order model for system (1) described by equations of the form

$$\begin{aligned} \dot{x}_M &= A_M x_M + B_M u, \\ y_M &= C_M x_M, \end{aligned} \tag{14}$$

where $x_M(t) \in \mathbb{R}^\nu$ and $y_M(t) \in \mathbb{R}$. To find a reduced order model with desired eigenvalues it is thus sufficient to apply the result in Lemma 2, and the construction in this section, to system (14).

3 Nonlinear Systems

3.1 The Notion of Moment

In this section we derive a nonlinear enhancement of the notion of moment. While most of the results in Section 2 do not have a direct nonlinear counterpart, we can use Lemma 1 to give a definition of moment.

To this end, consider a nonlinear, single-input, single-output, continuous-time system described by equations of the form[8]

$$\begin{aligned} \dot{x} &= f(x, u), \\ y &= h(x), \end{aligned} \tag{15}$$

with $x(t) \in \mathbb{R}^n$, $u(t) \in \mathbb{R}$, $y(t) \in \mathbb{R}$, a signal generator described by the equations

$$\begin{aligned} \dot{\omega} &= s(\omega), \\ \theta &= l(\omega), \end{aligned} \tag{16}$$

with $\omega(t) \in \mathbb{R}^\nu$ and $\theta(t) \in \mathbb{R}$ and the interconnected system

$$\begin{aligned} \dot{\omega} &= s(\omega), \\ \dot{x} &= f(x, l(\omega)), \\ y &= h(x). \end{aligned} \tag{17}$$

Suppose, in addition, that the mappings $f(\cdot,\cdot)$, $h(\cdot)$, $s(\cdot)$ and $l(\cdot)$ are smooth and that $f(0,0) = 0$, $h(0) = 0$, $s(0) = 0$ and $l(0) = 0$. The signal generator captures the requirement that one is interested in studying the behaviour of system (15) only in *specific circumstances.* However, for this to make sense and to provide a generalization of the notion of moment, we need the following assumptions and definitions.

[8] All functions and mappings are assumed sufficiently smooth.

Assumption 1. *There is a unique mapping $\pi(\omega)$, locally*[9] *defined in a neighborhood of $\omega = 0$, which solves the partial differential equation*

$$f(\pi(\omega), l(\omega)) = \frac{\partial \pi}{\partial \omega} s(\omega). \tag{18}$$

Assumption 1 implies that the interconnected system (17) possesses an invariant manifold, described by the equation $x = \pi(\omega)$. Note that the (well-defined) dynamics of the system restricted to the invariant manifold are described by

$$\dot{\omega} = s(\omega),$$

i.e. are a copy of the dynamics of the signal generator (16).

Assumption 2. *The signal generator (16) is observable,* i.e. *for any pair of initial conditions $\omega_a(0)$ and $\omega_b(0)$, such that $\omega_a(0) \neq \omega_b(0)$, the corresponding output trajectories $l(\omega_a(t))$ and $l(\omega_b(t))$ are such that*

$$l(\omega_a(t)) - l(\omega_b(t))) \not\equiv 0.$$

Definition 3. *Consider system (15) and the signal generator (16). Suppose Assumptions 1 and 2 hold. The function $h(\pi(\omega))$, with $\pi(\omega)$ solution of equation (18), is* the moment of system (15) at $s(\omega)$.

Definition 4. *Consider system (15) and the signal generator (16). Suppose Assumption 1 holds. Let the signal generator (16) be such that $s(\omega) = 0$ and $l(\omega) = \omega$. Then the function $h(\pi(\omega))$ is the 0-moment of system (15) at $s^\star = 0$.*

Definition 5. *Consider system (15) and the signal generator (16). Suppose Assumption 1 holds. Let the signal generator (16) be such that*

$$s(\omega) = \begin{bmatrix} 0 & \omega^\star \\ -\omega^\star & 0 \end{bmatrix} \omega, \qquad l(\omega) = \begin{bmatrix} L_1 \; L_2 \end{bmatrix} \omega,$$

with $\omega^\star \neq 0$ and $L_1^2 + L_2^2 \neq 0$. Then the function $h(\pi(\omega))$ is the 0-moment of system (15) at $s^\star = i\omega^\star$.

The above definitions allow to derive a nonlinear counterpart of Lemma 1.

Theorem 3. *Consider system (15) and the signal generator (16). Assume Assumption 2 holds. Assume the zero equilibrium of the system*

$$\dot{x} = f(x, 0)$$

is locally exponentially stable and system (16) is Poisson stable.

Then Assumption 1 holds and the moment of system (15) at $s(\omega)$ coincides with the (locally well-defined) steady-state response of the output of the interconnected system (17).

[9] All statements are local, although global versions can be easily given.

Proof. To begin with note that, under the stated hypotheses, Assumption 1 holds by the center manifold theory [8] and the results in [19]. Moreover, by [19], the steady-state response of the system is (locally) well-defined, and this is given by $\pi(h(\omega))$, hence the claim. □

Remark 3. While for linear systems it is possible to define k-moments for every $s^\star \in \mathbb{C}$ and for any $k \geq 0$, for nonlinear systems it may be difficult, or impossible, to provide general statements if the signal $\theta(t)$, generated by system (16), has unbounded trajectories. Therefore, if the signal generator is linear we consider only 0-moments for $s^\star \in \mathbb{C}^0$, whereas if the signal generator is nonlinear we assume that it generates bounded trajectories. ◁

Example 1. Consider a linear system described by equations of the form (1) with $x(t) \in \mathbb{R}^n$, $n > 3$, $u(t) \in \mathbb{R}$, $y(t) \in \mathbb{R}$ and the nonlinear signal generator (16) with $\omega = [\omega_1, \omega_2, \omega_3]'$,

$$s(\omega) = \begin{bmatrix} \dfrac{I_2 - I_3}{I_1}\omega_2\omega_3 \\ \dfrac{I_3 - I_1}{I_2}\omega_3\omega_1 \\ \dfrac{I_1 - I_2}{I_3}\omega_1\omega_2 \end{bmatrix},$$

with $I_1 > 0$, $I_2 > 0$, $I_3 > 0$, and $I_i \neq I_j$ for $i \neq j$, and

$$l(\omega) = L\omega = \begin{bmatrix} L_1 \; L_2 \; L_3 \end{bmatrix} \omega,$$

with $L_1 L_2 L_3 \neq 0$. This signal generator, which describes the angular velocities of a free rigid body in space, is Poisson stable and, under the stated assumption on L, observable [23, 5].

Suppose system (1) is asymptotically stable. The moment of system (1) at $s(\omega)$ can be computed as follows. Let

$$\pi(\omega) = \sum_i \pi_i(\omega),$$

with

$$\pi_i(\omega) = \begin{bmatrix} \pi_i^1(\omega) \\ \pi_i^2(\omega) \\ \vdots \\ \pi_i^n(\omega) \end{bmatrix}$$

and $\pi_i^j(\omega)$ a homogeneous polynomial of degree i in ω. Then equation (18) yields

$$\pi_1(\omega) = -A^{-1}BL\omega, \quad \pi_2(\omega) = -A^{-2}BL\dot{\omega}, \cdots \pi_i(\omega) = -A^{-i}BL\frac{d^{i-1}\omega}{dt^{i-1}}, \cdots .$$

Hence, the moment of system (1) at $s(\omega)$ is given by

$$C\pi(\omega) = -CA^{-1}[BL\omega + A^{-1}BL\dot{\omega}\cdots A^{-i+1}BL\frac{d^{i-1}\omega}{dt^{i-1}}\cdots],$$

which is a polynomial series in ω. △

Remark 4. The discussion in the previous sections allows to derive a nonlinear enhancement of the notion of frequency response of a linear system. This relies upon the notion of steady-state response of a nonlinear system, as developed in [18, 19].

Consider system (15) and the signal generator (16). Let the signal generator (16) be such that

$$s(\omega) = \begin{bmatrix} 0 & \omega^\star \\ -\omega^\star & 0 \end{bmatrix}\omega, \qquad l(\omega) = \begin{bmatrix} L_1 \; L_2 \end{bmatrix}\omega,$$

with $\omega^\star \neq 0$ and $L_1^2 + L_2^2 \neq 0$. Then, under the hypotheses of Theorem 3, for all $\omega^\star \in \mathbb{R}$, Assumptions 1 holds and the output of the interconnected system (17) converges towards a locally well-defined steady state response, which, by definition, does not depend upon the initial condition $x(0)$. Moreover, the steady-state response is periodic. Suppose, in addition, that the steady-state response has the same period of $l(\omega)$. This implies that it can be written in Fourier series as

$$h(\pi(\omega(t))) = \sum_{k=-\infty}^{\infty} c_k e^{ik\omega^\star t}.$$

Consider now the operator $\mathcal{P}_+$, which acts on a Fourier series as follows

$$\mathcal{P}_+(\sum_{k=-\infty}^{\infty} \alpha_k e^{ik\omega^\star t}) = \sum_{k=0}^{\infty} \alpha_k e^{ik\omega^\star t}.$$

With this definition we can define the frequency response of the nonlinear system (15) as

$$F(\omega(0), \omega^\star) = \frac{\mathcal{P}_+(h(\pi(\omega(t))))}{\mathcal{P}_+(l(\omega(t))}.$$

This function depends upon the frequency $\omega^\star$, just as in the linear case, and, unlike the linear case, upon the initial condition $\omega(0)$ of the signal generator. Note finally that if the system (15) were linear, hence described by the equations (1), then

$$F(\omega(0), \omega^\star) = |W(i\omega^\star)|e^{i\angle W(i\omega^\star)},$$

where $W(s) = C(sI - A)^{-1}B$. ◁

3.2 Model Reduction by Moment Matching

Analogously to the linear case, we now introduce the notion of reduced order model and characterize the solution of the model reduction problem by moment matching.

Definition 6. *The system*

$$\begin{aligned} \dot{\xi} &= \phi(\xi, u), \\ \psi &= \kappa(\xi), \end{aligned} \tag{19}$$

with $\xi \in I\!R^\nu$, *is a* model at $s(\omega)$ of system (15) *if system (15) has the same moment at* $s(\omega)$ *as (19). In this case, system (19) is said to* match *the moment of system (15) at* $s(\omega)$*. Furthermore, system (19) is a reduced order model of system (15) if* $\nu < n$.

Theorem 4. *Consider the system (15), the system (19) and the signal generator (16). Suppose Assumptions 1 and 2 hold. System (19)* matches *the moments of (15) at* $s(\omega)$ *if the equation*

$$\phi(p(\omega), l(\omega)) = \frac{\partial p}{\partial \omega} s(\omega) \tag{20}$$

has a unique solution $p(\omega)$ *such that*

$$h(\pi(\omega)) = \kappa(p(\omega)), \tag{21}$$

where $\pi(\omega)$ *is the solution of equation (18).*

Proof. The claim is a direct consequence of the definition of moment. □

3.3 Construction of an Asymptotically Stable Reduced Order Model

In this section we provide a nonlinear counterpart of the construction in Section 2.3. For, note that to construct a reduced order model it is necessary to determine mappings $\phi(\cdot, \cdot)$, $\kappa(\cdot)$ and $p(\cdot)$ such that equations (20) and (21) hold, where $\pi(\omega)$ is the solution of equation (18).

To solve this problem we make the following assumptions.

Assumption 3. *There exists mappings* $\kappa(\omega)$ *and* $p(\omega)$ *such that* $k(0) = 0$, $p(0) = 0$, $p(\omega)$ *is locally* C^1, *equation (21) holds and*

$$\det \frac{\partial p(\omega)}{\partial \omega}(0) \neq 0,$$

i.e. *the mapping* $p(\omega)$ *possesses a local inverse* $p^{-1}(\cdot)$.

Remark 5. Similarly to the linear case Assumption 3 holds selecting $p(\omega) = \omega$ and $k(\omega) = h(\pi(\omega))$. ◁

Assumption 4. *There is a mapping* $\tilde{\phi}_1(\cdot)$ *such that the zero equilibrium of the system*

$$\dot{z} = s(z) - \tilde{\phi}_1(z) l(z)$$

is locally asymptotically stable.

Remark 6. In the linear case Assumption 4 holds by observability of the pair (L, S). In the nonlinear case Assumption 4 holds if, for example, the pair

$$\left(\frac{\partial l(\omega)}{\partial \omega}(0), \frac{\partial s(\omega)}{\partial \omega}(0)\right)$$

is observable, or detectable. Note, however, that this is not necessary. ◁

A direct computation shows that a reduced order model, for which the zero equilibrium is locally asymptotically stable, achieving moment matching, provided equation (20) has a unique solution $p(\omega)$, is described by

$$\begin{aligned}\dot{\xi} &= \phi_0(\xi) + \frac{\partial p(\omega)}{\partial \omega}\phi_1(\xi)u, \\ \psi &= \kappa(\xi),\end{aligned}$$

where $\kappa(\xi)$ and $p(\omega)$ are such that Assumption 3 holds,

$$\phi_1(\xi) = \tilde{\phi}_1(p^{-1}(\xi)),$$

with $\tilde{\phi}_1(\cdot)$ as in Assumption 4, and

$$\phi_0(\xi) = \left[\frac{\partial p(\omega)}{\partial \omega}\left(s(\omega) - \phi_1(p(\omega))l(\omega)\right)\right]_{\omega = p^{-1}(\xi)}.$$

3.4 Model Reduction by 0-Moment Matching at $s^\star = 0$

In this section we focus on the model reduction problem with 0-moment matching at $s^\star = 0$. Such a problem can be solved, under specific assumptions, without the need to solve any partial differential equation, as detailed in the following statement.

Proposition 1 (0-moment matching at $s^\star = 0$). *Consider system (15) and the signal generator $\dot{\omega} = 0$, $\theta = \omega$. Assume the zero equilibrium of the system*

$$\dot{x} = f(x, 0)$$

is locally exponentially stable. Then the zero moment of system (15) is (locally) well defined and given by $h(\pi(\omega))$, with $\pi(\omega)$ the unique solution of the algebraic equation

$$f(\pi(\omega), \omega) = 0.$$

Finally, a reduced order model, for which the zero equilibrium is locally asymptotically stable is given by

$$\begin{aligned}\dot{\xi} &= -\phi_1(\xi)(\xi - u), \\ \psi &= h(\pi(\xi)),\end{aligned}$$

with $\phi_1(\xi)$ such that $\phi_1(0) > 0$.

Proof. We simply need to show that equation (20) has a unique solution. For, note that, in this case, equation (20) rewrites as

$$-\phi_1(p(\omega))\,(p(\omega)-\omega))=0,$$

which, by positivity of $\phi_1(\cdot)$, has indeed a unique solution. □

Example 2. The averaged model of the DC–to–DC Ćuk converter is given by the equations [24]

$$\begin{aligned} L_1\frac{d}{dt}i_1 &= -(1-u)\,v_2+E,\\ C_2\frac{d}{dt}v_2 &= (1-u)\,i_1+u\,i_3,\\ L_3\frac{d}{dt}i_3 &= -u\,v_2-v_4,\\ C_4\frac{d}{dt}v_4 &= i_3-G\,v_4,\\ y &= v_4, \end{aligned} \tag{22}$$

where $i_1(t)\in I\!\!R^+$ and $i_3(t)\in I\!\!R^-$ describe currents, $v_2(t)\in I\!\!R^+$ and $v_4(t)\in I\!\!R^-$ voltages, L_1, C_2, L_3, C_4, E and G positive parameters and $u(t)\in(0,1)$ a continuous control signal which represents the slew rate of a PWM circuit used to control the switch position in the converter.

The 0-moment of the system at $s^\star=0$ is

$$h(\pi(\omega))=\frac{\omega}{\omega-1}E,$$

and a locally asymptotically stable reduced order model achieving moment matching at $s^\star=0$ is

$$\begin{aligned} \dot\xi &= -\phi_1(\xi)(\xi-u),\\ \psi &= E\frac{\xi}{\xi-1}, \end{aligned}$$

with $\phi_1(0)>0$, which is well-defined for $\xi\neq1$. This is consistent with the fact that the 0-moment at $s^\star=0$ is defined for $\omega\neq1$. △

4 Summary

The model reduction problem by moment matching for linear systems has been revisited with the goal to provide the basic tools to develop a theory for nonlinear systems. In addition, for linear systems, a novel model reduction procedure, which allows to assign the eigenvalues of the reduced order model, has been developed. In the case of nonlinear system we have provided an enhancement of the notion of moment, thus paving the way for the development of a nonlinear model reduction theory based on the notion of moment matching.

Acknowledgments

The author would like to thank Prof. A.C. Antoulas for several illuminating discussions on model reduction problems.

References

1. A.C. Antoulas. *Approximation of Large-Scale Dynamical Systems.* SIAM Advances in Design and Control, 2005.
2. A.C. Antoulas. A new result on passivity preserving model reduction. *Systems & Control Letters*, pages 361–374, 2005.
3. A.C. Antoulas and A. Astolfi. H_∞ norm approximation. In *Unsolved problems in Mathematical Systems and Control Theory, V. Blondel and A. Megretski Editors*, pages 267–270. Princeton University Press, 2004.
4. A.C. Antoulas, J.A. Ball, J. Kang, and J.C. Willems. On the solution of the minimal rational interpolation problem. *Linear Algebra and Its Applications, Special Issue on Matrix Problems*, pages 511–573, 1990.
5. A. Astolfi. Output feedback control of the angular velocity of a rigid body. *Systems & Control Letters*, 36(3):181–192, 1999.
6. C. I. Byrnes and A. Isidori. Asymptotic stabilization of minimum phase nonlinear systems. *IEEE Trans. on Automat. Contr.*, 36:1122–1137, 1991.
7. C. I. Byrnes, A. Isidori, and J. C. Willems. Passivity, feedback equivalence, and the global stabilization of minimum phase nonlinear systems. *IEEE Trans. on Automat. Contr.*, 36:1228–1240, 1991.
8. J. Carr. *Applications of center manifold theory.* Springer Verlag, 1998.
9. P. Feldman and R.W. Freund. Efficient linear circuit analysis by Pade approximation via a Lanczos method. *IEEE Trans. on Computer-Aided Design*, pages 639–649, 1995.
10. K. Fujimoto and J.M.A. Scherpen. Nonlinear input-normal realizations based on the differential eigenstructure of Hankel operators. *IEEE Trans. on Automat. Contr.*, 50:2–18, 2005.
11. K. Glover. All optimal Hankel-norm approximations of linear multivariable systems and their L^∞-error bounds. *Int. J. of Control*, 39:1115–1193, 1984.
12. E.J. Grimme, D.C. Sorensen, and P. van Dooren. Model reduction of state space systems via an implicitly restarted Lanczos method. *Numerical Algorithms*, pages 1–31, 1995.
13. T. Hu, A.R. Teel, and Z. Lin. Lyapunov characterization of forced oscillations. *Automatica*, 41:1723–1735, 2005.
14. X.-X. Huang, W.-Y. Yan, and K.L. Teo. H_2 near optimal model reduction. *IEEE Trans. on Automat. Contr.*, 46:1279–1285, 2001.
15. A. Isidori. *Nonlinear Control Systems, Third Edition.* Springer Verlag, 1995.
16. A. Isidori. A tool for semiglobal stabilization of uncertain non-minimum-phase nonlinear systems via output feedback. *IEEE Trans. on Automat. Contr.*, 45:1817–1827, 2000.
17. A. Isidori and A. Astolfi. Disturbance attenuation and H_∞-control via measurement feedback in nonlinear systems. *IEEE Trans. on Automat. Contr.*, 37:1283–1293, 1992.

18. A. Isidori and C. I. Byrnes. Steady state response, separation principle and the output regulation of nonlinear systems. In *Proc. of the 28th IEEE Conf. on Decision and Contr.*, pages 2247–2251, 1989.
19. A. Isidori and C. I. Byrnes. Output regulation of nonlinear systems. *IEEE Trans. on Automat. Contr.*, 35(2):131–140, 1990.
20. I.M. Jaimoukha and E.M. Kasenally. Implicitly restarted Krylov subspace methods for stable partial realizations. *SIAM J. Matrix Anal. Appl.*, pages 123–132, 1997.
21. D. Kavranoglu and M. Bettayeb. Characterization of the solution to the optimal H_∞ model reduction problem. *Systems & Control Letters*, 20:99–108, 1993.
22. A.J. Krener. Model reduction for linear and nonlinear control systems. In *Proc. of the 45rd IEEE Conf. on Decision and Contr.*, 2006. Bode Lecture.
23. H. Nijmeijer and A. J. Van der Schaft. *Nonlinear Dynamical Control Systems.* Springer Verlag, 1989.
24. H. Rodriguez, R. Ortega, and A. Astolfi. Adaptive partial state feedback control of the DC-to-DC Ćuk converter. In *Proc. of the 2005 Amer. Contr. Conf.*, pages 5121–5126, 2005.
25. J.M.A. Scherpen. Balancing for nonlinear systems. *Systems & Control Letters*, 21:143–153, 1993.
26. J.M.A. Scherpen. H_∞ balancing for nonlinear systems. *Int. J. of Robust and Nonlinear Control*, 6:645–668, 1996.
27. J.M.A. Scherpen and A.J. van der Schaft. Normalized coprime factorizations and balancing for unstable nonlinear systems. *Int. J. of Control*, 60:1193–1222, 1994.

Adaptive Control of Nonlinear Systems with Unknown Parameters by Output Feedback: A Non-Identifier-Based Method

Hao Lei and Wei Lin*

Department of Electrical Engineering and Computer Science, Case Western Reserve University, Cleveland, OH 44106, USA.

Dedicated to our teacher and friend A. Isidori on the occasion of his 65th birthday

Summary. The problem of global state regulation by output feedback is investigated for a family of uncertain nonlinear systems that are bounded by a triangular system with unknown parameters. The bounding system is allowed to depend linearly on the unmeasurable states but nonlinearly on the systems output. Using the idea of universal control, together with the recent advance in nonlinear output feedback design, we develop a non-identifier based output feedback control scheme achieving, in the presence of nonlinear parameterization, global asymptotic state regulation as well as boundedness of the closed-loop system. The main contribution of the paper is two-fold: 1) the polynomial growth condition imposed in the previous work [6] is removed in this work; 2) an extension to uncertain cascade systems with zero-dynamics is carried out under mild structural and ISS conditions.

1 Introduction

We consider a family of nonlinear systems with unknown parameters described by equations of the form

$$\begin{aligned}
\dot{x}_1 &= x_2 + \phi_1(x, \bar{\theta}) \\
\dot{x}_2 &= x_3 + \phi_2(x, \bar{\theta}) \\
&\vdots \\
\dot{x}_n &= u + \phi_n(x, \bar{\theta}) \\
y &= x_1
\end{aligned} \tag{1}$$

* This work was supported in part by the NSF under grant ECS-0400413, and in part by the AFRL under grants FA8651-05-C-0110 and FA8650-05-M-3540.

where $x = (x_1, \cdots, x_n)^T \in \mathbb{R}^n$, $u \in \mathbb{R}$ and $y \in \mathbb{R}$ are the system state, input and output, respectively, $\bar{\theta} \in R^s$ is an unknown constant vector. The functions $\phi_i : R^n \times R^s \to R$, $i = 1, \ldots, n$ are locally Lipschitz continuous in x, with $\phi_i(0, \bar{\theta}) = 0$, and need not to be precisely known.

The purpose of this work is to investigate the problem of global state regulation via output feedback for the nonlinear system (1) with parameter uncertainty. To address such a control problem, we make the following assumption throughout this paper.

Assumption 1. *There exist a C^1 function $c(y) \geq 1$ and an unknown constant $\theta \geq 0$, such that for $i = 1, \cdots, n$,*

$$|\phi_i(x, \bar{\theta})| \leq (|x_1| + \cdots + |x_i|)c(y)\theta. \tag{2}$$

Under Assumption 1 with θ being a known constant, various results have been obtained for global stabilization of the uncertain system (1) by output feedback. In the work [12], global output feedback stabilization was proved, among the other things, to be possible for the two-dimensional system (1) (see Corollary 3.1 in [12]). For the $n-$dimensional nonlinear system (1), the same conclusion was obtained under the extra requirement that $c(y)$ in (2) be a *polynomial* function [11]. The polynomial restriction on $c(y)$ was removed later on in [5, 18], where global stabilization of (1) was shown to be possible by output feedback as long as $c(y)$ is a C^1 function of y. The results of [5, 18] were motivated by the work [10], where a novel idea was presented for updating the observer gain on line through a Riccati differential equation.

When θ in (2) is an *unknown constant,* the question of how to achieve global adaptive stabilization of the uncertain system (1) by output feedback was only investigated recently. Under Assumption 1 with $c(y)$ being a constant or a *polynomial* function of y, universal control schemes have been proposed in [7, 6], providing solutions to the problem of global state regulation of the nonlinearly parameterized system (1) by output feedback.

The objective of this paper is to show that the polynomial condition imposed on $c(y)$ [6] can be relaxed and Assumption 1 alone suffices to ensure the solvability of the global adaptive stabilization by output feedback for the nonlinear parameterized system (1). This will be done by constructing a reduced-order adaptive observer and a non-identifier based output feedback controller. Our controller takes advantage of both the output feedback control schemes proposed in [5, 18] and the universal control strategy [17, 3, 1, 13]. Notably, the robust output feedback design in this paper is substantially different from the one in [6], which relies crucially on the linear structure of the dynamic output compensator, and hence is hard to be extended to the uncertain system (1) satisfying Assumption 1.

In the case when $c(y)$ is not a polynomial function, one of the major difficulties to be faced is that the idea of building a full-order observer in which the observer gain L is dynamically tuned by the estimate error $e_1 = x_1 - \hat{x}_1$, as illustrated in [7, 6], is not feasible and cannot be adopted. This is because so far, only a reduced-order (instead of full-order) observer based control scheme has been developed for the output feedback stabilization of system (1) without parameter uncertainty [5, 18].

Therefore, in the presence of both the unknown parameter θ and the non-polynomial function $c(y)$, how to design a reduced-order adaptive observer as well as dynamic updated laws for the observer gains are two critical questions which need to be investigated. Another technical issue is: due to the use of a reduced-order observer, the proof of the boundedness of the system output $y = x_1$ is quite involved and requires a tedious argument, as indicated by Claim 3 in the appendix.

The main contribution of this paper is the following theorem which addresses not only the critical issues discussed above but also solves the problem of global state regulation of the uncertain system (1) via output feedback.

Theorem 1. *Under Assumption 1, there exists a C^1 universal output feedback controller of the form*

$$\begin{aligned} \dot{\hat{z}}_2 &= \hat{z}_3 + r^2 a_3 y - \dot{r} a_2 y - r a_2 \hat{z}_2 - r^2 a_2^2 y \\ \dot{\hat{z}}_3 &= \hat{z}_4 + r^3 a_4 y - 2r\dot{r} a_3 y - r^2 a_3 \hat{z}_2 - r^3 a_3 a_2 y \\ &\vdots \\ \dot{\hat{z}}_n &= u - (n-1) r^{n-2} \dot{r} a_n y - r^{n-1} a_n \hat{z}_2 - r^n a_n a_2 y \end{aligned} \tag{3}$$

$$\begin{aligned} u = -r^n \Big[& k_2 \frac{\hat{z}_2 + r a_2 y + N y c^2(y)}{r} + k_3 \frac{\hat{z}_3 + r^2 a_3 y}{r^2} \\ & + \cdots + k_n \frac{\hat{z}_n + r^{n-1} a_n y}{r^{n-1}} \Big] \end{aligned} \tag{4}$$

with the gains $r = LM$ and N being updated by

$$\dot{M} = -\alpha M + \Delta(y, N), \tag{5}$$

$$\dot{L} = M^2 y^2 c^2(y) + \frac{[\hat{z}_2 + r a_2 y + N y c^2(y)]^2}{r^2} \tag{6}$$

$$\dot{N} = y^2 c^2(y), \tag{7}$$

such that for every $(x(0), \hat{z}(0)) \in R^n \times R^{n-1}$ and $M(0) = L(0) = N(0) = 1$, all the states $(x, \hat{z}, M, L, N)$ of the closed-loop system (1)–(3)–(7) are well-defined and bounded $\forall t \in [0, +\infty)$. Moreover,

$$\lim_{t \to +\infty} (x(t), \hat{z}(t)) = (0, 0),$$

where $a_i > 0$ and $b_i > 0, i = 2, \cdots, n$ are the coefficients of the Hurwitz polynomial $s^{n-1} + h_2 s^{n-2} + \cdots + h_{n-1} s + h_n$ with $h_i = a_i$ or $h_i = k_{n-i+2}$, $\alpha > 0$ is a suitable constant and $\Delta(y, N) \geq \alpha$ is a smooth function, both of them can be explicitly determined, for instance, by (26)–(27).

Remark 1. For the sake of simplicity, we use the notation $\dot{r} = \dot{L}M + L\dot{M}$ in the compensator (3) directly. From the gain-update laws (5)-(7), it is clear that $\dot{r}$ is a function of $(L, M, N, \hat{z}_2, x_1)$ and hence (3) is implementable. ◁

Remark 2. The dynamic output compensator (3)–(7) is composed of the reduced-order observer (3), the controller (4) and the gain update laws (5)–(7). While the reduced-observer based feedback controller (3)–(4) and the gain updated law (5) are, as done in [18, 5], used to achieve global stabilization of the nonlinear system (1), the gain tuning laws (6)–(7) for L and N are motivated by the design of universal controllers [6, 4, 14, 2, 19], and employed to handle the parameter uncertainty of the controlled plant (1). In contrast to the previous work [6] in which the adaptive control scheme relied heavily on the *linear* structure of an output compensator and a *full-order* observer, here we design the reduced-order observer (3) and nonlinear output feedback control law (4) to deal with the non-polynomial growth condition (2). Another difference is that in addition to the observer gains $L(t)$ and $M(t)$, an extra gain $N(t)$ is also introduced and needs to be updated dynamically. As we shall see in the next section, the introduction of the two dynamic gain update laws (6)–(7) are key in proving Theorem 1. ◁

A preliminary version of this work [8] was presented at the American Control Conference, Minneapolis, in June, 2006.

2 Proof of the Main Result

In this section, we give a constructive proof of the main result of this paper: Theorem 1. The proof reveals the key idea and motivation behind the construction of the universal output feedback controller (3)–(7) which results in, under Assumption 1, a solution to the problem of global state regulation of the uncertain system (1) via output feedback.

To make the proof easy to follow, we break it up into four parts.

(1) A Reduced-Order Observer and the Closed-Loop System

We begin by observing that (3) can be viewed as a reduced-order observer that estimates the unmeasurable states $z_i = x_i - r^{i-1}a_i x_1$, $i = 2, \cdots, n$. With this observation in mind, define the estimated states $\hat{x}_i = \hat{z}_i + r^{i-1}a_i x_1$ for $i = 2, \cdots, n$. Then, it is easy to verify that

$$\begin{aligned} \dot{\hat{x}}_2 &= \hat{x}_3 + ra_2(x_2 - \hat{x}_2) + ra_2\phi_1(x, \bar{\theta}) \\ &\vdots \\ \dot{\hat{x}}_n &= u + r^{n-1}a_n(x_2 - \hat{x}_2) + r^{n-1}a_n\phi_1(x, \bar{\theta}) \end{aligned} \tag{8}$$

which is nothing but a reduced-order, "state estimator" for the partial state $(x_2, \cdots, x_n)$ of the uncertain system (1).

Let $e_i = x_i - \hat{x}_i = x_i - \hat{z}_i - r^{i-1}a_i x_1$, $i = 2, \cdots, n$ be the estimated errors. From (1) and (8), it is easy to see that the error dynamics are given by

$$
\begin{aligned}
\dot{e}_2 &= e_3 - ra_2e_2 + \phi_2(x,\bar{\theta}) - ra_2\phi_1(x,\bar{\theta}) \\
&\vdots \\
\dot{e}_n &= -r^{n-1}a_ne_2 + \phi_n(x,\bar{\theta}) - r^{n-1}a_n\phi_1(x,\bar{\theta})\,.
\end{aligned} \tag{9}
$$

As done in [5, 18], we introduce the change of coordinates

$$
\begin{aligned}
\xi_2 &= \tfrac{\hat{x}_2+Nyc^2(y)}{r}, \qquad & \varepsilon_2 &= \tfrac{e_2}{r} \\
\xi_i &= \tfrac{\hat{x}_i}{r^{i-1}}, & \varepsilon_i &= \tfrac{e_i}{r^{i-1}}, \quad i=3,\cdots,n.
\end{aligned} \tag{10}
$$

In the new coordinates, (8)–(9) can be rewritten as

$$
\begin{aligned}
\dot{\xi}_2 &= r\xi_3 + a_2r\varepsilon_2 + a_2\phi_1(\,\cdot\,) - \frac{\dot{r}}{r}\xi_2 + \frac{1}{r}\dot{N}yc^2(y) \\
&\qquad + \frac{N}{r}\frac{\partial(yc^2(y))}{\partial y}\Big(r\varepsilon_2 + r\xi_2 - Nyc^2(y) + \phi_1(\,\cdot\,)\Big) \\
&\vdots \\
\dot{\xi}_n &= \frac{u}{r^{n-1}} + a_nr\varepsilon_2 + a_n\phi_1(\,\cdot\,) - (n-1)\frac{\dot{r}}{r}\xi_n
\end{aligned} \tag{11}
$$

$$
\begin{aligned}
\dot{\varepsilon}_2 &= r\varepsilon_3 - ra_2\varepsilon_2 + \frac{\phi_2(\,\cdot\,)}{r} - a_2\phi_1(\,\cdot\,) - \frac{\dot{r}}{r}\varepsilon_2 \\
&\vdots \\
\dot{\varepsilon}_n &= -ra_n\varepsilon_2 + \frac{\phi_n(\,\cdot\,)}{r^{n-1}} - a_n\phi_1(\,\cdot\,) - (n-1)\frac{\dot{r}}{r}\varepsilon_n\,.
\end{aligned} \tag{12}
$$

In view of (4)–(11)–(12) and (10), the closed-loop system can be expressed in the following compact form

$$
\begin{aligned}
\dot{\varepsilon} &= rA\varepsilon + \varPhi(\,\cdot\,) - \mathbf{a}\phi_1(\,\cdot\,) - \frac{\dot{r}}{r}D_{n-1}\varepsilon \\
\dot{x}_1 &= r\varepsilon_2 + r\xi_2 - Nyc^2(y) + \phi_1(\,\cdot\,) \\
\dot{\xi} &= rB\xi + \mathbf{a}(r\varepsilon_2 + \phi_1(\,\cdot\,)) - \frac{\dot{r}}{r}D_{n-1}\xi \\
&\quad + \frac{\mathbf{b}}{r}\Big[N\frac{\partial(yc^2(y))}{\partial y}(r\varepsilon_2 + r\xi_2 - Nyc^2(y) + \phi_1(\,\cdot\,)) + y^3c^4(y)\Big]
\end{aligned} \tag{13}
$$

where $r = LM$ and the gains $L(t), M(t)$ and $N(t)$ are tuned by (5)–(7). In addition,

$\varPhi(\,\cdot\,) = [\frac{\phi_2(x,\bar{\theta})}{r}, \frac{\phi_3(x,\bar{\theta})}{r^2}, \cdots, \frac{\phi_n(x,\bar{\theta})}{r^{n-1}}]^T$

$\varepsilon = (\varepsilon_2, \cdots, \varepsilon_n)^T$, $\xi = (\xi_2, \cdots, \xi_n)^T$, $\mathbf{a} = (a_2, \cdots, a_n)^T$,

$\mathbf{b} = (1, 0, \cdots, 0)^T \in R^{n-1}$, $D_{n-1} = \mathrm{diag}\{1, \cdots, n-1\}$ and

$$A = \begin{bmatrix} -a_2 & 1 \cdots 0 \\ \vdots & \vdots \ddots \vdots \\ -a_{n-1} & 0 \cdots 1 \\ -a_n & 0 \cdots 0 \end{bmatrix}, \quad B = \begin{bmatrix} 0 & 1 & \cdots & 0 \\ \vdots & \vdots & \ddots & \vdots \\ 0 & 0 & \cdots & 1 \\ -k_2 & -k_3 & \cdots & -k_n \end{bmatrix}.$$

Observe that starting from every $(x_1(0), \xi(0), \varepsilon(0)) \in \mathbb{R} \times \mathbb{R}^{n-1} \times \mathbb{R}^{n-1}$ and $N(0) = L(0) = M(0) = 1$, the closed-loop system (13)–(5)–(7) satisfies a local Lipschitz condition in a neighborhood of the initial condition (by construction, $L(t) \geq 1, M(t) \geq 1$ and $N(t) \geq 1$ for some small $t > 0$), and hence has a unique solution $(N(t), L(t), M(t), x_1(t), \xi(t), \varepsilon(t))$ on a small time interval $[0, T)$. Without loss of generality, suppose that this solution can be extended to the maximal interval $[0, T_f)$ for some T_f, with $0 < T_f \leq +\infty$.

(2) Stability Analysis

Using the matrix lemma in [11, 5], we can suitably choose the coefficients a_i and k_i in the Hurwitz matrices A and B defined above, so that there exist $P = P^T > 0$ and $Q = Q^T > 0$ satisfying

$$\begin{aligned} A^T P + PA &\leq -I, & c_1 I &\leq DP + PD \leq c_2 I \\ B^T Q + QB &\leq -2I, & c_3 I &\leq DQ + QD \leq c_4 I \end{aligned} \tag{14}$$

where $D = D_{n-1} - \frac{1}{2}I$ and $c_i > 0,\ i = 1, \cdots, 4$ are real constants.

Now, consider the Lyapunov function

$$V_1 = mr\varepsilon^T P\varepsilon + r\xi^T Q\xi + \frac{1}{2}y^2 \tag{15}$$

where $m = \|Q\mathbf{a}\|^2 + 1$. A simple calculation gives

$$\begin{aligned} \dot{V}_1 \leq & -mr^2\|\varepsilon\|^2 + 2mr\varepsilon^T P\Phi(\cdot) - 2mr\varepsilon^T P\mathbf{a}\phi_1(\cdot) - m\dot{r}\varepsilon^T(DP + PD)\varepsilon \\ & -\dot{r}\xi^T(DQ + QD)\xi - 2r^2\|\xi\|^2 + 2\xi^T Q\Big(r^2\mathbf{a}\varepsilon_2 + r\mathbf{a}\phi_1(\cdot) + \mathbf{b}y^3c^4(y)\Big) \\ & +2\xi^T Q\mathbf{b}N\frac{\partial(yc^2(y))}{\partial y}\Big[r\varepsilon_2 + r\xi_2 - Nyc^2(y) + \phi_1\Big] \\ & +y\Big(r\varepsilon_2 + r\xi_2 - Nyc^2(y) + \phi_1(\cdot)\Big). \end{aligned} \tag{16}$$

From the gain updated laws (5)–(6), it is easy to prove that $\forall t \in [0, T_f)$, $L(t) \geq L(0) = 1$ and $M(t) \geq M(0) = 1$. Hence, $r(t) = L(t)M(t) \geq r(0) = 1$. Using this fact, one can deduce from (10) and Assumption 1 that

$$\begin{aligned} |\Phi_i(\cdot)| = |\frac{1}{r^{i-1}}\phi_i(x, \bar{\theta})| &\leq \frac{1}{r^{i-1}}\theta c(y)(|y| + N|y|c^2(y) + r|\xi_2| + \cdots \\ & \quad + r^{i-1}|\xi_i| + r|\varepsilon_2| + \cdots + r^{i-1}|\varepsilon_i|) \\ &\leq \theta c(y)\Big[\frac{1}{r}|y|(1 + Nc^2(y)) + (n-1)^{\frac{1}{2}}(\|\varepsilon\| + \|\xi\|)\Big] \end{aligned}$$

and

$$\|\Phi(\,\cdot\,)\| \le \frac{(n-1)^{\frac{1}{2}}}{r}\theta c(y)|y|(1+Nc^2(y)) + \theta(n-1)c(y)(\|\varepsilon\|+\|\xi\|).$$

The last inequality, in turn, leads to

$$\begin{aligned}|2mr\varepsilon^T P\Phi(\,\cdot\,)| &\le 2m\|P\|(n-1)^{\frac{1}{2}}\theta\|\varepsilon\|c(y)|y|(1+Nc^2(y)) \\ &\qquad +2mr\|P\|(n-1)\theta\|\varepsilon\|c(y)(\|\varepsilon\|+\|\xi\|) \\ &\le \theta_1\|\varepsilon\|^2 + \Big((1+Nc^2(y))^2 + c^4(y)\Big)\|\varepsilon\|^2 + \theta_1 c^2(y)y^2 \\ &\qquad + \frac{r^2}{8}\|\varepsilon\|^2 + \frac{r^2}{8}\|\xi\|^2 \qquad (17)\end{aligned}$$

where θ_1 is a suitable *unknown* constant depending on θ.

Without loss of generality, from now on we use $\theta_1 \ge 1$ to represent a *generic unknown* constant depending on θ, which may be implicitly changed from places to places.

By the completion of square, the following estimations can be obtained:

$$\begin{aligned}|2mr\phi_1(\,\cdot\,)\varepsilon^T P\mathbf{a}| &\le \frac{r^2}{8}\|\varepsilon\|^2 + \theta_1 c^2(y)y^2 \\ |2r^2\varepsilon_2\xi^T Q\mathbf{a}| &\le \|Q\mathbf{a}\|^2 r^2\|\varepsilon\|^2 + r^2\|\xi\|^2 \\ |2r\phi_1(\,\cdot\,)\xi^T Q\mathbf{a}| &\le \frac{r^2}{8}\|\xi\|^2 + \theta_1 c^2(y)y^2 \qquad (18)\end{aligned}$$

and

$$|2N\frac{\partial(yc^2(y))}{\partial y}\phi_1\xi^T Q\mathbf{b}| \le N^2(\frac{\partial(yc^2(y))}{\partial y})^2\|\xi\|^2 + \theta_1 c^2(y)y^2 \qquad (19)$$

$$\begin{aligned}|2N\frac{\partial(yc^2(y))}{\partial y}(r\varepsilon_2 + r\xi_2)\xi^T Q\mathbf{b}| &\le \frac{r^2}{8}(\|\xi\|^2+\|\varepsilon\|^2) \\ &\qquad +16\|Q\mathbf{b}\|^2\Big(N\frac{\partial(yc^2(y))}{\partial y}\Big)^2\|\xi\|^2. \qquad (20)\end{aligned}$$

Moreover, observe that

$$\begin{aligned}&\left|2\Big(y^3c^4(y) - N^2yc^2(y)\frac{\partial(yc^2(y))}{\partial y}\Big)\xi^T Q\mathbf{b}\right| \\ &\le \Big(y^4c^6(y) + N^4c^2(y)(\frac{\partial(yc^2(y))}{\partial y})^2\Big)\|\xi\|^2 + 2\|Q\mathbf{b}\|^2c^2(y)y^2 \qquad (21)\end{aligned}$$

and

$$y\Big(r\varepsilon_2 + r\xi_2 - Nyc^2(y) + \phi_1(\,\cdot\,)\Big) \le \frac{r^2}{8}(\varepsilon_2^2+\xi_2^2) - (N-\theta_1)c^2(y)y^2. \qquad (22)$$

Now, substituting the estimations (17)–(22) into (16) yields

$$\dot{V}_1 \leq -\frac{r^2}{2}\|\varepsilon\|^2 + h_1(y,N)\|\varepsilon\|^2 + \theta_1\|\varepsilon\|^2 - m\dot{r}\varepsilon^T(DP+PD)\varepsilon - \frac{r^2}{2}\|\xi\|^2$$
$$-\dot{r}\xi^T(DQ+QD)\xi - [N-\theta_1]c^2(y)y^2 + h_2(y,N)\|\xi\|^2 \qquad (23)$$

where

$$h_1(y,N) = (1+Nc^2(y))^2 + c^4(y) \qquad (24)$$
$$h_2(y,N) = (1+16\|Q\mathbf{b}\|^2)N^2(\frac{\partial(yc^2(y))}{\partial y})^2 + y^4c^6(y)$$
$$+N^4c^2(y)(\frac{\partial(yc^2(y))}{\partial y})^2. \qquad (25)$$

Using the gain update laws (5)–(6) and the fact that $\dot{r} = \dot{L}M + L\dot{M} \geq L\dot{M}$ (because $M(t) \geq M(0) = 1$) $\forall t \in [0,T_f)$, it is not difficult to deduce from (14) that

$$-m\dot{r}\varepsilon^T(DP+PD)\varepsilon \leq -mL\dot{M}\varepsilon^T(DP+PD)\varepsilon$$
$$\leq c_2mr\alpha\|\varepsilon\|^2 - c_1m\Delta(y,N)\|\varepsilon\|^2$$

and

$$-\dot{r}\xi^T(DQ+QD)\xi \;\leq\; c_4r\alpha\|\xi\|^2 - c_3\Delta(y,N)\|\xi\|^2.$$

With this in mind, it follows from (23) that $\forall t \in [0,T_f)$

$$\dot{V}_1 \leq -(\frac{r^2}{2} - \theta_1 - c_2m\alpha r)\|\varepsilon\|^2 - (\frac{r^2}{2} - c_4r\alpha)\|\xi\|^2 - (N-\theta_1)c^2(y)y^2$$
$$-(c_1m\Delta(y,N) - h_1(y,N))\|\varepsilon\|^2 - (c_3\Delta(y,N) - h_2(y,N))\|\xi\|^2.$$

By choosing the parameter α and $\Delta(y,N)$ as

$$0 < \alpha < \frac{1}{4(c_4 + 2\lambda_{\max}(Q))} \qquad (26)$$
$$\Delta(y,N) \geq \frac{1}{c_1m}h_1(\cdot) + \frac{1}{c_3}h_2(\cdot) + \alpha, \qquad (27)$$

we arrive at $\forall t \in [0,T_f)$

$$\dot{V}_1 \leq -(\frac{r}{2} - \theta_1)r\|\varepsilon\|^2 - \frac{r^2}{4}\|\xi\|^2 - (N-\theta_1)c^2(y)y^2$$
$$\leq -(\frac{L}{2} - \theta_1)r\|\varepsilon\|^2 - \frac{M^2}{4}\|\xi\|^2 - (N-\theta_1)c^2(y)y^2. \qquad (28)$$

With the aid of the Lyapunov inequality (28), it can be proved that all the states $(x_1, \xi, \varepsilon, L, M, N)$ of the closed-loop system (13)–(5)–(6)–(7) are bounded on $[0,T_f)$, as illustrated below.

(3) Boundedness of the Closed-Loop System

It is worth pointing out that the boundedness of the closed-loop system would follow immediately from the Lyapunov inequality (28), if θ was a known constant. For instance, one can pick constant gains $L = 2\theta_1 + \frac{1}{2}$ and $N = \theta_1 + \frac{1}{4}$ in (28), so that $\dot{V}_1 \leq -\frac{1}{4}(\|\varepsilon\|^2 + \|\xi\|^2 + y^2)$ (as $c(y) \geq 1$), which implies the boundedness of ε, ξ and y on $[0, T_f)$.

When θ is an *unknown constant,* the boundedness of the states, i.e. $(x_1, \xi, \varepsilon, L, M, N)$, does not follow straightforwardly from (28) and a delicate analysis based on the inequality (28) is needed. Clearly, one cannot find constant gains L and N making the coefficients $\frac{L}{2} - \theta_1$ and $N - \theta_1$ positive, because they depend on the unknown parameter θ.

Fortunately, with the help of the proposed gain update laws (5)–(7), we can deduce from the Lyapunov inequality (28) the following important conclusion.

Proposition 1. *All the state variables of the closed-loop system (13)–(5)–(6)–(7), i.e., the dynamic gains (L, M, N) and the states (y, ε, ξ), are bounded on the maximal interval $[0, T_f)$.*

The proof of Proposition 1 involves tedious contradiction arguments and hence is given in the appendix. Using Proposition 1, one concludes immediately that $T_f = +\infty$. If not, T_f would be the finite-escape time of the closed-loop system. Therefore, the dynamic system (13)–(5)–(6)–(7) would blow up at $t = T_f$, a contradiction to the fact that the system (13)–(5)–(6)–(7) is bounded on the maximal interval $[0, T_f)$, and hence by continuity also bounded at $t = T_f$.

In summary, from the Lyapunov inequality (28) it is concluded that the closed-loop system (13)–(5)–(6)–(7) has a unique solution and is bounded on $[0, +\infty)$.

(4) Convergence of the States (ε, y, ξ) of System (13)

Using the boundedness of N, L, M, x_1, ε, ξ on $[0, +\infty)$, one can show that y, ε and ξ are not only L_∞ but also L_2. In fact, the L_2 property of y follows from (7) and the fact that $c(y) \geq 1$. The boundedness of $\int_0^\infty \|\varepsilon\|^2 dt$ can be deduced from (10)–(44) and the inequality (50), together with the fact that $1 \leq M < +\infty$ and $1 \leq L < +\infty$. Similarly, the square integrability of ξ follows from (51) and (44).

Finally, it is easy to see from the closed-loop system (13)–(5)–(6)–(7) that $\dot{x}_1(t), \dot{\varepsilon}(t), \dot{\xi}(t) \in L_\infty$. By the Barbalat's Lemma, we have

$$\lim_{t \to +\infty} x_1(t) = 0, \quad \lim_{t \to +\infty} \varepsilon(t) = 0 \quad \text{and} \quad \lim_{t \to +\infty} \xi(t) = 0,$$

which, in turn, leads to

$$\lim_{t \to +\infty} (x(t), \hat{z}_2(t), \cdots, \hat{z}_n(t)) = 0.$$

3 Extension and Discussion

So far, we have addressed the problem of global state regulation by output feedback for the uncertain system (1) satisfying Assumption 1. The purpose of this section is to discuss briefly how Theorem 1 can be extended, under appropriate conditions, to a class of uncertain systems with zero-dynamics of the form

$$
\begin{aligned}
\dot{Z} &= f(Z, x, \bar{\theta}) \\
\dot{x}_1 &= x_2 + \phi_1(Z, x, \bar{\theta}) \\
\dot{x}_2 &= x_3 + \phi_2(Z, x, \bar{\theta}) \\
&\vdots \\
\dot{x}_n &= u + \phi_n(Z, x, \bar{\theta}) \\
y &= x_1
\end{aligned}
\tag{29}
$$

where $Z \in R^m$ and $x \in R^n$ are the system states, and the functions $f : \mathrm{I\!R}^{\mathrm{m}} \times \mathrm{I\!R}^{\mathrm{n}} \times \mathrm{I\!R}^{\mathrm{s}} \to \mathrm{I\!R}^{\mathrm{m}}$ and $\phi_i : \mathrm{I\!R}^{\mathrm{m}} \times \mathrm{I\!R}^{\mathrm{n}} \times \mathrm{I\!R}^{\mathrm{s}} \to \mathrm{I\!R}$, $\mathrm{i} = 1, \cdots, \mathrm{n}$, are locally Lipschitz continuous in (Z, x) with $f(0, 0, \bar{\theta}) = 0$ and $\phi_i(0, 0, \bar{\theta}) = 0 \ \forall \bar{\theta}$.

Remark 3. In view of the separation lemma introduced in [9] (see Lemma 2.5), it is not difficult to show that if $\phi_i(\cdot)$ is C^1 with respect to all the variables and $\phi_i(0, 0, \bar{\theta}) = 0$, $\phi_i(\cdot)$ can always be decomposed into the form

$$
\begin{aligned}
|\phi_i(Z, x, \bar{\theta})| &\leq \bar{\alpha}_i(\bar{\theta}, Z)||Z|| + \bar{\beta}_i(\bar{\theta}, x)||x|| \\
&\leq \alpha_i(\bar{\theta})\gamma_i(Z)||Z|| + \beta_i(\bar{\theta})b_i(x)(|x_1| + \cdots + |x_n|)
\end{aligned}
\tag{30}
$$

where $\alpha_i(\bar{\theta}) \geq 1$, $\gamma_i(Z) \geq 1$ $\beta_i(\bar{\theta}) \geq 1$ and $b_i(x) \geq 1$ are C^1 functions. ◁

Note that global stabilization of system (29) via output feedback is usually impossible without imposing certain growth conditions on $\phi_i(\cdot)$, particularly, on the unmeasurable states $(Z, x_2, \cdots, x_n)$. Keeping this in mind and in view of Remark 3 as well as the assumption of Theorem 1, it is natural to assume that system (29) satisfies the following conditions.

Assumption 2. *There exist smooth functions $\gamma : R^m \mapsto [1, +\infty)$, $c : R \mapsto [1, +\infty)$ and an unknown constant $\theta \geq 0$ such that for $i = 1, \cdots, n$,*

$$
|\phi_i(Z, x, \bar{\theta})| \leq \theta\gamma(Z)\|Z\| + \theta c(y)(|x_1| + \cdots + |x_i|)\,. \tag{31}
$$

Assumption 3. *There is a C^2 Lyapunov function $U_0(Z)$, which is positive definite and proper, such that*

$$
\frac{\partial U_0(Z)}{\partial z} f(Z, x, \bar{\theta}) \leq -\|Z\|^2 + \theta \hat{K}_0(y)y^2 \tag{32}
$$

where $\hat{K}_0(y) \geq 0$ is a known smooth function.

It is not difficult to see that Assumption 2 is a natural generalization of Assumption 1, while Assumption 3 is an ISS condition [16] imposed on the zero-dynamics.

Remark 4. Using the idea of changing supply rate [15], one can construct the Lyapunov function

$$V_0(Z) = \int_0^{U_0(Z)} \rho(s)ds \tag{33}$$

with $\rho(s) > 0$ being a C^0 nondecreasing function. By construction, $V_0(Z)$ is C^1, positive definite and proper. From Assumption 3 it follows that for a given smooth function $\gamma(z)$ in (31), there is a nondecreasing function $\rho(s)$ such that

$$\dot{V}_0 = \rho(U_0(Z))\frac{\partial U_0}{\partial Z}f(Z,x,\bar{\theta}) \le -\|Z\|^2 - \gamma^2(Z)\|Z\|^2 + \theta K_0(y)y^2 \tag{34}$$

where $K_0(y) \ge 1$ is a known smooth function. ◁

With the help of Remark 4 and $K_0(y)$ obtained in (34), the following result can be established.

Theorem 2. *Under Assumptions 2 and 3, the problem of global state regulation of the uncertain cascade system (29) is solvable by output feedback, in particular, by the dynamic output compensator (3)–(7) in which $c(y)$ is replaced by $c(y) + K_0(y)$.*

Proof. This result can be proved by using a constructive method that is in the spirit of Theorem 1. In what follows, we give only a sketchy proof.

First of all, using the same change coordinates as in (10) in which $c(y)$ is replaced by $c(y) + K_0(y)$, it is easy to see that the closed-loop system can be put in the form

$$\begin{aligned}
\dot{Z} &= f(Z,x,\bar{\theta}) \\
\dot{\varepsilon} &= rA\varepsilon + \Phi(\,\cdot\,) - \mathbf{a}\phi_1(\,\cdot\,) - \frac{\dot{r}}{r}D\varepsilon \\
\dot{x}_1 &= r\varepsilon_2 + r\xi_2 - Ny(c(y)+K_0(y))^2 + \phi_1(\,\cdot\,) \\
\dot{\xi} &= rB\xi + \mathbf{a}(r\varepsilon_2 + \phi_1(\,\cdot\,)) - \frac{\dot{r}}{r}D\xi + \mathbf{b}\frac{y^3(c(y)+K_0(y))^4}{r} \\
&\quad + \mathbf{b}N\frac{\partial(y(c(y)+K_0(y))^2)}{\partial y}\left(r\varepsilon_2 + r\xi_2 - Ny(c(y)+K_0(y))^2 + \phi_1(\,\cdot\,)\right).
\end{aligned} \tag{35}$$

Similar to section 2, without loss of generality, suppose that the closed-loop system (35) has a unique solution on the maximal interval $[0, T_f)$, with $0 < T_f \le +\infty$.

Now, choose the Lyapunov function

$$V_1 = \sigma V_0(Z) + mr\varepsilon^T P\varepsilon + r\xi^T Q\xi + \frac{1}{2}x_1^2$$

with $\sigma = 16m^2\|P\mathbf{a}\|^2 + 16\|Q\mathbf{a}\|^2 + \|Q\mathbf{b}\|^2$ and $m = \|Q\mathbf{a}\|^2 + 1$.

Following the stability analysis of Theorem 1, it is not difficult to arrive at $\forall t \in [0, T_f)$

$$\dot{V}_1 \leq -\|Z\|^2 - (\frac{L}{2} - \theta)r\|\varepsilon\|^2 - \frac{M^2}{4}\|\xi\|^2 - (N - \theta)(c(y) + K_0(y))^2 y^2.$$

Then, one can prove that the gains N and L are bounded on $[0, T_f)$ by using the same contradiction argument as done in Section 2. The boundedness of N on $[0, T_f)$ implies that Z and $\int_0^t \gamma^2(Z)\|Z\|^2 dt$ are bounded on $[0, T_f)$. The rest of the proof is almost identical to that of Theorem 1. In conclusion, it can be shown that all the states of closed-loop system are bounded on $[0, +\infty)$ and

$$\lim_{t\to+\infty} Z(t) = 0, \quad \lim_{t\to+\infty} x(t) = 0, \quad \text{and} \quad \lim_{t\to+\infty} \hat{z}(t) = 0,$$

thus completing the proof of Theorem 2. □

We conclude this section with a prototype example that demonstrates the effectiveness of Theorem 1.

Example 1. Consider the uncertain nonlinear system

$$\begin{aligned} \dot{x}_1 &= x_2 \\ \dot{x}_2 &= u + \theta_2(1 + |x_1|^{\theta_1})x_2 \\ y &= x_1 \end{aligned} \tag{36}$$

where θ_1 and θ_2 represent *unknown* constants with $\theta_1 \geq 0$. △

When θ_1 is known and only θ_2 is unknown, global state regulation of the nonlinear system (36) can be achieved by output feedback, for instance, using the work [6].

In the case when θ_1 and θ_2 are both unknown, how to globally regulate all the states of system (36) by output feedback is an unsolved problem and cannot be dealt with by [6]. However, observe that the uncertain system (36) satisfies Assumption 1. Indeed, no matter how big of θ_1, there exist an unknown constant $\theta = |\theta_2|e^{\frac{\theta_1^2}{2}} \geq 0$ and a smooth function $c(y) = e^{\frac{y^2}{2}}$, such that

$$\left|\theta_2(1 + |x_1|^{\theta_1})x_2\right| \leq \theta e^{\frac{y^2}{2}}|x_2|.$$

According to Theorem 1, one can construct the universal output feedback controller (3)–(4)–(5)–(6)–(7), with the observer parameter $a_2 = 1$, controller parameter $k_2 = 1$, $\alpha = 0.25$ and

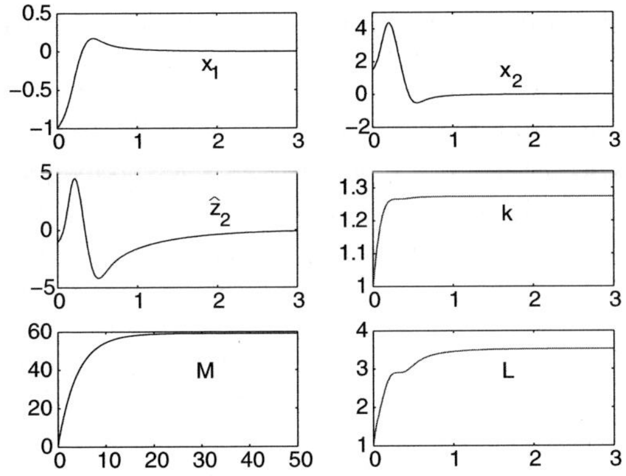

Fig. 1. Transient response of the closed-loop system (36)–(3)–(4)–(5)–(6)–(7)

$$\Delta(y, N) = 4(N^4 + 1)e^{y^2} + 0.25.$$

The simulation results shown in Figure 1 were obtained with the system parameters $\theta_1 = 2$ and $\theta_2 = 1$. The initial conditions are $(x_1(0), x_2(0), \hat{z}_2(0)) = (-1, 1.5, -1)$ and $N(0) = L(0) = M(0) = 1$.

4 Conclusion

In this paper, we have removed the restriction that $c(y)$ be a *polynomial* function – a crucial condition assumed in [6], and generalized the universal output feedback stabilization result obtained in [6] to a larger class of uncertain nonlinear systems such as (1) satisfying Assumption 1 or the cascade system (29) with zero dynamics. These were made possible by developing a non-identifier based output feedback control scheme which takes advantage of a reduced-order observer design and further elaborates the universal control idea proposed in [6] by introducing, in addition to the observer gains $L(t)$ and $M(t)$, an extra gain $N(t)$ that also needs to be updated dynamically.

A Proof of Proposition 1

We prove the Proposition 1 via a contradiction argument.

Suppose the closed-loop system (13)–(5)–(7) is *not bounded* on the maximal interval $[0, T_f)$. As a result,

$$\lim_{t\to T_f} \sup \|(N(t), L(t), M(t), x_1(t), \xi(t), \varepsilon(t))\| = +\infty. \tag{37}$$

In what follows, we shall show that (37) leads to a contradiction.

Claim 1: The dynamic gain N is bounded on the maximal interval $[0, T_f)$.

To see why, suppose that $\lim_{t\to T_f}\sup N(t) = \lim_{t\to T_f} N(t) = +\infty$. By construction, $M(t) \geq M(0) = 1$, $\forall t \in [0, T_f)$. As a consequence, $\dot{L} = M^2\dot{N} + \xi_2^2 \geq \dot{N}$, $\forall t \in [0, T_f)$. This, in turn, results in $\lim_{t\to T_f}\sup L(t) = \lim_{t\to T_f} L(t) \geq N(T_f) = +\infty$. Thus, there exists a finite time $t_1^* \in [0, T_f)$, such that

$$N(t) \geq \theta_1 + 1 \quad \text{and} \quad L(t) \geq 2\theta_1 + 2 \quad \forall t \in [t_1^*, T_f).$$

This, in view of (28), yields

$$\dot{V}_1 \leq -\|\varepsilon\|^2 - \frac{1}{4}\|\xi\|^2 - c^2(y)y^2, \qquad \forall t \in [t_1^*, T_f).$$

Consequently,

$$\int_{t_1^*}^{T_f} c^2(y)y^2 dt \leq V_1(r(t_1^*), x_1(t_1^*), \xi(t_1^*), \varepsilon(t_1^*)) = C. \tag{38}$$

From (38), it follows that

$$+\infty = N(T_f) - N(t_1^*) = \int_{t_1^*}^{T_f} \dot{N}(t)dt = \int_{t_1^*}^{T_f} c^2(y)y^2 dt \leq C$$

thus a contradiction.

In conclusion, the dynamic gain N is well-defined and bounded on $[0, T_f)$. From $\dot{N} = c^2(y)y^2$, it is deduced that $\int_0^{T_f} c^2(y)y^2 dt < +\infty$.

Claim 2: The dynamic gain L is bounded on the maximal interval $[0, T_f)$

If not, by construction, L is a monotone nondecreasing function. Thus, $\lim_{t\to T_f}\sup L(t) = \lim_{t\to T_f} L(t) = +\infty$. Then, there exists a finite time $t_2^* \in [0, T_f)$, such that

$$L(t) \geq 2\theta_1 + 2 \qquad \text{when} \quad t_2^* \leq t < T_f.$$

This, together with (28), yields

$$\dot{V}_1 \leq -\|\varepsilon\|^2 - \frac{1}{4}\|\xi\|^2 - (N - \theta_1)\dot{N}, \quad \forall t \in [t_2^*, T_f). \tag{39}$$

Using (39) and the boundedness of N on $[0, T_f)$, we arrive at

$$V_1(T_f) \leq -\int_{t_2^*}^{T_f} (N - \theta_1)dN + V_1(t_2^*) = \text{constant}, \tag{40}$$

$$\int_{t_2^*}^{T_f} \xi_2^2 dt \leq -4\int_{t_2^*}^{T_f} (N - \theta_1)dN + 4V_1(t_2^*) = C_1. \tag{41}$$

According to (15) and (40), it is clear that y is bounded on $[0, T_f)$. The boundedness of N and y on $[0, T_f)$ implies that the function $\varDelta(y, N)$ defined by (27) is bounded on $[0, T_f)$ by a constant, say β. With this in mind, it is deduced from (5) that

$$\dot{M} = -\alpha M + \varDelta(y, N) \leq -\alpha M + \beta, \quad \forall t \in [0, T_f) \tag{42}$$

which leads to the conclusion that $|M(t)| \leq C$, $\forall t \in [0, T_f)$, where $C > 0$ is a suitable constant.

Using this fact, together with (7) and (41), we deduce from (6) that

$$\begin{aligned} +\infty = L(T_f) - L(t_2^*) &= \int_{t_2^*}^{T_f} M^2 c^2(y) y^2 dt + \int_{t_2^*}^{T_f} \xi_2^2 dt \\ &\leq C^2(N(T_f) - N(t_2^*)) + C_1, \end{aligned}$$

which leads to a contradiction. Hence, L must be bounded on $[0, T_f)$. As a consequence, $\int_0^{T_f} (M^2 c^2(y) y^2 + \xi_2^2) dt < +\infty$.

Claim 3: The state x_1 is bounded on the maximal interval $[0, T_f)$.

The proof of boundedness of $y = x_1$ here is quite complicated and substantially different from the case when $c(y)$ is a polynomial function [6]. This is the price we have to pay due to the use of the reduced-order observer (3) instead of the full-order observer in [6]. Notably, the boundedness property of x_1 is not straightforward to establish even in the absence of parameter uncertainty.

In what follows, we shall prove claim 3 using a Lyapunov argument that is similar to the part (2) in Section 2.

Recall that the closed-loop system can be represented as (see (9), (11) and (13))

$$\begin{aligned} \dot{x}_1 &= e_2 + r\xi_2 - Nyc^2(y) + \phi_1(\cdot) \\ \dot{e}_2 &= e_3 - ra_2 e_2 + \phi_2(\cdot) - ra_2\phi_1(\cdot) \\ &\ \ \vdots \\ \dot{e}_n &= -r^{n-1} a_{n-1} e_2 + \phi_n(\cdot) - r^{n-1} a_n \phi_1(\cdot) \\ \dot{\xi} &= rB\xi + \mathbf{a}(r\varepsilon_2 + \phi_1(\cdot)) - \frac{\dot{r}}{r} D_{n-1}\xi + \frac{\mathbf{b}}{r}\Big[N\frac{\partial(yc^2(y))}{\partial y}\big(r\varepsilon_2 + r\xi_2 - Nyc^2(y) \\ &\qquad + \phi_1(\cdot)\big) + y^3 c^4(y)\Big]. \end{aligned} \tag{43}$$

Introduce the change of coordinates

$$\delta_1 = \frac{x_1}{r^*}, \quad \delta_i = \frac{e_i}{r^{*i}} \quad i = 2, \cdots, n, \quad \eta = \frac{\xi}{r^*} \tag{44}$$

where $r^* = L^* M$ with L^* being a sufficiently large constant.

In the new coordinates, (43) can be written as (by adding $r^*\mathbf{g}\delta_1$ and subtracting the same term)

$$\begin{aligned}\dot{\delta} &= r^* G\delta + r^*\mathbf{g}\delta_1 - r\Gamma\delta_2 + \Phi^*(\,\cdot\,) - \frac{\dot{M}}{M} D_n\delta \\ &\quad + (\mathbf{b}^T, 0)^T (r\eta_2 - N\frac{yc^2(y)}{r^*}) \\ \dot{\eta} &= rB\eta + \mathbf{a}(r^*\delta_2 + \frac{\phi_1}{r^*}) - \frac{\dot{r}}{r} D_{n-1}\eta - \frac{\dot{M}}{M}\eta \\ &\quad + \frac{\mathbf{b}}{rr^*}\Big[N\frac{\partial(yc^2(y))}{\partial y}(r^{*2}\delta_2 + rr^*\eta_2 \\ &\quad - Nyc^2(y) + \phi_1(\,\cdot\,)) + c^4(y)y^3\Big] \end{aligned} \tag{45}$$

where $D_n = \mathrm{diag}\{1, \cdots, n\}$, $\delta = (\delta_1, \cdots, \delta_n)^T$, $\eta = (\eta_2, \cdots, \eta_n)^T$, $\mathbf{g} = (g_1, \cdots, g_n)^T$, $\Gamma = [0, a_2, a_3\frac{L}{L^*}, \cdots, a_n(\frac{L}{L^*})^{n-2}]^T$ and

$$G = \begin{bmatrix} -g_1 & 1 \cdots 0 \\ \vdots & \vdots \ \ddots \ \vdots \\ -g_{n-1} & 0 \cdots 1 \\ -g_n & 0 \cdots 0 \end{bmatrix}, \qquad \Phi^*(\,\cdot\,) = \begin{bmatrix} \frac{1}{r^*}\phi_1(x, \bar{\theta}) \\ \frac{1}{r^{*2}}\phi_2(x, \bar{\theta}) + \frac{ra_2\phi_1(x,\bar{\theta})}{r^{*2}} \\ \vdots \\ \frac{1}{r^{*n}}\phi_n(x, \bar{\theta}) + \frac{r^{n-1}a_n\phi_1(x,\bar{\theta})}{r^{*n}} \end{bmatrix}.$$

The parameters $g_i > 0$, $i = 1, \cdots, n$ are chosen in such way that the polynomial $s^n + g_1 s^{n-1} + \cdots + g_{n-1}s + g_n$ is Hurwitz, and there exists a matrix $R = R^T > 0$ satisfying

$$G^T R + RG \le -I, \quad 0 < c_5 I \le \bar{D}R + R\bar{D} \le c_6 I \tag{46}$$

with $\bar{D} = D_n - \frac{1}{2}I$.

Now, consider the Lyapunov function

$$V_2(r, \delta, \eta) = mr^*\delta^T R\delta + r\eta^T Q\eta\,.$$

A straightforward but tedious calculation shows that the derivative of V_2 along system (45) satisfies

$$\begin{aligned}\dot{V}_2 \le &- mr^*\|\delta\|^2 + 2mr^{*2}\delta_1\delta^T R\mathbf{g} + 2mr^*\delta^T R\Phi^*(\,\cdot\,) - 2r^2\|\eta\|^2 \\ &+ 2rr^*\eta^T Q\mathbf{a}\delta_2 + 2mr^*(r\eta_2 - N\frac{yc^2(y)}{r^*})\delta^T R(\mathbf{b}^{\,T}, 0)^T \\ &- 2mr^* r\delta_2\delta^T R\Gamma + 2r\eta^T Q\mathbf{a}\frac{\phi_1(\,\cdot\,)}{r^*} - \dot{r}\eta^T(DQ + QD)\eta \\ &- m\dot{r}^*\delta^T(\bar{D}R + R\bar{D})\delta - 2L\dot{M}\eta^T Q\eta + 2\eta^T\frac{Q\mathbf{b}}{r^*}y^3c^4(y) \\ &+ 2\eta^T Q\mathbf{b}N\frac{\partial(yc^2(y))}{\partial y}(r^*\delta_2 + r\eta_2 - \frac{Nyc^2(y) - \phi_1(\,\cdot\,)}{r^*})\,. \end{aligned} \tag{47}$$

Following a similar procedure in the part (2) of Section 2 (see the estimations of each term in (16)), one can obtain

$$\begin{aligned}\dot{V}_2 \le& -(\frac{L^*}{2} - \theta_1)L^* M^2\|\delta\|^2 \\ &+ \theta_1(y^2 + c^2(y)y^2 + \xi_2^2) - \frac{r^2}{4}\|\eta\|^2 \\ &- (c_5 m L^* \Delta(y,N) - 2h_1(y,N))\|\delta\|^2\end{aligned}$$

where θ_1 is a unknown constant depending on θ.

With the relation (27) in mind, one can choose a sufficiently large constant L^*, for instance, $L^* \ge \max\{2\theta_1 + 2, \frac{2c_1}{c_5}\}$, such that

$$\dot{V}_2 \le -M^2(\|\delta\|^2 + \frac{1}{4}\|\eta\|^2) + \theta_1\big((1 + c^2(y))y^2 + \xi_2^2\big)\,. \tag{48}$$

From (48), it follows that

$$V_2(r(T_f),\delta(T_f),\eta(T_f)) \le \theta_1 \int_0^{T_f} \left(y^2 + \dot{N} + \xi_2^2\right)dt + V_2(0), \tag{49}$$

$$\int_0^{T_f} M^2\|\delta\|^2 dt \le \theta_1 \int_0^{T_f} \left(y^2 + \dot{N} + \xi_2^2\right)dt + V_2(0), \tag{50}$$

$$\int_0^{T_f} M^2\|\eta\|^2 dt \le 4(\theta_1 \int_0^{T_f} \left(y^2 + \dot{N} + \xi_2^2\right)dt + V_2(0). \tag{51}$$

Using the boundedness of $\int_0^{T_f} c^2(y)y^2 dt$, $\int_0^{T_f}(M^2c^2(y)y^2 + \xi_2^2)dt$ and $\int_0^{T_f} y^2 dt$ (as $c^2(y) \ge 1$) which are proved in Claim 1 and Claim 2, we concludes from (50) and (51) that $\int_0^{T_f} M^2\|\delta\|^2 dt$ and $\int_0^{T_f} M^2\|\eta\|^2 dt$ are bounded. Consequently, $\int_0^{T_f} \frac{e_2^2}{M^2}dt$ is bounded. From (49) and the definition of $V_2(\,\cdot\,)$, it is concluded that $M^{\frac{1}{2}}\|\delta\|$, $M^{\frac{1}{2}}\|\eta\|$ are also bounded on $[0, T_f)$.

Now, we are ready to prove the boundedness of x_1 on the interval $[0, T_f)$. Choose $V_3(x_1) = \frac{1}{2}x_1^2$. Then, it is easy to see that (by the completion of square)

$$\begin{aligned}\dot{V}_3 &\le x_1(e_2 + r\xi_2 - Nyc^2(y) + \phi_1(\,\cdot\,)) \\ &\le (L^2+1)M^2y^2 + \xi_2^2 + \frac{e_2^2}{M^2} - (N-\theta)c^2(y)y^2 \\ &\le (L^2+1)\dot{L} + \frac{e_2^2}{M^2} - (N-\theta)\dot{N}\,.\end{aligned} \tag{52}$$

By the boundedness of L and N on $[0, T_f)$, we have $\forall t \in [0, T_f)$

$$V_3(x_1(t)) - V_3(x_1(0)) \le (\bar{L}+1)\int_0^t \dot{L}dt + \int_0^t \frac{e_2^2}{M^2}dt - \int_0^t (N-\theta)dN,$$

which yields

$$V_3(x_1(t)) \leq \text{constant} \qquad \forall t \in [0, T_f).$$

As a consequence, x_1 is bounded on $[0, T_f)$.

Claim 4: The dynamic gain M is bounded on the maximal interval $[0, T_f)$.

Since both y and N are bounded on $[0, T_f)$, the boundedness of M on $[0, T_f)$ follows immediately from the inequality (42).

Claim 5: The states (ε, ξ) are bounded on the maximal interval $[0, T_f)$.

Using the fact that M and L are bounded on the maximal interval $[0, T_f)$, together with the boundedness of $M^{\frac{1}{2}}\|\delta\|$ and $M^{\frac{1}{2}}\|\eta\|$ concluded from (49), implies that ε and ξ are bounded on $[0, T_f)$.
Putting the five claims together, it is easy to see that Proposition 1 holds.

References

1. E. Bullinger and F. Allgöwer. Adaptive λ-tracking for nonlinear systems with higher relative degree. *Proc. of the 39th IEEE Conf. on Decision and Contr.*, pages 4771–4776, 2000.
2. Z. Ding. Universal output regulation for nonlinear systems in output feedback form. *Proc. of the 41st IEEE Conf. on Decision and Contr.*, pages 3837–3842, 2002.
3. A. Ilchmann. *Non-Identifier-Based High-Gain Adaptive Control*, volume 189 of *Lecture Notes in Control and Information Sciences.* Springer Verlag, Berlin, 1993.
4. A. Ilchmann and E.P. Ryan. On gain adaptation in adaptive control. *IEEE Trans. on Automat. Contr.*, 48:895–899, 2003.
5. P. Krishnamurthy and F. Khorrami. Dynamic high gain scaling: state and output feedback with application to systems with ISS appended dynamics driven by all states. *IEEE Trans. on Automat. Contr.*, 49:2219–2229, 2004.
6. H. Lei and W. Lin. A universal control approach for a family of uncertain nonlinear systems. *Proc. of the 44rd IEEE Conf. on Decision and Contr.*, pages 802–807, 2005.
7. H. Lei and W. Lin. Universal output feedback control of nonlinear systems with unknown growth rate. *Automatica*, 42:1783–1789, 2006.
8. H. Lei and W. Lin. Using a reduced-order observer for adaptive output feedback stabilization of uncertain cascade systems. *Proc. of the 2006 Amer. Contr. Conf.*, pages 4016–4017, 2006.
9. W. Lin and R. Pongvuthithum. Global stabilization of cascade systems by c^0 partial state feedback. *IEEE Trans. on Automat. Contr.*, 47:1356–1362, 2002.
10. L. Praly. Asymptotic stabilization via output feedback for lower triangular systems with output dependent incremental rate. *IEEE Trans. on Automat. Contr.*, 48:1103–1108, 2003.

11. L. Praly and Z.P. Jiang. Linear output feedback with dynamic high gain for nonlinear systems. *Systems & Control Letters*, 53:107–116, 2004.
12. C. Qian and W. Lin. Output feedback control of a class of nonlinear systems: a nonseparation principle paradigm. *IEEE Trans. on Automat. Contr.*, 47:1710–1715, 2002.
13. E.P. Ryan. A nonlinear universal servomechanism. *IEEE Trans. on Automat. Contr.*, 39:753–761, 1994.
14. E.P. Ryan. Universal stabilization of a class of nonlinear systems with homogeneous vector fields. *Systems & Control Letters*, 26:177–184, 1995.
15. E.D. Sontag and A.R. Teel. Changing supply functions in input/state stable systems. *IEEE Trans. on Automat. Contr.*, 40:1476–1478, 1995.
16. E.D. Sontag and Y. Wang. On characterizations of the input-to-state stability property. *Systems & Control Letters*, 24:351–359, 1999.
17. J. C. Willems and C.I. Byrnes. *Global Adaptive Stabilization in the Absence of Information on the Sign of the High Frequency Gain*, volume 62 of *Lecture Notes in Control and Information Sciences*. Springer-Verlag, Berlin, 1984.
18. B. Yang and W. Lin. Further results on global stabilization of uncertain nonlinear systems by output feedback. *Int. J. of Robust and Nonlinear Control*, 15:247–268, 2005.
19. X. Ye. Universal λ-tracking for nonlinearly-perturbed systems without restrictions on the relative degree. *Automatica*, 35:109–119, 1999.

Hybrid Feedback Stabilization of Nonlinear Systems with Quantization Noise and Large Delays

Claudio De Persis

Dipartimento di Informatica e Sistemistica, Università di Roma "La Sapienza", Via Eudossiana 18, 00184 Roma, Italy.

Summary. Control systems over networks with a finite data rate can be conveniently modeled as hybrid (impulsive) systems. For the class of nonlinear systems in feedforward form, we design a hybrid controller which guarantees stability, in spite of the measurement noise due to the quantization, *and* of an arbitrarily large delay which affects the communication channel. The rate at which feedback packets are transmitted from the sensors to the actuators is shown to be arbitrarily close to the infimal one.

This paper is dedicated to Professor Alberto Isidori on the occasion of his 65th birthday, with admiration and gratitude.

1 Introduction

The problem of controlling systems under communication constraints has attracted much interest in recent years. In particular, many papers have investigated how to cope with the finite bandwidth of the communication channel in the feedback loop. For the case of linear systems (cf. e.g. [2, 7, 9, 8, 23, 27, 3]) the problem has been very well understood, and an elegant characterization of the minimal data rate – that is the minimal rate at which the measured information must be transmitted to the actuators – above which stabilization is always possible is available. Loosely speaking, the result shows that the minimal data rate is proportional to the inverse of the product of the unstable eigenvalues of the dynamic matrix of the system. Controlling using the minimal data rate is interesting not only from a theoretical point of view, but also from a practical one, even in the presence of communication channels with a large bandwidth. Indeed, having control techniques which employ a small number of bits to encode the feedback information implies for instance that the number of different tasks which can be simultaneously carried out is maximized, results in explicit procedures to convert the analog information provided by the sensors into the digital form which can be transmitted, and improves the performance of the system ([15]). We refer the reader to [25] for an excellent survey on the topic of control under data rate constraints.

The problem for nonlinear systems has been investigated as well (cf. [16, 18, 24, 6, 13, 4]). In [16], the author extends the results of [2] on quantized control to nonlinear systems which are *input-to-state* stabilizable. For the same class, the paper [18] shows that the approach in [27] can be employed also for continuous-time nonlinear systems, although in [18] no attention is paid on the minimal data rate needed to achieve the result. In fact, if the requirement on the data rate is not strict, as it is implicitly assumed in [18], it is shown in [6] that the results of [18] actually hold for the much broader class of *stabilizable* systems. The paper [24] shows, among the other results, that a minimal data rate theorem for *local* stabilizability of nonlinear systems can be proven by focusing on linearized system. To the best of our knowledge, *non* local results for the problem of minimal data rate stabilization of nonlinear systems are basically missing. Nevertheless, the paper [4] has pointed out that, if one restricts the attention to the class of nonlinear *feedforward* systems, then it is possible to find the infimal data rate above which stabilizability is possible. We recall that feedforward systems represent a very important class of nonlinear systems, which has received much attention in recent years (see e.g. [29, 22, 12, 10, 19], to cite a few), in which many physical systems fall ([11]), and for which it is possible to design stabilizing control laws in spite of saturation on the actuators. When *no* communication channel is present in the feedback loop, a recent paper ([20], see also [21]) has shown that any feedforward nonlinear system can be stabilized regardless of an arbitrarily large delay affecting the control action.

In this contribution, exploiting the results of [20], we show that the minimal data rate theorem of [4] holds when an arbitrarily large delay affects the channel (in [4], instantaneous delivery through the channel of the feedback packets was assumed). Note that the communication channel not only introduces a delay, but also a quantization error and an impulsive behavior [26], since the packets of bits containing the feedback information are sent only at discrete times. Hence, the methods of [20], which are studied for continuous-time delay systems, can not be directly used to deal with impulsive delay systems in the presence of measurement errors. In addition, our result requires an appropriate redesign, not only of the parameters in the feedback law of [20], but also of the encoder and the decoder of [4]. See [17] for another approach to control problems in the presence of delays and quantization.

In the next section, we present some preliminary notions useful to formulate the problem. The main contribution is stated in Section 3. Building on the coordinate transformations of [28, 20], we introduce in Section 4 a form for the closed-loop system which is convenient for the analysis discussed in Section 5). For the sake of simplicity, not all the proofs are presented, and they can be found in [5]. In the conclusions, it is emphasized how the proposed solution is also robust with respect to packet drop-out. The rest of the section summarizes the notation adopted in the paper.

Notation. Given an integer $1 \leq i \leq \nu$, the vector $(a_i, \ldots, a_\nu) \in \mathbb{R}^{\nu-i+1}$ will be succinctly denoted by the corresponding uppercase letter with index i, i.e. A_i. For $i = 1$, we will equivalently use the symbol A_1 or simply a. I_i denotes the $i \times i$ identity matrix. $\mathbf{0}_{i\times j}$ (respectively, $\mathbf{1}_{i\times j}$) denotes an $i \times j$ matrix whose entries are all 0 (respectively, 1). When only one index is present, it is meant that the matrix is a (row or column) vector.

If x is a vector, $|x|$ denotes the standard Euclidean norm, i.e. $|x| = \sqrt{x^T x}$, while $|x|_\infty$ denotes the infinity norm $\max_{1\leq i\leq n} |x_i|$. The vector $(x^T \; y^T)^T$ will be simply denoted as (x, y). $\mathbb{Z}_+$ (respectively, $\mathbb{R}_+$) is the set of nonnegative integers (real numbers), $\mathbb{R}^n_+$ is the positive orthant of $\mathbb{R}^n$. A matrix M is said to be Schur stable if all its eigenvalues are strictly inside the unit circle.

The symbol $\text{sgn}(x)$, with x a scalar variable, denotes the sign function which is equal to 1 if $x > 0$, 0 if $x = 0$, and equal to -1 otherwise. If x is an n-dimensional vector, then $\text{sgn}(x)$ is an n-dimensional vector whose ith component is given by $\text{sgn}(x_i)$. Moreover, $\text{diag}(x)$ is an $n \times n$ diagonal matrix whose element (i, i) is x_i.

Given a vector-valued function of time $x(\cdot) : \mathbb{R}_+ \to \mathbb{R}^n$, the symbol $||x(\cdot)||_\infty$ denotes the supremum norm $||x(\cdot)||_\infty = \sup_{t\in\mathbb{R}_+} |x(t)|$. In the paper, two time scales are used, one denoted by the variable t in which the delay is θ, and the other one denoted by r, in which the delay is τ. Depending on the time scale, the following two norms are used: $||x_t|| = \sup_{-\theta\leq\varsigma\leq 0} |x(t+\varsigma)|$, $||x_r|| = \sup_{-\tau\leq\sigma\leq 0} |x(r + \sigma)|$. Moreover, $x(\bar{t}^+)$ represents the right limit $\lim_{t\to\bar{t}^+} x(t)$.

The saturation function [20] $\sigma : \mathbb{R} \to \mathbb{R}$ is an odd $\mathcal{C}^1$ function such that $0 \leq \sigma'(s) \leq 1$ for all $s \in \mathbb{R}$, $\sigma(s) = 1$ for all $s \geq 21/20$, and $\sigma(s) = s$ for all $0 \leq s \leq 19/20$. Furthermore, $\sigma_i(s) = \varepsilon_i \sigma(s/\varepsilon_i)$, with ε_i a positive real number.

2 Preliminaries

Consider a nonlinear system in feedforward form [29, 22, 12, 19], that is a system of the form

$$\dot{x}(t) = f(x(t), u(t)) := \begin{pmatrix} x_2(t) + h_1(X_2(t)) \\ \cdots \\ x_n(t) + h_{n-1}(X_n(t)) \\ u(t) \end{pmatrix}, \tag{1}$$

where $x_i(t) \in \mathbb{R}$, $X_i(t)$ is the vector of state variables $x_i(t), x_{i+1}(t), \ldots, x_n(t)$, $u(t) \in \mathbb{R}$, each function h_i is C^2, and there exists a positive real number $M > 0$ such that for all $i = 1, 2, \ldots, n-1$, if $|X_{i+1}|_\infty \leq 1$, then

$$|h_i(X_{i+1})| \leq M|X_{i+1}|^2 . \tag{2}$$

We additionally assume that a bound on the compact set of initial conditions is available to both the encoder and the decoder, namely a vector $\bar{\ell} \in \mathbb{R}^n_+$ is known for which

$$|x_i(t_0)| \le \bar{\ell}_i\ , \quad i = 1, 2, \ldots, n\ . \tag{3}$$

We investigate the problem of stabilizing the system above, when the measurements of the state variables travel through a communication channel. There are several ways to model the effect of the channel. In the present setting, we assume that there exists a sequence of strictly increasing transmission times $\{t_k\}_{k\in\mathbb{Z}_+}$, satisfying

$$T_m \le t_{k+1} - t_k \le T_M\ , \quad k \in \mathbb{Z}_+ \tag{4}$$

for some positive and known constants T_m, T_M, at which a packet of $N(t_k)$ bits, encoding the feedback information, is transmitted. The packet is received at the other end of the channel θ units of time later, namely at the times $\theta_k := t_k + \theta$. In problems of control under communication constraints, it is interesting to characterize how often the sensed information is transmitted to the actuators. In this contribution, as a measure of the data rate employed by the communication scheme we adopt the *average data rate* [27] defined as

$$R_{av} = \limsup_{k\to+\infty} \sum_{j=0}^{k} \frac{N(t_j)}{t_k - t_0}\ , \tag{5}$$

where $\sum_{j=0}^{k} N(t_j)$ is the total number of bits transmitted during the time interval $[t_0, t_k]$. An *encoder* carries out the conversion of the state variable into packets of bits. At each time t_k, the encoder first samples the state vector to obtain $x(t_k)$, and then determines a vector $y(t_k)$ of symbols which can be transmitted through the channel. We recall below the encoder which has been proposed in [4], inspired by [27, 18], and then propose a modification to handle the presence of the delay. The encoder in [4] is as follows:

$$\begin{aligned}
\dot{\xi}(t) &= f(\xi(t), u(t)) & \\
\dot{\ell}(t) &= \mathbf{0}_n & t \ne t_k \\
\xi(t^+) &= \xi(t) + g_{\mathcal{E}}(x(t), \xi(t), \ell(t)) & \\
\ell(t^+) &= \Lambda \ell(t) & t = t_k \\
y(t^+) &= \mathrm{sgn}(x(t) - \xi(t)) & t = t_k\ ,
\end{aligned} \tag{6}$$

where ξ, ℓ is the encoder state, y is the feedback information transmitted through the channel, Λ is a Schur stable matrix, and $g_{\mathcal{E}}(x, \xi, \ell) = 4^{-1}\mathrm{diag}\,[\mathrm{sgn}\,(x - \xi)]\,\ell$. Note that each component of y takes value in $\{0, \pm 1\}$, therefore y can be transmitted as a packet of bits of finite length. In particular, if ξ_i is on the left of x_i then $+1$ is transmitted, if it is on the right, then -1 is transmitted. The system above is an *impulsive* system ([1, 14]) and its behavior is easily explained. At $t = t_0$, given an initial condition $\xi(t_0), \ell(t_0)$, the updates $\xi(t_0^+), \ell(t_0^+)$ of the encoder state and $y(t^+)$ of the output are obtained. The former update serves as initial condition for the continuous-time dynamics, and the state $\xi(t), \ell(t)$ is computed over the interval $[t_0, t_1]$. At the endpoint of the interval, a new update $\xi(t_1^+), \ell(t_1^+)$ is obtained and the procedure can be iterated an infinite number of times to compute the solution $\xi(t), \ell(t)$ for all t.

At the other end of the channel lies a decoder, which receives the packets $y(t_k)$, and reconstructs the state of the system. The decoder is very similar to the encoder. In fact, we have:

$$\begin{aligned} \dot{\psi}(t) &= f(\psi(t), u(t)) & \\ \dot{\nu}(t) &= \mathbf{0}_n & t \neq t_k \\ \psi(t^+) &= \psi(t) + g_{\mathcal{D}}(y(t), \nu(t)) & \\ \nu(t^+) &= \Lambda \nu(t) & t = t_k \end{aligned} \tag{7}$$

with $g_{\mathcal{D}}(y, \nu) = 4^{-1}\mathrm{diag}(y)\nu$. The control law is

$$u(t) = \alpha(\psi(t)) \ , \tag{8}$$

where α is the nested saturated function specified later. Note that this control law is feasible because the decoder and the actuator are *co-located.* If the encoder and the decoder agree to set their initial conditions to the same value, then it is not hard to see ([4]) that $\xi(t) = \psi(t)$ and $\ell(t) = \nu(t)$ for all t. Moreover, one additionally proves that $\xi(t)$ is an asymptotically correct estimate of $x(t)$, and the latter converges to zero [4].

When a delay affects the channel, the decoder does not know the first state sample throughout the interval $[t_0, t_0 + \theta]$, and hence it can not provide any feedback control action. The control is therefore set to zero. As the successive samples $y(t_k)$ are all received at times $\theta_k = t_k + \theta$, the decoder becomes aware of the value of ξ θ units of time later. Hence, the best one can expect is to reconstruct the value of $\xi(t - \theta)$ (see Lemma 1 below), and to this purpose the following decoder is proposed:

$$\begin{aligned} \dot{\psi}(t) &= f(\psi(t), \alpha(\psi(t-\theta))) & \\ \dot{\nu}(t) &= \mathbf{0}_n & t \neq \theta_k \\ \psi(t^+) &= \psi(t) + g_{\mathcal{D}}(y(t-\theta), \nu(t)) & \\ \nu(t^+) &= \Lambda \nu(t) & t = \theta_k \\ u(t) &= \alpha(\psi(t)) \ . & \end{aligned} \tag{9}$$

We also need to modify the encoder. Indeed, as mentioned in the case with no delay, for the encoder to work correctly, the control law (8), and hence $\psi(t)$, must be available to the encoder. To reconstruct this quantity, the following equations are added to the encoder (6):

$$\begin{aligned} \dot{\omega}(t) &= f(\omega(t), \alpha(\omega(t-\theta))) & t \neq \theta_k \\ \omega(t^+) &= \omega(t) + g_{\mathcal{E}}(x(t-\theta), \xi(t-\theta), \ell(t-\theta)) & t = \theta_k \ . \end{aligned}$$

As in [28, 20], we shall adopt a *linear* change of coordinates in which the control system takes a special form convenient for the analysis. Differently from [4], this change of coordinates plays a role also in the encoding/decoding procedure. Indeed, denoted by Φ the nonsingular matrix which defines the change of coordinates, and which we define in detail in Section 4, the functions $g_{\mathcal{E}}$, $g_{\mathcal{D}}$ which appear in (11) and, respectively, (9) are modified as

$$g_{\mathcal{E}}(x,\xi,\ell) = (4\Phi)^{-1}\mathrm{diag}\,[\mathrm{sgn}\,(\Phi(x-\xi))]\,\ell\,, \quad g_{\mathcal{D}}(y,\nu) = (4\Phi)^{-1}\mathrm{diag}(y)\nu\,,$$

the initial conditions of the encoder and decoder are set as

$$\begin{aligned} ||\omega_{\theta_0}|| = 0\,, \quad \xi(t_0) = \mathbf{0}_n\,, \quad \ell(t_0) = 2\Phi\bar{\ell}\,, \\ ||\psi_{\theta_0}|| = 0\,, \quad \nu(\theta_0) = 2\Phi\bar{\ell}\,, \end{aligned} \tag{10}$$

and, finally, the vector y which is transmitted through the channel take the expression

$$y(t^+) = \mathrm{sgn}(\Phi(x(t)-\xi(t)))\,.$$

Overall, the equations which describe the encoder are:

$$\begin{array}{lll} \dot{\omega}(t) &= f(\omega(t),\alpha(\omega(t-\theta))) & t \neq \theta_k \\ \dot{\xi}(t) &= f(\xi(t),\alpha(\omega(t))) & \\ \dot{\ell}(t) &= \mathbf{0}_n & t \neq t_k \\ \omega(t^+) &= \omega(t) + g_{\mathcal{E}}(x(t-\theta),\xi(t-\theta),\ell(t-\theta)) & t = \theta_k \\ \xi(t^+) &= \xi(t) + g_{\mathcal{E}}(x(t),\xi(t),\ell(t)) & \\ \ell(t^+) &= \Lambda\ell(t) & \\ y(t^+) &= \mathrm{sgn}(\Phi(x(t)-\xi(t))) & t = t_k\,. \end{array} \tag{11}$$

The following can be easily proven.

Lemma 1. *In the above setting, we have: (i)* $\omega(t) = \psi(t)$ *for all* $t \geq t_0$*, (ii)* $\xi(t-\theta) = \psi(t)$ *and* $\nu(t-\theta) = \ell(t)$ *for all* $t \geq \theta_0$.

As anticipated, the encoder and decoder we introduced above are such that the internal state of the former is exactly reconstructed from the internal state of the latter. This also implies that in the analysis to come it is enough to focus on the equations describing the process and the decoder only.

3 Main Result

The problem we tackle in this paper is, given any value of the delay θ, find the matrices Λ, Φ in (11) and (9), and the control (8) which guarantee the state of the entire closed-loop system to converge to the origin. As recalled in the previous section, at times t_k, the measured state is sampled, packed into a sequence of $N(t_k)$ bits, and fed back to the controller. In other words, the information flows from the sensors to the actuators with an average rate

R_{av} given by (5). In this setting, it is therefore meaningful to formulate the problem of stabilizing the system *while* transmitting the minimal amount of feedback information per unit of time, that is using the minimal average data rate. The problem can be formally cast as follows.

Definition 1. *System (1) is semi-globally asymptotically and locally exponentially stabilizable using an average data rate arbitrarily close to the infimal one if, for any* $\bar{\ell} \in \mathbb{R}^n_+$, $\theta > 0$, $\hat{R} > 0$, *an encoder (11), a decoder (9), initial conditions (3), (10), and a controller (8) exist such that for the closed-loop system with state* $X := (x, \omega, \xi, \ell, \psi, \nu)$, *we have the following properties.*

(i) The origin is a stable equilibrium point;
(ii) There exist a compact set C *containing the origin, and* $T > 0$, *such that* $X(t) \in C$ *for all* $t \geq T$;
(iii) For all $t \geq T$, *for some positive real numbers* k, δ,

$$|X(t)| \leq k||X_T|| \exp(-\delta(t-T)) .$$

(iv) $R_{av} < \hat{R}$.

Remark 1. It is straightforward to verify that the origin is indeed an equilibrium point for the closed-loop system. Moreover, item (iii) explains what is meant by stabilizability using an average data rate arbitrarily close to the infimal one. As a matter of fact, (iv) requires that the average data rate can be made arbitrarily close to the zero, which of course is the infimal data rate. It is "infimal" rather than "minimal", because we could never stabilize an open-loop unstable system such as (1) with a zero data rate (no feedback). ◁

Compared with the papers [29, 22, 12, 19], concerned with the stabilization problem of nonlinear feedforward systems, the novelty here is due to the presence of impulses, quantization noise which affects the measurements and delays which affect the control action (on the other hand, we neglect parametric uncertainty, considered in [19]). In [30], it was shown robustness with respect to measurement errors for non-impulsive systems with no delay. In [20], the input is delayed, but neither impulses nor measurement errors are present. Impulses and measurement errors are considered in [4], where the minimal data rate stabilization problem is solved, but instantaneous delivery of the packets is assumed.
We state the main result of the paper.

Theorem 1. *System (1) is semi-globally asymptotically and locally exponentially stable with an average data rate arbitrarily close to the infimal one.*

Remark 2. The proof is constructive and explicit expressions for Λ, Φ, and the controller are determined. ◁

Remark 3. This result can be viewed as a nonlinear generalization of the well-known data rate theorem for linear systems. Indeed, the linearization of the feedforward system at the origin is a chain of integrators, for which the minimal data rate theorem for linear systems states that stabilizability is possible using an average data rate arbitrarily close to zero. ◁

4 Change of Coordinates

Building on the coordinate transformations in [20, 28], we put the system composed of the process and the decoder in a special form. Before doing this, we recall that for feedforward systems encoders, decoders and controllers are designed in a recursive way [28, 29, 22, 12, 20, 4]. In particular, at each step $i = 1, 2, \ldots, n$, one focuses on the last $n - i + 1$ equations of system (1), design the last $n - i + 1$ equations of the encoder and the decoder, the first i terms of the nested saturated controller, and then proceed to the next step, where the last $n - i$ equations of (1) are considered. To this end, it is useful to introduce additional notation to denote these subsystems. In particular, for $i = 1, 2, \ldots, n$, we denote the last $n - i + 1$ equations of (1) by

$$\dot{X}_i(t) = H_i(X_{i+1}(t), u(t)) = \begin{pmatrix} x_{i+1}(t) + h_i(X_{i+1}(t)) \\ \cdots \\ x_n(t) + h_{n-1}(X_n(t)) \\ u(t) \end{pmatrix} , \tag{12}$$

with $u(t) = \alpha(\psi(t))$, while for the last $n - i + 1$ equations of the decoder (9) we adopt the notation

$$\begin{array}{ll} \dot{\Psi}_i(t) = H_i(\Psi_{i+1}(t), u(t - \theta)) & \\ \dot{N}_i(t) = \mathbf{0}_{n-i+1} & t \neq \theta_k \; , \\ \Psi_i(t) = \Psi_i(t^-) + (4\Phi_i)^{-1}\mathrm{diag}(Y_i(t - \theta))N_i(t^-) & \\ N_i(t) = \Lambda_i N_i(t^-) & t = \theta_k \; , \end{array} \tag{13}$$

where N_i denotes the components from i to n of ν. Moreover, for given positive constants $L \leq M$, $\kappa \geq 1$, with M defined in (2), we define the *non singular positive* matrices[1] Φ_i as:

$$\Phi_i X_i := \begin{bmatrix} p_i\left(\frac{M}{L}\kappa^{i-1}x_i, \ldots, \frac{M}{L}\kappa^{n-1}x_n\right) \\ \cdots \\ p_n\left(\frac{M}{L}\kappa^{n-1}x_n\right) \end{bmatrix} , \quad i = 1, \ldots, n \; , \tag{14}$$

where the functions $p_i, q_i : \mathbb{R}^{n-i+1} \to \mathbb{R}$ are [28, 20]

$$p_i(a_i, \ldots, a_n) = \sum_{j=i}^{n} \frac{(n-i)!a_j}{(n-j)!(j-i)!} ,$$
$$q_i(a_i, \ldots, a_n) = \sum_{j=i}^{n} \frac{(-1)^{i+j}(n-i)!a_j}{(n-j)!(j-i)!} ,$$

[1] The matrix Φ_1 will be simply referred to as Φ.

with $p_i(q_i(a_i,\ldots,a_n),\ldots,q_n(a_n)) = a_i$, $q_i(p_i(a_i,\ldots,a_n),\ldots,p_n(a_n)) = a_i$. Finally, let us also introduce the change of *time scale*

$$t = \kappa r \ , \tag{15}$$

and the *input* coordinate change

$$v(r) = \kappa p_n\left(\frac{M}{L}\kappa^{n-1}u(\kappa r)\right) \ . \tag{16}$$

Then we have the following.

Lemma 2. *Let $i \in \{1,2,\ldots,n\}$ and*

$$\tau := \theta/\kappa \ , \quad r_k := t_k/\kappa \ , \quad \rho_k := \theta_k/\kappa \ . \tag{17}$$

The change of coordinates (15), (16), and

$$\begin{aligned} Z_i(r) &= \Phi_i X_i(\kappa r) \\ E_i(r) &= \Phi_i(\Psi_i(\kappa r) - X_i(\kappa(r-\tau))) \ , \\ P_i(r) &= N_i(\kappa r) \end{aligned} \tag{18}$$

transforms system (12)–(13) into

$$\begin{aligned} \dot{Z}_i(r) &= \Gamma_i Z_i(r) + \mathbf{1}_{n-i+1}v(r) + \varphi_i(Z_{i+1}(r)) \\ \dot{E}_i(r) &= \Gamma_i E_i(r) + \varphi_i(E_{i+1}(r) + Z_{i+1}(r-\tau)) \\ &\quad -\varphi_i(Z_{i+1}(r-\tau)) \\ \dot{P}_i(r) &= \mathbf{0}_{n-i+1} \qquad\qquad\qquad\qquad r \neq \rho_k \\ Z_i(r^+) &= Z_i(r) \\ E_i(r^+) &= E_i(r) + 4^{-1}\mathrm{diag}(\mathrm{sgn}(-E_i(r)))P_i(r) \\ P_i(r^+) &= \Lambda_i P_i(r) \qquad\qquad\qquad\qquad r = \rho_k \ , \end{aligned} \tag{19}$$

where

$$\Gamma_i := \begin{bmatrix} 0\,1\,1\ldots 1\,1 \\ 0\,0\,1\ldots 1\,1 \\ \vdots\;\vdots\;\vdots\;\;\vdots\;\;\vdots\;\vdots \\ 0\,0\,0\ldots 0\,1 \\ 0\,0\,0\ldots 0\,0 \end{bmatrix} , \qquad \varphi_i(Z_{i+1}) := \begin{bmatrix} f_i(Z_{i+1}) \\ f_{i+1}(Z_{i+2}) \\ \ldots \\ f_{n-1}(Z_n) \\ 0 \end{bmatrix} .$$

Proof. It is shown in [20] that, (15), (16) and $Z_i(r) = \Phi_i X_i(\kappa r)$ transforms (12) into

$$\begin{aligned} \dot{Z}_i(r) &= F_i(Z_{i+1}(r), v(r)) \\ &:= \begin{bmatrix} \sum_{j=i+1}^{n} z_j(r) + v(r) + f_i(Z_{i+1}(r)) \\ \sum_{j=i+2}^{n} z_j(r) + v(r) + f_{i+1}(Z_{i+2}(r)) \\ \vdots \\ v(r) \end{bmatrix} , \end{aligned} \tag{20}$$

where

$$|f_i(Z_{i+1})| \leq P|Z_{i+1}|^2 , \quad P = n^3(n!)^3 L\kappa^{-1} , \tag{21}$$

provided that $|Z_{i+1}|_\infty \leq (M\kappa)/(L(n+1)!)$. Clearly, the equation (20) is equal to the first equation of (19). Bearing in mind (20), and by differentiating E_i defined in (18), we obtain

$$\begin{aligned} \dot{E}_i(r) &= F_i(E_{i+1}(r) + Z_{i+1}(r-\tau), v(r-\tau)) - F_i(Z_{i+1}(r-\tau), v(r-\tau)) \\ &= \Gamma_i(E_i(r) + Z_i(r-\tau)) + \mathbf{1}_{n-i+1}v(r-\tau) + \varphi_i(E_{i+1}(r) + Z_{i+1}(r-\tau)) \\ &\quad -\Gamma_i(Z_i(r-\tau)) - \mathbf{1}_{n-i+1}v(r-\tau) - \varphi_i(Z_{i+1}(r-\tau)) \\ &= \Gamma_i E_i(r) + \varphi_i(E_{i+1}(r) + Z_{i+1}(r-\tau)) - \varphi_i(Z_{i+1}(r-\tau)) \end{aligned} \tag{22}$$

for $r \neq \rho_k$, while for $r = \rho_k$, we have:

$$\begin{aligned} E_i(\rho_k^+) &= \Phi_i(\Psi_i(\kappa\rho_k^+) - X_i(\kappa(\rho_k - \tau)^+)) \\ &= \Phi_i(\Psi_i(\theta_k^+) - X_i(t_k^+)) \\ &= \Phi_i(\Psi_i(\theta_k) + (4\Phi_i)^{-1}\mathrm{diag}(Y_i(t_k^+))N_i(\theta_k) - X_i(t_k)) \\ &= \Phi_i(\Psi_i(\theta_k) - X_i(t_k)) + 4^{-1}\mathrm{diag}(\mathrm{sgn}(\Phi_i[X_i(t_k) - \Xi_i(t_k)]))N_i(\theta_k) \\ &= \Phi_i(\Psi_i(\theta_k) - X_i(t_k)) + 4^{-1}\mathrm{diag}(\mathrm{sgn}(\Phi_i[X_i(t_k) - \Psi_i(\theta_k)]))N_i(\theta_k) , \end{aligned} \tag{23}$$

where the last equality descends from (ii) in Lemma 1, and implies

$$E_i(\rho_k^+) = E_i(\rho_k) + 4^{-1}\mathrm{diag}(\mathrm{sgn}(-E_i(\rho_k)))N_i(\theta_k) . \tag{24}$$

The thesis then follows if we observe that the variable P_i defined in (18) satisfies

$$\begin{aligned} \dot{P}_i(r) &= \mathbf{0}_{n-i+1} \;\; r \neq \rho_k \\ P_i(r^+) &= \Lambda_i P_i(r) \;\; r = \rho_k . \end{aligned} \tag{25}$$

□

Before ending the section, we specify the nested saturated controller $u(t) = \alpha(\psi(t))$ which is shown to stabilize the closed-loop system in the next section. In particular, we have

$$\begin{aligned} \alpha(\psi(t)) = -\frac{L}{M\kappa^n}\sigma_n\Bigg(p_n\left(\kappa^{n-1}\frac{M}{L}\psi_n(t)\right) + \sigma_{n-1}\Bigg(p_{n-1}\Bigg(\kappa^{n-2}\frac{M}{L}\psi_{n-1}(t), \\ \kappa^{n-1}\frac{M}{L}\psi_n(t)\Bigg) + \ldots + \sigma_i\left(p_i\left(\kappa^{i-1}\frac{M}{L}\psi_i(t), \ldots, \kappa^{n-1}\frac{M}{L}\psi_n(t)\right)\right. \\ +\lambda_{i-1}(t))\ldots) , \end{aligned}$$

with

$$\begin{aligned} \lambda_{i-1}(t) = \sigma_{i-1}\left(p_{i-1}\left(\kappa^{i-2}\frac{M}{L}\psi_{i-1}(t), \ldots, \kappa^{n-1}\frac{M}{L}\psi_n(t)\right)\right. \\ \left. + \ldots + \sigma_1\left(p_1\left(\frac{M}{L}\psi_1(t), \ldots, \kappa^{n-1}\frac{M}{L}\psi_n(t)\right)\right)\ldots\right) , \end{aligned}$$

and where the saturation levels ε_i of $\sigma_i(r) = \varepsilon_i\sigma(r/\varepsilon_i)$ are chosen as follows:

$$1 = 80\varepsilon_n = 80^2\varepsilon_{n-1} = \ldots = 80^n\varepsilon_1 \, . \tag{26}$$

In the new coordinates (15)–(16), (18), the controller takes the form

$$\begin{aligned} v(r) = -\sigma_n(e_n(r) + z_n(r-\tau) + \sigma_{n-1}(e_{n-1}(r) + z_{n-1}(r-\tau) + \\ \ldots + \sigma_i(e_i(r) + z_i(r-\tau) + \hat{\lambda}_{i-1}(r))\ldots)) \, , \end{aligned} \tag{27}$$

with $\hat{\lambda}_{i-1}(r) = \sigma_{i-1}(e_{i-1}(r) + z_{i-1}(r-\tau) + \ldots + \sigma_1(e_1(r) + z_1(r-\tau))\ldots)$.

5 Analysis

In the previous sections, we have introduced the encoder, the decoder and the controller. In this section, in order to show the stability property, we carry out a step-by-step analysis, where at each step i, we consider the subsystem (19) in closed-loop with (27). We first introduce two lemmas which are at the basis of the iterative construction. The first one, which, in a different form, was basically given in [4], shows that the decoder asymptotically tracks the state of the process under a boundedness assumption. The proof we present here is more straightforward than the original one.

Lemma 3. *Suppose (3) is true. If for some $i = 1, 2, \ldots, n$ there exists a positive real number $\bar{\mathrm{Z}}_{i+1}$ such that* [2]

$$||Z_{i+1}(\,\cdot\,)||_\infty \le \bar{\mathrm{Z}}_{i+1} \, ,$$

and, for all $r \ge \rho_0$,

$$|e_j(r)| \le p_j(r)/2 \, , \quad j = i+1, i+2, \ldots, n \, ,$$

with[3]

$$P_{i+1}(\rho^+) = \Lambda_{i+1}P_{i+1}(\rho) \quad \rho = \rho_k \, ,$$

and Λ_{i+1} a Schur stable matrix, then for all $r \ge \rho_0$,

$$|e_i(r)| \le p_i(r)/2 \, ,$$

with $p_i(r^+) = p_i(r)/2$, for $r = \rho_k$, if $i = n$, and

$$\begin{bmatrix} p_i(r^+) \\ P_{i+1}(r^+) \end{bmatrix} = \begin{bmatrix} 1/2 & * \\ \mathbf{0}_{n-i} & \Lambda_{i+1} \end{bmatrix} \begin{bmatrix} p_i(r) \\ P_{i+1}(r) \end{bmatrix} \; r = \rho_k \, , \tag{28}$$

if $i \in \{1, 2, \ldots, n-1\}$, where $*$ is a $1 \times (n-i)$ row vector depending on $\bar{Z}_{i+1}$, $\bar{\ell}$, and T_M.

---

[2] The conditions are void for $i = n$.

[3] In the statement, the continuous dynamics of the impulsive systems are trivial – the associated vector fields are identically zero – and hence omitted.

Proof. Recall first (22), (24). Furthermore, by (10), the definition of Φ, and (3), $|e_j(\rho_0)| \le p_j(\rho_0)/2$ for $j = i, i+1, \ldots, n$. For $i = n$, as $|e_n(\rho_0)| \le p_n(\rho_0)/2$, it is immediately seen that

$$|e_n(\rho_0^+)| = |e_n(\rho_0) + 4^{-1}\text{sgn}(-e_n(\rho_0))p_n(\rho_0)| \le 4^{-1}p_n(\rho_0)$$

which proves that $|e_n(\rho_0^+)| \le p_n(\rho_0^+)/2$, provided that $\Lambda_n = 1/2$. As $\dot{e}_n(r) = 0$, then $|e_n(r)| \le p_n(\rho_0^+)/2$ for $r \in [\rho_0, \rho_1)$. As $\dot{p}_n(r) = 0$, also $|e_n(\rho_1)| \le p_n(\rho_1)/2$, and iterative arguments prove that $|e_n(r)| \le p_n(\rho_k^+)/2$ on each interval $[\rho_k, \rho_{k+1})$. Notice that the single trivial eigenvalue of Λ_n is strictly less than the unity. The first equation of (22) writes as:

$$\begin{aligned}\dot{e}_i(r) &= \mathbf{1}_{n-i}E_{i+1}(r) + \varphi_i(E_{i+1}(r) + Z_{i+1}(r-\tau)) - \varphi_i(Z_{i+1}(r-\tau)) \\ &= \left(\mathbf{1}_{n-i} + \left[\frac{\partial \varphi_i(y_{i+1})}{\partial y_{i+1}}\right]_{\alpha(r)E_{i+1}(r)+Z_{i+1}(r-\tau)}\right) E_{i+1}(r)\,,\end{aligned}$$

with $\alpha(r) \in [0,1]$ for all r. As both E_{i+1} and Z_{i+1} are bounded, it is not hard to see [4] that there exists a positive real number F_i depending on $\bar{Z}_{i+1}$ and $\bar{\ell}$, such that, for $r \in [\rho_k, \rho_{k+1})$,

$$e_i(r) \le e_i(\rho_k^+) + F_i(\rho_{k+1} - \rho_k) \sum_{j=i+1}^{n} p_j(\rho_k^+)/2\,,$$

with $|e_i(\rho_0^+)| \le p_i(\rho_0)/4$. By iteration, the thesis is inferred provided that

$$\begin{aligned}p_i(\rho_k^+) &= \frac{1}{2}p_i(\rho_k) + F_i T_M \mathbf{1}_{n-i}\Lambda_{i+1}P_{i+1}(\rho_k) \\ &\ge \frac{1}{2}p_i(\rho_k) + F_i(\rho_{k+1} - \rho_k) \sum_{j=i+1}^{n} p_j(\rho_k^+)\,.\end{aligned}$$

Note that, by the definition of $p_i(\rho_k^+)$ above, $P_i(\rho_k^+) = \Lambda_i P_i(\rho_k)$, with Λ_i the matrix in (28), that shows Λ_i to be a Schur stable matrix provided that so is Λ_{i+1}. □

The following remark will be useful later on.

Remark 4. From the proof of the lemma, it is possible to see that, if $||z(\cdot)||_\infty \le Z$, for some $Z > 0$, then e and p in (19) (with $i = 1$) obey the equations[4]

$$\begin{array}{ll}\dot{e}(r) = A(r)e(r) & \\ \dot{p}(r) = \mathbf{0}_n & r \ne \rho_k \\ e(r) = e(r^-) + 4^{-1}\text{diag}[\text{sgn}(-e(r^-))]p(r^-) & \\ p(r) = \Lambda p(r^-) & r = \rho_k\,,\end{array} \qquad (29)$$

[4] Again, we adopt the symbol Λ rather than Λ_1.

with

$$A(r) := \begin{bmatrix} 0 & a_{12}(r) & a_{13}(r) & \dots & a_{1\,n-1}(r) & a_{1n}(r) \\ 0 & 0 & a_{23}(r) & \dots & a_{2\,n-1}(r) & a_{2n}(r) \\ \vdots & \vdots & \vdots & \vdots & \vdots & \vdots \\ 0 & 0 & 0 & \dots & 0 & a_{n-1\,n}(r) \\ 0 & 0 & 0 & \dots & 0 & 0 \end{bmatrix}, \tag{30}$$

and where the off-diagonal components of A, rather than as functions of $(r, e(r), z(r-\tau))$, are viewed as bounded (unknown) functions of r, whose absolute value can be assumed without loss of generality to be upper bounded by a positive constant depending on Z, $\bar{\ell}$ and T_M. ◁

The next statement, based on Lemma 10 in [20], shows that a controller exists which guarantees the boundedness of the state variables, a property required in the latter result. Note that the arguments of the proof in [20] hold even in the presence of a "measurement" disturbance e induced by the quantization, which can be possibly large during the transient but it is decaying to zero asymptotically.

Lemma 4. *Consider the system*

$$\dot{Z}(r) = -\varepsilon\sigma\left[\frac{1}{\varepsilon}(Z(r-\tau) + e(r) + \lambda(r))\right] + \mu(r)$$

where $Z \in \mathbb{R}$, ε is a positive real number, and additionally:

- *$\lambda(\cdot)$ and $\mu(\cdot)$ are continuous functions for which positive real numbers λ_* and μ_* exist such that, respectively, $|\lambda(r)| \le \lambda_*$, $|\mu(r)| \le \mu_*$, for all $r \ge r_0$;*
- *$e(\cdot)$ is a piecewise-continuous function for which a positive time r_* and a positive number e_* exist such that $|e(r)| \le e_*$, for all $r \ge r_*$.*

If

$$\tau \in \left(0, \frac{1}{24}\right], \quad \lambda_* \in \left(0, \frac{\varepsilon}{80}\right], e_* \in \left(0, \frac{\varepsilon}{80}\right], \quad \mu_* \in \left(0, \frac{\varepsilon}{80}\right],$$

then there exist positive real numbers Z_* and $R \ge 0$ such that $||Z(\cdot)||_\infty \le Z_*$, and for all $r \ge R$,

$$|Z(r)| \le 4(\lambda_* + \mu_* + e_*) .$$

Remark 5. The upper bounds on λ_*, e_*, μ_* could be lowered to $\varepsilon/40$ and the result would still hold. The more conservative bounds are needed in forthcoming applications of the lemma. ◁

To illustrate the iterative analysis in a concise manner, the following is very useful (cf. [20]).

Inductive Hypothesis There exists $\bar{Z}_i > 0$ such that $||Z_i(\cdot)|| \le \bar{Z}_i$. Moreover, for each $j = i, i+1, \dots, n$, $|e_j(r)| \le p_j(r)/2$, for all $r \ge \rho_0$, and there exists $R_i > \tau$ such that for all $r \ge R_i$,

$$|z_j(r)| \le \frac{1}{4}\varepsilon_j\,, \quad |e_j(r)| \le \frac{1}{2n}\cdot\frac{1}{80^{j-i+2}}\varepsilon_j\,.$$

Initial step ($i = n$) The initial step is trivially true, provided that $\tau \le 1/24$, and $\varepsilon_n = 1/80$. Indeed, consider the closed-loop system (19), (27) with $i = n$, to obtain:

$$\begin{array}{ll} \dot z_n(r) = -\sigma_n(z_n(r-\tau) + e_n(r) + \hat\lambda_{n-1}(r)) & \\ \dot e_n(r) = 0 & \\ \dot p_n(r) = 0 & r \ne \rho_k \\ z_n(r^+) = z_n(r) & \\ e_n(r^+) = e_n(r) + 4^{-1}\mathrm{sgn}(-e_n(r))p_n(r) & \\ p_n(r^+) = \Lambda_n p_n(r) & r = \rho_k\,, \end{array} \tag{31}$$

where we set $\Lambda_n := 1/2$. By Lemma 3 and (31), $|e_n(r)| \le \varepsilon_n/80$ from a certain time R_n' on. Applying Lemma 4 to the z_n sub-system, we conclude that $||z_n(\cdot)||_\infty \le \bar Z_n$, and there exists a time $R_n > R_n'$ such that $|z_n(r)| \le \varepsilon_n/4$, and $|e_n(r)| \le \varepsilon_{n-1}/(n\cdot 160)$ for all $r \ge R_n$, the latter again by Lemma 3.

Inductive step The inductive step is summarized in the following result.

Lemma 5. *Let*

$$P \le P_m \le [20\cdot(80)^n n]^{-1}\,, \quad \tau \le \tau_m \le [4\cdot 80^{n+1} n(n+2)]^{-1}\,. \tag{32}$$

If the inductive hypothesis is true for some $i \in \{2,\ldots,n\}$, then it is also true for $i-1$.

Applying this lemma repeatedly, one concludes that, after a finite time, the state converge to the linear operation region for all the saturation functions, and the closed-loop system starts evolving according to the equations (cf. Remark 4)

$$\begin{array}{ll} \dot z(r) = A_1 z(r) + A_2 z(r-\tau) + A_2 e(r) + \varphi(z(r)) & \\ \dot e(r) = A(r)e(r) & \\ \dot p(r) = \mathbf{0}_n & r \ne \rho_k \\ z(r^+) = z(r) & \\ e(r^+) = e(r) + 4^{-1}\mathrm{diag}[\mathrm{sgn}(-e(r))]p(r) & \\ p(r^+) = \Lambda p(r) & r = \rho_k\,, \end{array} \tag{33}$$

where:

(i) A_1, A_2 are matrices for which there exist $q = (1+n^2)^{n-1}$, $a = n$, and $Q = Q^T > 0$ such that

$$(A_1 + A_2)^T Q + Q(A_1 + A_2) \le -I\,,$$

with $||Q|| \le q$ and $||A_1||, ||A_2|| \le a$;

(ii) there exists $\gamma > 0$ such that $\varphi(z(r)) := [\, f_1(Z_2(r)) \;\ldots\; f_{n-1}(Z_n(r))\; 0\,]^T$ satisfies

$$|\varphi(z)| \leq \gamma |z| \,;$$

(iii) $A(r)$ is as in (30);
(iv) Λ is the Schur stable matrix designed following the proof of Lemma 3.

Remark 6. It can be shown that the same arguments used for the proofs of the Lemma 3 to 5 lead to the conclusion that there always exists a sufficiently small neighborhood of initial conditions for the system (19), (27), with $i = 1$, such that the entire state evolves in a set where all the saturation functions operate in their linear region. This remark is important to conclude Lyapunov stability of the closed-loop system. ◁

In [20] the authors investigate the stability property of

$$\dot{z}(r) = A_1 z(r) + A_2 z(r-\tau) + \varphi(z(r)) \,,$$

that is the first component of system (33), with $e = 0$ and no impulses. In the present case, e is due to the quantization noise and drives the z-subsystem. The "driver" subsystem is composed of the (e, p) equations of (33). Hence, we have to study the stability of a *cascade* system with impulses. To this end, concisely rewrite the (e, p) equations of the system above as ([1])

$$\begin{aligned} \dot{\epsilon}(r) &= B(r)\epsilon(r) \;\; r \neq \rho_k \\ \epsilon(r^+) &= g_k(\epsilon(r)) \;\; r = \rho_k \,, \end{aligned} \tag{34}$$

with $\epsilon = (e, p)$, $|g_k(\epsilon)| \geq |\epsilon|/2$, and notice the following consequence of Lemma 3 above, and [1], Theorem 15.2.

Corollary 1. *There exists a function $V(r, \epsilon) = V(r, e, p) : \mathbb{R}_+ \times \mathbb{R}^n \times \mathbb{R}^n \to \mathbb{R}_+$ such that, for all $r \in \mathbb{R}_+$ and for all $\epsilon = (e, p) \in \mathbb{R}^n \times \mathbb{R}^n$ for which $|e| \leq |p|/2$, satisfies*

$$\begin{aligned} c_1|\epsilon|^2 \leq V(r, \epsilon) &\leq c_2|\epsilon|^2 \\ \frac{\partial V}{\partial r} + \frac{\partial V}{\partial \epsilon} B(r)\epsilon(r) &\leq -c_3|\epsilon|^2 \;\; r \neq \rho_k \\ V(r^+, g_k(\epsilon)) &\leq V(r, \epsilon) \;\; r = \rho_k \\ \left| \frac{\partial V(r, \epsilon)}{\partial \epsilon} \right| &\leq c_4|\epsilon| \,, \end{aligned}$$

for some positive constants c_i, $i = 1, \ldots, 4$.

The corollary points out that there exists an exponential Lyapunov function for the system (34). Based on this function, one can build a Lyapunov-Krasowskii functional to show that the origin is exponentially stable for the entire cascade impulsive system (33), thus extending Lemma 11 in [20] in the following way.

Lemma 6. *Consider system (33), for which the conditions (i)–(iv) hold. If*

$$\gamma \leq \frac{1}{8q} \quad \text{and} \quad \tau \leq \min\left\{\frac{1}{16a^2(8aq+1)^2}, \frac{1}{32q^2a^4}, 2\right\},$$

then, for all $r \geq \rho_0$, for some positive real numbers k, δ, we have

$$|(z(r), \epsilon(r))| \leq k||(z,\epsilon)_{\rho_0}|| \exp(-\delta(r-\rho_0)) .$$

We can now state the following stability result for the system (19), (27).

Proposition 1. *Consider the closed-loop system (19), (27) and let $|e_i(\rho_0)| \leq p_i(\rho_0)/2$ for all $i = 1, 2, \ldots, n$. If (26) holds,*

$$L \leq \min\{M, \frac{M\kappa}{(n+1)!}\} \tag{35}$$

and

$$\begin{aligned} 0 \leq \tau \leq \tau_m &= [\max\{4\cdot(80)^{n+1}n(n+2), 16n^2(8n(1+n^2)^{n-1}+1)^2, \\ &\quad 32(1+n^2)^{2(n-1)}n^4\}]^{-1} \\ 0 \leq P \leq P_m &= [\max\{20\cdot 80^{n+1}n, 8(1+n^2)^{n-1}\sqrt{n(n-1)}\}]^{-1} , \end{aligned} \tag{36}$$

then the following properties hold.

(i) The origin of the closed-loop system is stable;
(ii) There exist a compact neighborhood $\hat{C}$ of the origin and $R > 0$ such that, for all $r \geq R$, the state belongs to $\hat{C}$;
(ii) For all $r \geq R$, for some positive real numbers $\hat{k}, \hat{\delta}$,

$$|(z(r), e(r), p(r))| \leq \hat{k}||(z,e,p)_R|| \exp(-\hat{\delta}(r-R)) . \tag{37}$$

Proof. Bearing in mind (21) and that $\varepsilon_i < 1$, for $i = 1, 2, \ldots, n$, and $(M\kappa)/(L(n+1)!) \geq 1$, then γ in (ii) after (33) is seen to be equal to $\sqrt{n(n-1)}P$, and the condition $P \leq [8(1+n^2)^{n-1}\sqrt{n(n-1)}]^{-1}$ in (36) actually implies $\gamma \leq 1/(8q)$. Analogously, one can check that the requirements on τ and P in (36) imply that all the conditions in Lemma 5 and 6 are true. These lemma (see also Remark 6) allow us to infer the thesis. □

The proof of the main result of the paper simply amounts to rephrase the proposition above in the original coordinates. This is straightforward and we omit it. We only discuss briefly the issue of the minimality of the data rate. By definition of R_{av}, it is always possible to guarantee that $R_{av} < \hat{R}$, provided that $T_m \geq 2n/\hat{R}$. Now the stability results we presented hold for a given value of T_m which may or may not fulfill the inequality above. Suppose it does not. Can we increase T_m above $2n/\hat{R}$ and still have stability? The answer is yes, for the value of T_m (and hence of $T_M \geq T_m$) affects the entries of $A(r)$ and Λ, but the exponential stability of the (e,p) equations (and therefore of system

(29)) remains true, as it is evident from the proof of Lemma 3. Hence, the arguments above still apply and minimality of the data rate is proven.

We stress that the proof of the result is constructive, that is we give the explicit expressions of the encoder, the decoder and the controller which solve the problem. As a matter of fact, the equations of encoder and the decoder are introduced in (11) and, respectively, (9). The matrices Λ and Φ appearing there are, respectively, designed in Lemma 3, and defined in (14). The parameters of the nested saturated controller are L, κ and the saturation levels ε_i. The latter are defined in (26). The former must be chosen in such a way that (35) and (36) are satisfied. Bearing in mind the definitions (17), (21), it is easy to see that, for any value of the delay θ, there exist a sufficiently large value of κ and a sufficiently small value of L such that (35) and (36) are true. These values will be, respectively, larger and smaller than the corresponding values given in [20], as the presence of the quantization error requires a stronger control action.

6 Conclusion

We have shown that minimal data rate stabilization of nonlinear systems is possible even when the communication channel is affected by an arbitrarily large transmission delay. The system has been modeled as the feedback interconnection of a couple of impulsive nonlinear control systems with the delay affecting the feedback loop. In suitable coordinates, the closed-loop system turns out to be described by a cascade of impulsive delay nonlinear control systems, and semi-global asymptotic plus local exponential stability has been shown. The proof relies, among other things, on the design of a Lyapunov-Krasowskii functional for an appropriate cascade impulsive time-delay system. If the encoder is endowed with a device able to detect abrupt changes in the rate of growth of x_n, or if a dedicated channel is available to inform the encoder about the transmission delays, then it is not difficult to derive the same kind of stability result for the case when the delays are time-varying and upper-bounded by θ. Similarly, by adjusting T_M in (4), it is possible to show that the solution proposed in this paper is also robust with respect to packet drop-outs. The same kind of approach appears to be suitable for other problems of control over communication channel with finite data rate, delays and packet drop-out.

References

1. D.D. Bainov and P.S. Simeonov. *Systems with impulse effect. Stability, theory and applications.* Ellis Horwood Limited, Chichester, UK, 1989.
2. R.W. Brockett and D. Liberzon. Quantized feedback stabilization of linear systems. *IEEE Trans. on Automat. Contr.*, 45:1279–1289, 2000.
3. A. Cepeda and A. Astolfi. Control of a planar system with quantized input/output. In *Proc. of the 2004 Amer. Contr. Conf.*, pages 3053–3058, 2004.

4. C. De Persis. n-bit stabilization of n-dimensional nonlinear systems in feedforward form. *IEEE Trans. on Automat. Contr.*, 53:299–311, 2005.
5. C. De Persis. Minimal data rate stabilization of nonlinear systems over networks with large delays. Preprint available at `www.dis.uniroma1.it/~depersis`, 2006.
6. C. De Persis and A. Isidori. Stabilizability by state feedback implies stabilizability by encoded state feedback. *Systems & Control Letters*, 53:249–258, 2004.
7. N. Elia and S.K. Mitter. Stabilization of linear systems with limited information. *IEEE Trans. on Automat. Contr.*, 46:1384–1400, 2001.
8. F. Fagnani and S. Zampieri. Stability analysis and synthesis for scalar linear systems with a quantized feedback. *IEEE Trans. on Automat. Contr.*, 48:1569–1584, 2003.
9. H. Ishii and B.A. Francis. Stabilizing a linear system by switching control with dwell time. *IEEE Trans. on Automat. Contr.*, 47:1962–1973, 2002.
10. A. Isidori. *Nonlinear control systems*, volume 2. Springer, London, 1999.
11. A. Isidori, L. Marconi, and A. Serrani. *Robust autonomous guidance: An internal model-based approach.* Springer Verlag London, Series Advances in Industrial Control, 2003.
12. M. Jankovic, R. Sepulchre, and P.V. Kokotovic. Constructive Lyapunov stabilization of nonlinear cascade systems. *IEEE Trans. on Automat. Contr.*, 41:1723–1735, 1996.
13. G. Kaliora and A. Astolfi. Nonlinear control of feedforward systems with bounded signals. *IEEE Trans. on Automat. Contr.*, 49:1975–1990, 2004.
14. V. Lakshmikantham, D.D. Bainov, and P.S. Simeonov. *Theory of impulsive differential equations.* World Scientific, Singapore, 1989.
15. M.D. Lemmon and R.J. Sun. Performance-rate functions for dynamically quantized feedback systems. In *Proc. of the 45rd IEEE Conf. on Decision and Contr.*, pages 5513–5518, 2006.
16. D. Liberzon. Hybrid feedback stabilization of systems with quantized signals. *Automatica*, 39:1543–1554, 2003.
17. D. Liberzon. Quantization, time delays, and nonlinear stabilization. *IEEE Trans. on Automat. Contr.*, 51:1190–1195, 2006.
18. D. Liberzon and J.P. Hespanha. Stabilization of nonlinear systems with limited information feedback. *IEEE Trans. on Automat. Contr.*, 50:910–915, 2005.
19. L Marconi and A. Isidori. Robust global stabilization of a class of uncertain feedforward nonlinear systems. *Systems & Control Letters*, 41:281–290, 2000.
20. F. Mazenc, S. Mondié, and R. Francisco. Global asymptotic stabilization of feedforward systems with delay in the input. *IEEE Trans. on Automat. Contr.*, 49:844–850, 2004.
21. F. Mazenc, S. Mondié, and S. Niculescu. Global asymptotic stabilization for chains of integrators with delay in the input. *IEEE Trans. on Automat. Contr.*, 48:57–63, 2003.
22. F. Mazenc and L. Praly. Adding integrations, saturated controls and stabilization for feedforward systems. *IEEE Trans. on Automat. Contr.*, 41:1557–1559, 1996.
23. G.N. Nair and R.J. Evans. Exponential stabilisability of finite-dimensional linear systems with limited data rates. *Automatica*, 39:585–593, 2003.

24. G.N. Nair, R.J Evans, I.M. Mareels, and W. Moran. Topological feedback entropy and nonlinear stabilization. *IEEE Trans. on Automat. Contr.*, 49:1585–1597, 2004.
25. G.N. Nair, F. Fagnani, S. Zampieri, and R.J Evans. Feedback control under data rate constraints: An overview. *Proceedings of the IEEE*, 95:108 137, 2007.
26. D. Nešić and A.R. Teel. Input-to-state stability of networked control systems. *Automatica*, 40:2121–2128, 2004.
27. S. Tatikonda and S.K. Mitter. Control under communication constraints. *IEEE Trans. on Automat. Contr.*, 49:1056–1068, 2004.
28. A.R. Teel. Global stabilization and restricted tracking for multiple integrators with bounded controls. *Systems & Control Letters*, 18:165–171, 1992.
29. A.R. Teel. A nonlinear small gain theorem for the analysis of control systems with saturations. *IEEE Trans. on Automat. Contr.*, 41:1256–1270, 1996.
30. A.R. Teel. On l_2 performance induced by feedbacks with multiple saturation. *ESAIM: Control, Optimization, Calculus of Variations*, 1:225–240, 1996.

Printing: Krips bv, Meppel, The Netherlands
Binding: Stürtz, Würzburg, Germany